Green Energy and Technology

Climate change, environmental impact and the limited natural resources urge scientific research and novel technical solutions. The monograph series Green Energy and Technology serves as a publishing platform for scientific and technological approaches to "green"—i.e. environmentally friendly and sustainable—technologies. While a focus lies on energy and power supply, it also covers "green" solutions in industrial engineering and engineering design. Green Energy and Technology addresses researchers, advanced students, technical consultants as well as decision makers in industries and politics. Hence, the level of presentation spans from instructional to highly technical.

Indexed in Scopus.

Indexed in Ei Compendex.

Panneerselvam Sathishkumar
Editor

Photocatalysis for Energy and Environmental Applications

Current Trends and Future Perspectives

Springer

Editor
Panneerselvam Sathishkumar
Department of Chemistry, School
of Advanced Sciences
Vellore Institute of Technology (VIT)
Vellore, Tamil Nadu, India

ISSN 1865-3529 ISSN 1865-3537 (electronic)
Green Energy and Technology
ISBN 978-981-97-1938-9 ISBN 978-981-97-1939-6 (eBook)
https://doi.org/10.1007/978-981-97-1939-6

This Springer imprint is published by the registered company Springer Nature Singapore Pte Ltd.
The registered company address is: 152 Beach Road, #21-01/04 Gateway East, Singapore 189721, Singapore

Preface

Photoelectrochemical water splitting in the presence of TiO_2 and ultraviolet light is considered as a significant discovery in energy and environmental applications of photocatalysis. Following that, around 50 years of research and development revolved around photocatalysis, and the various parameters causing the process to become expensive were identified and replaced by suitable components. In addition, various nanostructured materials were investigated to improve the efficiency of photocatalysts in the presence of solar irradiation. The inexhaustible availability of sunlight on the Earth's surface should be able to solve human demands for future needs of energy and environment. However, the present techniques available for the utilization of sunlight need improvement in order to achieve the 100% utilization of solar light efficiently.

Approximately 47% of visible light ($400 > \lambda < 700$ nm) is available as sunlight on the Earth's surface and theoretical calculations indicate that the Earth's surface receives 89,300 TW of solar irradiation. The development of imperative technologies based on the utilization of sunlight can reduce the cost required for energy and environmental applications of photocatalysis. Alternatively, many diseases are caused by the use of a polluted environment and the adulteration of chemicals/contaminants on the nanoscale level (whether naturally and/or artificially) tends to create negative consequences. Therefore, it is the responsibility of investigators to provide a clean, green, and safe environment to ensure the survival of the next generation on the planet.

The development of efficient nanotechnology-based catalysts which utilize solar light can be considered as a suitable and sustainable technology for energy and environmental applications in the future. The contents of the book discuss existing photocatalytic reactor designs and the developments needed for the future progress of both solar light-driven hydrogen generation and environmental purification. The proposed chapters discuss the renewable and commercial aspects of translucent polymer-linked heterojunction nanocomposites as visible light-responsive photocatalysts. The field is growing with great interest in black-TiO_2, perovskites, MXenes, and their numerous applications. The framework, structural features, and the need for mesocrystals for solar light-driven photocatalysis are also explained. Additional features of green chemistry-based synthesis of nanomaterials are discussed in order to reduce any secondary pollutants that may be released into the environment due to unsafe disposal. In addition, the critical views for the necessity of bisumuth-based nanophotocatalysts towards energy and environmental applications are comprised for their future developments to upraise as alternative photocatalysts for prevailing nanomaterials. The core research to generate green fuel Hydrogen is conferred through photocatalytic and photoelectrochemical technologies. Overall, this book reviews the contemporary developments transpired in the last decade and presents future perspectives of photocatalysis and its application towards energy and environment.

Vellore, India Panneerselvam Sathishkumar

Contents

Critical Review of Photocatalytic Reactor Designs for Environmental Applications

S. K. Sheik Moideen Thaha, Pugazhenthiran Nalandhiran, Selvam Kaliyamoorthy, Isao Mizota, Ramalinga V. Mangalaraja, and Panneerselvam Sathishkumar

Abstract Photocatalytic processes are a viable alternative to reduce fossil fuel use and global warming. The photocatalytic approach has the potential to restore the globe to its equilibrium state through a combination of energy production (solar cells), energy conversion (CO_2 to fuels, water splitting), and environmental remediation (degradation). Despite the rocketing research on photocatalytic materials for photocatalytic degradations, research on photocatalytic reactors is urgently required to match the explosive growth of research on photocatalytic materials. Although numerous innovative, futuristic designs are brought up by various research fields, the development of photocatalytic reactors to achieve commercialization objectives has not yet attained maturity. The photocatalytic treatment of contaminants is typically influenced by three factors: the pollutant (liquid), the catalyst (solid), and light (massless). It is difficult to advance the research further since all three components are in different phases. In order to decrease the restrictions brought on by photocatalysis, such as mass transfer and light penetration, numerous approaches have been proposed. As of now, many reactors have been constructed in a variety of sizes and shapes, from bench-top and laboratory designs to industrial pilot scales with different operational modes. The overall efficiency of the photocatalytic reactor is influenced by its working components. Reactor types classified according to their catalyst installation and geometry have been identified as the primary groups and their advantages and disadvantages were assessed. Additionally, several potential

S. K. S. M. Thaha · P. Sathishkumar (✉)
Department of Chemistry, School of Advanced Sciences, Vellore Institute of Technology (VIT), Vellore, India
e-mail: sathishkumar.p@vit.ac.in

P. Nalandhiran
Department of Chemistry, Universidad Técnica Federico Santa María, Av. España 1680, Valparaíso, Chile

S. Kaliyamoorthy · I. Mizota
Department of Applied Chemistry, Graduate School of Engineering, Mie University, Tsu 514-8507, Mie, Japan

R. V. Mangalaraja
Faculty of Engineering and Sciences, Universidad Adolfo Ibáñez, Diagonal Las Torres 2640, Peñalolén, Santiago, Chile

P. Sathishkumar (ed.), *Photocatalysis for Energy and Environmental Applications*, Green Energy and Technology, https://doi.org/10.1007/978-981-97-1939-6_1

modifications and designs for sunlight-driven photocatalytic reactors are discussed. Finally, conclusions and future improvement points were highlighted as well as a collective tabulated comparison of the reactor's advantages and disadvantages.

1 Introduction

It is well-known that photocatalysis is a versatile technique for eliminating environmental contaminants in addition to offering zero waste through renewable resources like solar energy. Technological advancements and environmental concerns have increased the need to explore sustainable methods for environmental remediation. Research into the use of light and heterogeneous catalysts has been expanding at a rapid pace since the publication of Fujishima Honda's work in 1972 [1]. But there is one research parameter that is still under development, limiting the industrial development of photocatalysis: photocatalytic reactor. Research on various heterogenous catalyst-assisted photocatalysis with more than 90% degradation efficiency has been widely tested and achieved [2–7]. Even though they had good efficiency, catalyst reusability, and cost-effectiveness, the majority of research advancements in this field remain confined to the laboratory scale. The reason that restricts it from reaching industrial applications is the suitable photocatalytic reactor. At the laboratory scale, each photocatalytic degradation research employed photocatalytic reactors with unique designs (light power, number of light sources, volume, homogenizer, etc.). However, for commercialization, uniform designs, multidisciplinary choices of pollutants, and standard efficiency-delivering photocatalytic reactors were expected (Fig. 1).

Globalization has accelerated the industry's diversity; developing a photocatalytic reactor that can remove contaminants from a wide variety of mixtures requires the design and development of a highly efficient reactor. Mass transfer, contact, catalyst installation, surface area to volume ratio, and illumination are the most challenging aspects of the scaling-up procedure [8], in which illumination of light is the most important variable for inducing the basic mechanism of photocatalysis. Since semiconductors have a lower bandgap energy, when light equal to or higher than the semiconductor catalyst shines on the surface, the electrons of the valence band are excited to the conduction band. This leads to the creation of voids in the valence band (void species are termed as holes (h^+)) and the accumulation of electrons (e^-) in the valence band. These two photogenerated species are termed as photogenerated charge carriers [9]. These are the species that react with the water and generate reactive oxygen species (ROS). The ROS commonly generated by photocatalysis are $OH^{\cdot}$, $O_2^{\cdot -}$, and $HOO^{\cdot}$ which are created by the redox reaction of charge carriers with water molecules. These generated ROS have a high oxidation potential and a high tendency to oxidize any form of contaminants. The overall reaction mechanism of the photocatalytic process is divided into four stages which are mentioned below, and all equally pay a definite contribution to overall degradation efficiency.

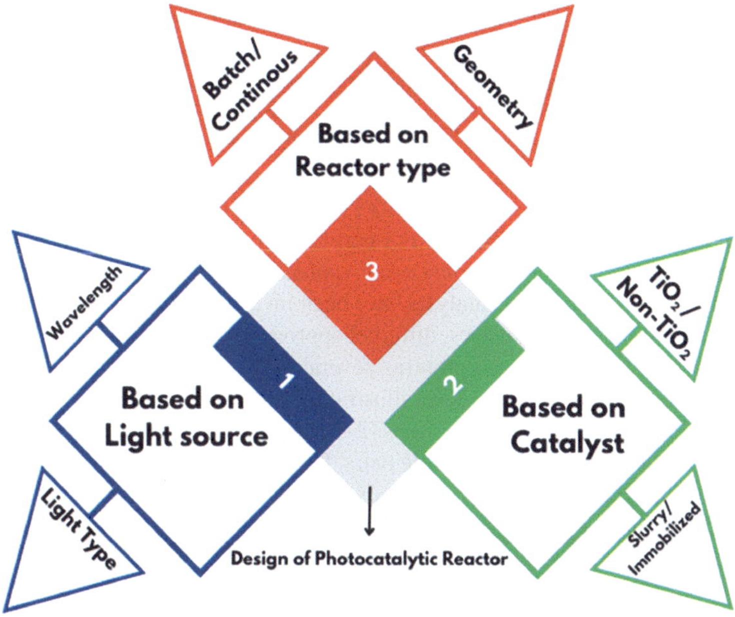

Fig. 1 Basic structure and components of photocatalytic reactor

Stage 1: Diffusion and Adsorption of pollutant molecules on the surface of the semiconductor catalyst. The rate of initial adsorption depends on the internal conditions of the reactor (pH, temperature, etc.)

Stage 2: Photoinduced excitation of catalyst; The electron present in the valence band (VB) shall excite to the conduction band (CB) of the semiconductor catalyst when the appropriate light energy is radiated.

Stage 3: Charge diffusion: Migration of produced charge carriers from the bulk to the photocatalyst's surface. A drop in the quantum efficiency of the photocatalyst will result from recombination if it occurs at this stage.

Stage 4: Redox reaction: Chemical reaction of charge carriers with the H_2O molecules present in the reactor medium to generate ROS.

For a typical good photocatalytic process, the quantity of ROS generated should be high. This depends on various factors like light transfer, mass transfer, catalyst ability, pH, catalytic dosage, adsorption, and suitable light energy. For a photocatalytic reactor to perform well in analysis, it must consider all of the parameters mentioned above. Catalyst installation methods in photocatalytic reactors were broadly classified into two categories: Suspended catalyst system (SCS) and immobilized catalyst system (ICS). The production of ROS and the recyclability of the catalyst differ between SCS to ICS. SCS is likely to generate more ROS compared to ICS, but SCS is more complicated when it comes to separating the catalyst phase. The problems associated with ICS are light transfer, catalyst thickness, and the proximity of catalyst and pollutant molecules. But these can be tackled by the appropriate design of photocatalytic ICS. One of the significant factors influencing the system's effectiveness is the type of photocatalyst. The photoactivity of the catalyst determines the production of essential ROS, including $OH^{\cdot}$, $O_2^{\cdot-}$, and $HOO^{\cdot}$. The absorption of light is dependent upon the bandgap of the photocatalyst; typically, the higher bandgap semiconductors absorb the UV light region, which increases the cost of illumination, whereas the lower bandgap semiconductors (i.e., the infrared range) undergo photocorrosion rather than photocatalysis. For commercial photocatalytic processes, it is preferable to generate charge carriers using visible or ultraviolet light. In the span of its evolution, numerous photocatalysts have been investigated. To reduce the recombination of photogenerated species, different approaches have been adopted, through the formation of heterojunctions granting several improvements. The heterojunction types, their formation, and band-level illustrations are classified in Table 1.

The above-mentioned heterojunctions (2–8) are employed to reduce the recombination of photogenerated electron–hole pairs. Each heterojunction possesses its own distinct mechanism, advantages, and disadvantages. The comprehensive elaboration of heterojunction is a challenging task that falls outside of the scope of the current chapter's theme. The following is a concise overview of the heterojunction catalyst.

- **Straddling gap heterojunction (Type 1)**: VB and CB of semiconductor 1, respectively, are higher and lower than the corresponding bands of semiconductor 2. Since generated electrons and holes migrate to the same semiconductor 2 which is lower in redox potential, this heterojunction exhibits low redox ability as well as a lesser effective separation of charge carriers [10].
- **Staggered gap heterojunction (Type 2)**: CB and the VB levels of semiconductor 1 are higher than the corresponding levels of semiconductor 2. The generated holes and electrons will be migrated to both semiconductors 1 & 2, respectively, leading to the effective separation of electron–hole pairs.
- **Broken gap heterojunction (Type 3)**: The architectural design of type 3 is very similar to that of a staggered heterojunction. However, the energy bands of the two semiconductors in this type do not overlap, making it unsuitable for effectively separating electron–hole pairs and minimizing recombination.
- **p–n heterojunction**: It is a combination of p-type and n-type semiconductors which leads to the formation of charge-diffused space known as the internal electric field. When the light energy equal to or higher than its bandgap is irradiated, VB electrons of p-type and n-type semiconductors are excited, generating electron–hole pairs in their band level. The photogenerated electrons and holes in the p-type and n-type semiconductors will migrate under the influence of the internal electric field to the CB of the n-type semiconductor and the VB of the p-type semiconductor, respectively, which results in the spatial separation of the electron–hole pairs [10].
- **Surface heterojunction**: The approach of electron–hole separation can also be achieved by utilizing the different crystal facets of a single semiconductor material. Different crystal-faceted semiconductors exhibit distinct band energies, hence enabling the creation of heterojunctions [11].

Table 1 Types of heterojunctions in photocatalysis

S. No	Type	Nature of formation	Band illustration
1.	Single photocatalyst	There is no formation. Existing molecular orbitals act as bands	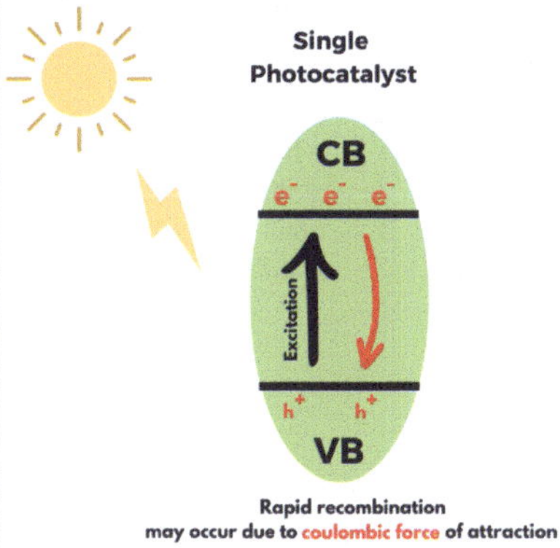
2.	Straddling gap (Type 1)	Formation of heterojunction by two semiconductors in which ones VB and CB congested to other	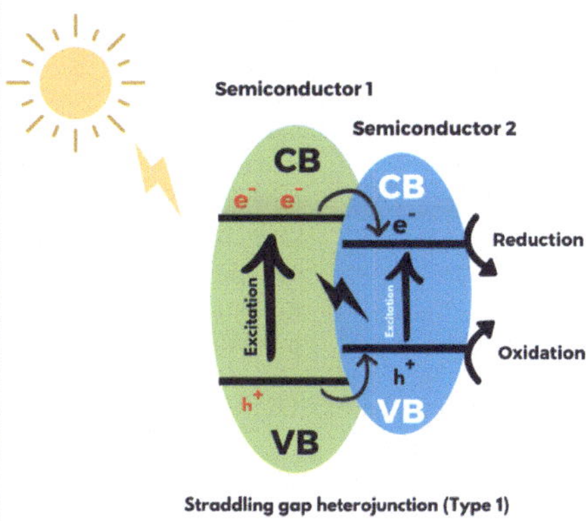
3.	Staggered gap (Type 2)	Formation of heterojunction by two semiconductors in which ones VB and CB higher to corresponding of other	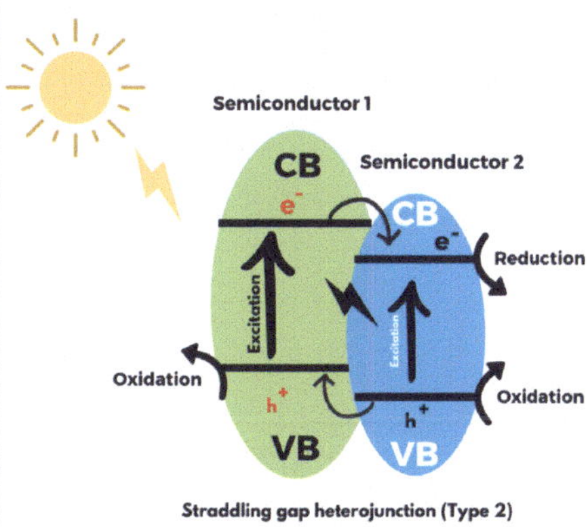

(continued)

Table 1 (continued)

S. No	Type	Nature of formation	Band illustration
4.	Broken bandgap (Type 3)	Formation of heterojunction by two semiconductors in which ones VB and CB higher to other with no overlap	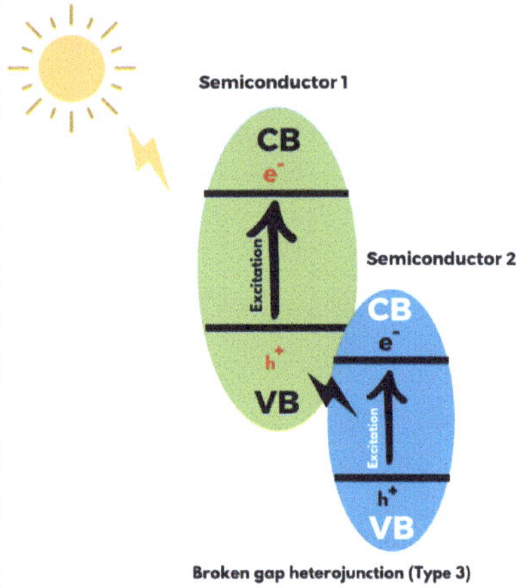
5.	P–n heterojunction	Formation of heterojunction by two different types of semiconductor (p-type and n-type) to form p–n interface	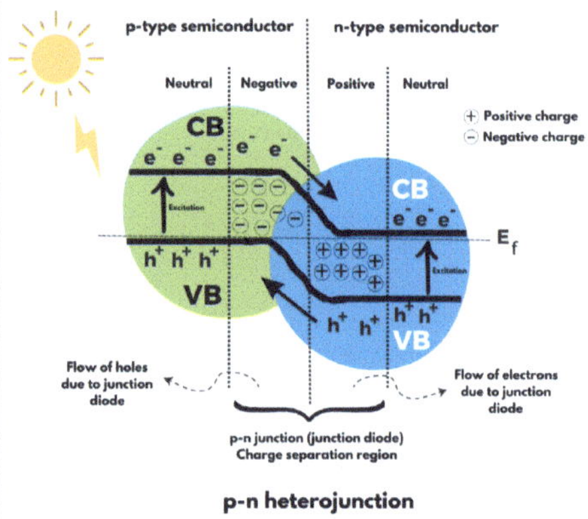

(continued)

Table 1 (continued)

S. No	Type	Nature of formation	Band illustration
6.	Surface heterojunction	Formation of heterojunction by two different crystal facets on a single semiconductor	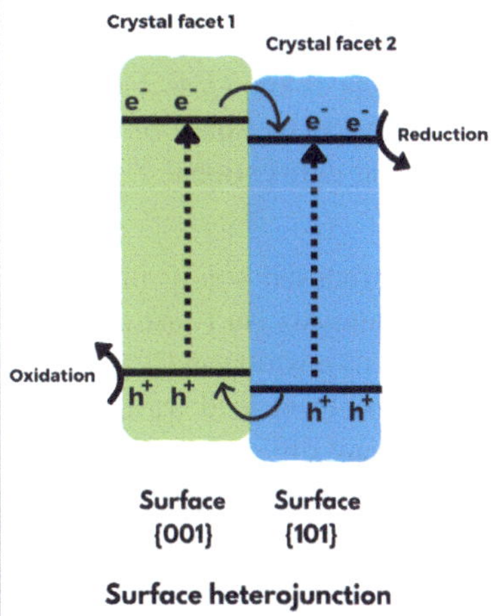
7.	Z-scheme	Formation of heterojunction by two different semiconductors in which ones excited electron will be transferred to others via a mediator to the valence band	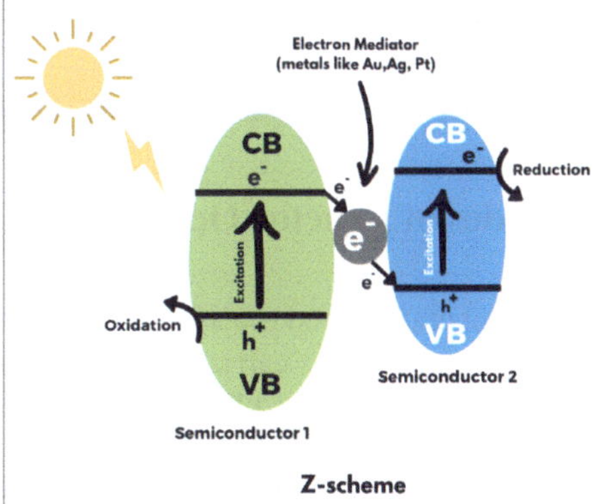
8.	Direct Z-scheme	Formation of heterojunction by two different semiconductors in which ones excited electron will be transferred to others to the valence band without any mediator	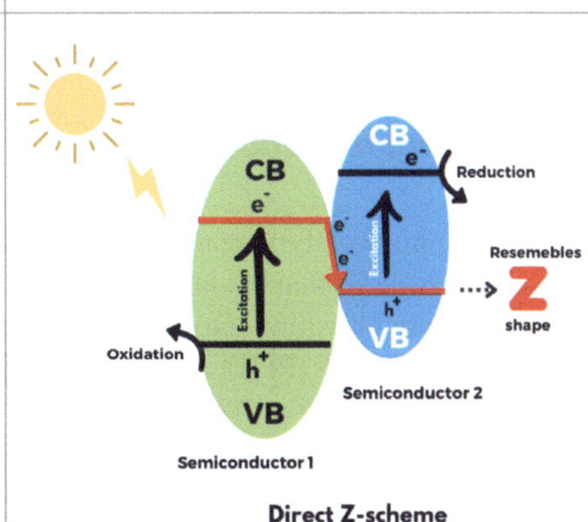

- **Z-scheme and Direct Z-scheme heterojunctions**: The separation of electron–hole pairs is carried out by merging two semiconductors (1 and 2) through an electron mediator material. The electrons from CB of semiconductor 1 are transferred to VB of semiconductor 2 via mediator material, which effectively separates the photogenerated charge carriers and enhances and retains the redox ability of the catalyst. This type is known as the Z-scheme since the resemblance of the electron transport channel to the shape of the letter Z. The direct Z-scheme heterojunction is similar to the conventional Z-scheme, in which the electron mediator is not needed.

So far, we've focused solely on the materials and process of photocatalysis. However, photocatalysis isn't just about the materials. Since light is so important here, playing with its position and wavelength could lead to various outcomes. Similar to our top load and front load washing appliance, photocatalytic reactors also can be modified in many different ways that have a significant impact on their performance. This book chapter comprehensively analyses photocatalytic reactors through a systematic classification, covering the reactor's proposed concepts, construction methods, and their merits and demerits. It also provides insights into the design optimization of photocatalytic reactors, as well as the various reactor configurations. Finally, it discusses the future prospects of photocatalytic reactors and their applications in environmental protection.

2 Modes of Reactor Operation

Different operating modes can be used to carry out the photocatalysis process. The way the procedure was carried out is referred to as the operative mode. There are numerous modes that use different technologies, such as batchwise, continuous flow, hybrid, sequential, and others. Since only photocatalysis applies continuous and batch operating modes, the other two are not covered in this section.

2.1 Continuous Mode

Continuous photocatalytic reactors are those in which the reactant liquid is pumped into a reaction vessel that has inlet and outlet tubes for the liquid flow on a continuous basis. These are classified into two types: continuous-stirred tank reactors (CSTR) and plug-flow reactors (PFR). As the name implies, CSTR are reactors whose reaction vessel is continuously stirred, the reactant liquid is fed through the inlet, and the product is gained through the outlet; therefore, the concentration and temperature of the output are equivalent to those of the input [12]. On the other hand, in plug-flow photoreactor modes, the reactant liquid enters the reactor vessel from one side and exits from the other side, but the fluid flow follows a straight channel (tubular) without any homogenizer. Slurry photoreactors frequently use CSTR modes, whereas immobilized systems usually operate in PFR modes [13]. The PFR outperforms the CSTR in terms of efficiency and duration. The main problem with CSTR is that the product and reactant are back mixed, which causes problems with adsorption, separation, and duration [14, 15]. These continuous operating mode reactors are feasible to scale up for commercialization considering their merits, such as being feasible for high reactant volumes, automation, and reduced man power [16].

2.2 Batch Mode

This type of reactor mode has no inlet or outlet for reactants or products during photocatalysis; this means that the reactants enter the reaction vessels and undergo the photocatalytic reactions over time; at the end of the time, the products and unreacted reactants are removed from the reactor by employing post-treatment methods. For this kind of operation, the homogenizer must thoroughly mix the reactant to ensure uniformity and maintain a constant reaction rate. In other words, the characteristics of the reaction mixture and the rate of the reaction are the same throughout the reactor vessel in stirred batch mode photocatalytic reactors [17]. Both slurry and immobilized reaction systems may be suitable for these batchwise operational modes. However, the efficiency and better kinetics depend upon the type of reactor and catalyst employed.

3 Types of Photocatalytic Reactor

A photocatalytic reactor is a reactor whereby photocatalysis is implemented to eliminate contaminants. Since its beginning, various photocatalytic reactor designs have been studied, and some of its notable progressions are tabulated year-wise in Table 2 [18]. The main components of the reactor are the catalyst, contaminant solution, and light. Some reactors may or may not have extra features such as the influence of lamps [19], air sparging [20], stirring, inert conditions [21, 22], a temperature–pressure controller [23–25], and more. The modeling and constructed architecture of the photocatalytic reactor are essential to achieving efficient degradation of environmental contaminants. Therefore, in designing the photoreactor, the operational mode, catalyst installation, catalyst type, and geometry of the reactor must be considered. On the basis of operational mode, the reactor is classified as batch and continuous type photoreactor. On the basis of catalyst installation, the reactor is classified as slurry photocatalytic reactor, immobilized photocatalytic reactor, and membrane-type reactor. Due to the significant number of variables concerned, the classification and comprehensive discussion of the diverse range of photocatalytic reactors is a challenging task. So, in a broader sense, photocatalytic reactors are categorized based on (i) light sources, (ii) catalyst, and (iii) reactor type.

Table 2 Year-wise notable progressions of photocatalytic reactors

S. No	Reactor	Description
1.	Fluidized bed reactor (1996)	Cylindrical reactor with lamp located at the center and the distance between the lamp to the fluidized bed is 2.15 cm. In addition, aluminum foil was used for covering the reactor wall. Cooling shell made of Pyrex glass (0.01 m thick) was located outside the UV lamp-containing tube. Thickness of fluidized layer was 0.5 cm and the total reactor surface irradiated is 0.04 m^2
2.	Multi-tube reactor (1998)	Reactor is a cylindrical vessel (ø0.056 m) with 54 hollow Quartz glass tubes (ø0.006 m) coated with photocatalyst. Tubes arranged in triangular pitch. There are four feed ports and four exit ports, where the reaction liquid flows outside the tube and light passes through the tube
3.	Tube-light reactor (1998)	The reactor is a rectangular cube (0.132 m × 0.016 m × 0.248 m) with 21 holes containing U-shaped lamps. It contains five inlet and exit ports
4.	Water fountain solar photocatalytic reactors (2003)	Two reactors of trapezoidal (300 × 600 mm) and circular (ø500 mm) configuration were developed, where fountain nozzle assembly was used to create a water fountain at the center of this solar photocatalytic reactor
5.	Optical fiber monolith reactor (2005)	Reactor components are an UV lamp, TiO_2-coated optical fiber tubes, TiO_2-coated 61-channeled monolith block (length 30 cm and ø3 mm) made of ceramic, reaction vessel, and a reservoir
6.	Magnetically agitated photocatalytic reactor (2005)	Barium ferrite/SiO_2/TiO_2 was controlled by dual magnetic component: permanent magnet to counteract gravitational force on catalyst particles and Solenoid to agitate the suspended catalyst
7.	Rayonet multi-lamp system (2006)	Multiple lamps are arranged around a quartz cylindrical photocatalytic reactor. Photocatalyst particles are retained by 425 μm nylon mesh screen held between steel plates with 58 holes of 5 mm diameter
8.	Micro-channel reactor (2006)	Cylindrical porous disks made of alumina ceramics with TiO_2 supported on it, where the dimensions of disk are ø5 mm × 0.2 mm and 1250 holes of ø50 μm
9.	CPCR (2008)	CPCR is an annular reactor; with a tube 90 mm × 500 mm (volume was 2.82 L). Centrally located lamp inside the quartz glass tube with the reactor wall to lamp clearance of 35 mm
10.	Oxygen purged photocatalytic reactor (2008)	Photocatalytic reactor consists of a quartz tube (ø3 cm) oxygen purge inlet during reaction. Oxygen was bubbled at a flow rate of 200 cc/min
11.	UV-LED Photocatalytic reactor (2009)	Reactor is a 125 mL glass beaker with 5 UVA-LED (ø5 mm) mounted in a circular pattern on an acrylic sheet, which is held 1 cm above the reaction mixture

(continued)

Table 2 (continued)

S. No	Reactor	Description
12.	Annular reactor (2009)	Annular reactor consists of UV lamp located at the center inside the quartz thimble with four sampling ports and an air sparging device
13..	TiO_2-coated rotating disk photocatalytic reactor (2010)	TiO_2-coated disk (ø6 cm) irradiated by 2 UV-C lamps, where the reactant is pumped onto the spinning disk at a specific feed flow rate
14.	Integrated membrane photocatalysis	Reactor consists of 7 quartz tubes, UVA lamp, feed tank, flat sheet polysulfone membrane of 15.2 cm^2, and pump of pressure 0.3 MPa
15.	Rotating corrugated drum reactor (2010)	Reactor consists of a drum made of stainless steel (6.4 cmø and 20.3 cm length) with 2 cm length of corrugations and a light source. Seven drum configurations were developed such as no corrugations, sixteen fins, twenty-eight fins, forty fins, thirty-degree angle, twenty-degree angle, and ten-degree angle
16.	Fixed-bed photocatalytic reactor (2010)	Continuous annular Pyrex photocatalytic reactor of outer diameter 72 mm and annulus gap 7 mm. A fixed bed of height 186 mm and void fraction 0.29 contains the catalyst coated on Pyrex beads
17.	UV-LED photocatalytic reactor (2011)	This novel photocatalytic reactor consists of 3 catalyst-coated Quartz tubes (120 mm × ø5 mm) inside the cylindrical vessel and 15 UV-LEDs mounted inside the reactor vessel
18.	Capillary array photocatalytic reactor (2012)	A complicated process was involved in reactor fabrication. Totally 18 capillaries (each 15 cm length and 0.012 cm^3 volume) were inserted into two tertrafluoroethylene tubes on both the sides
19.	Membrane-LED photoreactor (2012)	Reactor is operated in a continuous flow through mode, where TiO_2 suspensions are delivered to the membrane module and permeate is partially recycled to the photoreactor to maintain the working volume. Feed flow rate and permeate flux were closely monitored
20.	Closed circulation photocatalytic reactor with in-stream sensor (2012)	A batch glass reactor (height 70 mm and inner diameter 84 mm) was connected with a mini centrifugal pump, 15 UV-LEDs, and a coated wafer fixed to the reactor base. Maximum capacity of reactor volumeis 250 mL
21.	Multi-layer rotating disk photocatalytic reactor (2012)	Four level stacked aluminum disks (ø12 cm) coated with TiO_2 irradiated by two UV lamps each
22.	Corrugated wall Photocatalytic reactor (2013)	Photocatalytic reactor consists of an acrylic box with a folded plate coated with TiO_2 forming a set of triangular section channels, irradiated by two sets of actinic UV lamps. The pollutant air stream follows a zigzag pattern from inlet to outlet

(continued)

Table 2 (continued)

S. No	Reactor	Description
23.	Laminar falling film slurry photocatalytic reactor (2013)	Reactor glass column of length 525 mm and inner diameter 31 mm is surrounded by six 18 W UV lamps emitting 366 nm. This setup is placed in a cylindrical metal housing of length 875 mm and inner diameter 300 mm coated by a reflective aluminum foil
24.	Batch LED reactor (2013)	Cylindrical reaction vessel with 20 mmø and 50 mm height and cell holder made of Polyoxymethylene (POM) to avoid unwanted light absorption. UV-LED (5.4 mmø) and angular full width at half maximum 4° for optimal degradation compared to 120°
25.	Solar reactor (2013)	Three identical independent reactors were designed. Dimensions of the flat panel were 30 cm × 100 cm × 2 cm. The angle of inclination was 42° in the south direction
26.	Spiral-Shaped Photocatalytic reactor (2014)	Reactor consists of a spiral glass tube (overall spiral diameter 240 cm and length 55 cm) with inner and outer diameters as 0.6 cm and 1 cm. UV and blue LEDs are hexagonally arranged inside the spiral glass tube
27.	Immobilized Photocatalytic micro-reactors (2014)	Micro-reactor was fabricated by attaching a TiO_2-coated substrate to Polydimethylsiloxane (PDMS) slab containing a micro-channel, which is rectangular in shape and follows a zigzag motion. Dimensions of 50 μm × 500 μm × 5.96 cm (Volume 1.49 μL)
28.	Fluidized bed optical fiber photocatalytic reactor (2014)	Reactor consists of gas inlet segment, fluidizing segment, gas distribution segment with a support section, gas outlet segment, and 8 optical fibers placed at a longitudinal distance of 3.4 mm enclosed within an Aluminum jacket
29.	TiO_2/β-SiC Alveolar foam LED reactor (2014)	Tubular flow reactor of 40 mm diameter with 20 mm length β-SiC Alveolar foam where 300 mg TiO_2 was deposited on the foam with 7 LEDs placed uniformly in the direction of flow
30.	Continuous fixed-bed reactor (2015)	Reactor consists of feed tank with propeller for homogenization and two stainless steel tubes (ø1 cm) enclosed with two pyrex windows of thickness 2.5 cm, packed bed N-doped TiO_2 onto pyrex spheres, and 5 UV lamps irradiating the pollutant medium
31.	High-power LED reactor (2016)	LED Cylindrical photoreactor (Ø12 cm), where 6 LEDs were arranged at the inner side of steel vessel with perforations (for heat sink) in 2 rows at a specific distance. Computer case fan was used for cooling and ventilation
32.	TiO_2 immobilized mosquito net LED reactor (2016)	Hexagonally shaped 18 LEDs (350 nm) in total placed in the inner cylinder and the outer cylinder consist of TiO_2-coated mosquito net with dye solution

(continued)

Table 2 (continued)

S. No	Reactor	Description
33.	Spinning disk reactor (2017)	Cylindrical vessel made up of pyrex glass (length 35 cm and internal diameter 30 cm) consists of 8 UV-C lamps and horizontal rotating stainless steel disk (ø22 cm, thickness 1 mm), where the best results were obtained for disk with baffles (0.3 cm thickness,0.5 cm height, 2.16, 4.32, 6.87 cm as length for 1st, 2nd, and 3rd circles, respectively)
34.	Flat plate solar photocatalytic reactor (2017)	Panel reactor has 70 cm × 68 cm × 1 cm dimensions with a tilt angle of 20°, where TiO_2 is coated onto plastic granules with glass on the top and a mirror on the bottom of the panel
35.	Foam reactor (2018)	In this reactor, 40 or 56 LEDs were used. Disk-shaped foam of dimensions ø65 mm and height 40 mm made up of TiO_2/β-SiC with an inner fan for cooling PCB at the rate of 10 m^3/h
36.	Translucent packed bed reactor (2019)	Reactor consists of TiO_2 P25-coated borosilicate beads with a diameter of 3 mm coated beads, where 192 LEDs are used in the array structure to illuminate the reaction zone of 5 cm × 15 cm
37.	Packed bed photoreactor (2019)	The main component includes a cylindrical tube made of Pyrex glass (40 cm length, 5 cmø) and N-doped TiO_2 on polystyrene spheres as a photocatalyst bed. Visible LED strip was wrapped on the walls of the photoreactor
38.	3D-printed CPC photoreactor (2020)	The main component of the reactor includes a 3D-printed reflector for harvesting visible light and a quartz tube packed with immobilized graphite-like carbon nitride on chitosan hydrogel beads
39.	Wisconsin photoreactor platform (WPP) (2021)	The reactor consists of a 3D-printed vial holder, computer fan cooling, and a light source from the bottom. The WPP can be placed above the stirring plate for homogenization. It can be assembled in less than 24 h and costs approximately $100 USD
40.	Bristol Myers Squibb Photoreactor (BMS-PR) (2022)	The reactor's housing is made of austenitic steel, which resists corrosion. 21 reaction arrays and commercially accessible LED strips were designed. The reactor is set up on a standard stir plate and lacks an integrated stirring mechanism. LEDs are wrapped around the beaker's inner wall and the airflow tube's outer wall for enhanced photon exposure
41.	PhotoRedOx Box reactor (2023)	The reactor consists of several light source choices ranging from 365 to 808 nm (top-illuminated), can accommodate vials of various sizes (up to 32 vials), and also contains an integrated fan cooling system

ø—Diameter

3.1 Based on Light Sources

Light and catalyst are the key ingredients of the entire phenomena; the name of the process itself indicates the importance of light (i.e., photocatalysis). The rate of the photochemical reaction is significantly influenced by the transmission of light to the system's catalyst. The local volumetric rate of photon absorption (LVRPA) is the key factor in modeling the suitable photocatalytic reactor. It is the number of photons absorbed per unit reactor and per unit time at each location within the photocatalytic reactor space [26]. The distribution of light within a catalyst medium, such as slurry or suspended systems, is determined by two processes: elastic scattering and absorption. Absorption is a phenomenon in which the catalyst absorbs the light energy and proceeds with the electron excitation process; it depends only on the available local light intensity. Scattering is the redistribution or reflection of incident irradiated light in various directions; it depends on the refractive index, size, morphology, and composition of the catalyst [27]. Many photocatalytic reactors use the elastic scattering method such as reflectors to maximize the light distribution [28–30] (Fig. 2).

The photocatalytic reactors based on light are further categorized into two major categories: (i) wavelength and (ii) lamp type. Numerous studies have been conducted on ultraviolet-assisted photocatalysis based on the wavelength of light, due to its increased efficiency and rapid reaction time [31–34]. Utilization of TiO_2 and ZnO-based catalysts is another significant cause for employing this radiation since they have corresponding bandgap energy. However, the key issue with this form of radiation method is its cost-effectiveness and safety concerns. Visible light-driven photocatalysis is more practical and easier to commercialize, but the synthesis and design of the catalyst and photocatalytic reactor are, respectively, in stages of development. The sustainable, renewable, and green method to conduct the photocatalysis is by utilizing solar light-driven or artificial light-induced photocatalysis (Fig. 3).

The light source is further subdivided by lamp type into (i) conventional lamps and (ii) light-emitting diodes (LED) lamps. Photocatalytic light sources have evolved significantly from conventional lamps to LED lamps. Conventional lamps like incandescent lamps (lamps that produce light when heated) and gas discharge lamps are

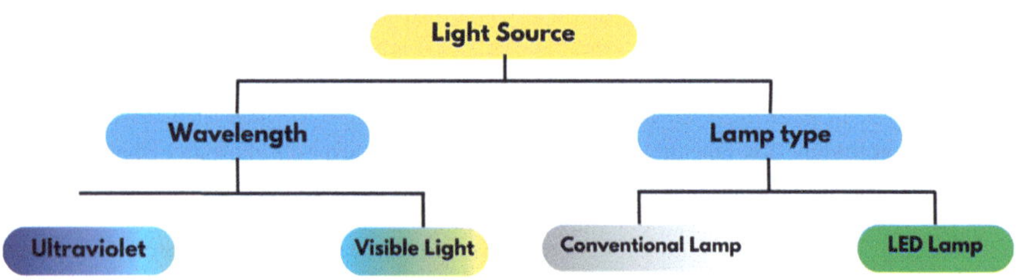

Fig. 2 Classification of photocatalytic reactors based on their light source

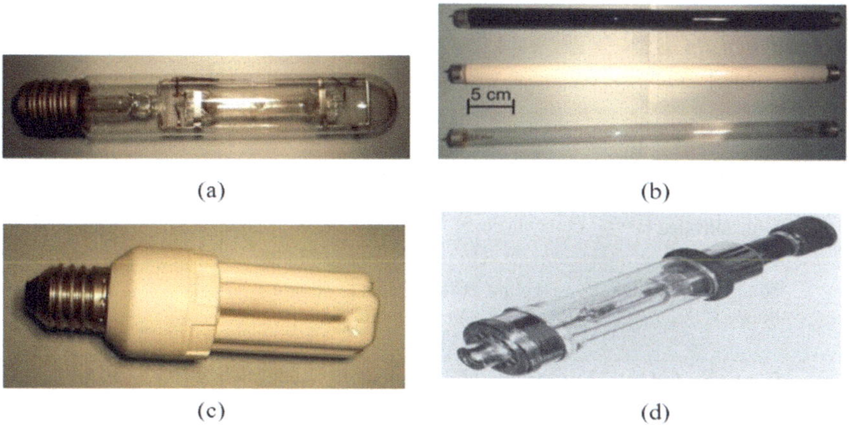

Fig. 3 Examples of light sources in artificial light reactor: **a** medium-pressure Hg lamp, **b** low-pressure Hg lamp, **c** CFL (compact fluorescent lamp), and **d** high-pressure Hg lamp [35]

poor at converting electrical energy to optical energy. The key problems associated with this type of lamp are toxicity, lifespan, and heat restriction in installations; the majority of the lamps contain toxic elements such as mercury and tungsten, their lifespan is very short, and they sometimes explode inside the reactor, causing further complications. These lamps emit thermal energy while glowing for an extended period of time, resulting in an abnormal temperature change in photocatalytic systems due to which they cannot be used in immersed lamp reactors [18]. The whole photocatalysis theme has been developed for environmentally friendly cleaning and production, so carrying out processes of the same theme only adds value. LED lamps can be developed with mono wavelength, non-toxic, durable, and capable of submerging features, making them more practical and energy efficient [18]. There are reports of numerous studies using UV-visible light sources to degrade various pollutants [36, 37]. Table 3 displays the degradation efficiency of tetracycline antibiotics by various light sources used in photocatalytic reactors. It was understood that photocatalytic reactor efficiency varies depending on the light source employed, but this may also be due to the catalyst's ability, which is another important component. Thus, it was discovered that both catalysts and light sources are joint-arm components for the effective design of photocatalytic reactors.

3.2 *Based on Catalyst Installation*

Based on the loading method of the catalyst, it has been further divided into three categories, respectively, Slurry, immobilized, and membrane systems. Over the years, many catalyst installation methods have been studied. The classification of photocatalytic reactors based on their catalyst installation is shown in Fig. 4.

Table 3 A comparison of the degradation efficiency of tetracycline using various light sources and photocatalysts

S. No	Catalyst	Pollutant	Light category	Light source	Degradation efficiency (%)	References
1.	TiO_2 decorated Ferroferric oxides with activated carbon	Tetracycline	LED lamp	UV-C lamp (6W, 254 nm)	96	[38]
2.	N-carbon quantum dots/g-C_3N_4	Tetracycline	Conventional Xenon lamp	Xenon lamp, 300 W	79.11	[39]
3.	C/g-C_3N_4–CeO_2	Tetracycline	Conventional Xenon lamp	Xenon lamp, 300 W	78.9	[40]
4.	N-defect/B-dopant g-C_3N_4	Tetracycline	Conventional Xenon lamp	Xenon lamp	62.8	[41]
5.	NiO/g-C_3N_4	Tetracycline	Conventional Xenon lamp	Xenon lamp, 500 W	74	[42]
6.	Phosphorus-doped CdS nanorods	Tetracycline	LED lamp	Blue LED lamp, 10 mW cm^{-2}	95.4	[43]
7.	TiO_2	Tetracycline	LED lamp	Blue LED lamp	80	[44]
8.	Phosphorus-doped /g-C_3N_4	Tetracycline	LED lamp	White LED lamp	80	[45]
9.	CdS nanorods	Tetracycline	LED lamp	White LED lamp	95.6	[46]

3.2.1 Slurry or Suspended Catalyst Systems (SCS)

Slurry-based photocatalysts are the most widely utilized photocatalytic reactor system. Slurry or suspended catalyst systems have been used for the majority of experimental investigations because of their simplicity of usage and adaptability to a wide range of catalysts. Through the use of a stirrer, the photocatalyst is dispersed throughout the reaction mixture. This type of installation is also referred to as a fluidized bed photoreactor due to its characteristic structure that prevents sedimentation, disperses the catalyst, and enables the catalyst's mobility. Haarstick et al. [47] constructed the cylindrical-shaped fluidized bed reactor in which they achieved optimum concentration of H_2O_2 production and $OH^{\cdot}$ radical production. The slurry

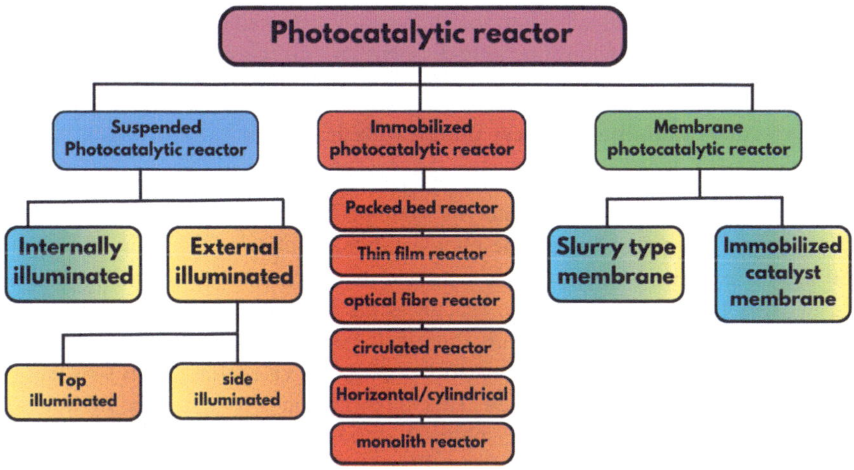

Fig. 4 Sub-classification of photocatalytic reactors based on catalyst

photocatalytic reactor design was the most utilized one [48–50] due to its high surface to volume ratio, good mass transfer, and adsorption mechanism [51]. In spite of having advantages, some of the drawbacks of photocatalytic reactors are reduced penetration of light and abrasion of the catalyst due to vigorous agitation. The slurry photocatalytic reactors can be classified on the basis of lamp position; (i) Internally illuminated and (ii) Externally illuminated as follows.

(i) **Internally illuminated slurry photocatalytic reactor**

In this type of photocatalytic reactor, a light source is installed inside the reactor vessel so that the light irradiation is enveloped by the reactants or pollutants to be illuminated. This kind of reactor is also referred to as lamp-submerged photocatalytic reactor. Due to its use of efficient light irradiation, tubular, cylinder-shaped photocatalytic reactor configurations are frequently employed in internal illumination methods. The two main reasons are that (1) light travels a relatively short distance and (2) the area of light that falls directly on the reactor is very large in comparison to externally lighted reactors. For example, Lu et al. [52] degraded isopropyl alcohol (IPA) photocatalytically using Ag/TiO_2 catalyst in an aqueous solution using 254 nm UV irradiation. The UV lamp is kept inside the photocatalytic reactor for internal illumination. The composition at 1.0 wt% Ag/TiO_2 gave the highest removal efficiency of 80% of the IPA. Metal halide lamp used as internal illumination source was studied by Shinde et al. [53]. They degraded orange M2R dye photocatalytically using a ZnO catalyst in an aqueous solution using a metal halide lamp. Almost 93% of the dye was degraded with 4 h of illumination of UV light. Even though, this type of reactor has some common advantages like any other slurry photocatalytic reactor, the internally illuminated slurry photocatalytic reactors are having three major drawbacks that restrict it from commercial applications. (i) Size restriction; only photocatalytic reactors of a certain size can be constructed using this technique.

(ii) Light distribution and adhesion; the photocatalyst may adhere to the walls of the illumination tube, thereby limiting light distribution. (iii) It is impossible to carry out sunlight-driven photocatalysis.

(ii) **Externally illuminated slurry photocatalytic reactor**

The externally illuminated slurry photocatalytic reactors allow many design systems to be used for this sort of photocatalytic system. Typically, an internal or external homogenizer is required to mix the photocatalytic solution in this type of reactor, and light irradiation is supplied from outside the reaction vessel, as illustrated in Fig. 5. This type of photocatalytic reactor is pretty good; when compared to the adverse remarks concerning internally illuminated photocatalytic reactors, externally illuminated slurry photocatalytic reactors are further categorized into two types: side-illuminated and top-illuminated. This classification is based on the limitations of the externally illuminated slurry photocatalytic reactor. The major drawback is the utilization of light. Assume that if we illuminate one side of the reaction vessel, the opposite side will remain dark or poorly illuminated. In order to achieve good efficiency, it is necessary to illuminate from both sides of the reaction vessel; however, that will result in an excessive amount of use of energy. Even though the majority of externally illuminated slurry photocatalytic reactors are cylindrical in shape to prevent catalyst segregation in the corners, if the catalyst is dark in color (i.e., black), the amount of light utilized by the catalyst in the center will be very low. The use of a homogenizer is required by the above; however, it makes the process uneconomical.

Side-illuminated Photocatalytic reactor: The positioning of light sources sideways to the reaction vessel facilitates the use of multiple light sources around the reaction vessel. The same is also considered a drawback of this reactor in terms of cost. The reflection of light is a bit low and the passing of light through multiple sides of the

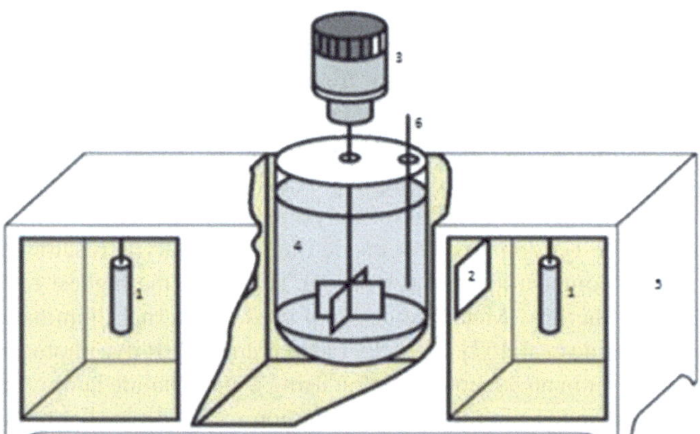

Fig. 5 Sideways Externally illuminated slurry photocatalytic reactor [(1) UV lamp, (2) square opening, (3) stirring module, (4) reactor, (5) box, and (6) sampling tube] [54]

vessel often leads to an increase in temperature that will lead to the alteration of particle size, particle aggregation, and kinetic collision of particles.

Top-illuminated Photocatalytic reactor: The positioning of light sources above the reaction vessel is a bit complicated when compared with a side-illuminated slurry photocatalytic reactor. Since the area of illumination is so narrow, the color of the catalyst will make light passage more difficult. As the absorption and scattering effects of light also contribute significantly [44], when light falls on the suspended catalyst of a side-illuminated system, it passes through it and very little light radiation is reflected back. In a top-illuminated system, when light falls to the bottom of the vessel from the top angle, the possibility of reflection increases, resulting in greater light energy utilization. Since this system imitates the natural solar light-driven photocatalytic system, this can be adopted for practical usage.

3.2.2 Immobilized Catalyst System (ICS)

Due to the limitations and demerits of slurry photocatalytic reactor setup, the catalyst should possess lower particle size, which makes fouling of light, reduction of light transfer, and additional cost and energy spend for removal/separation of catalyst. These limitations of slurry photocatalysts compelled the researchers to propose a new mode of catalyst installation called as immobilized or fixed-bed photocatalytic reactors [55–58]. In this type of photocatalytic reactor, the photocatalyst is placed as a static module (immobilized state) on the fixed supports and the photocatalysis process can be operated in continuous mode. Many new innovations like photoreactors followed by fixed-bed bioreactors are developed in recent days [59] as shown in Fig. 6. The important features that make fixed-bed photoreactors unique from slurry reactors are listed below.

1. **Reduction in the cost of separation**: Since the catalyst is immobilized on the support, there is no need for the catalyst from the resulting treated water.
2. **Light reachability**: Photons can directly reach the catalyst surface without any hindrance in immobilized reactors.
3. **Portability**: This immobilized catalyst reactor provides the photocatalytic system with some sort of portability. The catalyst can be fixed at various places in the reactor enabling portability.
4. **High penetration depth**: The problem of low penetration depth of photons can be solved in this type of reactor. Thin-film type of fixed bed can be employed to provide more surface area resulting in more absorption of light.

However, these photocatalytic reactors possess some demerits in their design. (1) Light distribution and (2) dependency on support. The light absorption in the fixed-bed photoreactors is good; however, the scattering of light and distribution of light through the bed limit the potential efficiency of the system. Since the catalyst is placed as static state with support, the light distribution to reach all of the catalyst molecules inside the bed is difficult. It makes the light to process conversion very

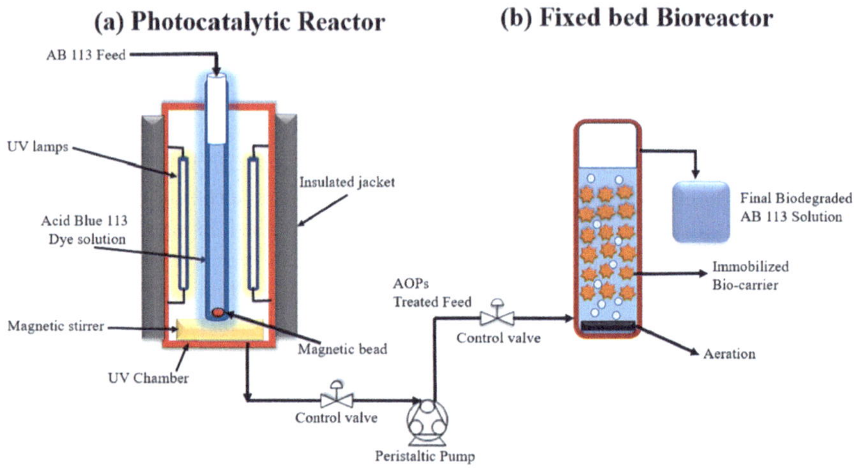

Fig. 6 Experimental hybrid reactor setup: **a** Photocatalytic reactor; **b** fixed-bed bioreactor [59]

low. These reactors are also further classified based on their geometry and type of loading [60–63]. The catalyst is placed in the bottom either as a powered form or placed in between the reaction vessels through some suitable supporting materials such as glass balls [64], monoliths [65], and optical fibers [66] to avoid the cost of separation of the catalyst. Various supporting materials were employed in this immobilized photocatalytic reactor system. Joseph and Vijayanandan [67] reviewed the various support materials that are employed in immobilized photoreactors along with their method of immobilization and pollutants. The compilation is reproduced here as Table 4, for a better understanding of the supporting materials for nanocatalyst in the immobilized systems [67]. However, the limitations in this type are that continuous usage is restricted due to the thermal energy generation in between the bed therefore leading to less efficiency, design, and optimization of thin catalytic bed having nanosized photocatalysts which makes the higher rate of light penetration.

3.2.3 Membrane Photoreactor

These reactors are hybrids that incorporate conventional technology with advanced oxidation techniques. Since photocatalysis is involved with three different phases (1) pollutant (liquid), (2) photocatalysis (solid), and (3) photon (massless), it makes the entire process complicated. Chester et al. [68] implemented the utilization of membrane separation and photocatalysis in their study. An annular membrane reactor was constructed for the purpose of degrading formic acid and atrazine. Later in this research, a few works were published on this regard on the degradation of dyes [69] and textile effluents [70, 71]. The current class of reactors differs by the integration of other technologies, leading to its further classification into various sub-categories based on the specific methodologies employed like ultrafiltration membrane module

Table 4 Studies on immobilized photocatalysts on suitable supporting materials [67]

Photocatalyst	Supporting material	Method of immobilization	Pollutant
Green BiOX	Ceramic fibers	Sol–gel method	Bisphenol A and Ampicillin
TiO_2	Cellulose	Sol–gel method	Phenol
CdS nanoparticles	Cellulose	Dip-coating	–
TiO_2	Diatomite	Modified sol–gel method	Rhodamine B
TiO_2	Borosilicate glass spheres	Dip-coating	Methylene blue
TiO_2	Silicone sealant/glass	Dip-coating	Rhodamine B
Z-scheme P3HT/ α-Fe_2O_3	α-Fe_2O_3 nano-array	Electrophoretic deposition	Tetracycline and Bisphenol A
TiO_2	Oxide films	Anodization	Rhodamine B
TiO_2	Carbon nanotube	Sol–gel method	Methyl orange
ZnO	Glass plate	Heat attachment method	Acid Red 88
TiO_2	UV-resistant Polymer	Solvent casting	Reactive Blue 8 (RB 8) and Reactive Blue 220
TiO_2	Stainless steel woven meshes	Dip-coating and electrophoretic deposition	–
TiO_2	Activated carbon	Solvent casting	Basic Red 18 and Basic Red 46
N-doped TiO_2	Polystyrene	Solvent-cast method	E Coli
N-doped TiO_2	Glass plates	Dip-coating	E Coli
TiO_2	Glass spheres	Dip-coating	15 emerging contaminants
TiO_2	Quartz	Dip-coating	Phenol
TiO_2	Quartz sand	Sol–gel	Propranolol Diclofenac Carbamazepine Ibuprofen
$AgBiO_3$	Cellulose acetate	Phase inversion	4-nitrophenol
TiO_2	Quartz tubes	Dip-coating	Amoxicillin
TiO_2	Cellulose fiber and Ceramic templates	Dip-coating	Methyl Orange, Cyanide
TiO_2	Stainless steel mesh and fiberglass strips	Sol suspension	Phenol and reactive blue 19
TiO_2	Black volcanic ashes	Hydrothermal decomposition	Methylene blue

(continued)

Table 4 (continued)

Photocatalyst	Supporting material	Method of immobilization	Pollutant
N-doped TiO_2	Ceramic membrane	Dip-coating	Diclofenac
TiO_2	Polypropylene copolymer surfaces	Electrostatic-heating coating	Methylene blue
TiO_2 & ZnO	Poly (vinylidene difluoride)–co-trifluoroethylene	Solvent casting	Methylene blue
TiO_2	Glass Raschig rings	Dip-coating & Sol–gel	Diclofenac
TiO_2	Poly vinylidene fluoride (PVDF) dual-layer hollow fiber membranes	Co-extrusion technique	Trimethoprim, Metoprolol, and Carbamazepine
TiO_2	Membrane—nano-engineered membrane	Sol–gel	Oxytetracycline
TiO_2/rGO/ g-C_3N_4	Polystyrene film	Solvent casting method	Remazol turquoise blue dye
TiO_2	Quartz fiber filters	Dip-coating/ thermal–chemical oxidation	Carbamazepine, venlafaxine, fluoxetine, atenolol, sulfamethoxazole, ibuprofen, atorvastatin, naproxen
NF-TiO_2	Nitrogen and Fluorine matrix	Sol–gel	1,4-dioxane
TiO_2	Polyethersulfone matrix	Phase inversion technique	Methyl orange dye
g-C_3N_4	Chitosan hydrogel beads	Sol–gel	Sulfamethoxazole and Carbamazepine
ZnO	Glass spheres	Dip-coating method	Real tannery wastewater
ZnO	Glass plates	Dip-coating method	Formaldehyde
TiO_2	Activated carbon	Hydrothermal method	Aflatoxin B1 (AFB1) and aflatoxin B2 (AFB2)
TiO_2	Posidonia Oceanica fibers	Sol–gel	Phenol
ZnO/Zn	Pt/C plate	Sonication and heat attachment method	Reactive Green 19
BiOCl	Calcium Alginate Hydrogel beads	Sol–gel	Rhodamine B
ZnO	Concrete	Water sealer method	Phenol

(continued)

Table 4 (continued)

Photocatalyst	Supporting material	Method of immobilization	Pollutant
TiO_2	Activated carbon	Solar irradiation	Amoxicillin, ampicillin, diclofenac, and paracetamol
ZnO	Stone	Dip-coating	Ciprofloxacin
BiOBr	Stainless steel wire mesh	Solvothermal process	Rhodamine B
TiO_2	Chitosan	Sol–gel	Reactive Red 4
Au@Ag core–shell nanoparticles	TiO_2 nanofibers	Dip-Coating	Methylene blue
Ag@AgCl	Chloridized g-C_3N_4 nanosheets	Ion exchange implanting approach	Tetracycline
TiO_2	Concrete	Water sealer method	Furfural
TiO_2	Borosilicate glass, refractory brick, and alumina foam	Dip-coating	Imazalil

[72], membrane distillation, membrane dialysis [73], etc. Since it is not appropriate to our context, we will concentrate on the basic idea and its classification based on catalyst installation.

Photocatalyst is deposited/suspended on porous membranes in this type of reactor. Larger pollutants may remain unfiltered by the membrane while the photocatalyst oxidizes the pollutant into smaller products (permeate) along with water that are capable of passing through the membrane shown in Fig. 7. The permeates can be reutilized to achieve a higher scale of efficiency. These are further classified based on the catalyst installation [74]. (1) Slurry or suspended photocatalytic membrane reactor (2) Fixed-bed or immobilized photocatalytic membrane reactor. Just as the slurry and immobilized reactors, these reactors also reported several disadvantages by various studies [75, 76]. The main limitation is the photocatalyst's fouling of the membrane's pores. Over the years, numerous anti-fouling photocatalytic membranes have been developed to reduce their fouling property [77–79]. The characteristic fouling mechanisms in membrane reactors [80] are shown in Fig. 8. This fouling feature is commonly observed in slurry-type membrane reactors. The obstructing effect of catalytic particles on the membrane surface pores [81] can be resolved by employing a fixed-bed membrane reactor in which the photocatalyst is fixed as a static module on the membrane, thereby reducing membrane fouling.

There are two methods to implement the fixed-bed membrane photocatalytic reactor: (1) mixed matrix membrane; (2) coated membrane, as shown in Fig. 9 [82]. Each membrane approach has benefits and drawbacks. When the bed gets thinner over time, the coated membrane-mediated photocatalytic system exhibits poor stability and fouling properties [83]. But this coated membrane has a simple preparation process (spin coating), only a few factors to take into consideration, and

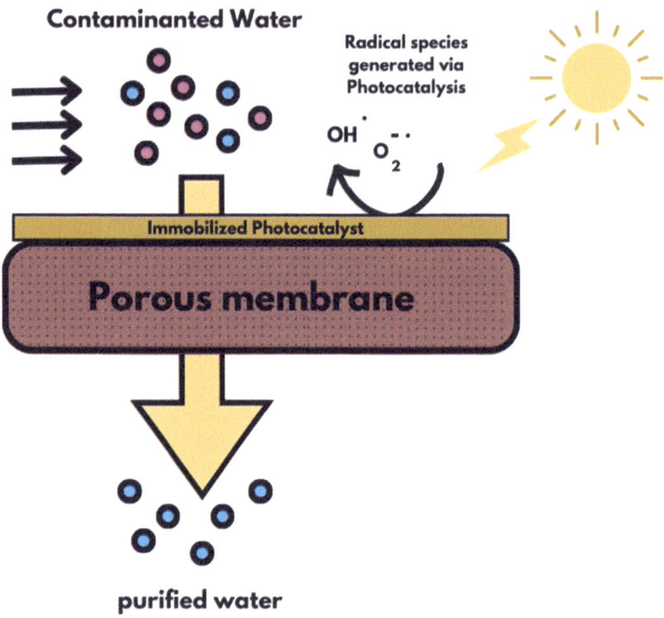

Fig. 7 Illustration of photocatalytic membrane reactor mechanism

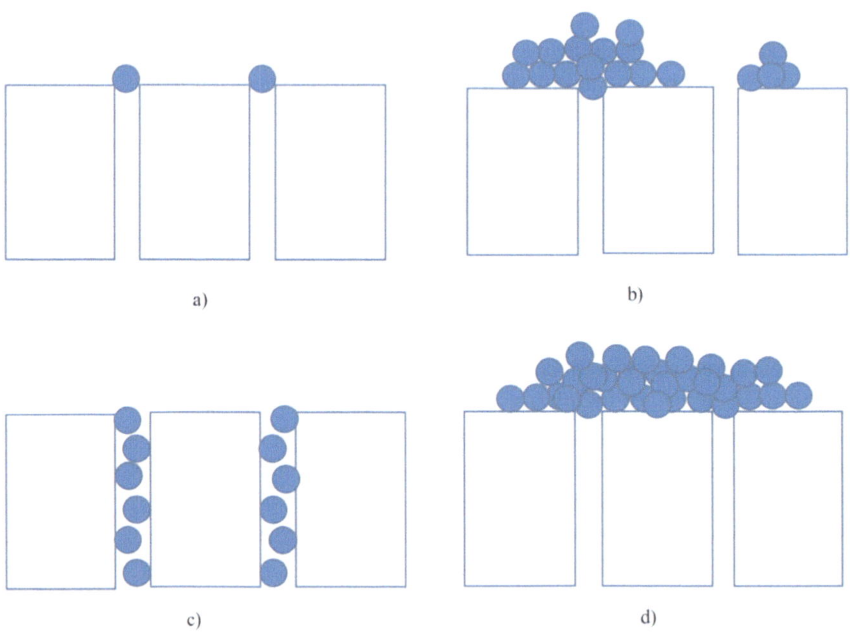

Fig. 8 The schematic diagram of membrane fouling mechanism, **a** complete blocking, **b** intermediate blocking, **c** standard blocking, and **d** cake formation [72]

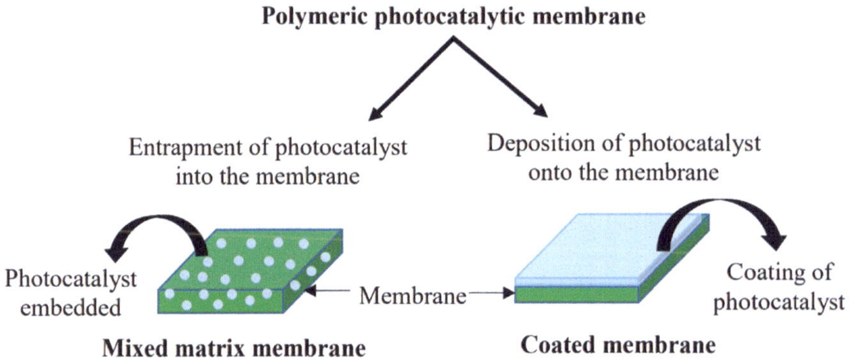

Fig. 9 Pictorial representation of fabrication of photocatalytic membrane [82]

very convenient to handle [84]. In the case of mixed matrix membranes, they exhibit good stability and anti-fouling properties over time; typically, the mixed matrix membranes are prepared using phase inversion methods, which makes the preparation quite different [82]; however, it is very complicated to obtain a uniformly distributed matrix, necessitating for attention to many variables like thickness, concentration, and porosity [85]. The diversity in the membrane separation concept, membrane preparation, and catalyst installation makes the photocatalytic membrane reactors quite different and has attracted more researchers from various fields. Table 5 summarizes some of the notable visible light-induced photocatalytic degradations from the last decade [86].

3.3 Based on Reactor Geometry/Design

A significant improvement in mass transfer [87], light–catalyst interaction [88], and adsorption kinetics was also made possible by the reactor's architecture. Regarding the subject, numerous works and reactors have been developed. The photocatalytic reactors received an immense number of geometrical modifications; however, it was discovered that the majority of them were designed for CO_2 photocatalytic reduction rather than degradation. We will therefore talk about the most popular photocatalytic reactors with geometric modifications in this part. These reactors could operate in continuous or batch modes, with installations for suspended or immobilized catalysts, and various light sources such that the reactors might have unique configurations with plenty of features we previously discussed.

Table 5 Photocatalytic membrane reactors (PMR) studied under visible light photocatalysis [86]

PMR configuration	Support	Photocatalyst	Pollutant	Main results	Years
HPC + MD	Microporous hydrophobic flat sheet PTFE membrane	Flower-like BiOBr microspheres	Methyl orange	High efficiency for MO photodegradation. High-quality permeate with constant flux. No membrane fouling The high energetic consumption of MD limited its coupling with HPC for water and wastewater treatment	2013
Photocatalyst coated on the membrane	Ceramic UF membranes	N–TiO_2 GO–TiO_2 Organic shell Layered TiO_2	Methylene blue (MB) and Methyl orange (MO)	29 and 15% MB and MO degradations by using the membrane coated with N-TiO_2 under visible light irradiation	2015
Photocatalyst coated on the membrane	Commercial α-Al_2O_3 photocatalytic membrane	N–TiO_2	Carbamazepine	Degradation rates of "flow through" the membrane > degradation rates "flow tangential to" the surface of the membrane Enhanced photoactivity of N-doped TiO_2-coated membranes under UV wavelengths, and activity under visible light A disadvantage of coated PMRs: the photocatalytic degradation is controlled by pollutant diffusion to the catalytic surface. The increase of the mass transfer with increasing water flux was limited by membrane properties	2016
Photocatalyst deposited on the membrane	α-Al_2O_3 membranes	TiO_2	MB	Complete MB degradation in only 40 min under solar light irradiation	2017

(continued)

Table 5 (continued)

PMR configuration	Support	Photocatalyst	Pollutant	Main results	Years
Photocatalyst coated on the membrane	PVDF membrane	$ZnIn_2S_4$	Tetracycline	Removal efficiency >92% was maintained for 36 h of continuous operation (under influent and effluent flux of 26.09 L m^{-2} h^{-1}) with 100 μg L^{-1} drug concentration Good membrane stability: the surface and structure of the PVDF membrane were not affected by the photocatalytic process	2018
Photocatalyst immobilized in the membrane	Polysulfone membrane	Novel mesoporous graphitic carbon nitride/titanium dioxide (mpg-C_3N_4/TiO_2) nanocomposite	Sulfamethoxazole	Sulfamethoxazole was degraded into seven non-toxic and pharmaceutically inactive by-products by the PMR technology Satisfactory sulfamethoxazole SMX removal efficiency was obtained by operating with the membrane named PSf-3 (with 1% mpg-C_3N_4/TiO_2 loading) for 30 h of consecutive irradiation Good membrane stability: membrane provided a stable support with high integrity and flexibility after solar irradiation The prepared photocatalytic membrane has a great potential to be applied in the water treatment industry	2018

(continued)

Table 5 (continued)

PMR configuration	Support	Photocatalyst	Pollutant	Main results	Years
SMPR with suspended photocatalyst	Polypropylene hollow fiber membrane	Ce–ZnO nanoparticles	Reactive orange 29	In the best conditions, 97.84% of dye removal was achieved in the continuous flow visible light SPMS reactor GC-Mass, COD, and TOC analyses demonstrated the degradation and mineralization of RO29 The Ce-ZnO nanocomposite showed a favorable antibacterial behavior against positive and negative bacteria	2019
SMPR with suspended photocatalyst	MF ceramic membrane	N–TiO_2	Diclofenac	The efficiency of the photocatalytic process decreased by increasing the initial concentration of the drug while it was improved by adding H_2O_2	2020
PVDF/ LDH@g-C3N4@PDA/ GO composite membrane	PVDF	C_3N_4	Methylene blue (MB), rhodamine b (RhB), gasoline, diesel, and petroleum	Rejection rates of methylene blue (MB), rhodamine B (RhB), gasoline, diesel, and petroleum ether were 100%, 94.61%, 96.74%, 93.22%, and 92.35%, respectively	2021

3.4 *Based on the Geometry and Its Configuration*

The architectural layout of the reactor and its configuration can have a significant impact on the efficiency and capabilities of photocatalytic processes. From bench-top to industrial scale, numerous reactors have been utilized for photocatalysis at various scales. It was understood by conducting a literature review that numerous reactor configurations and operating modes are proposed; however, the following are the most common variants and have been chosen for further discussion:

1. Rotating disk/drum photoreactor.
2. Thin-film photoreactor.
3. Spinning disk photoreactor.
4. Cascade-type photoreactor.
5. Solar light-driven photoreactor.

3.4.1 Rotating Disk/Drum Reactors

The Beer–Lamberts equation indicates that effective light utilization depends on the thickness and concentration of the solution [89]. The problem with general slurry or immobilized photocatalytic reactors is the utilization of light. Even though catalyst performance is optimized to its maximum level, the light loss in these reactor models is still high. To sort out the problem of light loss, rotating disk or drum-based photocatalytic reactors are developed. Dionysiou et al. [90] worked on the development of rotating disk photocatalytic reactors. They prepared a ceramic ball nanocomposite using commercially available TiO_2 and placed it as an immobilized photocatalyst. They conducted the research at various angular velocities for the degradation of various organic pollutants [90]. Now the researchers focused on the synthesis of efficient photocatalysts and coated them into a suitable support that will maximize the light absorption and reflection. Zangeneh et al. [88] prepared the L-histidine-TiO_2-CdS-coated glass disk and conducted the photocatalytic degradation of direct red 18 (DR18) on a rotating disk reactor as in Fig. 10. They found that DR18 completely degraded under optimized conditions of rough glass disk with 100 rpm with a light intensity of 13 lumens per m^2.

The architecture of this type of photocatalytic reactor consists mainly of a rotating disk, a light source, and a reactor vessel or path tube. They are built with consideration for the following factors: reactor capacity, distance between two disks, disk type, disk diameter, rotational speed, reactor path length, etc. [91]. The suitable disks are coated with photocatalytic substances and installed at certain distances. The disks carry the thin liquid layer, and the system is immersed in the reaction liquid while it's subjected to light. Contaminants that are present in the thin liquid film are destroyed by light. The key benefit of this type of reactor is its high light utilization and low recombination rates, which are caused by the angular velocity and centrifugal force generated by the disk's rotation [92]. This is further supported by multiple studies

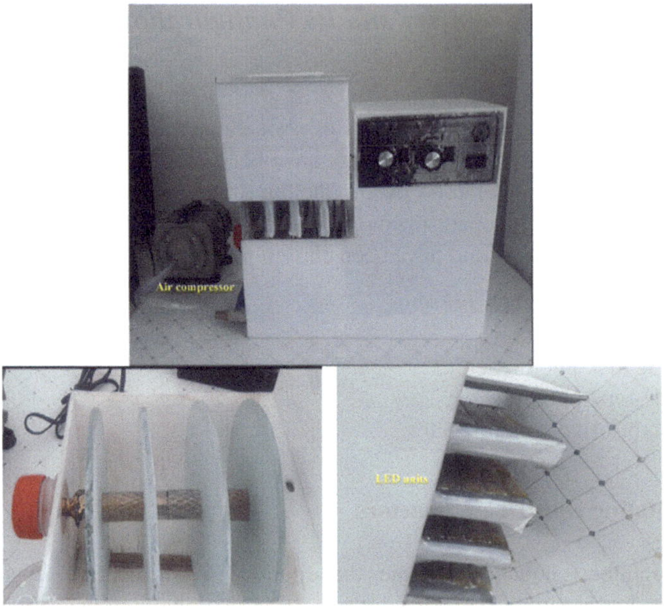

Fig. 10 Rotating glass disk photocatalytic reactor [88]

showing a difference in degradation or photocatalytic activity with disk rpm [93, 94]. This enables various alterations to be accomplished through modifying the rpm, light source, disk, etc. as shown in Fig. 11 [95].

3.4.2 Thin-Film Photocatalytic Reactor

This type of photocatalytic reactor falls under the category of immobilized catalytic reactors since the catalytic portion is present here as a thin-film plate. Usually, in fixed-bed photoreactors, only a small portion of the photocatalyst is illuminated, resulting in light loss and low cost-effectiveness. Therefore, thin-film photocatalytic reactors emerge as a result. The catalytic portion's broad dispersion accelerated faster light absorption and expanded surface area, which produced more reactive radical species. Typically, these thin-film reactors are made to work best with little dosages. The basic architecture of the reactor is that the photocatalyst is coated with a suitable binding agent on the surface of quartz or double-walled borosilicate glass/tile plates (Fig. 12). With an inlet flow rate of 0.1–1 mL/s, the contaminated water is allowed to enter the thin film; the liquid film's maximum thickness is in the range of 0.1–0.3 mm. Diverse outcomes in photocatalytic activity [96] can be accomplished by adjusting the flow rate, concentration, and thickness of the liquid film.

Saran et al. [98] developed and researched a thin-film plate photocatalytic reactor (TFPR) driven by solar light for the treatment of effluent from a sugar industrial

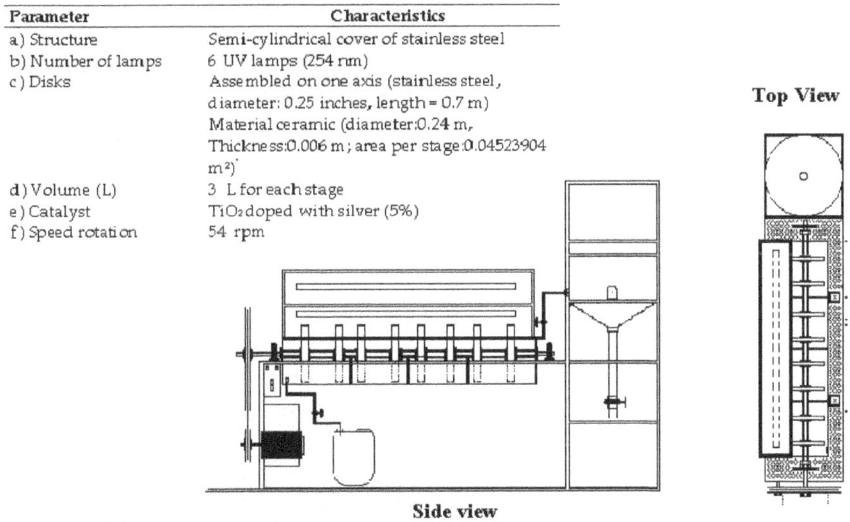

Fig. 11 Rotating disk photocatalytic reactor and its specification [95]

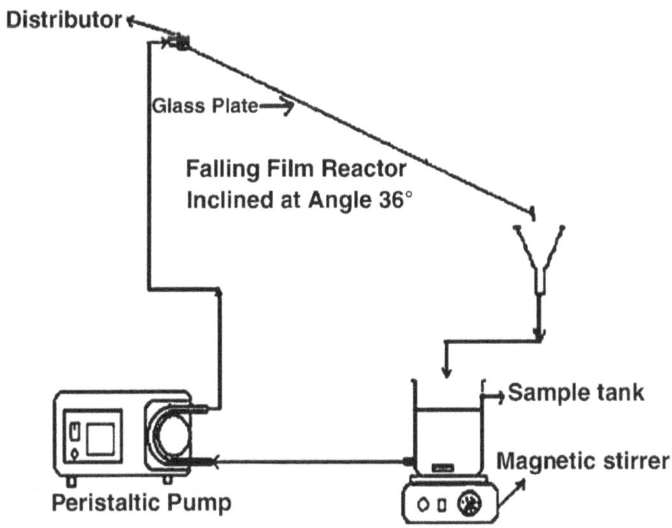

Fig. 12 The schematic diagram of thin-film immobilized surface reactor [97]

plant (Fig. 13). The ceramic tiles used as immobilized photocatalysts were coated with Ag/TiO_2 and placed at a specific angle for the water flow. The entire setup was kept in natural light, and the performance of the pure TiO_2 catalyst was compared to the results. The chemical oxygen demand (COD) results indicated that Ag/TiO_2 and pure TiO_2 were 95% and 86% efficient, respectively. It was discovered that efficiency

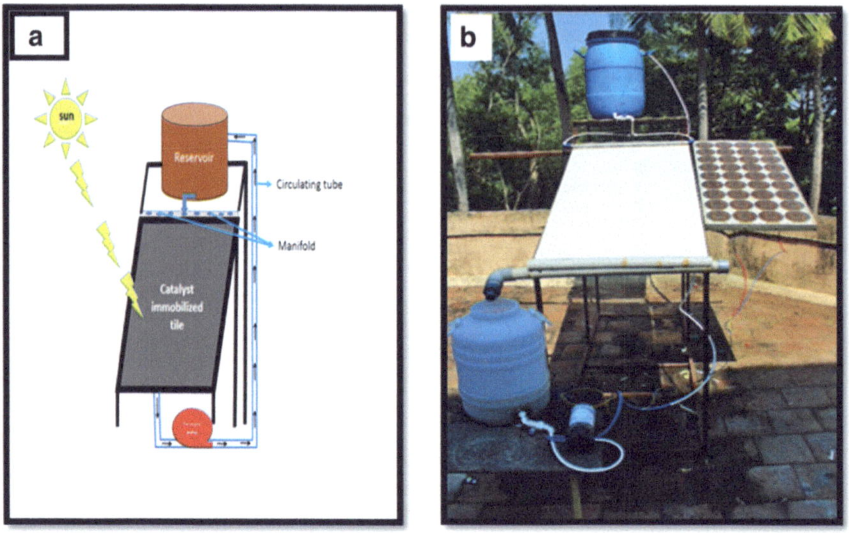

Fig. 13 Pilot scale Thin-film photocatalytic reactor (TFPR). **a** Schematic design of TFPR and **b** Actual image of TFPR [98]

up to 99% could be achieved with this sort of reactor by controlling the parameters like flow rate, initial pH, sunlight duration, and amount of H_2O_2.

3.4.3 Spinning Disk Photocatalytic Reactor

The rotating disk reactor and the spinning disk photocatalytic reactor are similar; however, the researchers focused on the process intensification to improve the efficiency and productivity of existing reactors. The spinning disk photoreactor's basic components are an immobilized catalyst-coated disk, a water injector setup, and a light source. The contaminated water is injected into the center of the spinning disk, which is also being illuminated by light (either internally or externally) while rotating at a specific speed. The spilled wastewater is collected at the reactor's bottom and can be injected repeatedly via a pump. This sort of reactor is effective since the treatment of the contaminated water can be repeated. After a period of time, the output flow is monitored, and the initial data is used for analyzing it.

Mirzaei et al. [99] modeled and evaluated the performance of a spinning disk photocatalytic reactor for the degradation of phenol. They employed TiO_2-coated stainless steel as the spinning disk photocatalyst module; the schematic setup is shown in Fig. 14. The findings showed that the phenol was completely destroyed away by the rough-baffled disk within 180 and 150 min. It was discovered that a spinning disk photocatalytic reactor could significantly speed up reaction time and enhance mass transfer. Similarly, studies based on the development of spinning disk photoreactors were conducted to improve their features and to know how they solve

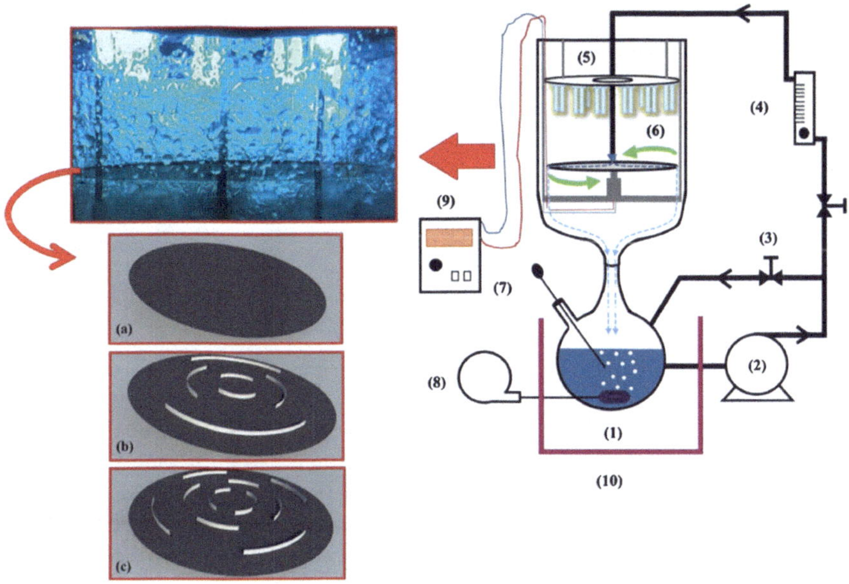

Fig. 14 Scheme of setup for photocatalytic experiments: (1) feed reservoir, (2) pump, (3) valve, (4) flow meter, (5) strip of UV lamps, (6) spinning disk, (7) temperature sensor, (8) compressor, (9) power supplier, and (10) circulating water bath. With different structures of disk surface: **a** smooth disk, **b** baffled disk-I, and **c** baffled disk-II [99]

the photon-transfer limitations [100–102]. Since the researchers worked on process intensification, different placement positions (Figs. 15 and 16) of the disk have been studied to evaluate the fluid dynamics, improved photon transfer, and optimal flow rate identification.

3.4.4 Cascade Photocatalytic Reactor

As the name itself suggests, cascade photocatalytic reactors are multistage reactors in which the contaminated water is allowed to enter multiple treatments sequentially. So, the output water from each stage is admitted for the next phase of treatment. If the total setup is designed in a closed-loop mode means, it is also referred to as closed-loop photocatalytic reactors. These reactors are evolved to fix the limitations of thin-film reactors as it is having prolong reaction times and slow kinetics. Stephen et al. [103] utilized a thin-film photocatalytic step reactor for the degradation of metolachlor and chlortoluron. The closed-loop thin-film reactor has a tank, pump, and spillway at the top and steps were designed with six identical steps covered with photocatalytic media as shown in Fig. 17.

TiO_2-coated stainless steel was used by Amiri et al. [87] as a photocatalyst component in a spinning disk photoreactor for the degradation of reactive yellow 81 shown

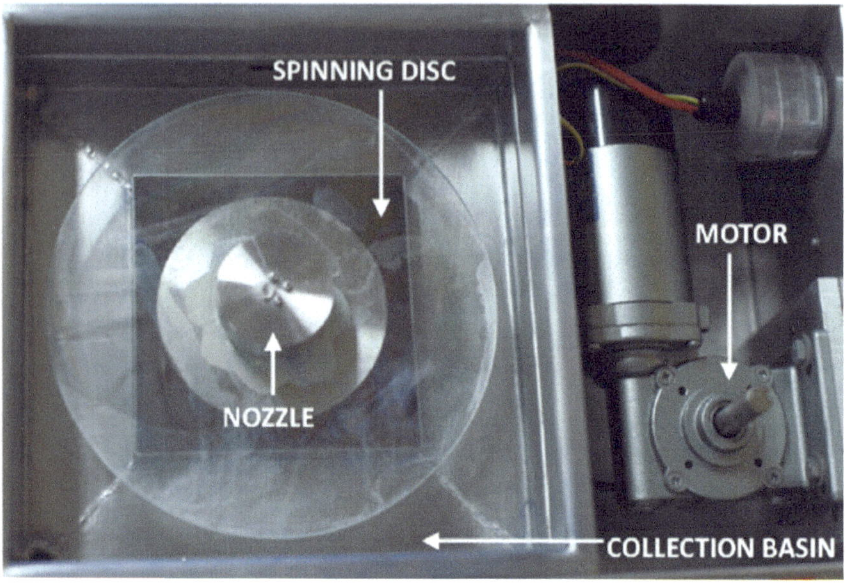

Fig. 15 Spinning disk photocatalytic reactor employed for the degradation of methylene blue [100]

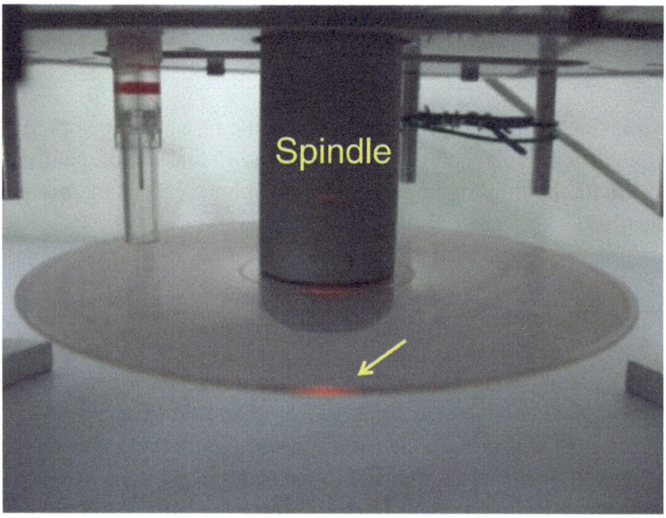

Fig. 16 Dual light circular plane disk photocatalytic reactor [101]

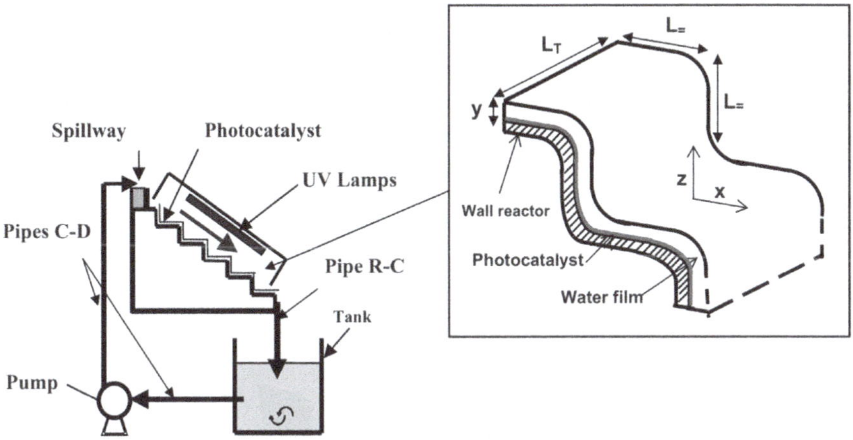

Fig. 17 Cascade thin-film photocatalytic reactor for the degradation of metolachlor and chlortoluron [103]

in Fig. 18. When compared to the typical immobilized reactor, it was discovered that the cascade photocatalytic reactor had good mass transfer and increased catalytic efficiency. The authors investigated the effect of roughness and flow rate on the system, and it was discovered that both variables strongly contributed to the mass transfer phenomenon. As the Reynolds number (Re) was raised from 328 to 822, it was clear that the spinning disk photocatalytic reactor improved fluid flow dynamics and mass transfer.

Recently, Fouad et al. [104] have proposed a novel closed-loop photocatalytic reactor for the degradation of sulfamethazine (SMZ) with W-TiO_2-coated plates shown in Fig. 19. Using polysiloxane, tungsten-doped TiO_2 (W-TiO_2) was used to immobilize four stainless steel plates in sequence in the reactor. The efficiency of SMZ degradation after 40 min without any scavengers was 98.44%, according to the

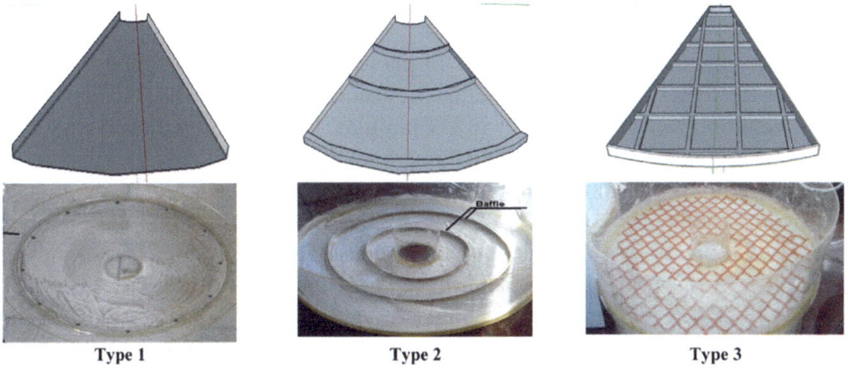

Fig. 18 Schematic design of roughness types studied via cascade photocatalytic reactor [87]

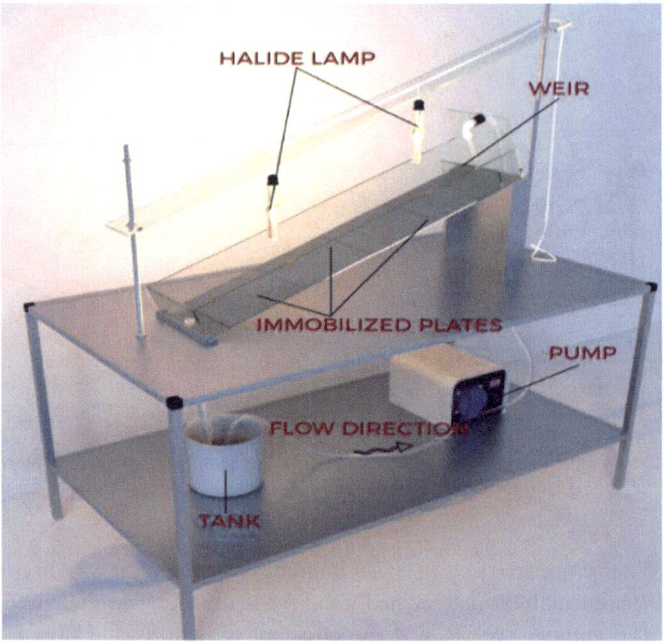

Fig. 19 Closed-loop photocatalytic reactor for the degradation of sulfamethazine [104]

authors. It was concluded that the repeated treatment of the contaminant in the loop mode produced significant results. Since the entire photocatalytic process takes place in the open air, no additional oxygen will be required, and the setup doesn't require any homogenization and air sparging setup, which lowers the cost of the instrument as well as the cost of energy.

3.4.5 Solar-Driven Photocatalytic Reactor

Here comes the actual green technique of photocatalysis, which is the use of solar energy as the light source. Solar radiation is cheaper than fossil fuels and, depending on the location, has a greater supply of UV photons. Radiation flow, however, fluctuates immensely and is only obtainable throughout the day [105]. The design components are simple: light collecting plates, a reservoir tank, and a catalyst module. The majority of research has been on artificial light, with very little attention paid to the utilization of natural sunlight. The fundamental goal of photocatalysis research is to make use of the plentiful natural sunlight; therefore, even though artificial light sources can mimic sunlight irradiations, they cannot replace them. Based on the incoming irradiation, photocatalytic reactors with a natural light source (solar photoreactors) are categorized into two categories: concentrating and non-concentrating light (Its merits and demerits are summarized in Table 6). The light collector is the main element of solar photoreactors that governs how efficiently the reactor operates. (1) Parabolic trough collectors (PTC) and (2) Flat plate collectors are the two main light collectors that have been proposed for solar photoreactors [106].

The concentration of the light energy for the photoexcitation is crucial for the entire reactor setup to produce photocatalytic activity. PTCs fall under the type of concentrated and non-concentrated solar reactors, and these are constructed of an assembly that holds a reflecting, concentrating parabolic surface. The collection window plane of this structure is continuously kept perpendicular to the sun's rays by one or more motors that are driven by a solar tracking system on one or more axes [107]. If it is placed statically with no tracking mechanism, then it will be considered a non-concentrated solar reactor. The primary benefit of this form of non-concentrated reactor is its simplicity, in addition to its lower cost than concentrated reactors. On an energy basis, they might capture sunlight less effectively for photocatalysis. Because of the limitations of current reactor designs, research on sunlight-driven reactors is lacking and rarely implemented. Some notable results for pollutant degradation using compound parabolic collector-assisted solar reactors are summarized by Tanveer and Guyer [108] and are reproduced below with permission as Table 7.

Since tubular [33] and annular [109] modified photocatalytic reactors are widely used and their performance is primarily determined by how and where the catalyst and light source are located (previously discussed), those reactors were not discussed in-depth; however, some of the positive reviews from recent years were cited here [66, 106, 110]. The schematic diagrams for the geometrically configured reactor designs are tabulated (Table 8).

Table 6 Advantages and disadvantages of solar reactor types

S. No	Type	Advantages	Disadvantages
1.	Non-concentrating solar reactors	• Cheaper • Easy to handle and simple mechanism • High quantum efficiency	• Static placement drives heat generation which may lead to evaporation of contaminants • Highly dependent on weather conditions • Regular monitoring required
2.	Concentrating solar reactors	• Highly efficient for collecting light radiation • Small reactor tube only required • More applicable for practical adaptations • Availability of direct and diffuse radiations for catalyst excitation • Regular monitoring is not needed, and it possesses easy maintenance to conduct photocatalysis	• High cost • Complex mechanism • Low quantum efficiency • High capital investment is needed

Table 7 CPC-assisted solar reactors for various wastewater treatments [108]

Pollutant types	Pollutant Conc	Catalyst type	Catalyst conc	pH	Oxidant type and conc	Efficiency	Time
AMBI	1 mM	Fe^{+3}	0.1 mM	N/A	10 mM H_20_2	90% COD removal	N/A
Remazol brilliant blue R-A	N/A	Fe^{+3}	0.5 mM	N/A	2281 mg/L of H_2O_2	100%color removal and 85% TOC removal	8 h
Azo-dye acid orange 24	N/A	Fe^{+2}	0.1 mM	N/A	5 mM of H_2O_2	85% color removal	65 min
4-Chloro-2-methylphenoxyacetic acid (MCPA)	186 mg/L	Fe^{+2}	1.0 mM	3	N/A	75% TOC reduction	2 h
Reactive blue 4	20 ppm	Fe^{+2}	7 ppm	2.5	120 ppm of H_2O_2	Complete removal of color and COD	1 h
Mixture of commercial pesticides	200 mg/L	Fe^{+2}	20 mg/L	2.8	16 mM H_2O_2	75% TOC reduction	4 h
Phenol	180–733 mg/L	Fe^{3+}	20 mg/L	7	H_2O_2, H_2O_2/phenol = 5.5	100% Degradation 85% COD reduction	1 h
Phenolic compounds	20 mg/L	Fe^{+2}	20 mg/L	4	H_2O_2, 40 mg/L	Total elimination of original pollutants and 94% TOC	5 h
Dihydroxy-benzene compounds	34 mg/L	Fe^{3+}	1 g/L	5	H_2O_2, 60 mg/L	Mineralization higher than 80%	6 h

(continued)

Table 7 (continued)

Pollutant types	Pollutant Conc	Catalyst type	Catalyst conc	pH	Oxidant type and conc	Efficiency	Time
Emerging contaminants in wastewater treatment plants	750 μg/L	TiO_2	50 mg/L	4	N/A	75% Degradation	100 min
Pesticides	1000 μgL	Fe^{+2}	140 mg/L	2.8	18 mM of H_2O_2	79% Mineralization and total pesticide removal	N/A
Ibuprofen	20–200 mg/L	TiO_2	0.1–1 g/L	4.5	N/A	50% TOC reduction	2 Days
4-Chlorophenol	3 mM	Fe^{2+}	0.75 mM	2.8	H_2O_2, 45 mM	90% Degradation	30 min
Cyanide-containing water	60 ppm	Fe^{2+}	8 ppm	3.3	H_2O_2, 2000 ppm	100% Degradation 90% TOC reduction	N/A
Orange II aqueous solutions	5 ppm	Fe^{2+}	2 ppm	3.5	H_2O_2, 225 ppm	100% COD reduction, 80% TOC reduction	180 min

Table 8 Summary of innovative reactor types and their schematic representation

S. No	Type of reactor	Schematic representation of model reactor
1.	Rotating drum photocatalytic reactor [111]	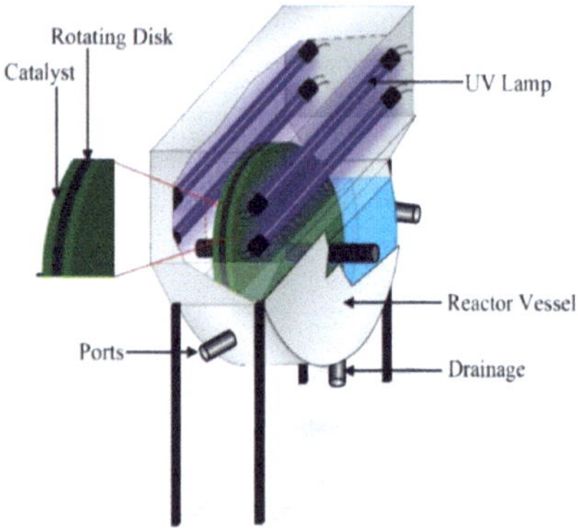
2.	Annular photocatalytic reactor [111]	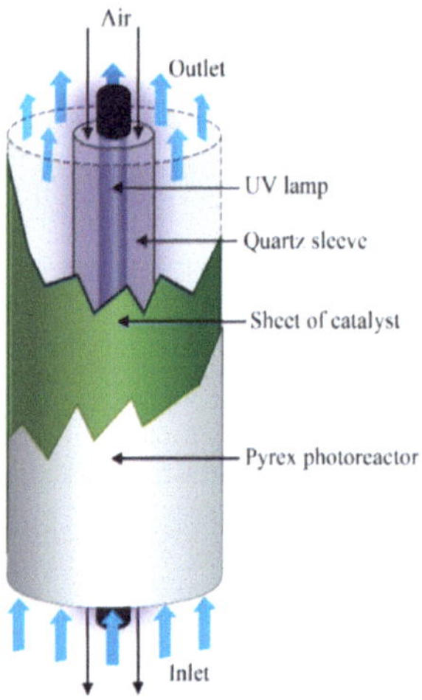

(continued)

Table 8 (continued)

S. No	Type of reactor	Schematic representation of model reactor
3.	Closed-loop photocatalytic reactor [111]	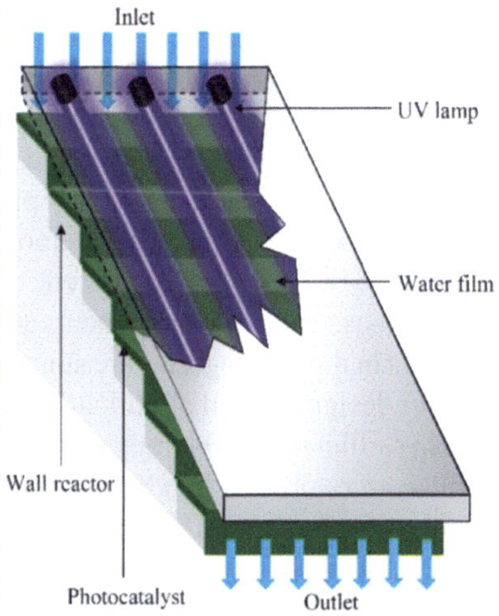
4.	Tubular photocatalytic reactor [112]	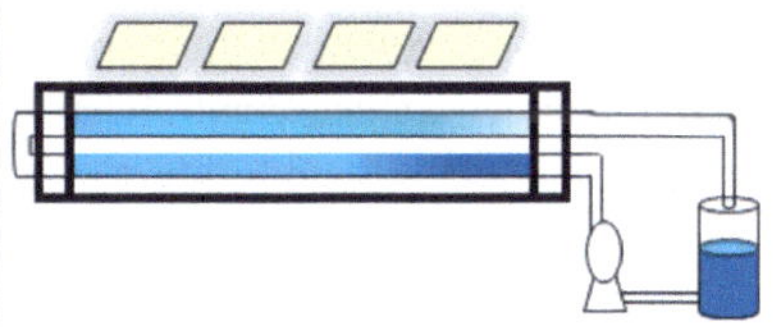
5.	Spinning disk reactor [113]	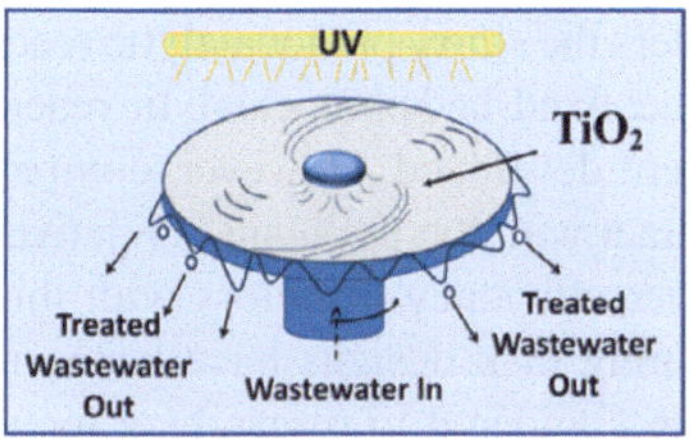
6.	Cylindrical photocatalytic reactor [114]	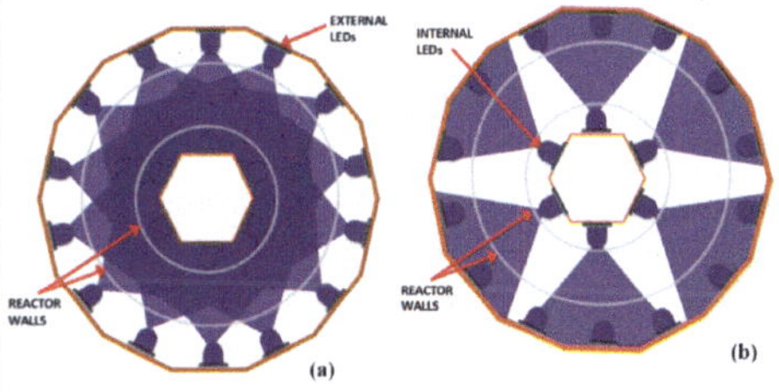

4 General Conclusion and Perspectives

In summary, it was understood by referring to various literature that the conversion of light energy into chemical energy via photocatalytic reactions can be significantly improved by modification of its components. Process intensification has been carried out by various research groups in this field, especially the fluid dynamics, photon-transfer limitations, mass transfer, and adsorption kinetics have been put into account for development by engineering and physics researchers. Since the variables to play with the photocatalytic reactors are more, it is difficult to address unique and customized reactors. However, photocatalytic reactors have been classified into three general categories based on their components and geometry. The design prospects and addressable issues in those categories are summarized neatly with visual illustrations of various reactor designs. Small modifications to the reactor designs, such as the use of modern energy-efficient LED lamps, catalyst-absorption-friendly targeted lamps, as well as others, were observed to be responsible for the significant efficiency and energy efficiency of the results. The essential design components of a photocatalytic reactor, as well as its systematic classification based on light, catalyst installation, and reactor shape, have been addressed in this chapter. Numerous variations, such as fluidized bed reactors, closed-loop reactors, rotating disk reactors, cascade, thin film, and spinning disks, as well as their specific properties, were addressed.

LED lamps have been an excellent alternative for reducing energy waste and as a good substitute for conventional lights. Suspended catalyst reactors feature a large surface area and more active sites, which increases efficiency but also adds sequential restrictions including catalyst aggregation, light penetration, and the requirement for post-treatment for recovering catalyst from the treated water stream. This uneconomical cause hinders the slurry photocatalytic reactor's commercialization potential. To address the issue, fixed-bed photocatalytic reactors with no post-treatment processing or filtration were developed. However, compared to slurry systems, the amount of catalyst in contact with the pollutant is relatively small in fixed-bed reactors, which results in poorer efficiency. Reactors with thin-film, membrane, cascade, rotating disk, and spinning disk designs have been created to get around the constraints. However, when compared to researchers focusing on photocatalytic degradations, the research groups dedicated to photocatalytic reactors are fairly small.

Table 9 summarizes the advantages and disadvantages of the several reactor designs, including some that haven't been covered in this chapter. Although we made an effort to compile a list of contemporary, prospective photocatalytic reactor configurations, it must be noted that photocatalytic reactors are by nowhere restricted to those covered in this chapter and that there will likely be plenty of room in the years to come for many new, innovative, practically adaptable design constructions.

Table 9 Summary of various photocatalytic reactors along with their merits and demerits

S. No	Reactor type	Catalyst installation type	Merits	Demerits	References
1.	Slurry-type photocatalytic reactor	Suspended	• High surface to volume ratio, good mass transfer, and adsorption mechanism	• Abrasion of the catalyst due to the agitation or stirring • Poor light penetration • Post-treatment like catalyst separation required	[51]
2.	Fixed-bed photocatalytic reactor	Immobilized	• High light penetration • No need for catalyst separation • Very portable • Cost-friendly	• Support material is required • Continuous usage can lead to less efficiency as it damages the catalytic bed	[106]
3.	Membrane photocatalytic reactor	Immobilized/ Slurry	• Integrated technology • Good efficiency within less time	• Membrane fouling • High cost	[84, 86, 115]
4.	Rotating disk photocatalytic reactor	Immobilized	• Extremely effective at degrading pollutants • Limitations on mass transfer and light penetration are sorted out • The proper mixing occurs • There is no need for external oxygenation or air sparging	• Expensive and complicated in design • Catalyst damage may result from excessively generated shear stresses	[90, 92, 94]

(continued)

Table 9 (continued)

S. No	Reactor type	Catalyst installation type	Merits	Demerits	References
5.	Spinning disk photocatalytic reactor	Immobilized	• Shorter reaction times with higher photocatalytic efficiency • An enhanced mass transfer	• More expensive and complicated design • Smooth and rough disks may produce various results; thus, optimization is essential	[73]
6.	Thin-film photocatalytic reactor	Immobilized	• Highly suitable for sunlight-driven photocatalysis • Suitable for large-scale commercialization • Coupling the system (considering the reactor is sunlight-driven) with solar cells may be a source of energy for the reactor	• Supporting materials were needed, most often expensive noble metals like gold, silver, and platinum • Works well at low concentrations of pollutants • Occupies a large space	[98]
7.	Cascade-type photocatalytic reactor	Immobilized/ Slurry	• Successive elimination of pollutants is possible and good efficiency is obtained quite often • System with closed loop usually eliminates more contaminants	• Relatively more expensive and utilizes more energy because a loop-wise pump is required • Considering the slurry type, catalyst separation is required in post-treatment processes	[87, 103]

(continued)

Table 9 (continued)

S. No	Reactor type	Catalyst installation type	Merits	Demerits	References
8.	Solar light-driven photocatalytic reactor	Immobilized/ Suspended	• Greener method, more sustainable • Unlimited free light source • Good quantum conversions • Possibility to construct a reactor that is independent of external energy supply	• High cost for light collectors • High capital to construct • Design of suitable catalyst is challenging	[108]

References

1. A. Fujishima, K. Honda, Electrochemical photolysis of water at a semiconductor electrode. Nature **238**(5358), 37–38 (1972)
2. S.S. Rezaei et al., Photocatalytic oxidation of tetracycline by magnetic carbon-supported TiO2 nanoparticles catalyzed peroxydisulfate: Performance, synergy and reaction mechanism studies. Sep. Purif. Technol. **258**, 117936 (2021)
3. S. Shekoohiyan et al., A novel CuO/Fe2O3/ZnO composite for visible-light assisted photocatalytic oxidation of Bisphenol a: kinetics, degradation pathways, and toxicity elimination. Sep. Purif. Technol. **242**, 116821 (2020)
4. J. Chen et al., Novel AgI/BiOBr/reduced graphene oxide Z-scheme photocatalytic system for efficient degradation of tetracycline. J. Alloy. Compd. **800**, 88–98 (2019)
5. T. Soltani, A. Tayyebi, B.K. Lee, Photolysis and photocatalysis of tetracycline by sonochemically heterojunctioned BiVO(4)/reduced graphene oxide under visible-light irradiation. J. Environ. Manag. **232**, 713–721 (2019)
6. M. Shariati, A. Babaei, A. Azizi, Synthesis of Ag2CrO4/Ag/Fe3O4/RGO nanocomposite as a suitable photocatalyst for degradation of methylene blue in aqueous media: RSM modeling, kinetic and energy consumption studies. Inorgan. Chem. Commun. **145** (2022)
7. S. Kumar, R.D. Kaushik, L.P. Purohit, RGO supported ZnO/SnO(2) Z-scheme heterojunctions with enriched ROS production towards enhanced photocatalytic mineralization of phenolic compounds and antibiotics at low temperature. J. Colloid Interface Sci. **632**(Pt A), 196–215 (2023)
8. A.K. Ray, A.A.C.M. Beenackers, Development of a new photocatalytic reactor for water purification. Catal. Today **40**(1), 73–83 (1998)
9. P.S. Sathish Kumar et al., Photocatalytic degradation of Acid Red 88 using Au–TiO2 nanoparticles in aqueous solutions. Water Res. **42**(19), 4878–4884 (2008)
10. J. Low et al., Heterojunction photocatalysts. Adv. Mater. **29**(20), 1601694 (2017)
11. J. Yu et al., Enhanced photocatalytic CO_2-reduction activity of anatase TiO_2 by coexposed 001 and 101 facets. J. Am. Chem. Soc. **136**(25), 8839–8842 (2014)
12. S.D.A. Zondag, D. Mazzarella, T. Noël, Scale-up of photochemical reactions: transitioning from lab scale to industrial production. Annu. Rev. Chem. Biomol. Eng. **14**(1), 283–300 (2023)
13. R. Molinari et al., Which configuration of photocatalytic membrane reactors has a major potential to be used at an industrial level in tertiary sewage wastewater treatment? Catalysts **13** (2023). https://doi.org/10.3390/catal13081204

14. A.K. Verma et al., Recent advances in the structural modifications of nanoparticles to enhance photocatalytic activity, in *Nanobiotechnology for Green Environment* (2021), pp. 67–81
15. S. Talwar, A.K. Verma, V.K. Sangal, Plug flow approaching novel reactor employing in-situ dual effect of photocatalysis and photo-Fenton for the degradation of metronidazole. Chem. Eng. J. **382**, 122772 (2020)
16. R. Binjhade, R. Mondal, S. Mondal, Continuous photocatalytic reactor: critical review on the design and performance. J. Environ. Chem. Eng. **10**(3), 107746 (2022)
17. O. Sacco, V. Vaiano, D. Sannino, Main parameters influencing the design of photocatalytic reactors for wastewater treatment: a mini review. J. Chem. Technol. Biotechnol. **95**(10), 2608–2618 (2020)
18. K.P. Sundar, S. Kanmani, Progression of photocatalytic reactors and it's comparison: a review. Chem. Eng. Res. Des. **154**, 135–150 (2020)
19. M. Martín-Sómer et al., Influence of light distribution on the performance of photocatalytic reactors: LED vs mercury lamps. Appl. Catal. B **215**, 1–7 (2017)
20. R. Sabouni, H.G. Gomaa, Comparative analysis of aeration and oscillation in a suspended catalyst photocatalytic membrane reactor. Chem. Eng. Res. Des. **173**, 55–62 (2021)
21. R. Priya, S. Kanmani, Batch slurry photocatalytic reactors for the generation of hydrogen from sulfide and sulfite waste streams under solar irradiation. Sol. Energy **83**(10), 1802–1805 (2009)
22. J. Wang et al., Reasons for the deactivation of Pt0/TiO2 photocatalyst treated by inert gas N2. Appl. Surf. Sci. **254**(17), 5375–5379 (2008)
23. Slamet et al., Photocatalytic reduction of CO2 on copper-doped Titania catalysts prepared by improved-impregnation method. Catal. Commun.**6**(5), 313–319 (2005)
24. M. Tahir, N.S. Amin, Photocatalytic CO2 reduction and kinetic study over In/TiO2 nanoparticles supported microchannel monolith photoreactor. Appl. Catal. A **467**, 483–496 (2013)
25. K. Kočí et al., Influence of reactor geometry on the yield of CO2 photocatalytic reduction. Catal. Today **176**(1), 212–214 (2011)
26. R.J. Brandi, O.M. Alfano, A.E. Cassano, Modeling of radiation absorption in a flat plate photocatalytic reactor. Chem. Eng. Sci. **51**(11), 3169–3174 (1996)
27. A. Visan et al., Photocatalytic reactor design: guidelines for kinetic investigation. Ind. Eng. Chem. Res. **58**(14), 5349–5357 (2019)
28. M. Martín-Sómer et al., Reflector design for the optimization of photoactivated processes in tubular reactors for water treatment. J. Environ. Chem. Eng. **11**(5), 110609 (2023)
29. T.F.C.V. Silva et al., Solar-driven heterogeneous photocatalysis using a static mixer as TiO2-P25 support: impact of reflector optics and material. Chem. Eng. J. **435**, 134831 (2022)
30. J. Moreno-SanSegundo, C. Casado, J. Marugán, Enhanced numerical simulation of photocatalytic reactors with an improved solver for the radiative transfer equation. Chem. Eng. J. **388**, 124183 (2020)
31. R. Rehman et al., Photocatalytic degradation of Alizarin Red S, Amaranth, Congo Red, and Rhodamine B dyes using UV light modified reactor and ZnO, TiO2, and SnO2 as catalyst. J. Chem. **2021**, 6655070 (2021)
32. M. Tanveer, G.T. Guyer, G. Abbas, Photocatalytic degradation of ibuprofen in water using TiO2 and ZnO under artificial UV and solar irradiation. Water Environ. Res. **91**(9), 822–829 (2019)
33. R. Wang et al., Photocatalytic degradation of Bisphenol A (BPA) using immobilized TiO2 and UV illumination in a horizontal circulating bed photocatalytic reactor (HCBPR). J. Hazard. Mater. **169**(1), 926–932 (2009)
34. S.K. Kaviya Piriyah Sundar, Comparative study on concentric cylindrical glass tube microreactor, UV-led strip photocatalytic reactor and classical annular reactor on Acid Red 87 removal. Desalin. Water Treat. **250**, 298–311 (2022)
35. W.S. Koe et al., An overview of photocatalytic degradation: photocatalysts, mechanisms, and development of photocatalytic membrane. Environ. Sci. Pollut. Res. **27**(3), 2522–2565 (2020)

36. N. Pugazhenthiran et al., Photocatalytic degradation of ceftiofur sodium in the presence of gold nanoparticles loaded TiO2 under UV-visible light. Chem. Eng. J. **241**, 401–409 (2014)
37. P. Sathishkumar et al., Synthesis of Fe3+ doped TiO2 photocatalysts for the visible assisted degradation of an azo dye. Colloids Surf. A **375**(1), 231–236 (2011)
38. B. Kakavandi et al., Enhanced sono-photocatalysis of tetracycline antibiotic using TiO(2) decorated on magnetic activated carbon (MAC@T) coupled with US and UV: a new hybrid system. Ultrason. Sonochem. **55**, 75–85 (2019)
39. F. Zhao et al., Photocatalytic degradation of tetracycline by N-CQDs modified S-g-C3N4 nanotubes and its product toxicity evaluation. Sep. Purif. Technol. **314** (2023)
40. F. Xu et al., Facile one-pot synthesis of carbon self-doped graphitic carbon nitride loaded with ultra-low ceric dioxide for high-efficiency environmental photocatalysis: Organic pollutants degradation and hexavalent chromium reduction. J. Colloid Interface Sci. **601**, 196–208 (2021)
41. L. Chen et al., Nitrogen defects/boron dopants engineered tubular carbon nitride for efficient tetracycline hydrochloride photodegradation and hydrogen evolution. Appl. Catal. B: Environ. **303** (2022)
42. K. Qi et al., Design of 2D–2D NiO/g-C3N4 heterojunction photocatalysts for degradation of an emerging pollutant. Res. Chem. Intermed. **46**(12), 5281–5295 (2020)
43. S. Das, Y.-H. Ahn, Preparation of P-doped CdS nanorods as an efficient photocatalyst for the degradation of the emerging pollutant tetracycline antibiotic under blue LED light irradiation. Dalton Trans. **51**(36), 13646–13656 (2022)
44. G. Jalloul et al., Fe-sensitized zeolite supported TiO2 for the degradation of tetracycline using blue LED irradiation. Front. Environ. Sci. **10** (2022)
45. X. Chi et al., Phosphorus-doped carbon nitride nanosheets as efficient white-LED-light-driven photocatalyst for hydrogen evolution and tetracycline degradation. Catal. Lett. **151**(12), 3592–3602 (2021)
46. S. Das, Y.-H. Ahn, Synthesis and application of CdS nanorods for LED-based photocatalytic degradation of tetracycline antibiotic. Chemosphere **291**, 132870 (2022)
47. A. Haarstrick, O.M. Kut, E. Heinzle, TiO2-assisted degradation of environmentally relevant organic compounds in wastewater using a novel fluidized bed photoreactor. Environ. Sci. Technol. **30**(3), 817–824 (1996)
48. Y. Wang et al., Nanocasting synthesis of chromium doped mesoporous CeO2 with enhanced visible-light photocatalytic CO2 reduction performance. J. Hazard. Mater. **372**, 69–76 (2019)
49. P. Praus et al., CdS nanoparticles deposited on montmorillonite: preparation, characterization and application for photoreduction of carbon dioxide. J. Colloid Interface Sci. **360**(2), 574–579 (2011)
50. K. Kočí et al., Effect of TiO2 particle size on the photocatalytic reduction of CO2. Appl. Catal. B **89**(3), 494–502 (2009)
51. R.J. Braham, A.T. Harris, A complete multi-scale simulation of light absorption within a fluidized bed photoreactor using integrated particle, fluid and photon behaviour models. Phys. Chem. Chem. Phys. **15**(29), 12373–12385 (2013)
52. K.-T. Lu et al., An internal-illuminated monolith photoreactor towards efficient photocatalytic degradation of ppb-level isopropyl alcohol. Chem. Eng. J. **296**, 11–18 (2016)
53. D.R. Shinde, I.S. Quraishi, R.A. Pawar, An efficient visible light driven photocatalytic removal of dyes from the dye effluent using metal halide lamp based slurry reactor. ES Energy Environ. (2021)
54. A. Ounnar et al., Photocatalytic degradation efficiency of hazardous macrolide compounds using an external UV-light irradiation slurry reactor. Water Sci. Technol. **82**(4), 695–703 (2020)
55. A.R. Soleymani, A.M. Tavassoli, H. Rezaei-Vahidian, Assessment of back-side activation of titania thin film using a fixed-bed photocatalytic-reactor: kinetic study, operating cost and ANN modeling. Chem. Eng. Res. Des. **190**, 759–776 (2023)
56. G. Xu et al., Continuous high-efficient degradation of organic pollutants based on sea urchin-like Fe3O4/ZnO/ZnSe heterostructures in photocatalytic magnetically fixed bed reactor. Colloids Surf. A **603**, 125198 (2020)

57. A. Manassero, O.M. Alfano, M.L. Satuf, Degradation of emerging pollutants by photocatalysis: radiation modeling and kinetics in packed-bed reactors. Water **14** (2022) https://doi.org/10.3390/w14223608
58. A.V. Mohod et al., Degradation of Rhodamine 6G dye using a novel glass-marble packed bed reactor-based cavitation. Chem. Eng. Process. Process Intensif. **184**, 109287 (2023)
59. H. Tiwari, R.K. Sonwani, R.S. Singh, A comprehensive evaluation of the integrated photocatalytic-fixed bed bioreactor system for the treatment of acid blue 113 dye. Biores. Technol. **364**, 128037 (2022)
60. O. Sacco, D. Sannino, V. Vaiano, Packed bed photoreactor for the removal of water pollutants using visible light emitting diodes. Appl. Sci. **9** (2019) https://doi.org/10.3390/app9030472
61. S. Horikoshi et al., Photodecomposition of a nonylphenol polyethoxylate surfactant in a cylindrical photoreactor with TiO2 immobilized fiberglass cloth. Appl. Catal. B **37**(2), 117–129 (2002)
62. J. Yang et al., Elemental mercury removal from flue gas over TiO2 catalyst in an internal-illuminated honeycomb photoreactor. Ind. Eng. Chem. Res. **57**(51), 17348–17355 (2018)
63. R.A. Damodar, T. Swaminathan, Performance evaluation of a continuous flow immobilized rotating tube photocatalytic reactor (IRTPR) immobilized with TiO2 catalyst for azo dye degradation. Chem. Eng. J. **144**(1), 59–66 (2008)
64. R. Foulady-Dehaghi, M. Behpour, Visible and solar photodegradation of textile wastewater by multiple doped TiO2/Zn nanostructured thin films in fixed bed photoreactor mode. Inorg. Chem. Commun. **117**, 107946 (2020)
65. M. Tahir et al., Enhanced photocatalytic carbon dioxide reforming of methane to fuels over nickel and montmorillonite supported TiO2 nanocomposite under UV-light using monolith photoreactor. J. Clean. Prod. **213**, 451–461 (2019)
66. A. Almansba et al., An engineering approach towards the design of an innovative compact photo-reactor for antibiotic removal in the frame of laboratory and pilot-plant scale. J. Photochem. Photobiol. A **418**, 113445 (2021)
67. A. Joseph, A. Vijayanandan, Review on support materials used for immobilization of nano-photocatalysts for water treatment applications. Inorg. Chim. Acta **545**, 121284 (2023)
68. G. Chester et al., A jacketed annular membrane photocatalytic reactor for wastewater treatment: degradation of formic acid and atrazine. J. Photochem. Photobiol. A **71**(3), 291–297 (1993)
69. S. Mozia, M. Tomaszewska, A.W. Morawski, Photocatalytic membrane reactor (PMR) coupling photocatalysis and membrane distillation—Effectiveness of removal of three azo dyes from water. Catal. Today **129**(1), 3–8 (2007)
70. K. Sopajaree et al., An integrated flow reactor-membrane filtration system for heterogeneous photocatalysis. Part I: experiments and modelling of a batch-recirculated photoreactor. J. Appl. Electrochem. **29**(5), 533–539 (1999)
71. K. Sopajaree et al., An integrated flow reactor-membrane filtration system for heterogeneous photocatalysis. Part II: experiments on the ultrafiltration unit and combined operation. J. Appl. Electrochem. **29**(9), 1111–1118 (1999)
72. A.L. Desa et al., Industrial textile wastewater treatment via membrane photocatalytic reactor (MPR) in the presence of ZnO-PEG nanoparticles and tight ultrafiltration. J. Water Process Eng. **31**, 100872 (2019)
73. D.S.M. Constantino et al., Intensification strategies for improving the performance of photocatalytic processes: a review. J. Clean. Prod. **340**, 130800 (2022)
74. C. Regmi et al., Comparison of photocatalytic membrane reactor types for the degradation of an organic molecule by TiO2-coated PES membrane. Catalysts **10** (2020). https://doi.org/10.3390/catal10070725
75. S. Sakarkar, S. Muthukumran, V. Jegatheesan, Factors affecting the degradation of remazol turquoise blue (RTB) dye by titanium dioxide (TiO2) entrapped photocatalytic membrane. J. Environ. Manag. **272**, 111090 (2020)
76. Q. Li et al., Enhanced catalytic degradation of amoxicillin with TiO2–Fe3O4 composites via a submerged magnetic separation membrane photocatalytic reactor (SMSMPR). RSC Adv. **9**(22), 12538–12546 (2019)

77. L. Zhu et al., Silicon carbide catalytic ceramic membranes with nano-wire structure for enhanced anti-fouling performance. Water Res. **226**, 119209 (2022)
78. C. Yang et al., A novel vertical immobilized photocatalytic membrane reactor based on Bi2WO6-g-C3N4/PVDF for enhanced removal of atrazine, anti-fouling performance and long-term stability. Chem. Eng. J. **471**, 144672 (2023)
79. S. Salehian et al., Photocatalytic TiO2@MIL-88A (Fe)/polyacrylonitrile mixed matrix membranes: characterization, anti-fouling properties, and performance on the removal of natural organic matter. Chemosphere **302**, 134893 (2022)
80. Y.G. Adewuyi, Sonochemistry: environmental science and engineering applications. Ind. Eng. Chem. Res. **40**(22), 4681–4715 (2001)
81. R. Sabouni, H. Gomaa, Photocatalytic degradation of pharmaceutical micro-pollutants using ZnO. Environ. Sci. Pollut. Res. **26**(6), 5372–5380 (2019)
82. S. Kundu, N. Karak, Polymeric photocatalytic membrane: an emerging solution for environmental remediation. Chem. Eng. J. **438**, 135575 (2022)
83. J.C. Espíndola et al., Photocatalytic membrane reactor performance towards oxytetracycline removal from synthetic and real matrices: suspended vs immobilized TiO2-P25. Chem. Eng. J. **378**, 122114 (2019)
84. Q. Li et al., Photocatalytic degradation of amoxicillin via TiO2 nanoparticle coupling with a novel submerged porous ceramic membrane reactor. J. Clean. Prod. **209**, 755–761 (2019)
85. T.E. Berger et al., Photocatalytic degradation of organic dye via atomic layer deposited TiO2 on ceramic membranes in single-pass flow-through operation. J. Membr. Sci. **604**, 118015 (2020)
86. R. Molinari, C. Lavorato, P. Argurio, The evolution of photocatalytic membrane reactors over the last 20 years: a state of the art perspective. Catalysts **11** (2021). https://doi.org/10.3390/catal11070775
87. H. Amiri, B. Ayati, H. Ganjidoust, Mass transfer phenomenon in photocatalytic cascade disc reactor: effects of artificial roughness and flow rate. Chem. Eng. Process. **116**, 48–59 (2017)
88. H. Zangeneh et al., Highly efficient azo dye degradation in a photocatalytic rotating disc reactor with deposited l-histidine-TiO2-CdS. Mater. Sci. Semicond. Process. **152**, 107071 (2022)
89. A. Chaudhuri et al., Scale-up of a heterogeneous photocatalytic degradation using a photochemical rotor-stator spinning disk reactor. Org. Process Res. Dev. **26**(4), 1279–1288 (2022)
90. D.D. Dionysiou et al., Rotating disk photocatalytic reactor: development, characterization, and evaluation for the destruction of organic pollutants in water. Water Res. **34**(11), 2927–2940 (2000)
91. O.A. Salu et al., Remediation of oily wastewater from an interceptor tank using a novel photocatalytic drum reactor. Desalin. Water Treat. **26**(1–3), 87–91 (2011)
92. A. Zhang et al., The combination of rotating disk photocatalytic reactor and TiO2 nanotube arrays for environmental pollutants removal. J. Hazard. Mater. **186**(2), 1374–1383 (2011)
93. Y. Cai et al., An Evaluation of metronidazole degradation in a plasma-assisted rotating disk reactor coupled with TiO2 in aqueous solution. Engineering **7**(11), 1603–1610 (2021)
94. Z. Jiang, K. Li, J. Jia, Recent developments of light-harvesting excitation, macroscope transfer and multi-stage utilization of photogenerated electrons in rotating disk photocatalytic reactor. Processes **11** (2023). https://doi.org/10.3390/pr11030838
95. C. Montalvo-Romero et al., A semi-pilot photocatalytic rotating reactor (RFR) with supported TiO2/Ag catalysts for water treatment. Molecules **23** (2018). https://doi.org/10.3390/molecules23010224
96. C. Bian et al., TiO2 nanocrystalline for enhanced hydrogen and oxygen generation of thin film photocatalyst: from catalytic mechanism and microstructural analysis. J. Mater. Sci.: Mater. Electron. **31**(13), 9961–9968 (2020)
97. R.A. Damodar, K. Jagannathan, T. Swaminathan, Decolourization of reactive dyes by thin film immobilized surface photoreactor using solar irradiation. Sol. Energy **81**(1), 1–7 (2007)

98. S. Saran et al., Pilot scale thin film plate reactors for the photocatalytic treatment of sugar refinery wastewater. Environ. Sci. Pollut. Res. **23**(17), 17730–17741 (2016)
99. M. Mirzaei et al., Evaluation and modeling of a spinning disc photoreactor for degradation of phenol: impact of geometry modification. J. Photochem. Photobiol. A **346**, 206–214 (2017)
100. I. Boiarkina, S. Pedron, D.A. Patterson, An experimental and modelling investigation of the effect of the flow regime on the photocatalytic degradation of methylene blue on a thin film coated ultraviolet irradiated spinning disc reactor. Appl. Catal. B **110**, 14–24 (2011)
101. H.J. Huang, K.-C. Huang, D.P. Tsai, Light absorption measurement of a plasmonic photocatalyst in the circular plane waveguide of a photocatalytic dual light source spinning disk reactor. Opt. Rev. **20**(2), 236–240 (2013)
102. I. Boiarkina, S. Norris, D.A. Patterson, Investigation into the effect of flow structure on the photocatalytic degradation of methylene blue and dehydroabietic acid in a spinning disc reactor. Chem. Eng. J. **222**, 159–171 (2013)
103. B. Stephan, L. Ludovic, W. Dominique, Modelling of a falling thin film deposited photocatalytic step reactor for water purification: pesticide treatment. Chem. Eng. J. **169**(1), 216–225 (2011)
104. K. Fouad et al., A novel photocatalytic reactor for the extended reuse of W-TiO2 in the degradation of sulfamethazine. Chemosphere **257**, 127270 (2020)
105. G.L. Puma et al., Scale-up and optimization for slurry photoreactors, in *Advanced Oxidation Processes*, ed. by B.-L. Ciro (IntechOpen, Rijeka, 2020), p. Ch. 8
106. A.G. Variar et al., Influence of various operational parameters in enhancing photocatalytic reduction efficiency of carbon dioxide in a photoreactor: a review. J. Ind. Eng. Chem. **99**, 19–47 (2021)
107. A. Fernández-García et al., Parabolic-trough solar collectors and their applications. Renew. Sustain. Energy Rev. **14**(7), 1695–1721 (2010)
108. M. Tanveer, G. Tezcanli Guyer, Solar assisted photo degradation of wastewater by compound parabolic collectors: review of design and operational parameters. Renew. Sustain. Energy Rev. **24**, 534–543 (2013)
109. T. Long et al., Fabrication of the annular photocatalytic reactor using large-sized freestanding titania-silica monolithic aerogel as the catalyst for degradation of glyphosate. Mater. Des. **159**, 195–200 (2018)
110. D. Wang et al., Engineering and modeling perspectives on photocatalytic reactors for water treatment. Water Res. **202**, 117421 (2021)
111. H. Zeghioud et al., Photocatalytic reactors dedicated to the degradation of hazardous organic pollutants: kinetics, mechanistic aspects, and design—A review. Chem. Eng. Commun. **203**(11), 1415–1431 (2016)
112. D. Olivo-Alanís et al., Generalized kinetic model for the photocatalytic degradation processes: validation for dye wastewater treatment in a visible-LED tubular reactor. Appl. Catal. B **317**, 121804 (2022)
113. I. Boiarkina, S. Norris, D.A. Patterson, The case for the photocatalytic spinning disc reactor as a process intensification technology: Comparison to an annular reactor for the degradation of methylene blue. Chem. Eng. J. **225**, 752–765 (2013)
114. A. Manassero, O.M. Alfano, M.L. Satuf, Radiation modeling and performance evaluation of a UV-LED photocatalytic reactor for water treatment. J. Photochem. Photobiol. A **436**, 114367 (2023)
115. Q. Wang et al., Submerged membrane photocatalytic reactor for advanced treatment of p-nitrophenol wastewater through visible-light-driven photo-Fenton reactions. Sep. Purif. Technol. **256**, 117783 (2021)

Renewable and Commercially Viable Porous Material-Supported Heterojunction Nanocomposites as UV-Visible Light-Responsive Photocatalysts for Environmental and Energy-Related Applications

Dhivya Jagadeesan, Denna Babu, Akhila Maheswari Mohan, and Prabhakaran Deivasigamani

Abstract The chapter deals with porous material-based nanocomposites as visible light active photocatalyst monoliths to remove lingering contaminants from the environment. Porous materials with well-defined, tailor-made structural and surface properties offer numerous applications related to catalysis, adsorption, separation, biomedicines, energy technologies, etc. In heterogeneous photocatalysis, porous materials such as zeolites, porous metal oxides, metal-organic frameworks, and porous polymer monoliths supported by nanocomposites contribute intensely to environmental and energy-related applications. Porous photocatalysts with a hierarchical stricture show great promise due to their distinct features compared to solid structures. These advantages encompass improved scattering and adsorption of light, abundance reaction sites, large specific surface areas, enhanced facilitation of mass and charge transfer, etc. These porous materials are efficient for hosting photoactive materials and are versatile owing to their lower production cost, high-performance efficiency, easy modification, and renewability. The chapter discusses the advances in heterogeneous photocatalysis using hierarchically porous materials as photocatalytic supports and their prospects in environmental decontamination and energy-related applications.

Keywords Porous materials · Nanocomposites · Visible light · Photocatalysis · Decontamination · Energy

D. Jagadeesan · D. Babu · A. M. Mohan · P. Deivasigamani (✉)
Department of Chemistry, School of Advanced Sciences, Vellore Institute of Technology (VIT), Vellore, Tamil Nadu 632014, India
e-mail: prabhakaran.d@vit.ac.in; prabhakarswordy@gmail.com

P. Sathishkumar (ed.), *Photocatalysis for Energy and Environmental Applications*, Green Energy and Technology, https://doi.org/10.1007/978-981-97-1939-6_2

1 Introduction

Environmental pollution from organics and their after-effects have become a global concern recently. This section highlights various compounds, some classified as contaminants of emerging concern (CECs), a class of pollutants being identified more frequently in various locales, lack regulation and oversight, and may have detrimental effects on human health and the environment. Even at low concentrations, they exhibit considerable toxicity in the aquatic medium and high diffusivity through water, which has a more significant detrimental impact on the ecosystem [1, 2]. In particular, they can lead to malfunctions in the endocrine and neurological systems, reproductive capacities, and hormonal balance, as well as various cancers (breast, ovary, prostate, testes, etc.). In the past few decades, pharmaceuticals, personal care products, and textile industries have been imperative in taking a significant share in the current environmental issues. Water pollution caused by these products is a significant concern due to their increasing consumption and utility, ending up as effluents, thus causing severe consequences to the aquatic ecosystem [3–5]. In recent years, worldwide concern has been to minimize these biologically toxic pollutants to maintain sustainability in the aquatic environment and human health. Nevertheless, only a minimal concentration of pollutants is evident during the initial stages of discharge. However, this concentration can escalate along the food chain, leading to cascading effects on human health and the ecosystem. These effects may manifest as alterations in enzyme activities, changes in growth rates and reproduction patterns, and a decrease in overall abundance and diversity [6–8]. Another difficulty involving CECs is the inability to achieve total elimination, as they typically present refractory characteristics, and conventional water treatment technologies cannot effectively obliterate these pollutants. Unfortunately, the traditional treatment methods are hampered because of some defects, such as complicated procedures, high costs, long cycles, and incomplete removal [9, 10].

The photocatalytic degradation process has emerged as a highly effective and promising approach for eliminating organic pollutants from industrial wastewater. In this context, the proficient separation of photogenerated charge carriers and the photocatalyst's successful absorption of organic contaminants are pivotal factors in the decontamination processes [11, 12]. As a result, advanced oxidation processes, dependent on producing hydroxyl radicals and other oxidizing species, are recognized as a viable and environmentally friendly option for treating water and wastewater. These processes demonstrate high efficiency in degrading Contaminants of Emerging Concern (CECs), contributing to the overall eco-compatibility of the treatment methodology [13, 14]. Among them, it is worth discussing heterogeneous photocatalysis due to their facile catalyst separation, recyclability, and critical assets for large-scale industrial technology implementation. To enhance the broader and more efficient application of photocatalytic technology, a pivotal challenge lies in developing a photocatalyst with a broad spectrum of light absorption wavelengths and superior charge separation capabilities [15]. Sunlight emission, with a radiant flux of (1353 W/m^2), surpasses humanity's total energy consumption and requirements. From an energy perspective, the distribution of sunlight can be categorized into approximately 3% in the high-energy UV range, 44% in the visible light range, and 53% in the lower energy infrared irradiation range. Compared to the highly energetic UV light, which often induces molecular bond fission, the visible light range offers distinct advantages, including a broad irradiation area and a sufficiently energetic level conducive to the photocatalytic reactions of environmental pollutants [16, 17]. Some notable semiconductor solids used in environmental heterogeneous photocatalysis include TiO_2, ZnO, CdS, WO_3, SnO_2, ZnS, CdTe, α-Fe_2O_3, $AgNbO_3$, and $SrTiO_3$ [18–27]. Nevertheless, the practical application of semiconductor nanoparticles faces challenges such as nanotoxicity, cost considerations, and non-recyclability. Additionally, many of these nanoparticles exhibit activity solely under UV–light exposure, limiting their efficiency in harnessing sunlight to its full potential. In water treatment, extensive research efforts have been directed toward identifying suitable support materials to address the shortcomings of semiconductor nanoparticles. Distinguished by their distinctive properties, these support materials have emerged as a critical focus in ongoing investigations. The exploration of visible light to facilitate chemical transformations represents an intriguing area of research, gaining particular significance in the current socioeconomic landscape [28, 29].

The heterogeneous photocatalysis processes have proved advantageous due to their chemical stability, mild operating conditions, no chemical requirements, absence of generation of secondary pollutants, low energy consumption, and cost-effectiveness. Many nanomaterials were developed for various photocatalytic applications over the years that are associated with recovery and reusability issues. Recently, there has been an increasing acknowledgment of the importance of hierarchically porous structures in stabilizing the dispersion of highly active sites and promoting the diffusion of reagents during catalytic reactions [30]. Typically, hierarchically porous materials exhibit organized networks characterized by a substantial surface area, adjustable porosity, spatial accessibility, recoverability, low density, and customizable chemical compositions specifically engineered to optimize light harvesting, mass loading/diffusion, and charge transfer/migration. These inherent physical attributes render such materials highly suitable for various applications, including energy storage/conversion, catalysis, photocatalysis, adsorption, separation, and gas sensing. Integrating macropores into meso-porous structures contributes to an augmentation of light scattering and harvesting capabilities, consequently elevating the efficiency of photocatalysis [31, 32]. Conversely, introducing macropores or mesopores into traditional microporous zeolites can diminish diffusion barriers and potentially enhance active site dispersion.

2 Overview of Nanocomposites in Heterogeneous Photocatalysis

Nanoparticles have been used as photocatalysts in the last four decades, as they offer considerable photocatalytic activity despite their moderate surface and porosity properties. Nanoparticles exhibit a high surface-to-volume ratio, thus dramatically changing their properties compared to their bulk-sized equivalents [33]. Besides, nanoparticles result in a quantum confinement effect that increases their energy bandgap, thus enhancing the photoinduced charge separation upon light irradiation. These nanomaterials can be synthesized into morphologies, while the properties of the specific morphology vary accordingly [34]. However, over the years, nanoparticles have evolved into heterostructure nanocomposites with the growing demand for newer applications. In the last few decades, several approaches have been devised to address the challenges associated with rapid charge recombination and restricted absorption of visible light in the context of engineering heterojunction photocatalysts. These nanocomposites featuring heterojunctions exhibit elevated photocatalytic activity, primarily attributed to the spatial segregation of photogenerated electron–hole pairs. Nanocomposites offer numerous advantages as they provide more flexible space for easy reconstruction, as their nanosize expands the limits, thus enhancing the confinement effect, mechanical stability, and large surface area, making them suitable for photocatalytic activities [35, 36]. Nanocomposites have garnered considerable attention owing to their versatility in diverse fields such as biomedical, catalysis, separation processes, chemical sensing, fuel cells, capacitors, microfabrication, tribology, resonant coupling, and high-flux gas transport [37, 38]. While a solitary photocatalytic semiconductor (either n-type or p-type) falls short in achieving the requisite performance for wastewater treatment, integrating heterostructures or nanocomposites comprising two or more photocatalytic materials emerges as a highly promising photocatalyst. These materials are inherent with charge separation capabilities, resilience against photo-corrosion, and propensity for harvesting visible light. The selection of photocatalytic components is contingent upon their valence band, conduction band potentials, and bandgap energy [39]. It is imperative to underscore the extensive application of nanocomposites in the realm of photocatalytic water treatment.

Nanostructured materials play a pivotal role in heterogeneous photocatalysis, demonstrating versatile applications across various processes such as oxidations, oxidative cleavages, reductions, isomerization, substitutions, condensations, and polymerizations [40, 41]. The application of nanotechnology principles to heterogeneous catalysis enhances our understanding of the transformations taking place at a molecular level on the catalyst's surface. The synthesis of nanomaterials not only aids in comprehending reaction mechanisms but also facilitates the design of innovative functional catalytic systems. Despite notable strides in developing reproducible materials through novel methods, several challenges persist. From a materials perspective, photocatalysts necessitate specific properties tailored to their intended applications, encompassing particle size, specific surface area, and the spacing between electronic levels, among other factors [42, 43]. Fundamentally, two approaches are employed for nanostructure synthesis: top-down and bottom-up. The former involves grinding, cutting, slicing, and milling processes to derive nanostructures from bulk materials.

On the contrary, the bottom-up methodology organizes atoms and molecules through diverse techniques, including physical methods such as sonication, physical vapor deposition, and thermal evaporation, chemical methods like sol–gel, hydrothermal, chemical vapor deposition, precipitation, and solvothermal, as well as biological methods like controlled deposition and growth, to craft nanostructures [44]. In contrast, the top-down approach proves unsuitable for achieving optimal photocatalytic performance due to the severe damage inflicted upon the crystal structure of nanomaterials, which substantially impacts their overall properties. Consequently, bottom-up techniques emerge as the preferred choice for fabricating photocatalytically active functional nanomaterials. The conventional technique for manufacturing metal oxides relies on the solid-state approach, where thermal energy triggers the formation of the intended phase characterized by high crystallinity. However, this method frequently results in agglomerated large grain forming, leading to a reduction in specific surface area. Mechanochemical approaches, such as ball milling, replace thermal energy with mechanical energy, though these methods have limited applicability and insufficient control over morphology and nanostructure [45, 46]. Notably, the creation of mixed oxide compositions generally requires elevated temperatures to synthesize the desired single-phase materials. Furthermore, attaining a uniform blending of precursors in solid-state and mechanochemical methodologies poses a considerable challenge. As a result, diverse chemical strategies have been formulated, encompassing a variety of precursors, templates, and reaction media. These approaches involve complexations, precipitation, hydrothermal, and templating routes [47].

Nanostructured photocatalytic materials find widespread applications in water and wastewater purification, renowned for their efficacy in degrading recalcitrant organic compounds. Photocatalysis facilitates diverse chemical transformations under mild conditions, minimizing energy consumption and the generation of contaminants. Semiconductor materials provide a platform for inducing "photo-redox" chemical pathways, allowing tailored reductive and oxidative reactions and radical formation based on band positions. While these materials are commonly used in dispersed or dissolved phases, immobilizing semiconductors on solid surfaces at the macroscale presents an appealing option for scalability. However, a notable challenge arises in the disintegration of nanocomposites, a concern effectively addressed by immobilizing them on solid supports. This approach prevents the photocatalyst from washing or leaching out, enhancing its reusability [48]. Nevertheless, immobilization introduces a significant issue: a decline in the degradation rate of pollutants due to the reduced surface area-to-volume ratio. The coming sections will summarize photocatalyst-supported porous networks from synthesis and application perspectives.

3 Role of Porous Materials as Host Templates in Heterogeneous Photocatalysis

In recent years, porous matrices have been employed as support materials or host templates for immobilizing nanocomposites that depend on their homogeneous distribution across the template matrix. Nonetheless, achieving a robust and straightforward immobilization of nanomaterials, accomplished by fine-tuning the nature of the interface, facilitates the circumvention of laborious purification processes. The reason is that simple filtration can easily separate the supported catalyst from the reaction mixture. Generally, a more vital interaction between the support and the photocatalyst reduces the likelihood of leaching [49, 50]. It is noteworthy that repetitive experiments involving the novel photocatalysts demonstrated the sustained stability of the porous templates without a significant decline in photocatalytic efficiency. In this context, contemplating chemical reactions occurring on the surface of porous catalytic materials, the efficacy of heterogeneous catalysis hinges on efficient diffusion and adsorption processes. Hence, substantial surface areas and open pore structures contribute significantly to heightened reaction rates.

3.1 Porous Organic Polymers

Porous organic polymers (POPs) comprise a group of complex porous matrices assembled through covalent linkages that are robust enough to hold the organic elements, which are the building bricks with distinct configurations and morphologies. The need for advanced development in porous materials is essential for various applications [51]. The salient features of POPs, including perpetual and tunable porosity, low density, diverse composition, facile functionalization, and high thermal/chemical stability, have accelerated the research where POPs have a significant role in sorption/separation, energy storage, biomedical applications, optical devices, and catalysis [52]. POPs with excellent Brunauer–Emmett–Teller surface area and rich structural flexibility offer a thorough analysis of current developments and show a lot of promise in heterogeneous and homogeneous catalysis. The pores in the POPs at the nanoscale level could effectively activate reactants, thus boosting catalytic activities. As a result, the catalytic performance of these reactants would be far better than their homogeneous match. POPs fall into two categories: crystalline and amorphous. Hyper-crosslinked polymers (HCPs), polymers of intrinsic microporosity (PIMs), conjugated microporous polymers (CMPs), and porous aromatic frameworks (PAFs) are examples of amorphous POPs, whereas covalent organic frameworks (COFs) are an example of crystalline POPs [53–58]. Compared to other amorphous POPs, CMPs have longer π conjugated skeletons, giving them an appealing light-harvesting capacity, an essential feature in photocatalysis.

The ordered and periodic arrangement of COFs with uniform and permanent nanopores allows precise integration of various functional building blocks. The photoelectric properties of the organic semiconductors incorporated into the COF framework can retain or even amplify the functional properties and catalytic abilities. Thus, the bottom-up strategy used for COF synthesis makes it more photocatalytically active and can be used for photocatalysis [59]. Another class of COFs, i.e., covalent triazine frameworks (CTFs), usually formed by the trimerization of aromatic nitriles, contains triazine rings as the functional unit in their architecture. The major hindrance to the development of CTFs when compared with COFs having imine, boroxine, or boronate ester, and azine linkages are the uncompromising synthetic conditions (e.g., elevated temperate over 400 °C) and irremediable generation of aromatic 1,3,5-triazine rings. CITs reported so far have features that make them attractive for photocatalytic applications, such as poor crystallinity but exceptional chemical and thermal stability and rich nitrogen content. POPs with enormous surface area, tunable chemical composition, designable pore walls, and high stabilities have been successfully developed [60].

POPs also show insolubility in different reaction media, and their wettability can be adjusted, which is significant for improving the mass transfer and the catalyst's performance. When the photocatalytic molecules are attached to or within the polymer network, they enable easy species retrieval and reduce photobleaching. The scattering of light from the surface of polymers due to their high optical density is a major challenge faced by polymers because the fraction of light available for photocatalytic reactions will be reduced, making it inefficient [61]. Therefore, further exploration in this field is necessary to maximize the utilization of light reaching the surface and enhance the photocatalytic systems' light-harvesting capability. Polymers can also be produced by adding a secondary monomer with multiple reactive centers where the coupling of various linear chains forms a polymer network. These systems are more resilient and durable and can resist wear-offs and degradation [62, 63]. The crosslinked polymer networks can also be targeted as heterogeneous photocatalysts because their facile recovery after the reaction offers a promising future for converting harmful substances to sustainable and biodegradable products. Catalytically active semiconductor junctions can be incorporated into the networks of POPs, and the catalytic activity can be enhanced by tuning the electronic properties and pore structures of solids, making them distinguishable from their homologous analogs. The incorporation can reduce the nanotoxicity and the cost of the particles involved and can be widely used for various leading environmental applications. The unique functionalities of POPs can be controlled precisely by modulating the building components and optimizing the structure–property relationship for specific applications [64].

POPs appear as an encouraging platform for effective photocatalysis accordant with their excellent degree of polymerization and strong chemical link connection, which makes most POPs interact with several solvents, making separation and recycling simple. Porous organic polymer monoliths can also be used as a template to support heterogenous photocatalysts to enhance their activity and reduce their nanotoxicity [65, 66]. However, most POPs can only be used photo-catalytically with a favorable redox potential. Therefore, POPs with very high redox potentials are preferred to investigate the range of their applications, which impact the efficiency of visible light capture. The adsorption selectivity and porous characteristics help POPs to be used for photocatalytic treatment of water pollutants [67].

In summary, with uniform porous structure and reusability characteristics, POPs (especially CMPs and COFs) guarantee enhanced photocatalytic activity. Moreover, POPs have a promising future, with more efficient heterogenous photocatalytic materials loaded to increase the durability and reusability features. POPs connected through covalent linkage have immense conjugation and suitable energy gaps, making them active in photocatalytic reactions. The skeletal backbone and surficial characteristics affect the morphological and functional properties of POPs, including light-harvesting capability, energy bandgap, and charge transfer efficiency. Photogenerated electrons and holes can be separated more easily in POPs where the donor–acceptor (D-A) system comprises electron-donating and electron-accepting monomers [68]. The customizable porosity, simplicity in preparation, and tunable textural properties make POPs unique and valuable for environmental remediations. The controllable structure and good thermal conductivity are also crucial in photocatalytic processes [69].

3.2 *Zeolites*

Zeolites, aluminosilicates, or permutites are hydrated aluminosilicates of alkaline and alkaline-earth metals. These crystalline microporous aluminosilicate materials are widely recognized for their diverse industrial applications in catalysis, adsorption, and ion exchange. Certain aluminosilicates release water upon heating, resulting in a porous structure characterized by a substantial surface area. Over the past two centuries, approximately 40 natural zeolites have been identified, with analcime, chabazite, clinoptilolite, erionite, ferrierite, heulandite, laumontite, mordenite, and phillipsite being the most prevalent [70]. Synthetic and natural zeolites are viable alternatives to activated carbon in municipal and industrial pollution control. Numerous studies have demonstrated the suitability of zeolites as effective adsorbents for organic compounds. The introduction of alkyl trimethylammonium surfactants (C_nH_{2n+1}–$(CH_3)_3NBr$) with varying alkyl chain lengths has been employed to generate intercrystalline mesoporosity in zeolites. Additionally, organosilanes are utilized to reduce the crystal size of zeolites, enhancing their applicability in various contexts [71, 72]. Zeolites with two-dimensional (2D) nanosheet structures were intensively developed to demonstrate remarkable catalytic activity relative to conventional three-dimensional (3D) zeolitic structures.

Typically, 2D structures contain more excellent accessible external surface area per mass/volume unit than 3D structures, thereby addressing the limitations of mass transport, enabling bulk molecules to undergo faster catalytic reactions and shortening the diffusion pathway [73–75]. Besides, 2D zeolites exhibit excellent thermal stability and acidic properties necessary for catalytic reactions. Zeolites, characterized by an orderly arrangement of molecularly sized micropores, have demonstrated their efficacy as suitable catalytic supports. Comprising TO_4 tetrahedra (where T, for instance, may be tetrahedrally coordinated Si or Al), zeolites form interconnected structures through oxygen-sharing vertices, giving rise to rings, channels, cages, and three-dimensional tetrahedral frameworks [76]. Constructed from tetrahedral $[SiO_4]^{4-}$ and $[AlO_4]^{5-}$ units, these frameworks are linked through corners. The negatively charged aluminosilicate lattice is counterbalanced by cations within the pore channels or cages, typically of alkaline and alkaline-earth origins, establishing electroneutrality. These cations, electrostatically interacting with the aluminosilicate lattice, are mobile and quickly replaced by other cations from water media. This ion exchangeability renders zeolites adaptable for diverse applications [77].

Zeolites are considered an attractive and significant solid catalyst in petroleum-based sectors in refining and separation due to their distinctive shape-selective catalytic prowess, adjustable acidity, excellent (hydro)thermal stability, and cost-effective manufacturing. Large surface area possessed by nanocomposites on zeolite surface plays a dominate role in adsorption. The selectivity of zeolites for removing organic pollutants can be adjusted by functionalizing them with suitable units. The tetrahedral unit of AlO_4 shows an overall negative charge, which can be compensated by adding positive charges (H^+, Na^+, K, Ca^{2+}, Mg^{2+}, etc.) [78]. Besides, the added cations can be swapped under nominal conditions because they are not tightly held. The structural characteristics like porosity, volume, and channel diameter are influenced by the zeolite framework that constitutes connected tetrahedral units of SiO_4 and AlO_4. By the latest study of the Structure Commission of the International Zeolite Association (IZA-SC), around 248 types of zeolite frameworks have been reported, with FAU, MFI, MOR, BEA, and HEU being the frequently studied types for the photocatalytic activity, as delineated in Table 1 [79].

Zeolites have amphoteric properties and acids, and primary sites are well known. Zeolites have emerged as potential carriers for photocatalytic particles, with a particular focus on synthetic zeolites such as ZSM-5, Zeolite A, and Zeolite Y. Recently, there has been a noteworthy exploration of the use of natural zeolites, notably clinoptilolite, which is the most abundant among them [80]. Various studies have investigated zeolites functionalized with photocatalytic materials like TiO_2 or ZnO, revealing promising outcomes with enhanced photocatalytic efficiency compared to bulk photocatalysts. The semiconductors are impregnated to the surface of zeolites by the hydrothermal method by transferring the materials to a Teflon-lined autoclave operated at a temperature stretching from 100 to 250 °C. The closed system endures elevated pressure and temperature for the consistent synthesis of the particular crystals on the zeolite surface in water. Several studies illustrate the synthesis of zeolite-based photocatalysts using metal oxides like ZnO. The crystals of ZnO are incorporated on the outer surface of NaA zeolite, and for NaY and USY zeolites, it is impregnated inside the cavities. The suitable energy bandgap of cadmium sulfide (CdS), i.e., 2.4 eV, makes it an attractive candidate for the betterment of visible light capture. The major challenge CdS faces is the decomposition when activated by the light, which could destabilize the system. Including CdS in the zeolite pockets can improve the photocatalyst's stability. The design and size of the semiconductor can be modulated by changing the reaction condition for hydrothermal treatment [81–85].

Table 1 Structural properties of five kinds of zeolite frameworks

Zeolite framework types	Framework structures	Maximum diameter of a sphere		Typical zeolite materials
		That can be included (Å)	That can diffuse along a × b × c (Å)	
FAU	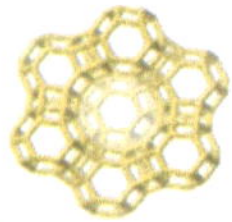	11.24	7.35 × 7.35 × 7.35	Zeolite X Zeolite Y
MFI	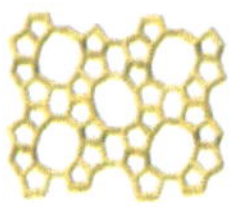	6.36	4.7 × 4.6 × 4.46	ZSM-5
MOR	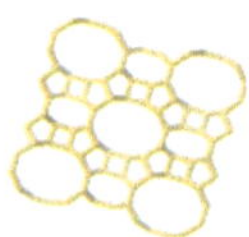	6.7	1.57 × 2.95 × 6.45	Mordenite
BEA	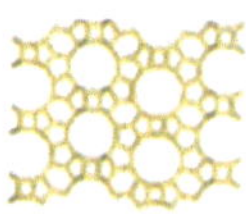	6.68	5.95 × 5.95 × 5.95	Beta
HEU	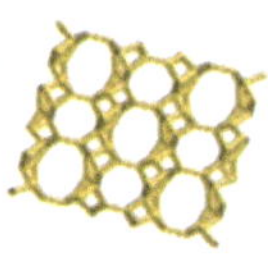	5.97	3.05 × 1.34 × 3.67	Clinoptilolite

The skeletal structure and the ratio between Si and Al in zeolite determine the main features like the adsorption capability, photocatalysis of hybrid composites, and elimination of byproducts formed during the reaction. Aluminosilicates come under the acid catalyst category; they do not intrinsically own photocatalytic capability [86]. The solution to this problem is the uncomplicated substitution of cations with metal ions that leads to the doping of metal ions on the surface of zeolites by ion exchange and ion conduction and intensifies the photocatalytic activity [87]. Metal ions can also be placed inside the cavities of the zeolite framework and can even amplify the performance of the zeolite-based photocatalyst. The principle involved in this approach is called surface plasmon resonance, where the disunion of light-induced charge carriers takes place when noble metals like Au and Ag are embedded in the framework because of the number of active sites they create. The process will also increase the visible light capture and enhance the photocatalytic ability [88–90]. The doping of non-noble metals, such as iron, manganese, copper, titanium, and nickel, generates active charge carriers and helps improve the acidity of zeolites, thereby enhancing the photocatalytic treatment. Literature reports reveal that the engineering applications that utilize semiconductor/zeolite materials in powder form for photocatalysis show inherent limitations. The reuse of photocatalytic composite can be possible by extracting the particles after the treatment [91, 92]. Zeolites are used as pellets in fixed-bed photoreactors as support for photocatalysts. Its advantages include robust mechanical characteristics, broad surface area, and high capacity for adsorption, making them the perfect materials for photocatalysis. Another possible advantage of zeolites in photocatalysis is their ion exchange property, which can incorporate transition metal ions, which show critical photocatalytic properties due to vacant d-orbitals.

3.3 Metal–Organic Frameworks

Metal–organic frameworks (MOFs) are newly derived highly porous crystalline materials with wide surface area, tunable functional groups, acceptable stability, and efficiency. They consist of metal ion clusters associated with organic linkers and are studied for various applications, including the degradation of organic pollutants [93–95]. These linkers trap the photoelectrons and act as a vulnerable photocatalyst unit. Endurable MOFs can be prepared by using the HSAB principle. The metals with high valency (Ti^{4+}, Al^{3+}, Fe^{3+}) are treated with hard Lewis bases, and metals with low valency (Co^{2+}, Ni^{2+}) are treated with soft Lewis bases. MIL (Material Institute Lavoisier) comes under the former, and ZIF (Zeolitic imidazolate frameworks) falls under the latter category [96]. Combining nanocomposites with MOFs has received particular attention in decontaminating toxic organic pollutants due to their significant catalytic properties and large surface area, leading to many active sites. Photocatalysts templated by Metal–Organic Frameworks (MOFs) have gained significant popularity owing to their substantial surface area. These catalysts not only uncover the active sites effectively but also facilitate the synthesis of MOF-templated photocatalysts characterized by an elevated surface area. The above-mentioned feature enhances the exposure of active sites and promotes the mass transfer and transport of photogenerated charge carriers [97].

MOFs are also investigated under numerous applications, including gas purification and storage, catalytic mechanisms, medicines, etc. Further research on MOF goes on extensively, revealing the innovations in drug delivery, heterogeneous catalysis, water treatment, and degradation of toxic materials [98]. The viable way to generate visible light active photocatalysts is by modifying the organic linkers. Amine group functionalized linkers promote photocatalytic activity more than the other linkers. For instance, G. Capano et al., in their article about the electronic and optical properties of MOFs, deal with a case study about MIL-125 and MIL-125-NH_2. The recombination rate of electron–hole pairs in MIL-125-NH_2 is stunted than MIL-125 because, upon excitation, the holes and the electron get localized in the orthogonal regions of MIL-125-NH_2 while they significantly overlap in the case of MIL-125 [99]. However, reactions with long linkers result in reduced active centers and porosity.

Hence, refining the crystallinity can make the porosity better. This section primarily deals with using organic polymers and photoactive MOF nanomaterials to decontaminate organic pollutants for a sustainable environment. However, the drawbacks of MOF include the necessity of systematic protocol for the synthesis that owes high porosity and crystallinity. The insolubility of MOF makes it challenging to establish synthesis reproducibility and aqueous stability. The polymer-MOF monolith fabrication is a reasonable approach to stimulate charge migration and thus enhances the photocatalytic activity of MOF. Polymers like polyaniline, polypyrrole, and polyphenylene are conductive materials that stimulate the mobility of charge carriers and reduce the recombination rate, thus broadening the photocatalytic potentiality [100]. By increasing the conductivity, the number of charge carriers can be increased, and thus it can prevail over the energy barriers, accelerating the photocatalysis. Different techniques for introducing polymers to MOF include placing them straight into the MOF by stirring the polymers and MOFs in their molten state or with appropriate solvents that can be eliminated by evaporation. Adsorption of polymers to the nanopores of MOF highly relies on the solvent used. Their affinity toward solvents and MOFs can adjust the polymeric adsorption, so desorption of the solvent after the polymer insertion would be uncomplicated.

Graphitic carbon nitride (g-C_3N_4) is a 2D layered polymeric compound having conjugated π bonds extensively used as a photocatalytic candidate due to its excellent thermal stability, facile synthesis strategy, easiness in the availability, distinctive bandgap structure, and tremendous redox ability [101, 102]. However, they are deficient in the number of active centers to trap the photogenerated electrons and have a small surface area that diminishes their photocatalytic performance. One of the effective strategies to improve the photocatalytic ability is modifying the carbon–nitrogen vacancies. Creating N vacancies can more effectively induce photocatalysis by enhancing the separation between the charge carriers and increasing the specific surface area. In his paper, Sun et al. [103] developed RuL_2L'@ C_3N_4/MOF, where he loaded Co-MOF-74 and Ruthenium complex on g-C_3N_4 to study about photocatalytic reduction of CO_2. The UV–vis spectra and CO_2 adsorption isotherms have been presented in Fig. 1. The CO_2 absorptions for g-C_3N_4, RuL_2L'@ C_3N_4/MOF and Co-MOF-74 were reported as 1.1, 8.1, 18.8 cm^3g^{-1}. A strong red-shifted band is obtained for RuL_2L'@ C_3N_4/MOF than g-C_3N_4. The selectivity of CO_2 evolution was 86.6% with RuL_2L'@200 C_3N_4/MOF.

The hybridization of MOF with MXenes garnered immense interest from researchers as they light up toward a new area where the structural diversity of MOF is well explored. MXenes are transition metal carbides, nitrides, or both with 2D layered structures with significant flexibility and catalytic properties. Xiao et al., in their paper, *incorporated* CdS with Ti_3C_2 MXene to study the efficient photocatalytic hydrogen generation. Ti_3C_2 MXene unconvincingly degrades, and the oxidation rate decreases in organic solvents rather than water. The low hydrogen evolution of CdS is attributed to its self-corrosion property, but it became seven times higher

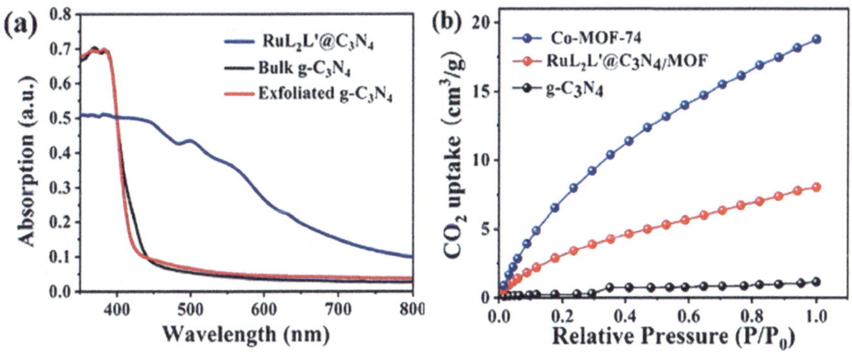

Fig. 1 **a** UV-vis DRS Spectra for RuL_2L' @ C_3N_4, bulk and exfoliated g-C_3N_4. **b** CO_2 adsorption isotherms for g-C_3N_4, RuL_2L' @ C_3N_4/MOF & Co-MOF-74. Reprinted with permission from Sun et al. [103], Copyright (2021) Royal Society of Chemistry

when Ti_3C_2 MXene was employed. After illumination, the photoexcited electrons from CdS migrate to MXene, reacting with the absorbed electrons and producing hydrogen [104].

Besides tailoring the energy gap of MOF for photocatalysis, they lack charge transfer and have a fast recombination rate. Consequently, integrating MOF with different heterojunctions that include metal clusters and oxides of semiconductors as guest molecules can improve their optical properties [105–107]. TiO_2 is prominent in photocatalysis because of its biocompatible characteristics, low toxicity, and high chemical and thermal stability. T. Zeng et al. demonstrated how a phosphonate-based MOF can positively impact photocatalytic behavior when doped with TiO_2 [108]. Unlike TiO_2, P-TiO_2 generates a redshift because the P atoms replace the Ti atom in the TiO_2 crystallite. The 3p electronic configuration of P has a crucial role in forming electron–hole pairs, increasing the efficiency of visible light absorption. MOF/P-TiO_2 shows higher absorption capacity with RhB molecule than pure P-TiO_2. About 97.6% of RhB was degraded in under 30 min; the results are presented in Fig. 2a. The HR-TEM, FFT, STEM, and EDX mapping images of MOF1.0/P–TiO_2 have been presented in Fig. 2b–e. The reduction of Cr(VI) was also reviewed; the results show 76% degradation within 60 min of irradiation. It was also concluded that by increasing the amount of MOF, the separation between the charge carriers can be improved and thus enhance the photocatalytic performance of the photocatalyst synthesized. Wang et al. reported the encapsulation of Pt nanoparticles in zeolitic imidazole frameworks, which is an efficient strategy because imidazole can serve as stabilizing agent and organic linker [109].

There are still issues when heavy metals get incorporated into MOF, including causing minor pollution to treated water bodies and the scarcity of noble metals. Different kinds of MOFs were integrated to form a hybrid MOF which is an eminent method rather than going with one that is insufficient in its function or bringing out multifunctional MOF materials. Keen attention should be taken while preparing MOF–MOF hybrid that even a minute neglectance can cause completely different

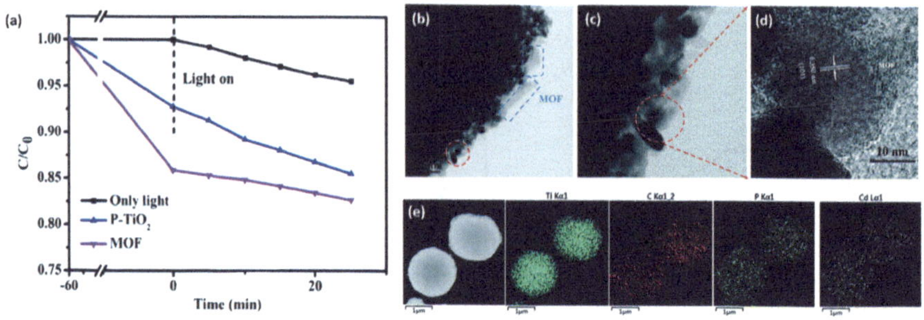

Fig. 2 **a** Photocatalytic degradation of RhB over different catalysts, HR-TEM images **b** and **c** and FFT image **d** of MOF1.0/P–TiO_2, and **e** STEM image and EDX mapping of Ti, C, P, and Cd in MOF0.5/P–TiO_2. Reprinted with permission from Zeng et al. [108], Copyright (2020) Royal Society of Chemistry

structures and properties. S. Kampouri et al. reported the photocatalytic activity of MOF–MOF heterojunction [110] between MIL-125-NH_2 and MIL-167. The generation of this heterojunction enhances the photocatalytic activity and exhibits light absorption from 500 to 750 nm. The UV-vis absorption spectra results and hydrogen evolution rate of MIL-125-NH_2 and MIL-167 are given in Fig. 3a, b. Inserting halogen groups into MOF is another expeditious technique that brings favorable characteristics for progressing photocatalytic applications. Yu et al. reported that incorporating a layered structure of bismuth oxyhalides into an MOF network promotes the separation of charge carriers and further gives beneficial advantages in the photocatalytic process. Incorporating BiOBr into UiO-66-NH_2 to form the heterojunction has efficiently removed hexavalent chromium and Rhodamine B from wastewater. The eradication abilities of UiO-66-NH_2, BiOBr, and UiO-66-NH_2/BiOBr were 56.3, 75.8, and 98.9% which can be attributed to the existence of interlayer surface and dispersed alignment of BiOBr on UiO-66-NH_2 which can adsorb extra $Cr_2O_7^{2-}$ [111].

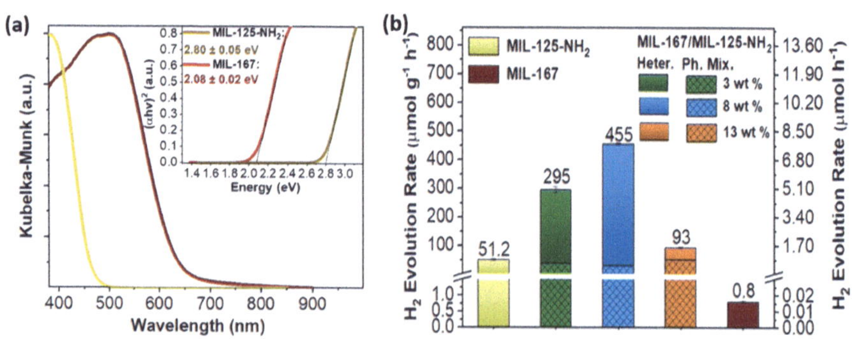

Fig. 3 **a** UV-vis absorption spectra and **b** hydrogen evolution rate of MIL-125-NH_2 and MIL-167. Reprinted with permission from Kampouri et al. [110], Copyright (2021) American Chemical Society

The application area in the MOF field is expanding to nanophotonics and intelligent sensors. The luminescent properties of MOF make it a compelling candidate for photocatalysis, and it functions as a shield for the guest molecules accommodated in its nanopores. Zeolites are microporous and have tetrahedral units with aluminosilicate skeletons. The uniform porosity, regeneracy, tunability of active centers, and high thermal stability make zeolites an active catalyst. Further encapsulating zeolites to MOF elevates the physical and chemical properties because they share standard features and thus improve photocatalytic phenomena. Zhu G. et al., in their study [112], demonstrate zeolite @ MOF by using zeolite socony mobil-5 (ZSM-5) and $Zr_6O_4(OH)_4(O_2C–C_6H_4–CO_2)_6$ (UiO-66) as a bifunctional catalyst in one-pot cascade reactions. Michael et al., in their study [113], illustrate the influence of SDA (structure directing agent) during the hydrothermal and solvothermal synthesis of porous zeolitic frameworks. The schematic diagram of Indium Zeolitic Metal–Organic Framework (In-ZMOF) topologies is shown in Scheme 1, describing the reaction between indium(II) nitrate hydrate and 4,5-imidazole dicarboxylic acid (H_3ImDC) giving raise to different ZMOFs in the presence of SDAs.

The structural features of MOF play a vital role in the field of photocatalysis. Nevertheless, the bonds are weak, and their instability is a significant concern. COFs

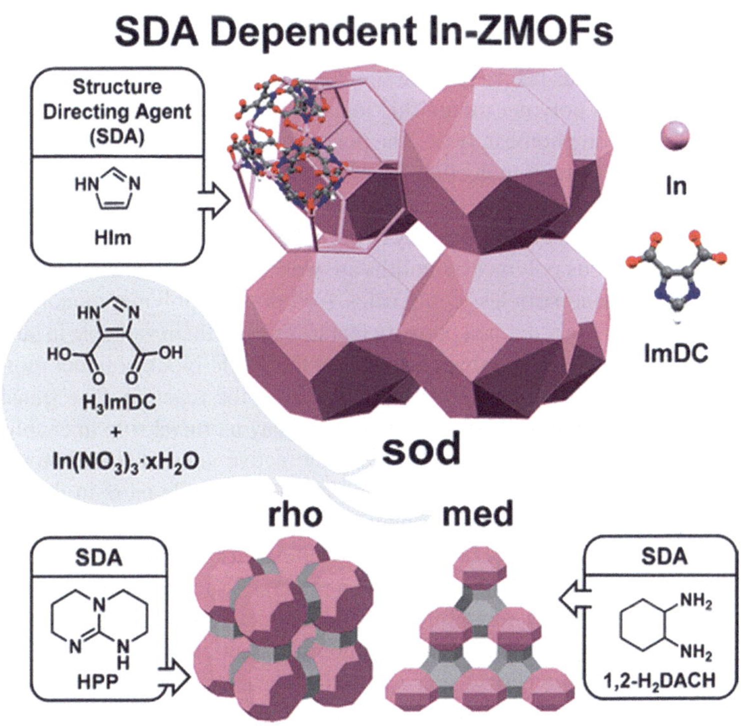

Scheme 1 Structure-Directing Agent (SDA) influence on indium zeolitic MOF topologies. Reprinted with permission from Sinnwell [113], Copyright (2020) American Chemical Society

contain C, N, and O with covalent bonding and have significant characteristics similar to MOF. The light intensity can be increased by introducing nitrogen-containing groups to COFs. However, COF cannot meet the charge separation efficiency and the large surface area. So, it would be practicable if a combination of MOF-COF is used for the applications. In particular, the facile modification of MOFs and COFs enables their intermesh into MOF/COF hybrids to enhance performance and extend scope toward diverse applications. Significant progress in the design and synthesis of MOF/COF hybrids, including MOF@COF, COF@MOF, MOF + COF, C-MOF, and COF-in-MOF, and their various applications in catalysis, gas adsorption, sensing, energy storage, and photodynamic therapy have been reported [114]. In the heterojunction of MOF-COF, the COF has a weak forbidden bandgap, mainly absorbing light and generating photoelectrons. The covalent bond between MOF and COF transfers the electrons and improves charge separation. The critical applications include photocatalytic hydrogen evolution (PHE) and photocatalytic degradation of organic pollutants [115].

4 Polymer-Supported Nanocomposite Photocatalysts for Decontamination of Emerging Pollutants

In recent years, there has been a notable surge in the interest surrounding translucent macro-/meso-porous polymer monolithic materials, particularly in their application for light-harvesting in photocatalysis. The porous monolithic structures facilitate the seamless diffusion of target molecules from the bulk solution to the photoactive sites within the anchored monolithic framework of nanocomposites. This enhancement significantly contributes to the efficiency of the process and accelerates disintegration kinetics. Macro-porous polymer monoliths are well suited for the facile and uniform immobilization of nanoparticles/-composites. Additionally, polymer monoliths' high surface area and precisely defined porosity features render them valuable in heterogeneous catalysis. Moreover, the shape and size of the single-block polymer monoliths are customizable and allow for swift recovery after the reaction. The translucent optical properties of polymer monolith materials play a crucial role in enabling the unimpeded diffusion of visible light toward the active sites of the photocatalyst, thereby fostering the voluminous generation of electron–hole pairs in the conduction band/valence band states of the nanocomposite. Furthermore, polymer monolith materials offer the advantage of being easily synthesized and functionalized at a relatively lower cost. In contrast to other porous materials such as zeolites, silica, and activated carbons, polymer monoliths provide distinctive feature variations. Additionally, implementing porous polymer monoliths is anticipated to enhance photocatalytic efficiency by offering greater adsorption capacity, thus addressing recovery and reusability concerns.

Sompalli et al. fabricated a homogeneously dispersed heterojunction Cr_2O_3–Ag_2O nanocomposite on a polymer monolith through solvothermal-assisted polymerization. The characterization data elucidates a substantial macro-/meso-porous monolithic structure exhibiting enhanced surface properties, facilitating photocatalytic activity facilitated by the uniformly distributed Cr_2O_3–Ag_2O nanocomposites. The morphological analysis reveals the uniform dispersion of Cr_2O_3–Ag_2O composites across the surface, creating homogeneous patterns within the macro-/meso-porous polymer monolith, as depicted in Fig. 4. The findings indicate that utilizing the Cr_2O_3–Ag_2O (60:40) nanocomposite-dispersed polymer monolith exhibits superior visible-light-induced photocatalytic performance, completely mineralizing the RB-10 dye pollutant within 1 h [116]. Asu et al. emphasized the synthesis of CaO-doped CeO2 nanocomposites (NCs) prepared through a sonochemical-assisted temperature-controlled hydrothermal method to disintegrate moxifloxacin, an antimicrobial drug. The surface topography of the prepared (20:80) CaO/CeO_2 NC- and NCs-loaded polymer monolith materials was examined using FE-SEM and HR-TEM analysis, as depicted in Fig. 5. Optimal results were achieved with the monolithic photocatalyst dispersed with 20:80 CaO/CeO_2 NC, achieving complete dissipation of moxifloxacin drug molecules (99% degradation) at a pH of 6.0. With 100 mg of photocatalyst, 15 ppm of moxifloxacin was dissipated entirely in less than 2 h of visible light irradiation. The proposed photocatalyst is cost-effective, durable, reusable, and efficient for the real-time dissipation of environmental pollutants [117].

Jagadeesan et al. presented a distinctive ZrO_2–Ag_2O heterojunction nanocomposite and an innovative p-In_2S_3 and n-BiOCl heterojunction nanocomposite uniformly dispersed on a macro-/meso-porous polymer monolithic template, serving as efficient visible-light-driven heterogeneous plasmonic photocatalysts for water decontamination. The structural properties and surface morphology of the optimized (60:40) ZrO_2–Ag_2O nanocomposite embedded in the polymer monolith are illustrated in Fig. 6. The photocatalyst exhibited complete degradation of the norfloxacin drug within 1 h of visible light irradiation [118]. In a subsequent report, the same research group detailed synthesizing a novel p-In_2S_3 and n-BiOCl heterojunction nanocomposite with varying β-In_2S_3 stoichiometry (10–50%) produced through hydrothermal and combustion methods. Using 20 wt.% of In_2S_3 to BiOCl nanocomposite, referred to as IB-20 NC, the structural and surface features of the photocatalyst were characterized through high-resolution transmission electron microscopy, selected area electron diffraction, and field-emission scanning electron microscopy-energy-dispersive X-ray spectroscopy, with the results presented in Figs. 7 and 8. Numerous analytical parameters were systematically optimized, including solution pH, nanocomposite ratio, photocatalyst dosage, dye concentration, photosensitizers, oxidizers, and light intensity. Under the established conditions of pH 2.0, 100 mg photocatalyst dosage, 30 ppm dye concentration, 1.0 mM $KBrO_3$, and 300 W/cm^2 visible light intensity, the monolithic photocatalyst demonstrated a dissipation efficiency of $\geq$98.6% for Rhodamine B in $\geq$0.3 h [119].

Kuppusamy et al. successfully synthesized a dual (UV-visible) light-responsive heterocomposite photocatalyst comprising NH2-MIL-125 metal–organic framework

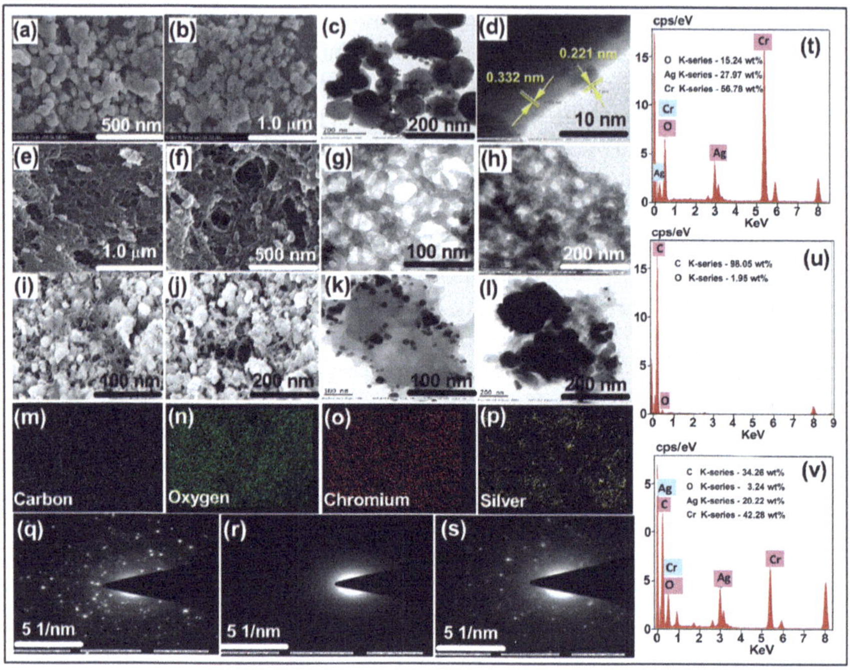

Fig. 4 **a**, **b** FE-SEM and **c**, **d** HR-TEM images of (60:40) Cr_2O_3–Ag_2O nanocomposite, **e**, **f** FE-SEM and **g**, **h** HR-TEM images of bare poly-EGDMA monolith, **i**, **j** FE-SEM and **k**, **l** HR-TEM images of (60:40) Cr_2O_3–Ag_2O nanocomposite impregnated polymer monolith, **m**–**p** elemental mapping of (60:40) Cr_2O_3–Ag_2O nanocomposite impregnated polymer monolith, **q**–**s** SAED pattern, and **t**–**v** EDAX spectra of (60:40) Cr_2O_3–Ag_2O nanocomposite, pristine poly-EGDMA monolith and (60:40) Cr_2O_3–Ag_2O nanocomposite impregnated polymer monolith, respectively. Reprinted with permission from Sompalli et al. [116], Copyright (2021) Elsevier

(MOF) integrated into a transparent bimodal macro-/meso-porous polymer monolith to decontaminate pharmaceutical pollutants from aqueous environments. The morphological characteristics of this material were thoroughly examined, and the outcomes are presented in Fig. 9. Optimal drug degradation was observed under UV and visible light illumination when utilizing 50 mg of the photocatalyst within 0.5 h, maintaining a pH range of 6.0–7.0. The photocatalyst underwent systematic characterization, and its dual light-responsive behavior was verified by monitoring its photocatalytic degradation efficiency under UV and visible light irradiation, a feature rarely observed in most photocatalysts. Notably, the MOF-poly (EGDMA) photocatalyst exhibited exceptional reusability and durability and superior performance in environmental clean-up applications [120].

In recent years, the quest for polymer-supported nanocomposites as a new generation of heterogeneous photocatalysts for developing sustainable chemical processes has been mainly on energy/environment-related applications. Enhancing the semiconductor nanocomposites across the porous network of polymer framework can

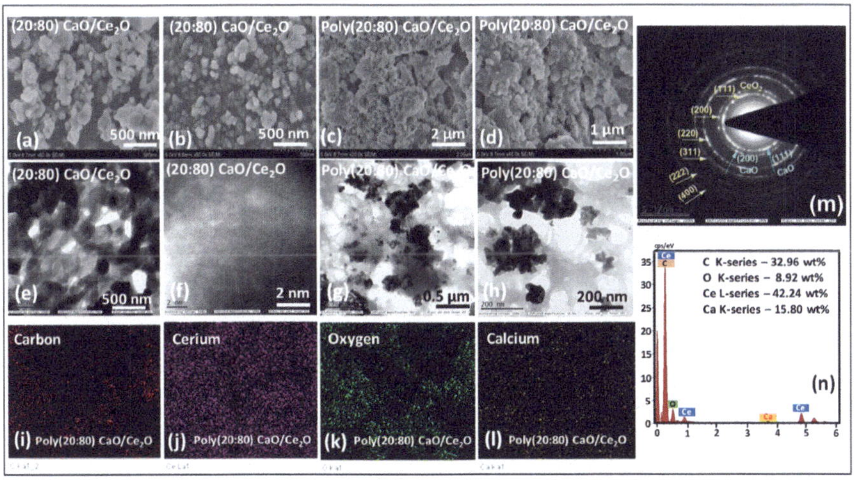

Fig. 5 **a**, **b** FE-SEM images of (20:80) CaO/CeO_2 nanocomposite, **c**, **d** FE-SEM images of (20:80) CaO/CeO_2 nanocomposite imprinted polymer monolith, **e**, **f** HR-TEM images of (20:80) CaO/CeO_2 nanocomposite, **g**, **h** HR-TEM images of (20:80) CaO/CeO_2 nanocomposite imprinted polymer monolith, **i–l** Elemental mapping analysis, **m** SAED pattern, and **n** EDAX spectra of (20:80) CaO/CeO_2 nanocomposite imprinted polymer monolith. Reprinted with permission from Asu et al. [117], Copyright (2022) Elsevier

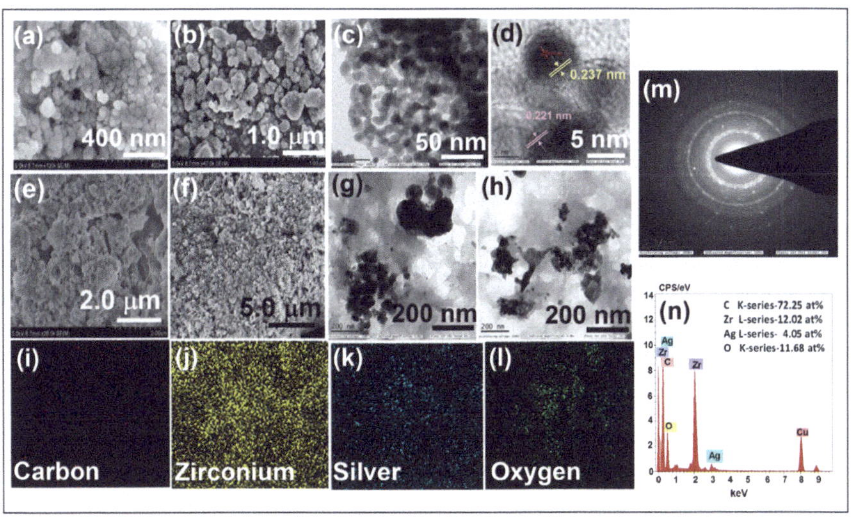

Fig. 6 **a**, **b** FE-SEM images of 60:40 (ZrO_2–Ag_2O) nanocomposites, **c**, **d** HR-TEM images of 60:40 (ZrO_2–Ag_2O) nanocomposites, **e**, **f** FE-SEM images 60:40 (ZrO_2–Ag_2O) nanocomposites imprinted poly(EGDMA) monolith, **g**, **h** HR-TEM images of 60:40 (ZrO_2–Ag_2O) nanocomposites imprinted poly(EGDMA) monolith, **i–l** Elemental mapping analysis, **m** SAED pattern, and **n** EDAX spectra of 60:40 (ZrO_2–Ag_2O) nanocomposites imprinted poly(EGDMA) monolith. (Reprinted with permission from Jagadeesan et al. [118], Copyright (2022) Elsevier

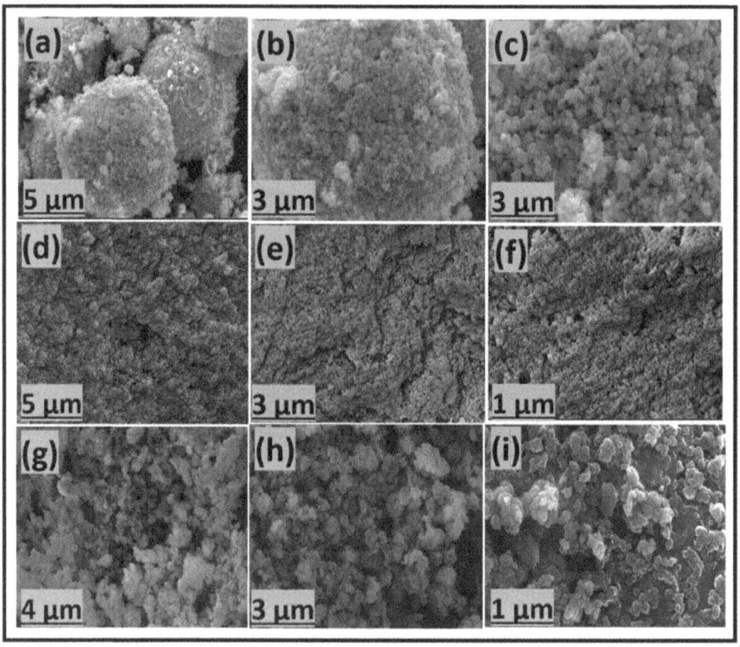

Fig. 7 FE-SEM images of **a–c** IB-20 nanocomposite, **d–f** Bare poly(EGDMA) monolith and **g–i** IB-20 nanocomposite loaded poly(EGDMA) monolith. Reprinted with permission from Jagadeesan and Deivasigamani [119], Copyright (2023) Elsevier

result in improved photoactivities. The quest for eco-friendly monomers to construct the framework of photoactive porous polymers is imperative. We propose that forthcoming research endeavors should explore the application boundaries of photoactive composites, extending their capabilities to address challenges like the removal of a combination of pesticides, personal care products, and chromium reduction. Preferably, these investigations should be conducted in authentic water matrices, simulating precise water/wastewater quality parameters.

5 Conclusion and Future Perspectives/Challenges

The contributions in this chapter can lead newcomers to new lines of inquiry on the structural features and surface morphology of porous support material for photocatalytic applications. The overview reveals that the decoration of nanocomposite heterojunctions on porous supports can produce recyclable, cheaper, and eco-friendly heterogeneous photocatalysts. In the context of heterogeneous supported catalysis, it is imperative to evaluate various crucial experimental parameters conscientiously. Particular attention must be given to factors such as the porosity range and

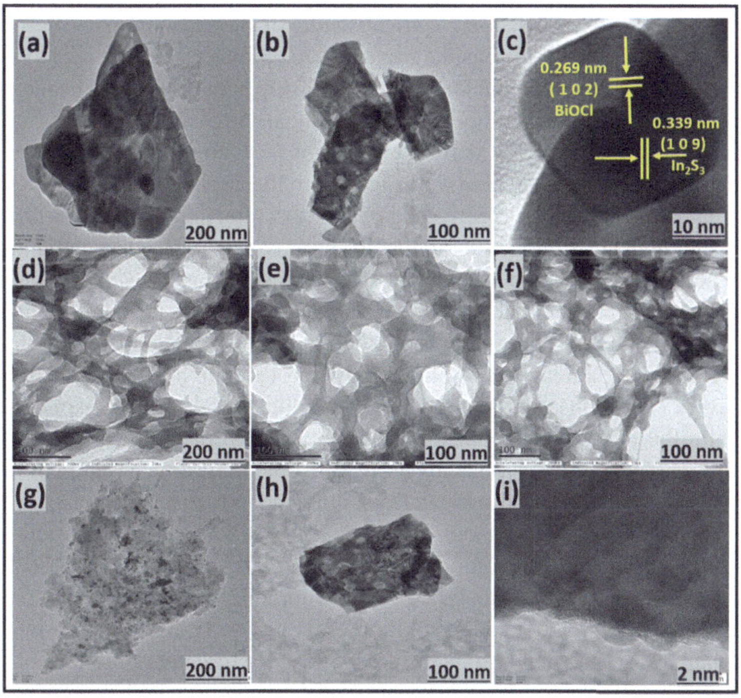

Fig. 8 HR-TEM images of **a–c** IB-20 nanocomposite, **d–f** bare poly (EGDMA) monolith, and **g–i** IB-20 nanocomposite loaded poly (EGDMA) monolith. Reprinted with permission from Jagadeesan and Deivasigamani [119], Copyright (2023) Elsevier

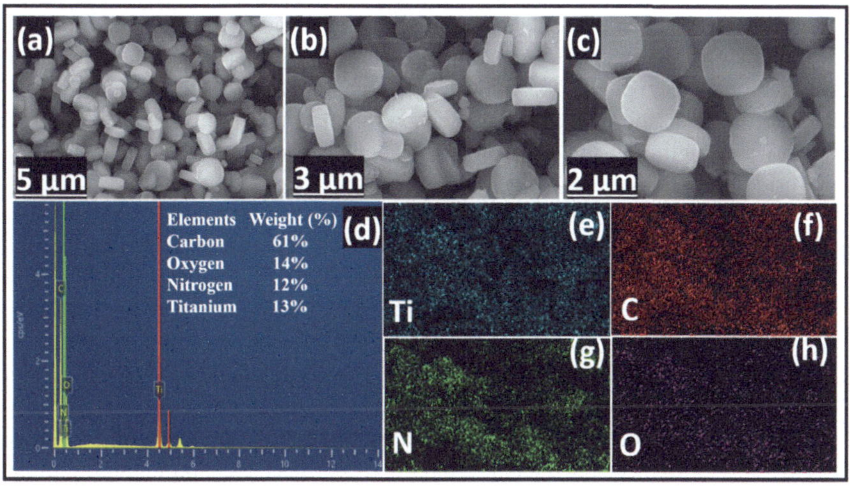

Fig. 9 **a–c** FE-SEM images of the synthesized NH_2-MIL-125 MOF **d–h** EDAX and elemental mapping for the synthesized NH_2-MIL-125 MOF. Reprinted with permission from Kuppusamy et al. [120], Copyright (2023) Elsevier

morphology of the materials, along with the nature of chemical moieties presented at the pore surface. This careful consideration aims to optimize the interactions between the support and the metal (precursor). The relationship between hierarchically porous structures and properties, according to their chemical composition and physical characteristics, proved crucial. The current evaluation seeks to provide a novel pathway for readers to acquire comprehensive insights into the applications of hierarchically porous materials. It aims to clearly understand the selection and design processes involved in choosing suitable hierarchically porous materials for specific applications. In addition to delving into the applications of such materials, this comprehensive overview has the potential to inspire researchers to embark on synthesizing innovative hierarchically porous solids. Nevertheless, their sustainability and efficiency pose a substantial challenge, given that many porous materials tend to disintegrate under severe conditions, and adjusting their photophysical properties proves challenging. Furthermore, their application introduces concerns related to the retrieval and reutilization of costly photocatalysts and the potential contamination of organic products by minute traces of photocatalysts, often laden with heavy metals in homogeneous conditions. Heterogeneous systems have been implemented to tackle these issues by immobilizing nanocomposites within organic or inorganic matrices. This approach aims to prolong the catalytic activity's durability and facilitate the recovery of photocatalysts from the reaction solution.

The chapter overview highlights the urgent need for researchers to develop a new class of visible light-responsive heterogeneous photocatalysts with promising water compatibility, fast response time, reusability, and durability. Fabricating high-quality photocatalysts on large scale for real-time applications remains to be a challenging task. Development is also necessary to fabricate an effective, sustainable, economically viable, and facile photocatalyst. Hence, further exploration is essential to delve into additional dimensions that have surfaced in response to the disclosed findings. Conducting an in-depth investigation into the environmental factors influencing porous support-based photocatalysis is imperative. Parameters such as aqueous pH, catalyst dosage, dissolved organic matter, pollutant concentration, and the impact of light wavelength and intensity on photocatalysis are crucial for understanding real-world conditions.

In summary, there are still many challenges before realizing their practical applications. Conducting a comprehensive life cycle assessment and identifying viable recycling pathways for photoactive crosslinked polymers are imperative to avoid landfilling and minimize waste generation. The future holds substantial potential for semiconductor-infused polymer networks and the resulting photoactive hybrid materials to spearhead industrial transformation. By consistently addressing the aforementioned challenges, we are optimistic that hierarchically porous materials based on polymers will offer a promising future trajectory. Moreover, most photocatalytic studies (>90%) have been conducted on dyes and pharmaceutical pollutants as model organic contaminants, which can extend to decontaminate healthcare products and energy-related applications such as hydrogen evolution reactions, nitrate and carbon dioxide reduction products.

References

1. M.J. Martín de Vidales, R. Prieto, G. Galán-Lucarelli, E. Atanes-Sánchez, F. Fernández-Martínez, Removal of contaminants of emerging concern by photocatalysis with a highly ordered TiO_2 nanotubular array catalyst. Catal. Today **413–415**, 1–8 (2023). https://doi.org/10.1016/j.cattod.2023.01.002
2. M. Guo, Z. Zhou, S. Yan, P. Zhou, F. Miao, S. Liang, J. Wang, X. Cui, Bi_2WO_6–BiOCl heterostructure with enhanced photocatalytic activity for efficient degradation of oxytetracycline. Sci. Rep. **10**, 1–13 (2020). https://doi.org/10.1038/s41598-020-75003-x
3. S.P. Tripathy, S. Subudhi, A. Ray, P. Behera, A. Bhaumik, K. Parida, Mixed-valence bimetallic Ce/Zr MOF-based nanoarchitecture: a visible-light-active photocatalyst for ciprofloxacin degradation and hydrogen evolution. Langmuir **38**, 1766–1780 (2022). https://doi.org/10.1021/acs.langmuir.1c02873
4. M.L. Xu, X.J. Jiang, J.R. Li, F.J. Wang, K. Li, X. Cheng, Self-assembly of a 3D hollow BiOBr@Bi-MOF heterostructure with enhanced photocatalytic degradation of dyes. ACS Appl. Mater. Interfaces **13**, 56171–56180 (2021). https://doi.org/10.1021/acsami.1c16612
5. F.K. Naqvi, M. Faraz, S. Beg, N. Khare, Synthesis and phase transformation studies of dysprosium-doped $Bi_4V_2O_{11}$ nanoparticles and their application in visible light photocatalytic degradation of tetracycline drug. ACS Omega **3**, 11300–11306 (2018). https://doi.org/10.1021/acsomega.8b01012
6. H. Santoke, W. Song, W.J. Cooper, J. Greaves, G.E. Miller, Free-radical-induced oxidative and reductive degradation of fluoroquinolone pharmaceuticals: kinetic studies and degradation mechanism. J. Phys. Chem. A **113**, 7846–7851 (2009). https://doi.org/10.1021/jp9029453
7. E. Pierini, G. Famiglini, F. Mangani, A. Cappiello, Fate of enrofloxacin in swine sewage. J. Agric. Food Chem. **52**, 3473–3477 (2004). https://doi.org/10.1021/jf049865c
8. S.K. Fanourakis, J. Peña-Bahamonde, P.C. Bandara, D.F. Rodrigues, Nano-based adsorbent and photocatalyst use for pharmaceutical contaminant removal during indirect potable water reuse. NPJ Clean Water **3** (2020). https://doi.org/10.1038/s41545-019-0048-8
9. M. Bodzek, M. Rajca, Photocatalysis in the treatment and disinfection of water. Part I. Theoretical backgrounds. Ecol. Chem. Eng. S. **19**, 489–512 (2012). https://doi.org/10.2478/v10216-011-0036-5
10. M. Ren, J. Di, W. Chen, Recent progress and application challenges of wearable supercapacitors. Batteries Supercaps **4**(8), 1279–1290 (2021). https://doi.org/10.1002/batt.202000333
11. B. Zhu, D. Song, T. Jia, W. Sun, D. Wang, L. Wang, J. Guo, L. Jin, L. Zhang, H. Tao, Effective Visible light-driven photocatalytic degradation of ciprofloxacin over flower-like Fe_3O_4/Bi_2WO_6 composites. ACS Omega **6**, 1647–1656 (2021). https://doi.org/10.1021/acsomega.0c05616
12. F. Quddus, A. Shah, F.J. Iftikhar, N.S. Shah, A. Haleem, Environmentally benign nanoparticles for the photocatalytic degradation of pharmaceutical drugs. Catalysts **13**, 1–24 (2023). https://doi.org/10.3390/catal13030511
13. K. Selvakumar, Y. Wang, Y. Lu, B. Tian, Z. Zhang, J. Hu, A. Raja, M. Arunpandian, M. Swaminathan, H. Dai, M. Sui, Single metal atom oxide anchored Fe_3O_4-ED-rGO for highly efficient photodecomposition of antibiotic residues under visible light illumination. Appl. Catal. B Environ. **300**, 120740 (2022). https://doi.org/10.1016/j.apcatb.2021.120740
14. X. Nie, G. Li, S. Li, Y. Luo, W. Luo, Q. Wan, T. An, Highly efficient adsorption and catalytic degradation of ciprofloxacin by a novel heterogeneous Fenton catalyst of hexapod-like pyrite nanosheets mineral clusters. Appl. Catal. B Environ. **300**, 120734 (2022). https://doi.org/10.1016/j.apcatb.2021.120734
15. R. Abazari, A.R. Mahjoub, Amine-Functionalized Al-MOF#@ $yxSm_2O_3$-ZnO: a visible light-driven nanocomposite with excellent photocatalytic activity for the photo-degradation of amoxicillin. Inorg. Chem. **57**, 2529–2545 (2018). https://doi.org/10.1021/acs.inorgchem.7b02880

16. S. Ghasimi, Conjugated porous polymers for visible-light photocatalysis, 157 (2016). https://d-nb.info/1123047847/34
17. E. Fernandes, P. Mazierski, M. Miodyńska, T. Klimczuk, M. Pawlyta, A. Zaleska-Medynska, R.C. Martins, J. Gomes, Carbon nitride exfoliation for photocatalysis and photocatalytic ozonation over emerging contaminants abatement. J. Environ. Chem. Eng. **11** (2023). https://doi.org/10.1016/j.jece.2023.110554
18. F.T. Geldasa, M.A. Kebede, M.W. Shura, F.G. Hone, Experimental and computational study of metal oxide nanoparticles for the photocatalytic degradation of organic pollutants: a review. RSC Adv. **13**, 18404–18442 (2023). https://doi.org/10.1039/d3ra01505j
19. N. Mzimela, S. Tichapondwa, E. Chirwa, Visible-light-activated photocatalytic degradation of rhodamine B using WO_3 nanoparticles. RSC Adv. **12**, 34652–34659 (2022). https://doi.org/10.1039/d2ra06124d
20. Y. He, L. Yang, F. Zhang, B. Zhang, G. Zou, Tunable electron-injection channels of heterostructured ZnSe@CdTe nanocrystals for surface-chemistry-involved electrochemiluminescence. J. Phys. Chem. Lett. **9**, 6089–6095 (2018). https://doi.org/10.1021/acs.jpclett.8b02645
21. D.A. Reddy, J. Choi, S. Lee, R. Ma, T.K. Kim, Self-assembled macro porous ZnS-graphene aerogels for photocatalytic degradation of contaminants in water. RSC Adv. **5**, 18342–18351 (2015). https://doi.org/10.1039/c4ra16494f
22. D. Wang, T. Kako, J. Ye, New series of solid-solution semiconductors $(AgNbO_3)$ 1−x$(SrTiO_3)$x with modulated band structure and enhanced visible-light photocatalytic activity. J. Phys. Chem. C **113**, 3785–3792 (2009). https://doi.org/10.1021/jp807393a
23. M. Niu, F. Huang, L. Cui, P. Huang, Y. Yu, Y. Wang, Hydrothermal synthesis, structural characteristics, and enhanced photocatalysis of $SnO_2/\alpha\text{-}Fe_2O_3$ semiconductor nanoheterostructures. ACS Nano **4**, 681–688 (2010). https://doi.org/10.1021/nn901119a
24. C. Kang, K. Xiao, Y. Wang, D. Huang, L. Zhu, F. Liu, T. Tian, Synthesis of $SrTiO_3$–TiN nanocomposites with enhanced photocatalytic activity under simulated solar irradiation. Ind. Eng. Chem. Res. **57**, 11526–11534 (2018). https://doi.org/10.1021/acs.iecr.8b01203
25. R. Das, S. Sarkar, S. Chakraborty, H. Choi, C. Bhattacharjee, Remediation of antiseptic components in wastewater by photocatalysis using TiO_2 nanoparticles. Ind. Eng. Chem. Res. **53**, 3012–3020 (2014). https://doi.org/10.1021/ie403817z
26. J.A. Nasir, Z.U. Rehman, S.N.A. Shah, A. Khan, I.S. Butler, C.R.A. Catlow, Recent developments and perspectives in CdS-based photocatalysts for water splitting. J. Mater. Chem. A. **8**, 20752–20780 (2020). https://doi.org/10.1039/d0ta05834c
27. W.H. Lin, Y.H. Chiu, P.W. Shao, Y.J. Hsu, Metal-particle-decorated ZnO nanocrystals: photocatalysis and charge dynamics. ACS Appl. Mater. Interfaces **8**, 32754–32763 (2016). https://doi.org/10.1021/acsami.6b08132
28. Y. Mingmongkol, A. Polnok, P. Phuinthiang, D. Channei, K. Ratananikom, A. Nakaruk, W. Khanitchaidecha, Photocatalytic degradation mechanism of the pharmaceutical agent salbutamol using the Mn-doped TiO_2 nanoparticles under visible light irradiation. ACS Omega **8**, 17254–17263 (2023). https://doi.org/10.1021/acsomega.3c01776
29. C. Dai, B. Liu, Conjugated polymers for visible-light-driven photocatalysis. Energy Environ. Sci. **13**, 24–52 (2020). https://doi.org/10.1039/c9ee01935a
30. K. Vijayan, S.P. Vijayachamundeeswari, Improving the multifunctional attributes and photocatalytic dye degradation of MB and RhB dye—A comparative scrutiny. Inorg. Chem. Commun. **144**, 109940 (2022). https://doi.org/10.1016/j.inoche.2022.109940
31. M.H. Sun, S.Z. Huang, L.H. Chen, Y. Li, X.Y. Yang, Z.Y. Yuan, B.L. Su, Applications of hierarchically structured porous materials from energy storage and conversion, catalysis, photocatalysis, adsorption, separation, and sensing to biomedicine. Chem. Soc. Rev. **45**, 3479–3563 (2016). https://doi.org/10.1039/c6cs00135a
32. R.A.C. Amoresi, R.C. Oliveira, N.L. Marana, P.B. De Almeida, P.S. Prata, M.A. Zaghete, E. Longo, J.R. Sambrano, A.Z. Simões, CeO_2 nanoparticle morphologies and their corresponding crystalline planes for the photocatalytic degradation of organic pollutants. ACS Appl. Nano Mater. **2**, 6513–6526 (2019). https://doi.org/10.1021/acsanm.9b01452

33. A.A. AbdelHamid, A. Mendoza-Garcia, J.Y. Ying, Advances in and prospects of nanomaterials' morphological control for lithium rechargeable batteries. Nano Energy **93**, 106860 (2022). https://doi.org/10.1016/j.nanoen.2021.106860
34. P. Moroz, A. Boddy, M. Zamkov, Challenges and prospects of photocatalytic applications utilizing semiconductor nanocrystals. Front. Chem. **6**, 1–7 (2018). https://doi.org/10.3389/fchem.2018.00353
35. J. Cai, X. Wu, S. Li, F. Zheng, Synthesis of TiO_2@WO_3/Au nanocomposite hollow spheres with controllable size and high visible-light-driven photocatalytic activity. ACS Sustain. Chem. Eng. **4**, 1581–1590 (2016). https://doi.org/10.1021/acssuschemeng.5b01511
36. N. Zhang, S. Liu, Y.J. Xu, Recent progress on metal core@semiconductor shell nanocomposites as a promising type of photocatalyst. Nanoscale **4**, 2227–2238 (2012). https://doi.org/10.1039/c2nr00009a
37. L.S. Gomez-Villalba, C. Salcines, R. Fort, Application of inorganic nanomaterials in cultural heritage conservation, risk of toxicity, and preventive measures. Nanomaterials **13** (2023). https://doi.org/10.3390/nano13091454
38. P. Samanta, A.V. Desai, S. Let, S.K. Ghosh, Advanced porous materials for sensing, capture and detoxification of organic pollutants toward water remediation. ACS Sustain. Chem. Eng. **7**, 7456–7478 (2019). https://doi.org/10.1021/acssuschemeng.9b00155
39. V. Harish, M.M. Ansari, D. Tewari, M. Gaur, A.B. Yadav, M.L. García-Betancourt, F.M. Abdel-Haleem, M. Bechelany, A. Barhoum, Nanoparticle and nanostructure synthesis and controlled growth methods. Nanomaterials **12**, 1–30 (2022). https://doi.org/10.3390/nano12183226
40. Q. Zhang, E. Uchaker, S.L. Candelaria, G. Cao, Nanomaterials for energy conversion and storage. Chem. Soc. Rev. **42**, 3127–3171 (2013). https://doi.org/10.1039/c3cs00009e
41. M.C. Orilall, U. Wiesner, Block copolymer based composition and morphology control in nanostructured hybrid materials for energy conversion and storage: solar cells, batteries, and fuel cells. Chem. Soc. Rev. **40**, 520–535 (2011). https://doi.org/10.1039/c0cs00034e
42. A. Ahmadi, M. Hajilou, S. Zavari, S. Yaghmaei, A comparative review on adsorption and photocatalytic degradation of classified dyes with metal/non-metal-based modification of graphitic carbon nitride nanocomposites: synthesis, mechanism, and affecting parameters. J. Clean. Prod. **382**, 134967 (2023). https://doi.org/10.1016/j.jclepro.2022.134967
43. S. Malato, J. Giménez, I. Oller, A. Agüera, J.A. Sánchez Pérez, Removal and degradation of pharmaceutically active compounds (PhACs) in wastewaters by solar advanced oxidation processes. Handb. Environ. Chem. **108**, 299–326 (2021). https://doi.org/10.1007/698_2020_688
44. Y. Khan, H. Sadia, S.Z. Ali Shah, M.N. Khan, A.A. Shah, N. Ullah, M.F. Ullah, H. Bibi, O.T. Bafakeeh, N. Ben Khedher, S.M. Eldin, B.M. Fadhl, M.I. Khan, Classification, synthetic, and characterization approaches to nanoparticles, and their applications in various fields of nanotechnology: a review. Catalysts **12** (2022). https://doi.org/10.3390/catal12111386
45. A.M. El-Khawaga, A. Zidan, A.I.A.A. El-Mageed, Preparation methods of different nanomaterials for various potential applications: a review. J. Mol. Struct. **1281** (2023). https://doi.org/10.1016/j.molstruc.2023.135148
46. N. Abid, A.M. Khan, S. Shujait, K. Chaudhary, M. Ikram, M. Imran, J. Haider, M. Khan, Q. Khan, M. Maqbool, Synthesis of nanomaterials using various top-down and bottom-up approaches, influencing factors, advantages, and disadvantages: a review. Adv. Colloid Interface Sci. **300**, 102597 (2022). https://doi.org/10.1016/j.cis.2021.102597
47. S. Rani, M. Sharma, D. Verma, A. Ghanghass, R. Bhatia, I. Sameera, Two-dimensional transition metal dichalcogenides and their heterostructures: role of process parameters in top-down and bottom-up synthesis approaches. Mater. Sci. Semicond. Process. **139**, 106313 (2022). https://doi.org/10.1016/j.mssp.2021.106313
48. D. Maity, S.R. Sahoo, S. Saha, Synthesis and characterization of nanomaterials for electrochemical sensors. ACS Symp. Ser. **1437**, 193–222 (2023). https://doi.org/10.1021/bk-2023-1437.ch009

49. C.T.J. Ferguson, K.A.I. Zhang, Classical polymers as highly tunable and designable heterogeneous photocatalysts. ACS Catal. **11**, 9547–9560 (2021). https://doi.org/10.1021/acscatal.1c02056
50. Q. Sun, Z. Dai, X. Meng, L. Wang, F.S. Xiao, Task-specific design of porous polymer heterogeneous catalysts beyond homogeneous counterparts. ACS Catal. **5**, 4556–4567 (2015). https://doi.org/10.1021/acscatal.5b00757
51. T.X. Wang, H.P. Liang, D.A. Anito, X. Ding, B.H. Han, Emerging applications of porous organic polymers in visible-light photocatalysis. J. Mater. Chem. A **8**, 7003–7034 (2020). https://doi.org/10.1039/d0ta00364f
52. T. Zhang, G. Xing, W. Chen, L. Chen, Porous organic polymers: a promising platform for efficient photocatalysis. Mater. Chem. Front. **4**, 332–353 (2020). https://doi.org/10.1039/c9qm00633h
53. Z. Fu, S. Han, J. Huang, Y.N. Liu, Comparison of hyper-cross-linked polystyrene/polyacryldiethylenetriamine (HCP/PADETA) interpenetrating polymer networks (IPNs) with hyper-cross-linked polystyrene (HCP): structure, adsorption and separation properties. RSC Adv. **6**, 32340–32348 (2016). https://doi.org/10.1039/c6ra01932c
54. S. Fajal, S. Dutta, S.K. Ghosh, Porous organic polymers (POPs) for environmental remediation. Mater. Horizons. (2023). https://doi.org/10.1039/d3mh00672g
55. D. Taylor, S.J. Dalgarno, Z. Xu, F. Vilela, Conjugated porous polymers: incredibly versatile materials with far-reaching applications. Chem. Soc. Rev. **49**, 3981–4042 (2020). https://doi.org/10.1039/c9cs00315k
56. M.Z. Ahmad, R. Castro-Munõz, P.M. Budd, Boosting gas separation performance and suppressing the physical aging of polymers of intrinsic microporosity (PIM-1) by nanomaterial blending. Nanoscale **12**, 23333–23370 (2020). https://doi.org/10.1039/d0nr07042d
57. C.C. Gu, F.H. Xu, W.K. Zhu, R.J. Wu, L. Deng, J. Zou, B.C. Weng, R.L. Zhu, Recent advances on covalent organic frameworks (COFs) as photocatalysts: different strategies for enhancing hydrogen generation. Chem. Commun. 7302–7320 (2023). https://doi.org/10.1039/d3cc01970e
58. H. Ma, H. Ren, X. Zou, F. Sun, Z. Yan, K. Cai, D. Wang, G. Zhu, Novel lithium-loaded porous aromatic framework for efficient CO_2 and H_2 uptake. J. Mater. Chem. A. **1**, 752–758 (2013). https://doi.org/10.1039/c2ta00616b
59. B. Dong, Y. Wan, Q. Cheng, H. Zhou, Z. Pan, Construction of novel MoS_2@COF-Ph heterojunction photocatalysts for boosted photocatalytic efficiency and hydrogen production performance under sunlight. Environ. Sci NANO **9**, 2799–2814 (2022). https://doi.org/10.1039/d2en00278g
60. M. Nasrollahzadeh, R. Akbari, S. Sakhaei, Z. Nezafat, S. Banazadeh, Y. Orooji, G. Hegde, Polymer supported copper complexes/nanoparticles for treatment of environmental contaminants. J. Mol. Liq. **330**, 115668 (2021). https://doi.org/10.1016/j.molliq.2021.115668
61. C. Li, W. Wang, L. Yan, Y. Wang, M. Jiang, Y. Ding, Phosphonium salt and ZnX_2–PPh_3 integrated hierarchical POPs: tailorable synthesis and highly efficient cooperative catalysis in CO_2 utilization. J. Mater. Chem. A **4**, 16017–16027 (2016). https://doi.org/10.1039/c6ta05823j
62. S. Raja, L.H.C. Mattoso, Functionalized polymer-based composite photocatalysts, 167–188 (2020). https://doi.org/10.1007/978-3-030-15608-4_7
63. L. Tan, K. Wang, Q. Li, Y. Yang, Y. Liu, B. Tan, Organic porous polymer materials: design, preparation, and applications (2017). https://doi.org/10.1007/978-3-319-57003-7_4
64. R. Gogoi, S. Dutt, P.F. Siril, Conjugated polymer—Based nanocomposites as photocatalysts (2021). https://doi.org/10.1002/9783527820115.ch8
65. Q. Sun, B. Aguila, Y. Song, S. Ma, Tailored porous organic polymers for task-specific water purification. Acc. Chem. Res. **53**, 812–821 (2020). https://doi.org/10.1021/acs.accounts.0c00007
66. P. Kaur, J.T. Hupp, S.T. Nguyen, Porous organic polymers in catalysis: opportunities and challenges. ACS Catal. **1**, 819–835 (2011). https://doi.org/10.1021/cs200131g

67. P. Ahuja, S.K. Ujjain, R. Kanojia, P. Attri, Transition metal oxides and their composites for photocatalytic dye degradation. J. Compos. Sci. **5**, 1–27 (2021). https://doi.org/10.3390/jcs5030082
68. N. Sun, C. Wang, H. Wang, X. Gao, J. Jiang, Photonic switching porous organic polymers toward reversible control of heterogeneous photocatalysis. ACS Appl. Mater. Interfaces **12**, 56491–56498 (2020). https://doi.org/10.1021/acsami.0c18062
69. L. Shao, Y. Li, J. Huang, Y.N. Liu, Synthesis of triazine-based porous organic polymers derived n-enriched porous carbons for CO_2 capture. Ind. Eng. Chem. Res. **57**, 2856–2865 (2018). https://doi.org/10.1021/acs.iecr.7b04533
70. G. Hu, J. Yang, X. Duan, R. Farnood, C. Yang, J. Yang, W. Liu, Q. Liu, Recent developments and challenges in zeolite-based composite photocatalysts for environmental applications. Chem. Eng. J. **417**, 129209 (2021). https://doi.org/10.1016/j.cej.2021.129209
71. O. Sacco, V. Vaiano, M. Matarangolo, ZnO supported on zeolite pellets as efficient catalytic system for the removal of caffeine by adsorption and photocatalysis. Sep. Purif. Technol. **193**, 303–310 (2018). https://doi.org/10.1016/j.seppur.2017.10.056
72. E.C. Umejuru, T. Mashifana, V. Kandjou, M. Amani-Beni, H. Sadeghifar, M. Fayazi, H. Karimi-Maleh, N.T. Sithole, Application of zeolite based nanocomposites for wastewater remediation: evaluating newer and environmentally benign approaches. Environ. Res. **231**, 116073 (2023). https://doi.org/10.1016/j.envres.2023.116073
73. D. Xu, X. Lu, Y. Zhang, P.R. Shearing, S. Zhang, D.J.L. Brett, S. Wang, Insights into in-situ catalytic degradation of plastic wastes over zeolite-based catalyst from perspective of three-dimensional pore structure evolution. Chem. Eng. J. **450**, 138402 (2022). https://doi.org/10.1016/j.cej.2022.138402
74. L. Wang, J. Li, Z. Du, M. Jin, J. Yao, Z. Zhang, $MnFe_2O_4$/zeolite composite catalyst for activating peroxymonosulfate to efficiently degrade antibiotic. Mater. Lett. **344**, 134460 (2023). https://doi.org/10.1016/j.matlet.2023.134460
75. J. Liu, X. Yu, L. Wang, M. Guo, W. Zhu, J. Chen, Study on degradation of marine product processing waste water by Zeolite/Bi/ZnO catalyst. J. Environ. Chem. Eng. **7**, 102863 (2019). https://doi.org/10.1016/j.jece.2018.102863
76. U. Khalil, Z. Liu, C. Peng, N. Hikichi, T. Wakihara, J. García-Martínez, T. Okubo, S. Bhattacharya, Ultrafast surfactant-templating of *BEA zeolite: an efficient catalyst for the cracking of polyethylene pyrolysis vapours. Chem. Eng. J. **412**, 128566 (2021). https://doi.org/10.1016/j.cej.2021.128566
77. G.T.M. Kadja, N.J. Azhari, S. Mardiana, N.T.U. Culsum, A. Maghfirah, Recent advances in the development of nanosheet zeolites as heterogeneous catalysts. Results Eng. **17**, 100910 (2023). https://doi.org/10.1016/j.rineng.2023.100910
78. L. Feyzi, N. Rahemi, S. Allahyari, Efficient degradation of tetracycline in aqueous solution using a coupled S-scheme ZnO/g-C_3N_4/zeolite P supported catalyst with water falling film plasma reactor. Process. Saf. Environ. Prot. **161**, 827–847 (2022). https://doi.org/10.1016/j.psep.2022.03.076
79. G. Hu, J. Yang, X. Duan, R. Farnood, C. Yang, J. Yang, W. Liu, Q. Liu, Recent developments and challenges in zeolite-based composite photocatalysts for environmental applications. Chem. Eng. J. **417**, 129209 (2021). https://doi.org/10.1016/j.cej.2021.129209
80. S. El-Nahas, A.I. Osman, A.S. Arafat, H. Ala'a, H.M. Salman, Facile and affordable synthetic route of nano powder zeolite and its application in fast softening of water hardness. J. Water Process Eng. **33**, 101104 (2020)
81. T.W. Duan, B. Yan, Novel luminescent hybrids prepared by incorporating a rare earth ternary complex into CdS QD loaded zeolite y crystals through coordination reaction. CrystEngComm **16**, 3395–3402 (2014). https://doi.org/10.1039/c3ce42479k
82. J. Shi, J. Chen, Z. Feng, T. Chen, X. Wang, P. Ying, C. Li, Time-resolved photoluminescence characteristics of subnanometer ZnO clusters confined in the micropores of zeolites. J. Phys. Chem. B **110**, 25612–25618 (2006). https://doi.org/10.1021/jp060439z
83. D.G. Boer, J. Langerak, P.P. Pescarmona, Zeolites as selective adsorbents for CO_2 separation. ACS Appl. Energy Mater. **6**, 2634–2656 (2023). https://doi.org/10.1021/acsaem.2c03605

84. T. Gong, L. Qin, J. Lu, H. Feng, ZnO modified ZSM-5 and y zeolites fabricated by atomic layer deposition for propane conversion. Phys. Chem. Chem. Phys. **18**, 601–614 (2016). https://doi.org/10.1039/c5cp05043j
85. J. Chen, Z. Feng, P. Ying, C. Li, ZnO clusters encapsulated inside micropores of zeolites studied by UV Raman and laser-induced luminescence spectroscopies. J. Phys. Chem. B **108**, 12669–12676 (2004). https://doi.org/10.1021/jp048746x
86. A. Corma, H. Garcia, Zeolite-based photocatalysts. Chem. Commun. **13**, 1443–1459 (2004). https://doi.org/10.1039/B400147H
87. S. Narayanan, P. Tamizhdurai, V.L. Mangesh, C. Ragupathi, P. Santhana krishnan, A. Ramesh, Recent advances in the synthesis and applications of mordenite zeolite—Review. RSC Adv. **11**, 250–267 (2020). https://doi.org/10.1039/d0ra09434j
88. L. Hlekelele, P.J. Franklyn, F. Dziike, S.H. Durbach, Novel synthesis of Ag decorated TiO_2 anchored on zeolites derived from coal fly ash for the photodegradation of bisphenol-A. New J. Chem. **42**, 1902–1912 (2018). https://doi.org/10.1039/c7nj02885g
89. H. Yang, X. Fang, Z. Li, H. Sun, H. Chen, Copper-doped small pore zeolites for CO_2 capture by honeycomb rotor with low temperature regeneration. ACS Sustain. Chem. Eng. **10**, 1759–1764 (2022). https://doi.org/10.1021/acssuschemeng.1c08347
90. Q. Zhang, S. Gao, J. Yu, Metal sites in zeolites: synthesis, characterization, and catalysis. Chem. Rev. **123**, 6039–6106 (2023). https://doi.org/10.1021/acs.chemrev.2c00315
91. M.E. Mahmoud, M.F. Amira, S. Daniele, A. El Nemr, M.E. Abouelanwar, B.M. Morcos, Adsorptive removal of Ag/Au quantum dots onto covalent organic frameworks@magnetic zeolite@arabic gum hydrogel and their catalytic microwave-Fenton oxidative degradation of Rifampicin antibiotic. J. Colloid Interface Sci. **624**, 602–618 (2022). https://doi.org/10.1016/j.jcis.2022.05.132
92. S.K. Lee, S.W. Han, G.Y. Cha, J.M. Park, H. Park, R. Ryoo, U.H. Lee, Base-type nitrogen doping in zeolite-templated carbon for enhancement of carbon dioxide sorption. J. CO2 Util. **62**, 102084 (2022). https://doi.org/10.1016/j.jcou.2022.102084
93. Z. Li, P. Liu, C. Ou, X. Dong, Porous metal-organic frameworks for carbon dioxide adsorption and separation at low pressure. ACS Sustain. Chem. Eng. **8**, 15378–15404 (2020). https://doi.org/10.1021/acssuschemeng.0c05155
94. A. Chatterjee, L. Wang, P. Van Der Voort, Metal-organic frameworks in photocatalytic Z-scheme heterojunctions: an emerging technology. Chem. Commun. **59**, 3627–3654 (2023). https://doi.org/10.1039/d2cc05819g
95. X. Zhang, J. Wang, X.X. Dong, Y.K. Lv, Functionalized metal-organic frameworks for photocatalytic degradation of organic pollutants in environment. Chemosphere **242**, 125144 (2020). https://doi.org/10.1016/j.chemosphere.2019.125144
96. S. Yuan, L. Feng, K. Wang, J. Pang, M. Bosch, C. Lollar, Y. Sun, J. Qin, X. Yang, P. Zhang, Q. Wang, L. Zou, Y. Zhang, L. Zhang, Y. Fang, J. Li, H.C. Zhou, Stable metal-organic frameworks: design, synthesis, and applications. Adv. Mater. **30**, 1–35 (2018). https://doi.org/10.1002/adma.201704303
97. S. Soni, P.K. Bajpai, C. Arora, A review on metal-organic framework: synthesis, properties and application. Charact. Appl. Nanomater. **2**, 1–20 (2018). https://doi.org/10.24294/can.v2i2.551
98. V. Kumar, V. Singh, K.H. Kim, E.E. Kwon, S.A. Younis, Metal-organic frameworks for photocatalytic detoxification of chromium and uranium in water. Coord. Chem. Rev. **447**, 214148 (2021). https://doi.org/10.1016/j.ccr.2021.214148
99. G. Capano, F. Ambrosio, S. Kampouri, K.C. Stylianou, A. Pasquarello, B. Smit, On the electronic and optical properties of metal-organic frameworks: case study of MIL-125 and MIL-125-NH_2. J. Phys. Chem. C **124**, 4065–4072 (2020). https://doi.org/10.1021/acs.jpcc.9b09453
100. L. Tao, J. Wang, Z. Luo, J. Ren, D. Yin, Fabrication of an S-Scheme heterojunction photocatalyst MoS_2/PANI with greatly enhanced photocatalytic performance. Langmuir **39**, 11426–11438 (2023). https://doi.org/10.1021/acs.langmuir.3c01295

101. M.D. Goudarzi, N. Khosroshahi, A. Hamlehdar, V. Safarifard, Construction of S-scheme heterojunction via incorporating g-C_3N_4 into Ce-based MOFs for promotion of charge-transfer in photocatalytic Cr(VI) detoxification. J. Environ. Chem. Eng. **11**, 110169 (2023). https://doi.org/10.1016/j.jece.2023.110169
102. S. Cao, J. Yu, G-C_3N_4-based photocatalysts for hydrogen generation. J. Phys. Chem. Lett. **5**(12), 2101–2107 (2014). https://doi.org/10.1021/jz500546b
103. W. Sun, X. Chen, X. Meng, Y. Gao, Highly efficient photocatalytic CO_2 reduction by a ruthenium complex sensitizing g-C_3N_4/MOF hybrid photocatalyst. New J. Chem. **45**, 8965–8970 (2021). https://doi.org/10.1039/d1nj01010g
104. R. Xiao, C. Zhao, Z. Zou, Z. Chen, L. Tian, H. Xu, H. Tang, Q. Liu, Z. Lin, X. Yang, In situ fabrication of 1D CdS nanorod/2D Ti_3C_2 MXene nanosheet Schottky heterojunction toward enhanced photocatalytic hydrogen evolution. Appl. Catal. B Environ. **268**, 118382 (2020). https://doi.org/10.1016/j.apcatb.2019.118382
105. M. Elazar, D. Ph, Optimization of metal organic framework (Mof) synthesis for use in drug delivery. Technium**4**, 53–61 (2022). https://doi.org/10.47577/technium.v4i10.7730
106. N.C. Chiu, M.T. Nord, L. Tang, L.S. Lancaster, J.S. Hirschi, S.K. Wolff, E.M. Hutchinson, K.A. Goulas, W.F. Stickle, T.J. Zuehlsdorff, C. Fang, K.C. Stylianou, Designing dual-functional metal-organic frameworks for photocatalysis. Chem. Mater. **34**, 8798–8807 (2022). https://doi.org/10.1021/acs.chemmater.2c02089
107. M.X. Wu, C. Wei, X.H. Wang, Q.Q. Xia, H. Wang, X. Liu, Construction and sensing amplification of raspberry-shaped MOF@MOF. Inorg. Chem. **61**, 4705–4713 (2022). https://doi.org/10.1021/acs.inorgchem.1c04027
108. T. Zeng, D. Shi, Q. Cheng, G. Liao, H. Zhou, Z. Pan, Construction of novel phosphonate-based MOF/P-TiO_2 heterojunction photocatalysts: enhanced photocatalytic performance and mechanistic insight. Environ. Sci. NANO **7**, 861–879 (2020). https://doi.org/10.1039/c9en01180c
109. P. Wang, J. Zhao, X. Li, Y. Yang, Q. Yang, C. Li, Assembly of ZIF nanostructures around free Pt nanoparticles: efficient size-selective catalysts for hydrogenation of alkenes under mild conditions. Chem. Commun. **49**, 3330–3332 (2013). https://doi.org/10.1039/c3cc39275a
110. S. Kampouri, F.M. Ebrahim, M. Fumanal, M. Nord, P.A. Schouwink, R. Elzein, R. Addou, G.S. Herman, B. Smit, C.P. Ireland, K.C. Stylianou, Enhanced visible-light-driven hydrogen production through MOF/MOF heterojunctions. ACS Appl. Mater. Interfaces **13**, 14239–14247 (2021). https://doi.org/10.1021/acsami.0c23163
111. F. Yu, M. Jin, Y. Zhang, C. Lei, L. Zhou, H. Zhu, B. Yu, Visible-light-driven Zr-MOF/BiOBr heterojunction for the efficient synchronous removal of hexavalent chromium and rhodamine B from wastewater. ACS Omega **7**, 25066–25077 (2022). https://doi.org/10.1021/acsomega.2c01298
112. G. Zhu, R. Graver, L. Emdadi, B. Liu, K.Y. Choi, D. Liu, Synthesis of zeolite@metal-organic framework core-shell particles as bifunctional catalysts. RSC Adv. **4**, 30673–30676 (2014). https://doi.org/10.1039/c4ra03129f
113. M.A. Sinnwell, Q.R.S. Miller, L. Palys, D. Barpaga, L. Liu, M.E. Bowden, Y. Han, S. Ghose, M.L. Sushko, H.T. Schaef, W. Xu, M. Nyman, P.K. Thallapally, Molecular intermediate in the directed formation of a zeolitic metal-organic framework. J. Am. Chem. Soc. **142**, 17598–17606 (2020). https://doi.org/10.1021/jacs.0c07862
114. Y. Li, M. Karimi, Y.N. Gong, N. Dai, V. Safarifard, H.L. Jiang, Integration of metal-organic frameworks and covalent organic frameworks: design, synthesis, and applications. Matter. **4**, 2230–2265 (2021). https://doi.org/10.1016/j.matt.2021.03.022
115. P. Xue, X. Pan, J. Huang, Y. Gao, W. Guo, J. Li, M. Tang, Z. Wang, In situ fabrication of porous MOF/COF hybrid photocatalysts for visible-light-driven hydrogen evolution. ACS Appl. Mater. Interfaces **13**, 59915–59924 (2021). https://doi.org/10.1021/acsami.1c18238
116. N.K. Sompalli, A. Mohanty, A.M. Mohan, P. Deivasigamani, Heterojunction Cr_2O_3–Ag_2O nanocomposite decorated porous polymer monoliths a new class of visible light fast responsive heterogeneous photocatalysts for pollutant clean-up. J. Environ. Chem. Eng. **9**, 104846 (2021). https://doi.org/10.1016/j.jece.2020.104846

117. S.P. Asu, N.K. Sompalli, S. Kuppusamy, A.M. Mohan, P. Deivasigamani, CaO/CeO_2 nanocomposite dispersed macro-/meso-porous polymer monoliths as new generation visible light heterogeneous photocatalysts. Mater. Today Sustain. **19**, 100189 (2022). https://doi.org/10.1016/j.mtsust.2022.100189
118. D. Jagadeesan, N.K. Sompalli, A.M. Mohan, C.V.S.B. Rao, S. Nagarajan, P. Deivasigamani, ZrO_2–Ag_2O nanocomposites encrusted porous polymer monoliths as high-performance visible light photocatalysts for the fast degradation of pharmaceutical pollutants. Photochem. Photobiol. Sci. **21**, 1273–1286 (2022). https://doi.org/10.1007/s43630-022-00218-y
119. D. Jagadeesan, P. Deivasigamani, Facile fabrication of novel In_2S_3–BiOCl nanocomposite-supported porous polymer monolith as new generation visible-light-responsive photocatalyst for decontaminating persistent toxic pollutants. Mater. Today Sustain. **23**, 100428 (2023). https://doi.org/10.1016/j.mtsust.2023.100428
120. S. Kuppusamy, D. Jagadeesan, A.M. Mohan, A. Pavoor Veedu, A.E. Jiji, A.M. John, P. Deivasigamani, NH_2-MIL-125 MOF integrated translucent mesoporous polymer monolith as dual-light responsive new-generation photocatalyst for the expeditious decontamination of perennial pharmaceuticals. J. Environ. Chem. Eng. **11**, 110355 (2023). https://doi.org/10.1016/j.jece.2023.110355

Contemporary Developments of Black TiO_2 Nanophotocatalysts for the Degradation of Agrochemical Pollutants

Resmy R. Nair, Kanagavel Deepan Kumar, Kumaresan Ramanathan, and Panneerselvam Sathishkumar

Abstract Pesticide-contaminated wastewater treatment has become a hotly debated topic in recent years. Titanium dioxide is the most widely used photocatalyst that effectively uses UV radiation. The recent advancement in black TiO_2 nanomaterials clearly demonstrated that structural and morphological traits and appropriate electrical properties are widely used in many research areas. The associated drawbacks of white TiO_2 and its modified forms were addressed largely after the discovery of black titania (B-TiO_2) because it not only absorbs photons in a broad spectral range (UV to IR region) but also significantly boosts the catalytic performance. The photocatalytic properties of titanium oxide are closely related to material defect disorders, such as oxygen vacancy and Ti^{3+} defect. Defective nanoparticles with superior properties to white TiO_2 and efficient solar harvest in the photocatalytic application were synthesized by various reduction and oxidation approaches and various characterization techniques such as XRD, DRS, XPS, SEM, TEM, and so on were applied to investigate the properties of the defective TiO_2. The enhanced photoactivity can be explained on the basis of various defects associated with the black TiO_2. Photocatalytic activity of black titanium oxide under visible light was over 90% of imidacloprid degradation for example and can be associated with the specific surface area, bandgap energy smaller than white TiO_2, a large amount of surface defect Ti^{3+}, and oxygen vacancies that promote charge carriers' separation.

Keywords Advanced oxidation process · Photocatalysis · Black TiO_2 · Ti^{3+} defects · Oxygen vacancies · Agrochemical pollutants

R. R. Nair · P. Sathishkumar (✉)
Department of Chemistry, School of Advanced Sciences, Vellore Institute of Technology (VIT), Vellore, India
e-mail: sathishkumar.p@vit.ac.in

K. Deepan Kumar
Department of Biotechnology, School of Bio Sciences and Technology, Vellore Institute of Technology (VIT), Vellore, India

K. Ramanathan
Department of Biomedical Sciences, Faculty of Medical Sciences, Institute of Health, Jimma University, Jimma, Ethiopia

P. Sathishkumar (ed.), *Photocatalysis for Energy and Environmental Applications*, Green Energy and Technology, https://doi.org/10.1007/978-981-97-1939-6_3

1 Introduction

Agriculture has been and will continue to be the primary chauffeur of human progresses. Agricultural harvest is critical for population, health, and economic balance (import and export), which directly reflects on the country's development and security. However agriculture today has a significant negative impact on climate change because of its high energy consumption and human consumerism [1]. Agrochemicals are used to manage insects, pests, and weeds, which contaminate water resources and encourage soil erosion and they are the primary causes of agricultural and environmental pollution. Among the agrochemicals, pesticides are frequently found in water sources due to their recurrent application in the intensive agricultural sectors. Unused chemical reagents, spray machines, and pesticide container washing are well-known causes of pollution from agricultural activities. Numerous studies have shown that heterogeneous photocatalysis is an efficient method for the complete removal of pesticides from the total environment [2, 3].

Photocatalysis has accomplished the interest of researchers all over the world in addressing energy and environmental issues through contaminated air and water, hydrogen production via water splitting, organic synthesis, and other applications [4]. Photocatalysis provides a green, renewable method of producing clean hydrogen fuel through water splitting. The ability of photocatalytic systems to operate at room temperature and the potential to use sunlight as a light source has been made them appealing. Recently, black-TiO_2 and ZnO-based photocatalysts have received a lot of attention because of their excellent light absorption properties, which allow even the NIR (Near Infrared) region of the solar spectrum to be beneficial. Since their discovery in 2011 by Chen et al., who synthesized black TiO_2 by introducing surface disorders in nanophase TiO_2 via hydrogenation, black-TiO_2 became an essential photocatalyst among the research community. The photocatalytic processes offer many advantages for the removal of pollutants even at the low concentration from water [5]. These include the following:

- Complete oxidation of organic pollutants within few hours.
- No formation of polycyclic products.
- Availability of highly active and cheap catalysts that are capable of adapting to specially designed reactor systems.
- Oxidation of pollutants in the ppb range, etc.

TiO_2 photocatalyst has gained significant attention in recent decades as a perfect model for the exploration of photocatalysis in a variety of applications, particularly water splitting for hydrogen production, pollutant removal for environmental protection, and CO_2 reduction for solar fuels. These elements are critical to the economy and for the long-term growth of human society. However, due to weak optical absorption in the UV-Vis spectrum and the quick electron-hole recombination, the solar energy conversion efficiency is extremely low in the case of TiO_2. A promising technique to enhance the visible light absorption of the nanocatalysts is its structural engineering, which primarily results in TiO_2 surface structure alteration with a colour change from white to black. The technique can potentially address both issues, as black TiO_2 nanoparticles frequently demonstrate strong absorption of visible light and enhanced quantum efficiency.

In 2011, hydrogenation was reported as an effective method for changing the bandgap of TiO_2 to around 1.5 eV, with an optical onset of around 1.0 eV and a colour of black. Naturally, the discovery of black TiO_2 nanoparticles has sparked the global research interest. Indeed, it is widely known that heating TiO_2 under a vacuum or in a reducing environment causes colour changes. Ti^{3+} or oxygen vacancies transform white TiO_2 into yellow, blue, or black nanomaterial with the tiny band gap (1.54 eV) which enables the light absorption even in the near-infrared region. Since, black TiO_2 has been a popular research material for a variety of applications, including photocatalytic pollutant degradation, hydrogen production via water splitting, photocatalytic CO_2 reduction, dye-sensitized solar cells (DSSCs), lithium batteries, and other photoelectrochemical applications [6]. Increased photocatalytic efficiency and lower recombination of photogenerated charge carriers have been regularly reported as advantages of the novel class of black semiconductor-based nanocatalysts. Many advanced approaches have been used to analyse the black TiO_2 nanocatalysts in order to better understand their outstanding physicochemical properties. The current chapter aims to address the various properties as well as the synthesis methods of black TiO_2 along with the degradation of agrochemicals and their future perspectives.

1.1 Advanced Oxidation Process

Many toxic substances have been detected in the environment in recent years due to the excessive use of agrochemicals in the fields which affects the soil as well as groundwater [7]. These compounds have the potential to cause catastrophic harm to both humans and the aquatic environment. As a result, there is an urgent need to remediate the existing harmful contaminants using appropriate treatment methods. Advanced oxidation processes (AOPs) are the most appealing and advantageous method for effective removal of organic contaminants. AOPs are the potential technologies for partial or total mineralization of pollutants by extremely reactive hydroxyl, hydroperoxyl, superoxide, and sulfate radicals and are considered as effective novel approaches for hastening the oxidation process [8–11]. All AOPs are distinguished by their ability to take advantage of the greater reactivity of hydroxyl radicals while driving oxidation processes. As demonstrated in Fig. 1, various types of chemical, photochemical, sonochemical, and electrochemical approaches have been performed for the degradation of organic pollutants. Comprehensive reviews of conventional AOPs, including O_3/UV, O_3/H_2O_2, UV/H_2O_2, Fenton, and electrochemical oxidation processes, are demonstrated in Fig. 1. The AOPs have been categorized principally based on the mode of radical formation, which includes chemical, photochemical, radiation-induced, cavitation, and electrochemical methods. Advanced oxidation technologies have greatly benefited from the advances in materials science and technology. Novel materials are continuously being identified

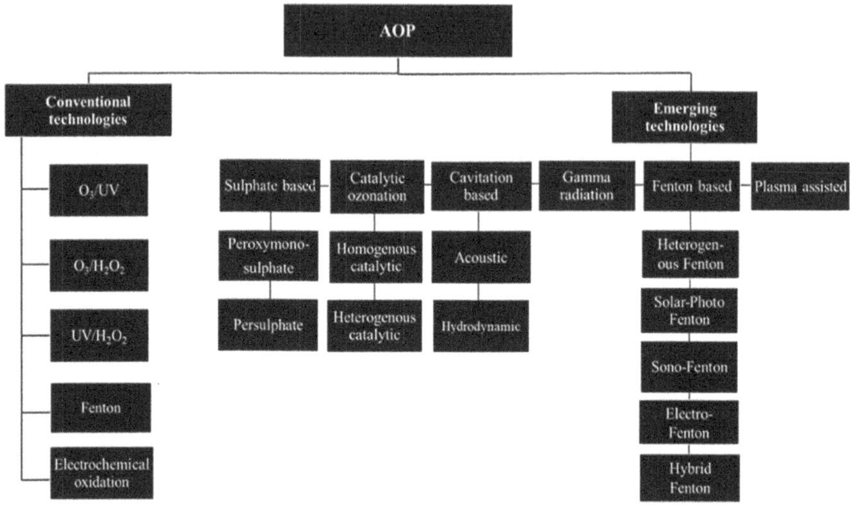

Fig. 1 Classification of conventional and emerging AOP-based technologies adopted from Ref. [12]

with better properties. Therefore, using AOPs for environmental applications is a topic in engineering, science, and many other disciplines. This is a new, impressive, and somewhat untapped concept.

The chemistry of AOP could be divided into 3 parts:

1. Formation of OH radicals
2. Initial attacks on target molecules by OH radicals and their breakdown into fragments
3. Subsequent attacks by ·OH until the complete mineralization

The mechanism of ·OH production (Part 1) highly depends on the sort of AOP technique adopted for the degradation of environmental contaminants. For example, ozonation, UV/H_2O_2, photocatalytic oxidation, and Fenton's oxidation rely on different mechanisms of ·OH generation [13].

UV/H_2O_2 based AOP:

$H_2O_2 + UV \rightarrow 2\cdot OH$ (homolytic bond cleavage of the O–O bond of H_2O_2 leads to formation of 2·OH radicals).

UV/HOCl based AOP:

$HOCl + UV \rightarrow \cdot OH + Cl\cdot$.

Ozone-based AOP:

$O_3 + HO^- \rightarrow HO_2^- + O_2$ (reaction between O_3 and a hydroxyl ion leads to the formation of H_2O_2 in charged form)
$O_3 + HO_2^- \rightarrow HO_2\cdot + O_3^{-\cdot}$ (a second O_3 molecule reacts with the HO_2^- to produce the ozonide radical)

$O_3^{-}\cdot + H^+ \rightarrow HO_3\cdot$ (this radical gives to ·OH upon protonation)
$HO_3\cdot \rightarrow \cdot OH + O_2$.

Fenton based AOP [9]

$Fe^{2+} + H_2O_2 \rightarrow Fe^{3+} + HO\cdot + OH^-$ (initiation of Fenton's reagent)
$Fe^{3+} + H_2O_2 \rightarrow Fe^{2+} + HOO\cdot + H^+$ (regeneration of Fe^{2+} catalyst)
$H_2O_2 \rightarrow HO\cdot + HOO\cdot + H_2O$ (Self scavenging and decomposition of H_2O_2).

Photocatalytic oxidation with TiO_2:

$TiO_2 + UV \rightarrow e^- + h^+$ (irradiation of the photocatalytic surface leads to an excited electron (e^-) and electron gap (h^+))
$Ti(IV) + H_2O \rightleftharpoons Ti(IV)\text{–}H_2O$ (water adsorbs onto the catalyst surface)
$Ti(IV)\text{–}H_2O + h^+ \rightleftharpoons Ti(IV)\text{–}\cdot OH + H^+$ (the highly reactive electron gap will react with water).

1.2 Photocatalysis

Developments at the nanoscale are leading to promising solutions for contamination in distant areas. The development of nanoscale metal oxides in recent years has substantially boosted catalytic activity due to the high specific surface area available for reactivity of the smaller particles. Furthermore, changing the band gap of the catalyst is a significant technique because it controls the percentage absorption of the solar spectrum of the catalysts. Photocatalysts with high activity have been demonstrated using visible light wavelengths (380–500 nm) where potentially bigger amounts of energy are available in the solar spectrum (solar peak energy is approximately 460 nm).

The use of different photocatalysts in photocatalysis is a useful alternative and an attractive option in advanced research. The photocatalyst must have higher photochemical stability and be able to make efficient use of solar radiation for wastewater treatment. The word "photocatalyst" is made up of two parts: photo and catalyst. The word "photo" means light and "catalyst" refers to the process in which the rate of oxidation is enhanced by a catalyst. In many situations, the catalyst is a solid material, when exposed to UV or visible light, creates electron-hole pairs, which produce free radicals. Photocatalysts are classified into three types: heterogeneous, homogeneous, and plasmonic antenna-reactor catalysts. The use of each catalyst is determined by the desired application in catalysis.

Heterogeneous catalysis was the widely studied catalysis which offer ecological, efficient, and flexible solutions to a number of environmental problems. Among others, the removal of harmful organic pollutants and the generation of H_2 through water separation are the well-explored applications. The process is based on the separation of charges induced by the excitation of the semiconductor photocatalyst through the absorption of photons. Due to the strong development of materials science, apart from the well-known TiO_2 and ZnO, several new semiconductor materials have been designed and synthesized to enhance the efficiency of heterogeneous photocatalysis based on the utilization of sunlight and/or visible light.

Figure 2 represents the various steps involved in the photocatalysis processes. Light with energy greater than or equal to the band gap energy is absorbed by the semiconductor used, and an electrons are excited from the valence band to the conduction band, leaving the holes in the valence band (Eq. 1). The photogenerated electron-hole pair is called an exciton. The excited electrons react with the oxidants to form the reduced product and the positive holes react with the reductants to form the oxidized product. Electrons can react with O_2 to give superoxide radicals ($O_2^{\cdot-}$) whereas the holes can react with water to yield hydroxyl radicals (·OH) as per Eqs. 2–4. The free radicals can later react with a variety of different species by participating in redox processes in numerous applications. At the present stage, the challenge of limiting its development is mainly material problems. These questions are complex and difficult because they involve simultaneously two distant fields of optoelectronics and catalysis. However, these challenges also provide a fertile research platform for testing different material design hypotheses and principles [14]. The whole mechanism of semiconductor photocatalysis can be represented in Fig. 2.

$$\text{Photocatalyst} + \text{hv} \quad h_{VB}^{+} + e_{CB}^{-} \tag{1}$$

$$h_{VB}^{+} + e_{CB}^{-} \qquad \text{energy (heat)} \tag{2}$$

$$H_2O + h_{VB}^{+} \qquad \text{energy (heat)} \rightarrow \cdot OH\ \text{(hydroxyl radical)} + H^{+} \tag{3}$$

$$O_2 + e_{CB}^{-} \rightarrow O_{2^{\cdot-}}\ \text{(superoxide radical)} \tag{4}$$

$$\cdot OH + \text{pollutant} \rightarrow \text{Intermediates} \rightarrow H_2O + CO_2 \tag{5}$$

$$O_{2^{\cdot-}} + \text{pollutant} \rightarrow \text{Intermediates} \rightarrow H_2O + CO_2 \tag{6}$$

The acceleration of a photoreaction in the presence of a photocatalyst, the excited state of which "repeatedly interacts with the reaction partners forming reaction intermediates and regenerates itself after each cycle of such interactions". Photocatalysis is defined as the change in the rate of a chemical reaction caused by a photocatalyst (metal oxide substance) in the presence of a light source, such as ultraviolet (UV), visible (VIS), a combination of both UV + VIS or solar radiation.

Heterogeneous photocatalysis has significant advantages over traditional water treatment methods such as chemical-based procedures, membrane technologies, adsorption, and so on, which have been plagued by a slew of issues when dealing with organic contaminants in wastewater [16, 17]. Titanium Dioxide (TiO_2) and TiO_2-based photocatalysts have dominated the research studies due to the earth's

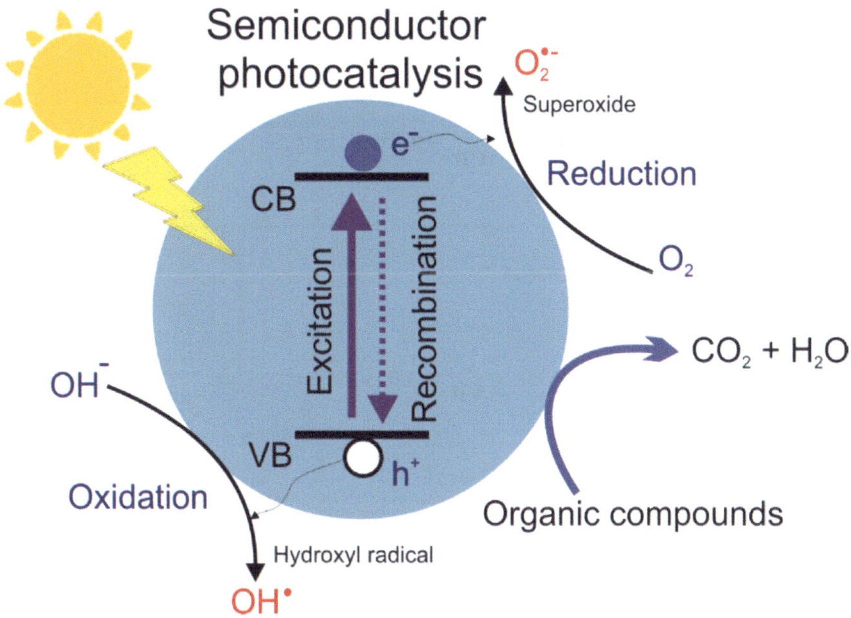

Fig. 2 Mechanism of semiconductor photocatalysis adopted from Ref. [14, 15]

abundance, chemical stability, good optical transparency, high refractive index, low cost, and non-toxicity of TiO_2 [18]. But white TiO_2 is active only in the ultraviolet (UV) spectrum of light which makes it less effective in using the entire solar spectra. Also, the rapid recombination of photogenerated charge carriers drastically reduce the efficiency of the whole process [19]. Black photocatalysts have recently received a lot of interest due to their remarkable light absorption capabilities, which make even the NIR (Near Infrared) part of the solar spectrum usable. This was synthesized by introducing surface defects in the nanophase TiO_2 through hydrogenation. Because of the tiny band gap (1.54 eV), light was absorbed even in the near-infrared range.

As illustrated in Fig. 3, the oxygen vacancy defects introduced mid-gap states in the black TiO_2 synthesized by Song et al. [16, 20], increasing the likelihood of electrons being excited at lower energy. This resulted in greater visible light absorption and, as a result, higher photocatalytic efficiency. The photoluminescence spectra further demonstrated that excellent separation of photogenerated charge carriers occurred. Figure 3 shows a simplified energy band structure of TiO_2 nanostructures and black TiO_2 nanostructures. The strong UV absorption of TiO_2 results from the electronic transition from VB to CB. Hydrogenation could create electronic state bands of oxygen vacancies in TiO_2 nanostructures. The visible and infrared light absorption is caused by electron transitions from VB to oxygen vacancy levels and/or from oxygen vacancies to VB [20].

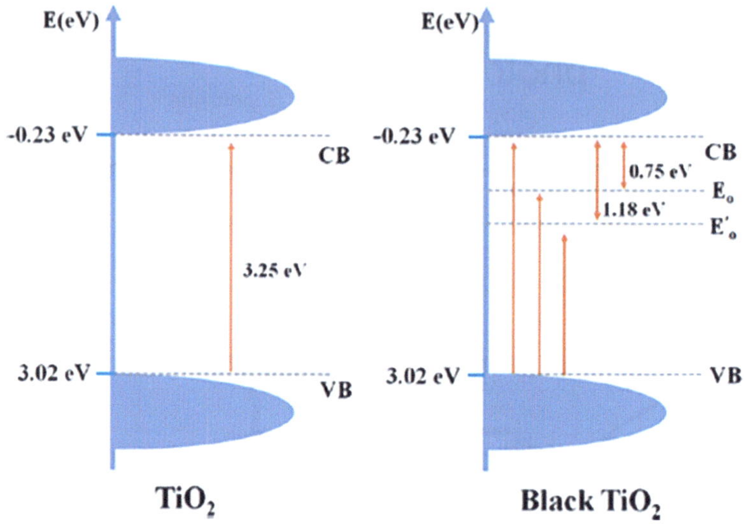

Fig. 3 The energy band structure of TiO_2 nanostructures and black TiO_2 nanostructures. Eo and E′o are considered as the energy levels of the oxygen vacancies located at 0.73 and 1.18 eV below the CB of TiO_2. Arrows are the possible electron transitions within the different energy levels in TiO_2 nanostructures and black TiO_2 nanostructures [16]

Bandgap narrowing is an excellent example of how the band structure of black TiO_2 changes. Because of the complexity of surface defects/disorders and functional groups, the origin of bandgap narrowing has been debated for years. One notion is that Ti^{3+} species in bulk TiO_2 are responsible for band gap narrowing and oxygen-vacancy midgap states further boost photon energy absorption below the direct bandgap via indirect electron transitions. Chen et al. hypothesized that bandgap narrowing resulted mostly from disorder-induced midgap states rather than Ti^{3+} species, significantly upshifting the valence band (VB) edge of black TiO_2 and that the putative conduction band (CB) tail states emerging from the surface disorder could only be observed.

1.3 Black TiO_2

When white TiO_2 is reduced in various ways, it turns black by creating defects such as Ti^{3+}, oxygen vacancies and Ti–H bonds in Black TiO_2. These structural modifications are responsible for the dark coloration, and superior optical, electronic, and catalytic properties [6]. A precise definition of "Black TiO_2" is more difficult due to the reported properties in various literature. According to the literature, the degree of blackness varies with different procedures and also with changes in synthesis procedure and process parameters. As a result, the term "black" in black TiO_2 can be hazy because "black" can symbolize anything from dark grey to black. Furthermore, terminology such as 'Hydrogenated TiO_2' and 'Reduced TiO_2', 'Defective, coloured TiO_2/TiO_{2-x}', and other terms have also been used extensively. The resulting colour is attributed to the formation of varying amounts of Ti^{3+} and VOs. Oxygen vacancies have been studied since a very long time and one of the earliest studies were done by Cronemeyer et al. [21] who observed a blue coloration of the TiO_2 rutile crystal upon reduction at 700 °C. It was proven that this reduction took place along with a loss of oxygen from the crystal. Increasing the "level" of reduction leads, in general, to a higher density of defects (e.g., Ti^{3+} and VO concentration) and consequently to "darker" TiO_2 powders.

Some studies recorded a wide range of coloured samples with white and black at the extremes and other colours such as brown, grey, yellow, and blue in the middle [22]. Furthermore, the coloured samples might have gone black with further changes in synthesis conditions (typically increasing the level of reduction). Nonetheless, all of these samples had characteristics such as higher visible light absorption, the presence of one or more defect species and superior activity as compared to pristine TiO_2. The colourful TiO_2 with better absorption properties and improved photocatalytic activities compared with white TiO_2 have been designed through (i) metal doping in which metal replaces Ti^{4+} ions in the TiO_2 lattice, (ii) non-metal doping to replace O_{2-} ions in the TiO_2 lattice, (iii) to replace partial Ti^{4+} and O_2 ions in the TiO_2 lattice. Because of the difference in charge states and ionic radii, substituting Ti^{4+} in TiO_2 with any cation is more accessible than substituting O_2 with anions (such as nitrogen, carbon, and sulphur). As shown in Fig. 4, black TiO_2 nanomaterials have the potential to absorb the entire range of solar spectra. In the development of black TiO_2, lattice disorder is generated, resulting in the formation of mid-gap energy levels within the band gap. The broad range of absorption in these nanoparticles as a result of this mid-gap energy level distribution is critical for optical and catalytic activity. The improved performance of black TiO_2 was attributed to higher light absorption caused by lattice disorder and H-doping, which lowered the optical bandgap of black TiO_2 to 1.54 eV by introducing electronic states producing substantial VB and CB tailing.

1.4 Defects

According to Chen et al., hydrogen thermal treatment of TiO_2 nanoparticles can produce an amorphous layer near the surface, resulting in defective black TiO_{2-x} nanoparticles. One of the most prevalent flaws in metal oxides, including TiO_2 nanomaterials, is oxygen vacancy [23, 24]. It significantly impacts the activity and kinetics of reactions occurring on the surface of metal oxides. Oxygen vacancy shortens the TiO_2 band gap to increase visible light absorption by creating energy levels below the conduction band [25, 26]. X-ray diffractometer, electron paramagnetic resonance (EPR) or electron spin resonance (ESR) [27], Raman spectroscopic analyses have been used to detect the oxygen vacancies present in black TiO_2 nanomaterials. Furthermore, high-concentration crystal defects are favourable for trapping

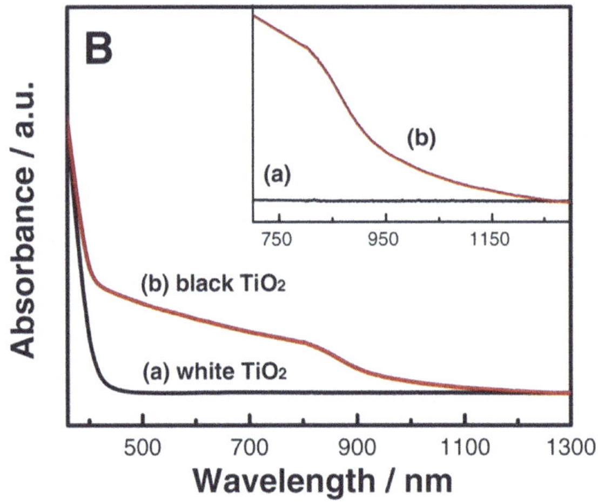

Fig. 4 Absorbance spectra of **a** white and **b** black TiO_2. Inset shows the extended absorption of black TiO_2 [6]

photogenerated carriers and inhibiting recombination [28]. Ti^{3+} species and oxygen vacancies are the most common types of defects. The defect generation mechanism seems to depend strongly on the synthesis route and conditions. The various compositions of the widely reported core-shell structure and the potential roles it could play in the enhanced properties of black TiO_2, as illustrated in Fig. 5. The discovered Ti^{3+} species may be a part of a considerably smaller band gap Magneli phase such as Ti_2O_3, Ti_4O_7, and so on, where they can be stabilized as opposed to Ti^{3+} in stoichiometric TiO_2 [27]. This could explain the disparities in conclusions concerning the existence of Ti^{3+} at the surface.

In Fig. 6, the XRD patterns of the acquired TiO_{2x} samples show diffraction peaks at 25.4°, 37.9°, 48.1°, and 53.1°, showing that the TiO_{2x} is in a pure anatase phase. The particle size of highly crystallized TiO_2 core steadily reduces from 50 to 10 nm as the amount of reducing agent increases, according to TEM analysis of the obtained defective TiO_2x samples. The TiO_2 core crystal lattices can be clearly detected in all samples, indicating that the TiO_2 core is strongly crystallized. Because the particles are overlapping, the disordered TiO_{2x} outer layer cannot be distinguished in TEM, but its presence can be indirectly confirmed by further sintering of the defective black TiO_{2x} at 500 °C in N_2 atmosphere, given that high vacancy concentrations can significantly reduce TiO_2 melting temperature. Theoretical studies show that high vacancy concentrations can create a vacancy band of electronic states just below the conduction band, reducing the band gap to 1.0 eV. Aside from its optical and electrical properties, the performance of defective TiO_2x is heavily influenced by its morphological and structural qualities [28].

The quantitative amount of Ti^{3+} in the samples after blackening was quantified using X-ray photoelectron spectroscopy (XPS). The Ti 2p XPS spectra of the as-prepared TiO_2 nanotubes are shown in Fig. 7 (Fig. 7a). It can be noted that both TiO_2 nanotubes and black-TiO_2 exhibit the tetravalent Ti characteristic peak having a binding energy of 463.9 eV (Ti^{4+} 2p1/2) and 458.2 eV (Ti^{4+} 2p3/2) [29]. The

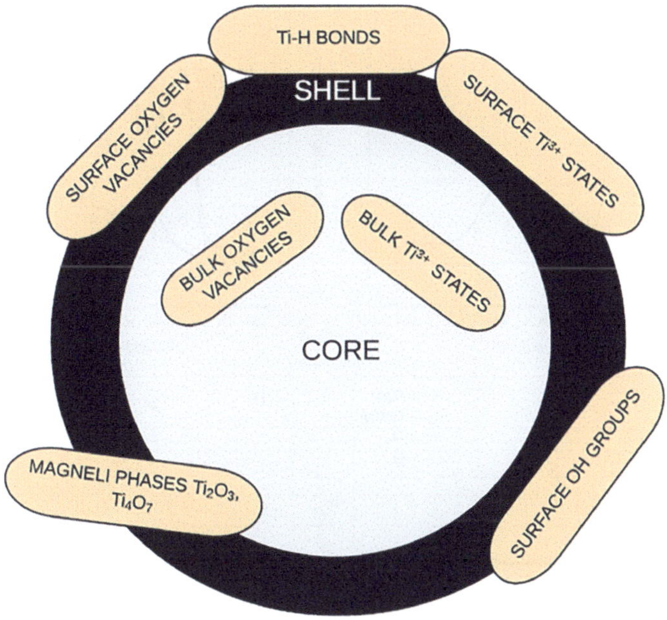

Fig. 5 Various defects and their location reported in the core shell structure of Black TiO_2 [27]

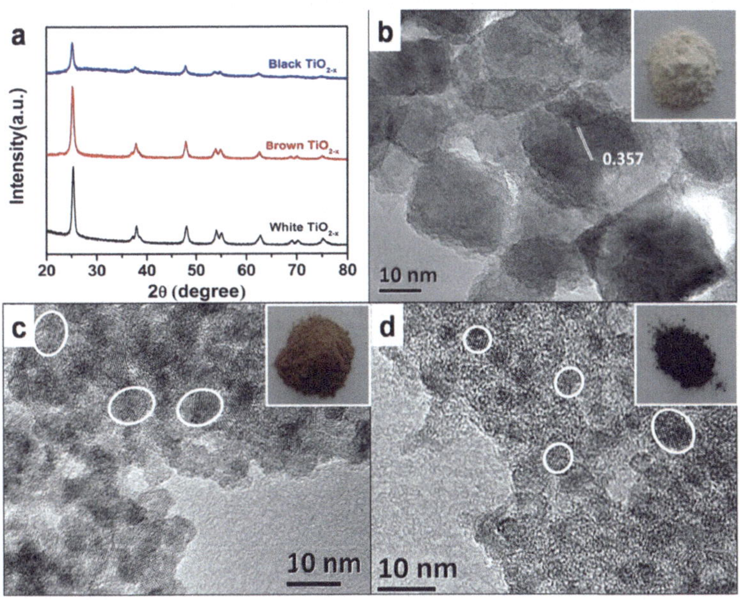

Fig. 6 XRD patterns (**a**), TEM images (white TiO_{2-x} (**b**), brown TiO_{2-x} (**c**), black TiO_{2-x} (**d**)), of the defective TiO_{2-x} nanocrystals [28]

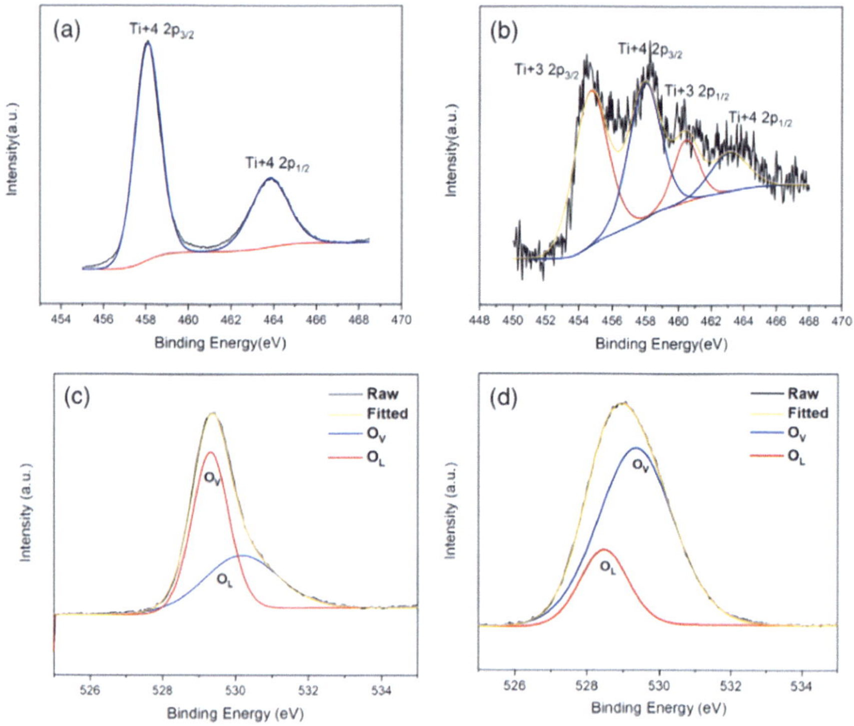

Fig. 7 Ti 2p XPS spectra of the as-prepared TiO_2 nanotubes (**a**), black TiO_2 nanotubes (**b**) and O 1 s XPS spectra of the as-prepared TiO_2 nanotubes (**c**) and black TiO_2 nanotubes (**d**) [31]

black TiO_2 Ti 2p XPS peaks have been greatly expanded, with four peaks clearly overlapping (Fig. 7b). It can be divided into two peaks at 463.9 eV (Ti^{4+} 2p1/2), 458.2 eV (Ti^{4+} 2p3/2), and 460.4 eV (Ti^{3+} 2p1/2), 454.7 eV (Ti^{3+} 2p3/2) [30]. By integrating the peaks area of two valence states, the Ti^{3+} content in the nanotubes is dominant (more than 70%). It implies that the surface layer of TiO_2 nanotubes is lowered by aluminium with high activity and a high concentration of oxygen vacancies (Vo) and trivalent Ti. Figure 7d, e show the O 1 s peak of an as-prepared TiO_2 nanotube and a black TiO_2 nanotube. Both O 1 s peaks may be divided into OV (oxygen vacancy) and OL (lattice oxygen) peaks, with the relative ratio of OV in black TiO_2 nanotubes clearly higher than in the as-prepared sample. This suggests that Ti is lowered following the aluminothermic reduction and many oxygen vacancies emerge on the surface.

1.4.1 Enhanced Photoactivity

Due to inadequate and ineffective light absorption and the presence of multiple carrier-recombination centers, TiO_2's visible light harvesting remains insufficient, resulting in diminished photocatalytic activity. To get around these constraints, a hydrogenated black TiO_2 material with a narrow bandgap of 1.5 eV has been created, which can boost the absorption of the full electromagnetic spectrum and has high photocatalytic properties when exposed to solar light. This discovery sparked international scientific interest in black TiO_2 (B-TiO_2) nanoparticles. B-TiO_2 absorbs all visible light wavelengths and even those in the near-infrared region as well. The constricted electronic band structure of black TiO_2 nanocrystals enables for enormous visible light absorption and conversion to chemical energy, resulting in significantly increased photocatalytic activity for photocatalytic pollutant removal and hydrogen generation from water [32]. The increased photocatalytic activity of black TiO_2 is also attributable to a surface lattice disorder and the existence of oxygen vacancies, Ti^{3+} ions, Ti–OH, and Ti–H groups[33] Cu doping can improve TiO_2's photocatalytic antibacterial activity under visible light irradiation by producing oxygen vacancies in the TiO_2 lattice by exchanging Ti^{4+} ions with Cu^{2+} ions and decreasing the bandgap around 2.8 eV [34].

Metal nanoparticle/semiconductor heterojunctions are thought to be a potent nanomaterilas for enabling electron-hole pair separation. It is generally known that at metal-semiconductor interfaces, a Schottky barrier forms which reducing the rate of recombination between photogenerated electrons transported to metal and holes. Furthermore, noble metals, such as Au, Ag, and Cu, can generate localized surface plasmon resonance (LSPR) when exposed to a wavelength of visible light. The noble metals could boost the photocatalytic activity of semiconductors by visible light-induced charge transfer from metals to semiconductors and/or LSPR-induced electromagnetic fields in the area of the plasmonic nanostructure when coupled with semiconductors.

1.4.2 Synthesis Routes

The routine manufacture of black TiO_2 has been a difficult and tough process because it necessitates extreme pressure and temperature variations, as well as inert and vacuum atmospheric conditions. In the recent past, hydrogenation was thought to be the most regularly used procedure for producing B-TiO_2. To create the black TiO_2 nanoparticles, several synthetic techniques have been devised. The majority of these techniques fall into two categories: partial reduction from TiO_2 and incomplete oxidation from low-valence-state Ti species.

The reduction approach can be generally described in Eq. (7):

$$TiO_2 + A \rightarrow TiO_{2-X} + AO_x \quad (7)$$

where A represents the reductant. If H_2 was utilized as the reductant, H-doping could also be encountered, which can be described by Eq. (8):

$$TiO_2 + H_2 \rightarrow TiO_{2-X}H_y + H_2O \quad (8)$$

Furthermore, balck-TiO_2 has been synthesized using a variety of other methods, including an inert atmosphere in the presence of argon, hydrogen, nitrogen, hydrogen/argon, nitrogen/argon environments, oxidation and reduction methods which includes reduction via metals such as Al, Zn, Mg and the reduction using NaBH4. In addition hydroxylation, pulsed laser, solvothermal, ionothermal and sol–gel methods were utilized for the successful synthesis of black TiO_2.

1.4.3 Reduction Approaches

(a) Hydrogen reduction

Chen et al. reported the synthesis of B-TiO_2 nanomaterial in 2011 by hydrogenating white TiO_2 crystal at 20 bar pressure for 5 days while maintaining a temperature of 200 °C [35]. The procedure initiates with the preparation of white nano TiO_2 from the hydrolysis of titanium tetraisopropoxide in hydrochloric acid/water/ethanol solution with the addition of pluronic F127, followed by calcined at 500 °C [35]. Even at the ambient conditions high-pressure hydrogen is able to convert commercial P25 powder into black TiO_2 nanoparticles [30]. When the hydrogenation treatment time was extended the colour of the hydrogenated TiO_2 nanoparticles gradually changed according to Lu et al. [36] After 3 days of hydrogenation, the white P25 turned a pale yellow colour and as on time progresses the TiO_2 colour will be changed to pale yellow (15 days) followed by the formation of grey TiO_2 (after 15 days) and greater than 15 days of hydrogenation leads to the formation of black TiO_2. In addition, pure H_2 (ambient conditions) or mixed gas of Ar/H_2, N_2/H_2 was also used to prepare the black TiO_2, however a high temperature treatment (400–700 °C) was required for the efficient conversion. It has been reported that slightly hydrogenated TiO_2 (white in color) showed better photocatalytic performance as compared to highly hydrogenated black TiO_2. It was concluded that higher concentration of surface defects along with higher ratio of trapped holes (O^- centers) helped reduce the recombination rate whereas for highly hydrogenated samples, a very high concentration of bulk defects acted as recombination centers where most of the holes were consumed through significantly enhanced non radiative recombination. This was also verified by the temperature changes during the degradation reactions [37]. In general, atmospheric H_2 cannot convert crystalline white TiO_2 nanoparticles to black [37, 38]. Nevertheless the direct utilisation of hydrogen gas is somewhat dangerous and difficult to synthesize black TiO_2. In this regard, the decomposition of hydrides to generate the molecular hydrogen under mild experimental conditions was the adopted strategy for the preparation of black TiO_2, for example $NaBH_4$ [25], N_2H_4 [39], CaH_2, [40] are attractive reductants for the simple and safe synthesis of black TiO_2. For hydride reduction, three ways were observed: (1) Dry procedure, in which the original TiO_2 and hydride were thoroughly mixed, followed by annealing at 300–400 °C in an Ar or N_2 environment; [40, 41] (2) A wet method in which the original TiO_2 was immersed in a hydride solution for the variable amount of time and temperature [39, 42, 43]. Other techniques, such as microwave or hydrothermal treatment can also be used to aid the synthetic process. (3) Sol-gel method. For example, Fang et al. used the NaBH4 aided sol–gel technique to prepare Ti^{3+} self-doped TiO_2. Solution A (tetrabutyl titanate/EtOH) and Solution B (HNO_3/EtOH) were combined to produce a sol, which was then reconstituted with a little amount of $NaBH_4$. The gel was placed in a ceramic dish and calcined for 180 min to produce the black TiO_2 nanoparticles.

(b) **Argon Annealing**

Various reports are available in the literature for the preparation of black TiO_2 under argon atmosphere [32, 44], and usually special starting TiO_2 was needed. For example, Myung et al. prepared black TiO_2 nanoparticles by annealing a yellow TiO_2 gel in Ar gas at 400–600 °C for 5 h. The TiO_2 gel was obtained by hydrolysis of TiCl4 in a water-ethanol mixture with HF and urea at 0 °C, followed by slow evaporation of the solution [45]. The colour of the annealed TiO_2 nanoparticles depended on the annealing temperature and the amount of HF added in the preparation of the TiO_2 gel [45].

(c) **Metal Reduction**

Manufacturing of B-TiO_2 by metal reduction is a cost-effective method because it involves the use of cheap and abundant metals (such as Al, Mg, Zn) for the reduction. For example, Wang et al. typically synthesized B-TiO_2 by pre-annulation of the TiO_2 precursor with Al and Pt as the reducing agent at 800 °C for 6 h and 500 °C for 8 h, respectively, followed by post-annealing at 800 and 900 °C. The crystalline core and amorphous shell-structured TiO_{2-X} obtained after aluminum reduction underwent a colour change starting from white to black which originated from the oxygen vacancies. The wide optical absorption was recorded ranging from visible to the IR region with an absorption edge at 440 nm having a bandgap of 2.18 eV for the resultant nanoparticles [23, 46].

(d) **Organic Reductant**

Imidazole and ascorbic acid have been used as reductants to produce black titania. Feng et al. found imidazole or 2-ethyl imidazole [46] could effectively introduce oxygen/Ti^{3+} vacancy concentration in the TiO_2, leading to the reduction of TiO_{2-x}. Moreover, grey TiO_2 nanoparticles were prepared by heating the mixture of porous amorphous TiO_2 precursors, imidazole, and hydrochloric acid in a preheated furnace (450 °C) for 6 h [20]. In another synthesis process, reduced TiO_{2-X} nanoparticles were prepared by burning ethanol/2-ethyl imidazole/HCl/titanium(IV) isopropoxide mixture at 500 °C for 5 h [42].

1.4.4 Oxidation Approaches

Low valence Ti species (such as TiH_2 [47, 48], TiO [49–51]$TiCl_3$ TiN [52] etc.) are incompletely oxidised to TiO_{2-X}, substantial oxygen vacancy or Ti^{3+} doping can be expected across the bulk Ti–O matrix. The resulting defective TiO_{2-X} materials are always different from the white colour and significantly absorb visible light and the defective TiO_{2-X} materials can be effectively utilised as the visible light-driven catalysts for the degradation of environmental contaminants. Few attempts made in the literature for the preparation of defective TiO_2 are discussed as follows.

Pei et al. prepared grey TiO_{2-X} by facile hydrothermal treatment of TiO in HCl solution[53]. The prepared grey TiO_2 showed enhanced visible light activity and good stability [44]. The oxidation process of TiO can be described by the following three reactions (Eqs. 9–11).

$$TiO + HCl \rightarrow Ti^{3+} + H_2 + H_2O \tag{9}$$

$$Ti^{3+} + H_2O \rightarrow TiOH_2^+ + H^+ \tag{10}$$

$$TiOH_2\ ++ O_2 \rightarrow Ti(IV)\ \text{oxo species} + O_2 \rightarrow TiO_2^{\cdot} \tag{11}$$

$TiCl_3$ and TiF_4 are also used as the starting materials for the one-step solvothermal process to produce self-doped blue TiO_2 [51]. The inclusion of Ti^{3+} into the TiO_2 matrix is directly derived from the $TiCl_3$ precursor. The presence of Ti^{4+} in the reaction system prevents Ti^{3+} oxidation during the solvothermal treatment. Because Ti^{3+} is not entirely oxidised to Ti^{4+} during a solvothermal process, oxygen vacancy generation is dominated by Ti^{3+} resulting in a high concentration of Ti^{3+} doping in bulk TiO_2 [51].

1.4.5 Other Approaches

Fan et al. discovered that intense ultrasonic treatment will convert the original white amorphous TiO_2 turned black while retaining its immaculate amorphous character [50]. The blackness of ultrasonic-treated TiO_2 will become deeper when the ultrasonic time is increased. Chen et al. introduced laser-irradiation to the TiO_2 nanoparticle suspension and discovered that the colour of the suspended TiO_2 solution changed from white to grey after 5 min of laser irradiation [4]. The colour of the TiO_2 suspension became darker as the duration of the laser irradiation increased, eventually turning black after 120 min of laser irradiation. Similarly there are various approaches reported in the literature for the preparation of black TiO_2, a simplified view of various approaches are presented in Fig. 8 and the brief summary of recent methodologies adopted for black TiO_2 synthesis is given in Table 1 [30, 54–58].

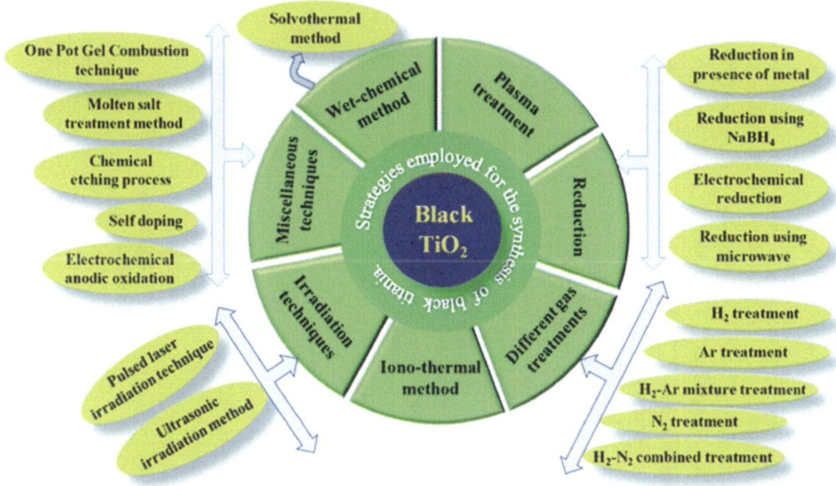

Fig. 8 Different preparation techniques adopted for fabricating Black TiO_2 [56]

1.5 Degradation of Pesticides in the Presence of Black TiO_2 Photocatalysts

Because of widespread intensive agricultural activities, the agrochemical sector has grown dramatically in the previous few decades. The increased use of pesticides and poor wastewater disposal procedures are particularly concerning for the freshwater (surface and groundwater), coastal, and marine habitats. The ecological impacts or toxicity of the pesticides and their breakdown products in different aquatic animals and plants are revealed by the assessment of the agrochemicals effects on the environment, even though the pesticide is not meant to kill a target species. It has an impact on soil, water, and air in such a way that it reduces biodiversity in soil and water, and agrochemicals pollute groundwater by seeping into the soil. A phenomenon known as pesticide drift occurs when these chemicals are easily transported by wind to other non-agricultural areas. The level of exposure is calculated in environmental concentration and anticipated in the field to those levels that cause hazardous effects. Based on the above facts, it is prudent to develop technologies that facilitate the easy breakdown of such bio-recalcitrant organic molecules. The use of advanced oxidation processes (AOPs) is a potential method for the mineralization of these types of compounds. These AOPs are distinguished by the production of highly reactive and oxidizing free radicals in aqueous solutions, which have high destructive power. Among the various AOP technologies suggested, heterogeneous photocatalysis effectively destroys organic pollutants in water.

Methomyl is a carbamate insecticide used mainly for tomato crops. It is produced by reacting S-methyl-N-hydroxylthio acetamidate (MHTA) in methylene chloride with gaseous methyl isocyanate at 30–35 °C. Methomyl has been classified as a

Table 1 Recently reported synthetic approaches for the preparation of black TiO_2

Synthesis route	Precursor	Synthesis conditions	Phase and morphology	Defect/ disorder	Application
Induction suspense on plasma spraying (ISPS)	Anatase TiO_2 nanopowders (10–30 nm)	A 40 kW inductively coupled radio-frequency plasma spraying system with Ar as plasma gas and H_2^+ Ar as sheath gas	30–200 nm spherical particles containing rutile phase/Ti_4O_7 Magnéli phases	Oxygen vacancies and Ti^{3+} species	Photothermal therapy of bone tumor
High-energy proton ion-implantation	Anatase TiO_2 nanotube arrays	Implantation of protons (p^+) at an energy of 30 keV with a dose of 10^{16} ions/cm^2	Anatase nanotube arrays encircled with amorphous rims	Ti^{3+} species	Photocatalytic H_2-evolution
Hollow cathode H_2 RF (radio frequency) plasma	TiO_2 thin films	Immersion of TiO_2 films in a hollow cathode H_2 RF plasma for 15–60 min	Densely packed clusters of small round particles 40–50 nm long with anatase phase	TiO, Ti_2O_3 and surface hydroxyl groups	Degradation of methylene blue (MB) under UV light
Continuous aerosol synthesis	Titanium-(IV) tetraisopropoxide (TTIP)	Decomposition of TTIP and subsequent reduction by H_2 at 1100 °C	Almost spherical particles of 20–160 nm with phase mixture of anatase, rutile, Ti_3O_5, Ti_4O_7 and Ti_5O_9	Magneli Phases	Not studied. But can be used for both water splitting and decomposition of organic pollutants
Suspension plasma spray fabrication	Degussa P25 suspension	Plasma spray coating system of 90 ml/min Spray distance of 80 mm with Ar and H_2 flow rate of 15 ml/min and 1.8 ml/min respectively	Composite structure comprising anatase, rutile, metastable brookite, and Ti_2O_3	Multiple phases detected due to lattice disorder. No defect species detected	Degradation of methylene blue (MB) under visible light

(continued)

Table 1 (continued)

Synthesis route	Precursor	Synthesis conditions	Phase and morphology	Defect/ disorder	Application
Pulsed laser irradiation in liquid (PLAL)	Titanium metal foil	1064 nm wavelength laser at 10 Hz, 532 nm wavelength laser at low repetition rate and 532 nm wavelength laser at high repetition rate	Complex branched structure for 1064 nm laser. Plate like structure with cracks for 532 nm laser	Sub oxides like Ti_2O_3 and TiO along with bound OH groups	Photocatalytic water treatment
Pulsed Laser Ablation in Liquid (PLAL)	Pristine anatase TiO2	Laser of 355 nm with a pulse duration of 3.6 ns and pulse repetition rate of 20 Hz; each pulse delivering 120 mJ/pulse. Irradiation time: 5–120 min	Core shell structure of hydrogenated TiO_2 with anatase crystal core along with TiO_2 micrsopheres with rutile shells on top of Lattice Expanded Planes (LEP) in the core	Ti^{3+} species and oxygen vacancies	Photocatalytic MB degradation
UV illumination of anatase in aqueous methanolic solution	Anatase particles (nanotube layers)	365 nm LED (90 mW/cm^2) illumination of anatase in oxygen free aq. methanol solution (50% vol)	Nanotube morphology with average particle size of 25 nm consisting of anatase phase	Ti^{3+} species	Photocatalytic H_2 evolution
Hydrothermal method involving controlled hydrolysis of TTIP	TTIP	Hydrolysis of TTIP in ethanol/acetonitrile medium using CTAB (hexadecyltrimethylammonium bromide) as structure-directing agent. Pressurized at 8 bars at 180 °C under H_2/N_2 flow for 1–5 days	Particles with lattice disorder interconnected with nanofibers. Decreased anatase content due to amorphous phase	Ti^{3+} species	Photocatalytic degradation of Orange G
Spark discharge method	Titanium rods	Sparks were generated at a discharge voltage of 1.5 kV on titanium rods of 5 mm diameter kept at a distance of 1 mm under Ar gas flow. Oxidation of Ti nanoparticles produced black titanium oxide particles	Particles of 5 nm diameter mostly in amorphous phase	Amorphous and crystalline TiO with crystalline Ti_2O_3	Not studied

very toxic and hazardous insecticide by the WHO (World Health Organization), EPA (Environmental Protection Agency, USA) and EC (European Commission). Methomyl is highly soluble in water (57.9 g L^{-1}), has a low sorption affinity to soil therefore it can easily cause groundwater contamination in agricultural areas. The effectiveness of the heterogeneous photocatalysis for the decontamination of water polluted by methomyl is studied' and the kinetics of disappearance of methomyl and the intermediates were identified in order to propose a suitable photocatalytic degradation pathway [57].

Methomyl concentration was decreased in dark with TiO_2 (Fig. 9), with 18% of the initial pollutant disappearing after 30 min of continuous stirring. This adsorption causes a considerable fall in concentration on the surface of TiO_2. On the other hand, the total disappearance of methomyl (1.23×10^{-4} mol/L; Irradiation at $\lambda >$ 290 nm) was resulted within 45 min in the presence of TiO_2. Organic carbon present in the methomyl is converted into CO_2, whereas nitrogen and sulphur heteroatoms are converted into inorganic ions such as nitrate or ammonium and sulphate ions, respectively. Because the initial concentration of methomyl is 1.23×10^{-4} mol/L, a final concentration of nitrogen ions of 2.46×10^{-4} mol/L which can be expected during mineralization (the nitrogen released is frequently measured as a combination of ammonia and nitrate ions) [57]. It was discovered that, under experimental conditions, methomyl degradation follows a first-order kinetic rate, the total disappearance of methomyl seen in less than 45 min. A thorough examination of parameters, including TOC disappearance, ion monitoring of the development of sulphate and ammonium ions were identified by chromatographic techniques. The evolution of intermediates were identified by HPLC-UV and HPLC-ESI-MS analyses. A better knowledge of the degradation mechanism and a possible degradation mechanism has been proposed in the reference [59].

The potential utilization of as-prepared black TiO_2 for the photodegradation of pesticide as an environmentally friendly technique was assessed after various time intervals of UV and UV-Vis irradiation. For example, the detection of imidacloprid (IMD) was accomplished by the use of a series of standard solutions. In the photolysis experiments, IMD degradation was less than 3% indicating that IMD is stable under UV and simulated visible light irradiation. A greater number of active centers at the catalyst surface were available for more pollutant molecules at an appropriate photocatalyst concentration. As a result, the degradation rate increases, which is typical characteristics of heterogeneous photocatalysis. The effect of initial catalyst concentration was studied at 0.6 and 1 g/L. For example, using 0.6 g/L photocatalysts results in the greatest photodegradation efficiency of IMD under UV light; increasing this amount to 1 g/L results in a fall in efficiency from 73 to 68%.

As illustrated in Fig. 10, the degradation efficiency of IMD in conjunction with B1, B2, and B3 photocatalysts reached 83, 90, and 86% (under UV irradiation), were achieved in 360 min, compared to 68% in the presence of white TiO_2 photocatalyst. However, the distinction between white and black TiO_2 is overlooked. Only the B2 sample was visible under simulated visible irradiation with 67% greater degradation efficiency [33].

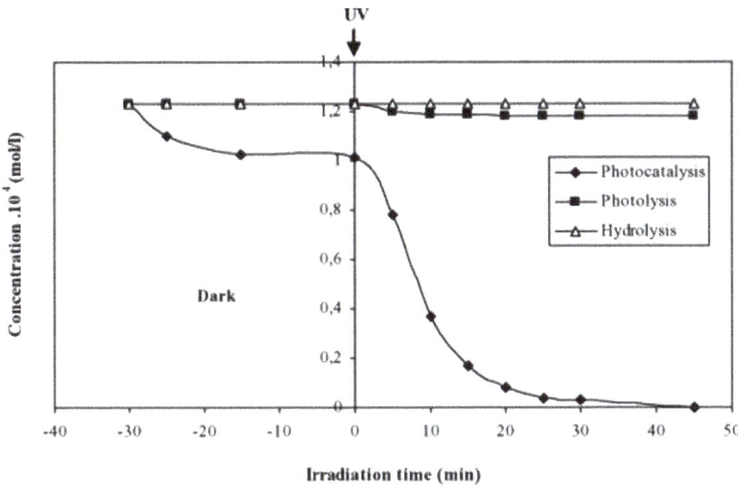

Fig. 9 Degradation of methomyl by photocatalysis (TiO_2-UV), direct photolysis (UV) and hydrolysis [57]

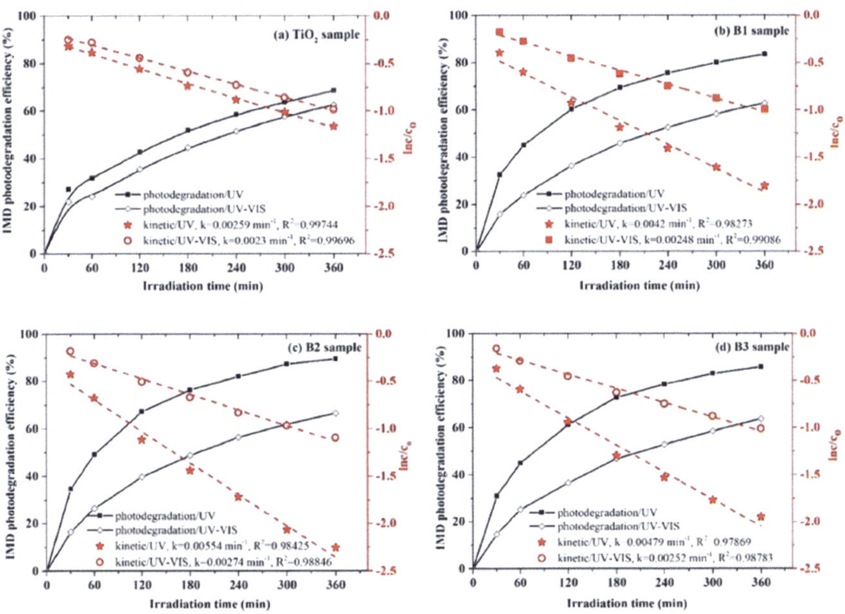

Fig. 10 The photocatalytic efficiency (black line) and the first-order degradation kinetic (red line) of materials. **a** TiO_2 sample, **b** B1 sample, **c** B2 sample, and **d** B3 sample. (Experimental conditions: photocatalyst 1 g/L, IMD 20 mg/L)

1.6 Developments of Black TiO_2 Photocatalyst in Recent years and Future Perspectives

Many more novel techniques for synthesis of black TiO_2 are likely to be developed in the forthcoming years due to its immense potential for harvesting visible light of solar spectrum. The diverse synthesis routes uncovered so far definitely widen the horizon on how black TiO_2 can be synthesized. As we discussed in the previous sections, the properties of black TiO_2 change immensely with the synthesis route and synthetic conditions like reduction atmosphere, reduction temperature and pressure, reductant dosage, precursor used etc. This diversity in the preparation methodologies offer numerous ways to induce defects in pristine TiO_2 and alter the physicochemical properties so as to benefit the intended applications. However, the biggest challenge in understanding the intricacies of black TiO_2 probably lies in comprehending the diverse behaviour reported in those varying experimental setups. Because of its enormous potential for harvesting visible light in the solar spectrum, many more unique ways for synthesis of black TiO_2 are anticipated to be developed in the future. The various synthesis techniques discovered up to now significantly broaden the scope on how black TiO_2 can be synthesised. However, recognising the complexities of black TiO_2 is likely to be the most difficult challenge in the futuristic synthetic methodologies.

2 Conclusion

In this chapter, the growing need of the scientific community towards the degradation of agrochemicals through a green, cost-effective method is described. Given this, it is prudent to create technologies that facilitate the easy breakdown of bio-recalcitrant organic molecules. A promising method for the mineralization of these compounds is produced through the use of advanced oxidation processes (AOPs). Although heterogeneous photocatalysis has been labelled as a costly treatment option, the cost can be reduced and optimised by implementing approaches for catalyst separation and recycling. The arrival of black TiO_2 which has superior optical, electronic, and photocatalytic properties than white TiO_2, with reduced band gap value can be able to degrade the organic pollutants, especially agrochemicals. The various synthesis methods and the defects associated with their superior properties are also explained in this chapter. The degradation of pesticides using black TiO_2 is an emerging area and more research will be concentrated in this field in the future. The results were found that these photocatalysts can effectively degrade selected pesticides as well as wide spectrum of environmental contaminants.

References

1. E. Borsato, P. Tarolli, F. Marinello, Sustainable patterns of main agricultural products combining different footprint parameters. J. Clean. Prod. **179**, 357–367 (Elsevier Ltd., 2018). https://doi.org/10.1016/j.jclepro.2018.01.044
2. H.D. Burrows, M. Canle, J.A. Santaballa, S. Steenken, Invited review reaction pathways and mechanisms of photodegradation of pesticides (2002). www.elsevier.com/locate/jphotobiol
3. S. Malato, J. Blanco, J. Cáceres, A.R. Fernández-Alba, A. Agüera, A. Rodríguez, Photocatalytic treatment of water-soluble pesticides by photo-Fenton and TiO_2 using solar energy (2002)
4. J. Liu, Z. Wei, W. Shangguan, Defects engineering in photocatalytic water splitting materials. ChemCatChem **11**(24), 6177–6189 (Wiley Blackwell, 2019). https://doi.org/10.1002/cctc.201901579
5. I.K. Konstantinou, T.A. Albanis, TiO2-assisted photocatalytic degradation of azo dyes in aqueous solution: kinetic and mechanistic investigations: a review. Appl. Catal. B: Environ. **49**(1), 1–14 (Elsevier, 2004). https://doi.org/10.1016/j.apcatb.2003.11.010
6. S.G. Ullattil, S.B. Narendranath, S.C. Pillai, P. Periyat, Black TiO2 nanomaterials: a review of recent advances. Chem. Eng. J. **343**, 708–736 (Elsevier B.V., 2018). https://doi.org/10.1016/j.cej.2018.01.069
7. M.H. El-Saeid, A. Baqais, M. Alshabanat, Study of the photocatalytic degradation of highly abundant pesticides in agricultural soils. Molecules **27**(3) (2022). https://doi.org/10.3390/molecules27030634
8. P. Sathishkumar, R.V. Mangalaraja, S. Anandan, Review on the recent improvements in sonochemical and combined sonochemical oxidation processes—A powerful tool for destruction of environmental contaminants. Renew. Sustain. Energy Rev. **55**, 426–454 (Elsevier Ltd., 2016). https://doi.org/10.1016/j.rser.2015.10.139
9. J. Madhavan, P.S.S. Kumar, S. Anandan, M. Zhou, F. Grieser, M. Ashokkumar, Ultrasound assisted photocatalytic degradation of diclofenac in an aqueous environment. Chemosphere **80**(7), 747–752 (2010). https://doi.org/10.1016/j.chemosphere.2010.05.018
10. V.K. Saharan, A.B. Pandit, P.S. Satish Kumar, S. Anandan, Hydrodynamic cavitation as an advanced oxidation technique for the degradation of Acid Red 88 dye. Ind. Eng. Chem. Res. **51**(4), 1981–1989 (2012). https://doi.org/10.1021/ie200249k
11. D. Ma et al., Critical review of advanced oxidation processes in organic wastewater treatment. Chemosphere **275** (Elsevier Ltd., 2021). https://doi.org/10.1016/j.chemosphere.2021.130104
12. M. Priyadarshini, I. Das, M.M. Ghangrekar, L. Blaney, Advanced oxidation processes: performance, advantages, and scale-up of emerging technologies. J. Environ. Manag. **316** (Academic Press, 2022). https://doi.org/10.1016/j.jenvman.2022.115295
13. Q.Q. Cai, L. Jothinathan, S.H. Deng, S.L. Ong, H.Y. Ng, J.Y. Hu, Fenton- and ozone-based AOP processes for industrial effluent treatment, in *Advanced Oxidation Processes for Effluent Treatment Plants* (Elsevier, 2020), pp. 199–254. https://doi.org/10.1016/B978-0-12-821011-6.00011-6
14. W.J. Chung et al., A magnetically separable and recyclable Ag-supported magnetic TiO2 composite catalyst: fabrication, characterization, and photocatalytic activity. J. Environ. Manag. **213**, 541–548 (2018). https://doi.org/10.1016/j.jenvman.2018.02.064
15. D. Nunes, A. Pimentel, R. Branquinho, E. Fortunato, R. Martins, Metal oxide-based photocatalytic paper: A green alternative for environmental remediation. Catalysts **11**(4) (MDPI, 2021). https://doi.org/10.3390/catal11040504
16. H. Song, C. Li, Z. Lou, Z. Ye, L. Zhu, Effective formation of oxygen vacancies in black TiO2 nanostructures with efficient solar-driven water splitting. ACS Sustain. Chem. Eng. **5**(10), 8982–8987 (2017). https://doi.org/10.1021/acssuschemeng.7b01774
17. Y. Ma, X. Wang, Y. Jia, X. Chen, H. Han, C. Li, Titanium dioxide-based nanomaterials for photocatalytic fuel generations. Chem. Rev. **114**(19), 9987–10043 (American Chemical Society, 2014). https://doi.org/10.1021/cr500008u

18. M. Kapilashrami, Y. Zhang, Y.S. Liu, A. Hagfeldt, J. Guo, Probing the optical property and electronic structure of TiO2nanomaterials for renewable energy applications. Chem. Rev. **114**(19), 9662–9707 (American Chemical Society, 2014). https://doi.org/10.1021/cr5000893
19. Y. Fu et al., An amine-functionalized titanium metal-organic framework photocatalyst with visible-light-induced activity for CO2 reduction. Angew. Chem. Int. Ed. **51**(14), 3364–3367 (2012). https://doi.org/10.1002/anie.201108357
20. M. Dahl, Y. Liu, Y. Yin, Composite titanium dioxide nanomaterials. Chem. Rev. **114**(19), 9853–9889 (American Chemical Society, 2014). https://doi.org/10.1021/cr400634p
21. C. Di Valentin, G. Pacchioni, A. Selloni, Reduced and n-type doped TiO2: nature of Ti3+ species. J. Phys. Chem. C **113**(48), 20543–20552 (2009). https://doi.org/10.1021/jp9061797
22. I. Zada et al., Superior photothermal black TiO2 with random size distribution as flexible film for efficient solar steam generation. Appl. Mater. Today **20** (2020). https://doi.org/10.1016/j.apmt.2020.100669
23. S.T. Myung et al., Black anatase titania enabling ultra high cycling rates for rechargeable lithium batteries. Energy Environ. Sci. **6**(9), 2609–2614 (2013). https://doi.org/10.1039/c3ee41960f
24. C. Yang et al., Core-shell nanostructured 'black' rutile titania as excellent catalyst for hydrogen production enhanced by Sulfur doping (2013). http://pubs.acs.org
25. A. Naldoni et al., Effect of nature and location of defects on bandgap narrowing in black TiO2 nanoparticles. J. Am. Chem. Soc. **134**(18), 7600–7603 (2012). https://doi.org/10.1021/ja3012676
26. J. Lao, S. Wu, J. Gao, A. Dong, G. Li, J. Luo, Electricity generation based on a photothermally driven Ti3C2Tx MXene nanofluidic water pump. Nano Energy **70** (2020). https://doi.org/10.1016/j.nanoen.2020.104481
27. T.S. Rajaraman, S.P. Parikh, V.G. Gandhi, Black TiO2: a review of its properties and conflicting trends. Chem. Eng. J. **389** (Elsevier B.V., 2020). https://doi.org/10.1016/j.cej.2019.123918
28. M. Wajid Shah et al., Facile synthesis of defective TiO2-x nanocrystals with high surface area and tailoring bandgap for visible-light photocatalysis. Sci. Rep. **5** (2015). https://doi.org/10.1038/srep15804
29. J. Wan et al., Defect effects on TiO2 nanosheets: stabilizing single atomic site au and promoting catalytic properties. Adv. Mater. **30**(11) (2018). https://doi.org/10.1002/adma.201705369
30. B. Wang, S. Shen, S.S. Mao, Black TiO2 for solar hydrogen conversion. J. Mater. **3**(2), 96–111 (Chinese Ceramic Society, 2017). https://doi.org/10.1016/j.jmat.2017.02.001
31. B. Li, Y. Jiang, J. Liu, K. Hui, L. Guo, Photocatalytic and eletrocatalytic activity of black TiO2 nanotube arrays by aluminothermic reduction. Micro Nano Lett. **18**(1) (2023). https://doi.org/10.1049/mna2.12152
32. M.C. Wu, K.C. Hsiao, Y.H. Chang, S.H. Chan, Photocatalytic hydrogen evolution of palladium nanoparticles decorated black TiO2 calcined in argon atmosphere. Appl. Surf. Sci. **430**, 407–414 (2018). https://doi.org/10.1016/j.apsusc.2017.08.071
33. L. Andronic, A. Enesca, Black TiO2 synthesis by chemical reduction methods for photocatalysis applications. Front. Chem. **8** (Frontiers Media S.A., 2020). https://doi.org/10.3389/fchem.2020.565489
34. T. Perarasan, I.J. Peter, A.M. Kumar, N. Rajamanickam, K. Ramachandran, C.R. Mohan, Copper doped titanium dioxide for enhancing the photovoltaic behavior in solar cell. in *Materials Today: Proceedings* (Elsevier Ltd., 2019), pp. 66–68. https://doi.org/10.1016/j.matpr.2019.06.377
35. X. Chen, L. Liu, F. Huang, Black titanium dioxide (TiO2) nanomaterials chemical society reviews black titanium dioxide (TiO2) nanomaterials
36. H. Lu et al., Safe and facile hydrogenation of commercial Degussa P25 at room temperature with enhanced photocatalytic activity. RSC Adv. **4**(3), 1128–1132 (2014). https://doi.org/10.1039/c3ra44493g
37. Y. Yan et al., Slightly hydrogenated TiO2 with enhanced photocatalytic performance. https://doi.org/10.1039/c0xx00000x

38. Y. Wang et al., Enhancing catalytic activity of titanium oxide in Lithium–Sulfur batteries by band engineering. Adv. Energy Mater. **9**(24) (2019). https://doi.org/10.1002/aenm.201900953
39. Z. Zhao et al., Reduced TiO2 rutile nanorods with well-defined facets and their visible-light photocatalytic activity. Chem. Commun. **50**(21), 2755–2757 (2014). https://doi.org/10.1039/c3cc49182j
40. C. Mao, F. Zuo, Y. Hou, X. Bu, P. Feng, In situ preparation of a Ti3+ Self-doped TiO2 film with enhanced activity as photoanode by N2H4 reduction. Angew. Chem. Int. Ed. **53**(39), 10485–10489 (2014). https://doi.org/10.1002/anie.201406017
41. A. Sinhamahapatra, J.P. Jeon, J.S. Yu, A new approach to prepare highly active and stable black titania for visible light-assisted hydrogen production. Energy Environ. Sci. **8**(12), 3539–3544 (2015). https://doi.org/10.1039/c5ee02443a
42. L. Li et al., Synthesis of Ti3+ self-doped mesoporous TiO2 cube with enhanced visible-light photoactivity by a simple reduction method. J. Alloys Compd. **845** (2020). https://doi.org/10.1016/j.jallcom.2020.156138
43. L. Liu et al., Engineering Coexposed 001 and 101 facets in oxygen-deficient TiO2 nanocrystals for enhanced CO2 photoreduction under visible light. ACS Catal. **6**(2), 1097–1108 (2016). https://doi.org/10.1021/acscatal.5b02098
44. Q. Kang, J. Cao, Y. Zhang, L. Liu, H. Xu, J. Ye, Reduced TiO2 nanotube arrays for photoelectrochemical water splitting. J. Mater. Chem. A Mater. **1**(18), 5766–5774 (2013). https://doi.org/10.1039/c3ta10689f
45. G. Chai, W. Huang, Q. Shi, S. Zheng, D. Wei, Preparation and characterization of λ-Ti3O5 by carbothermal reduction of TiO2. J. Alloys Compd. **621**, 404–410 (2015). https://doi.org/10.1016/j.jallcom.2014.10.011
46. G. Wang et al., Hydrogen-treated TiO2 nanowire arrays for photoelectrochemical water splitting. Nano Lett. **11**(7), 3026–3033 (2011). https://doi.org/10.1021/nl201766h
47. Z. Zhang, M.N. Hedhili, H. Zhu, P. Wang, Electrochemical reduction induced self-doping of Ti3+ for efficient water splitting performance on TiO2 based photoelectrodes. Phys. Chem. Chem. Phys. **15**(37), 15637–15644 (2013). https://doi.org/10.1039/c3cp52759j
48. X. Liu et al., Green synthetic approach for Ti3+ self-doped TiO2-x nanoparticles with efficient visible light photocatalytic activity. Nanoscale **5**(5), 1870–1875 (2013). https://doi.org/10.1039/c2nr33563h
49. G. Panomsuwan, A. Watthanaphanit, T. Ishizaki, N. Saito, Water-plasma-assisted synthesis of black titania spheres with efficient visible-light photocatalytic activity. Phys. Chem. Chem. Phys. **17**(21), 13794–13799 (2015). https://doi.org/10.1039/c5cp00171d
50. Z. Pei et al., Facile synthesis of defect-mediated TiO2-x with enhanced visible light photocatalytic activity. J. Mater. Chem. A Mater. **1**(35), 10099–10102 (2013). https://doi.org/10.1039/c3ta12062g
51. M. Li et al., Mesoporous palladium-copper bimetallic electrodes for selective electrocatalytic reduction of aqueous CO2–CO. J. Mater. Chem. A Mater. **4**(13), 4776–4782 (2016). https://doi.org/10.1039/c6ta00487c
52. Q. Zhu et al., Stable blue TiO2-x nanoparticles for efficient visible light photocatalysts. J. Mater. Chem. A Mater. **2**(12), 4429–4437 (2014). https://doi.org/10.1039/c3ta14484d
53. X. Zhou et al., TiO2 nanotubes: N-ion implantation at low-dose provides noble-metal-free photocatalytic H2-evolution activity. Angew. Chem. Int. Ed. **55**, 3763 (2016). https://doi.org/10.1002/anie.201511580/abstract
54. M. Domaschke, L. Strunz, W. Peukert, Single-step aerosol synthesis of oxygen-deficient blue titania. Chem. Eng. Sci. **206**, 327–334 (2019). https://doi.org/10.1016/j.ces.2019.05.044
55. E. Wierzbicka et al., Self-enhancing H2 Evolution from TiO2 nanostructures under illumination. Chemsuschem **12**(9), 1900–1905 (2019). https://doi.org/10.1002/cssc.201900192
56. H. Hamad, E. Bailón-García, F.J. Maldonado-Hódar, A.F. Pérez-Cadenas, F. Carrasco-Marín, S. Morales-Torres, Synthesis of TixOy nanocrystals in mild synthesis conditions for the degradation of pollutants under solar light. Appl. Catal. B **241**, 385–392 (2019). https://doi.org/10.1016/j.apcatb.2018.09.016

57. S.S. Sahoo, S. Mansingh, P. Babu, K. Parida, Black titania an emerging photocatalyst: review highlighting the synthesis techniques and photocatalytic activity for hydrogen generation. Nanosc. Adv. **3**(19), 5487–5524 (Royal Society of Chemistry, 2021). https://doi.org/10.1039/d1na00477h
58. M. Tamimi, S. Qourzal, A. Assabbane, J.M. Chovelon, C. Ferronato, Y. Ait-Ichou, Photocatalytic degradation of pesticide methomyl: determination of the reaction pathway and identification of intermediate products. Photochem. Photobiol. Sci. **5**(5), 477–482 (2006). https://doi.org/10.1039/b517105a
59. N. Pugazhenthiran et al., Graphene modified "black {001}TiO2" nanosheets for photocatalytic oxidation of ethylene: The implications of chemical surface characteristics in the reaction mechanism, Separ, Purifi. Technol. **292**, 121008 (2022). https://doi.org/10.1016/j.seppur.2022.121008
60. F. Zuo, L. Wang, T. Wu, Z. Zhang, D. Borchardt, P. Feng, Self-doped Ti3+ enhanced photocatalyst for hydrogen production under visible light. J. Am. Chem. Soc. **132**(34), 11856–11857 (2010). https://doi.org/10.1021/ja103843d
61. C. Fan et al., Black hydroxylated titanium dioxide prepared via ultrasonication with enhanced photocatalytic activity. Sci. Rep. **5**(1) (2015). https://doi.org/10.1038/SREP11712
62. M. Zhai et al., Efficient suspension plasma spray fabrication of black titanium dioxide coatings with visible light absorption performances. Ceram. Int. **45**(1), 930–935 (2019). https://doi.org/10.1016/j.ceramint.2018.09.268
63. M. Zimbone et al., Hydrogenated black-TiOx: a facile and scalable synthesis for environmental water purification. Catal. Today **321–322**, 146–157 (2019). https://doi.org/10.1016/j.cattod.2018.03.040
64. W. Zhang et al., A hydrogenated black TiO2 coating with excellent effects for photothermal therapy of bone tumor and bone regeneration. Mater. Sci. Eng. C **102**, 458–470 (2019). https://doi.org/10.1016/j.msec.2019.04.025
65. N. Liu et al., 'Black' TiO2 nanotubes formed by high-energy proton implantation show noble-metal-co-catalyst free photocatalytic H2-evolution. Nano Lett. **15**(10), 6815–6820 (2015). https://doi.org/10.1021/acs.nanolett.5b02663

Constructive Routes of Size Dependent Photocatalytic Nanomaterials and Its Application in Energy and Environmental Fields

Rajalakshmi Padmanabhan, Perarasu Thangavelu, and Saravanathamizhan Ramanujam

Abstract The emergence of nanotechnology has instigated a profound transformation in the field of photocatalysis, ushering in unparalleled prospects for addressing urgent energy and environmental dilemmas. This chapter delves into the domain of size-dependent photocatalytic nanomaterials, unveiling the intricate connection between nanoscale characteristics and photocatalytic efficacy. Through an exploration of synthesis techniques governing conversion and environmental rehabilitation, this chapter illuminates the constructive pathways steering the advancement of cutting-edge nanomaterials. Drawing upon an extensive compilation of comprehensive articles and illustrative case studies, this compendium offers a holistic perspective on how size-dependent photocatalytic nanomaterials are propelling innovation toward a sustainable future. This undertaking aspires to serve as a comprehensive reference for researches, engineers, and policymakers engaged in the forefront of size-dependent photocatalytic nanomaterials and their applications within the domains of energy and the environment. It encapsulates the pivotal role that nanotechnology plays in catalyzing progress toward cleaner energy production and environmental preservation, showcasing its potential to shape a more sustainable and ecologically balanced future. However, challenges related to nanoparticle synthesis, stability, and potential impacts must be addressed for the successful integration of this technology into practical applications. Continued research and innovation in size-dependent photocatalysis hold the promise of transformative solutions for a more sustainable and cleaner future.

Keywords Energy and environment · Nanomaterials · Photocatalysis · Size-dependent properties · Sustainability

R. Padmanabhan · P. Thangavelu (✉) · S. Ramanujam
Department of Chemical Engineering, AC Tech Campus, Anna University, Chennai 600 025, Tamil Nadu, India
e-mail: vtparasu@annauniv.edu

P. Sathishkumar (ed.), *Photocatalysis for Energy and Environmental Applications*, Green Energy and Technology, https://doi.org/10.1007/978-981-97-1939-6_4

1 Introduction

In recent years, photocatalysis has emerged as a promising and versatile technology for addressing a wide range of environmental and energy-related challenges. The size of photocatalytic materials, particularly semiconductor nanoparticles, has been identified as a crucial factor governing their performance in photocatalytic applications. This chapter provides a comprehensive overview of the size-dependent photocatalytic applications, challenges, and future prospects in this rapidly evolving field. Photocatalysis [31], the process of utilizing light to drive chemical reactions on the surface of semiconductors, has garnered immense attention for its potential to address global issues such as water purification, air pollution mitigation, and renewable energy production.

Semiconductor nanoparticles, including but not limited to titanium dioxide (TiO_2), zinc oxide (ZnO), and various other metal oxides, have proven to be highly effective photocatalyst (Fig. 1). The size of these nanoparticles plays a pivotal role in determining their photocatalytic activity. One of the most notable advantages of size-dependent photocatalysis is the tunability of catalytic performance (Fig. 2). Small nanoparticles exhibit a higher surface area-to-volume ratio, leading to enhances light absorption and increased active sites for photo-generated charge carriers. Consequently, they demonstrate superior photocatalytic activity compared to their bulk counterparts. This effect has been extensively explored in the degradation of organic pollutants, such as dyes and pharmaceuticals, as well as in the removal of heavy metals from water. Moreover, the size-dependent properties of photocatalytic materials have been harnessed for energy conversion applications. Quantum dots (QDs) and other nanoscale semiconductors have exhibited remarkable potential in photocatalytic hydrogen production, where their size-dependent bandgap enables efficient utilization of a broader spectrum of light. This has profound implications for the development of sustainable and clean energy sources.

Despite the numerous advantages, size-dependent photocatalysis also faces several challenges. One major concern is the synthesis and stability of these nanoparticles. Precise control over particle size and size distribution is essential, and various synthesis methods, including sol-gel, hydrothermal, and microwave-assisted techniques, have been developed. However, maintaining the stability of small nanoparticles in real-world applications, particularly under harsh environmental conditions, remains a significant challenge. Furthermore, the potential environmental and health risks associated with the release of nanoparticles during photocatalytic processes cannot be overlooked.

Research efforts have focused on understanding the fate and transport of nanoparticles in natural systems and developing strategies to minimize their impact on the environment and human health. The integration of size-dependent photocatalysis into practical applications has gained momentum in recert years. Photocatalytic coatings and films have been employed for self-cleaning surfaces, air purification, and water treatment systems.

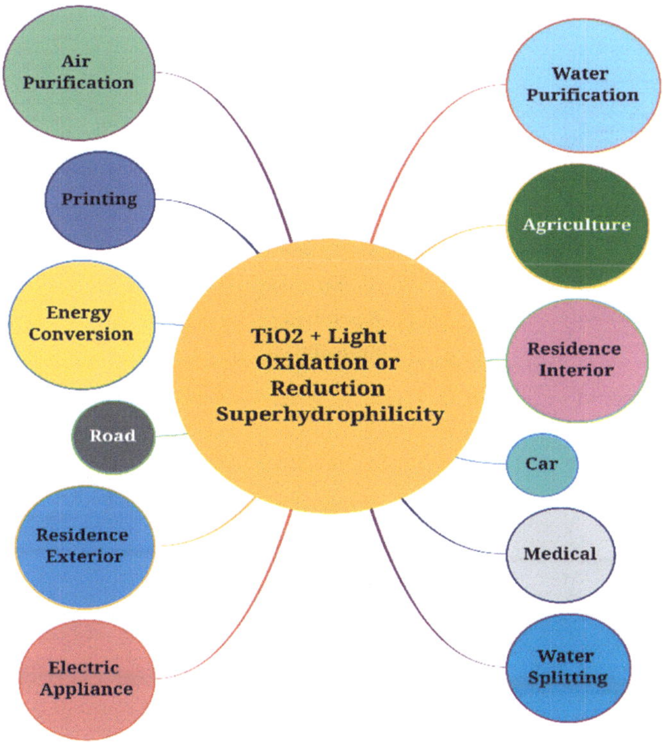

Fig. 1 Photocatalytic applications of semiconductor nanomaterials

Additionally, the development of photocatalytic reactors with immobilized nanoparticles allows for continuous and efficient pollutant degradation and water purification. In summary, size-dependent photocatalysis represents a promising avenue for addressing pressing environmental and energy challenges. The ability to finely tune photocatalytic performance through nanoparticle size manipulation offers exciting opportunities for clean water, air, and sustainable energy generation. However, overcoming challenges related to nanoparticle synthesis, stability, and potential environmental impacts is crucial for the successful implementation of this technology. Continued research and innovation in size-dependent photocatalysis are expected to yield groundbreaking solutions for a more sustainable and cleaner future.

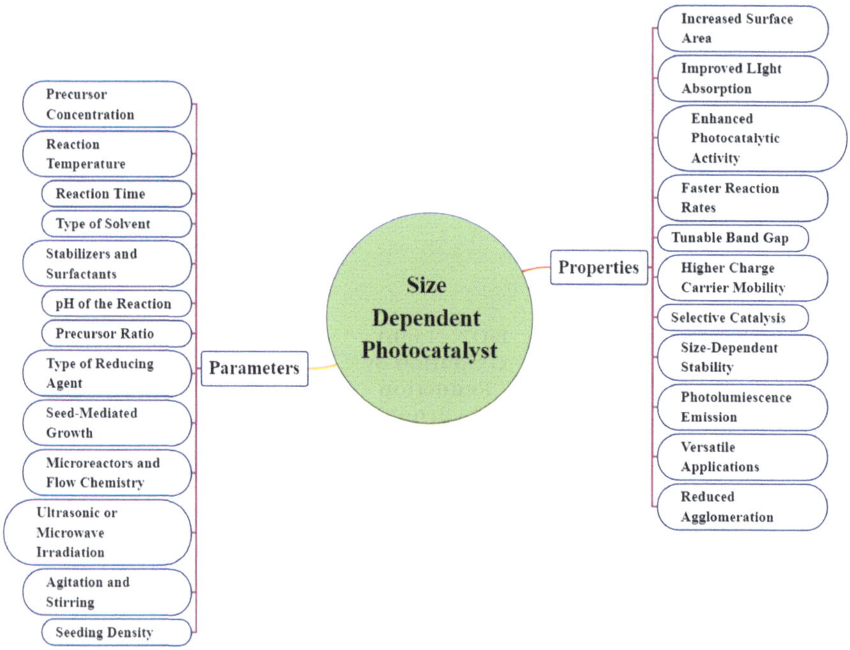

Fig. 2 Parameters affecting the size dependent photocatalyst and the enhancement in their properties

2 Nanomaterials Revolution

Nanomaterials, typically ranging from 1 to 100 nm in size, exhibit an array of extraordinary properties due to their high surface area-to-volume ratio and quantum confinement effects. These properties are a result of their size, shape, composition, and structure. The utilization of nanomaterials as photocatalysts has led to significant advancements in terms of efficiency, selectivity, and versatility. Efficiency and Selectivity.

One of the most striking advantages of nanomaterials in photocatalysis is their exceptional efficiency in utilizing light energy. With a higher surface area, more active sites are available for photon absorption, leading to increased reaction rates. Moreover, the tunable size and composition of nanomaterials allow for precise control over the bandgap, enabling them to absorb a broader spectrum of light, including visible and even infrared radiation. This expanded absorption range greatly enhances their potential for solar-driven photocatalysis. In terms of **selectivity**, nanomaterials can be tailored to exhibit specific surface facets or crystal structures, which can dictate reaction pathways. This level of control ensures that undesirable side reactions are minimized, resulting in higher product yields and purity. The improved selectivity also enables the design of photocatalysts for applications such as pollutant degradation and selective organic transformations. Versatility in Applications

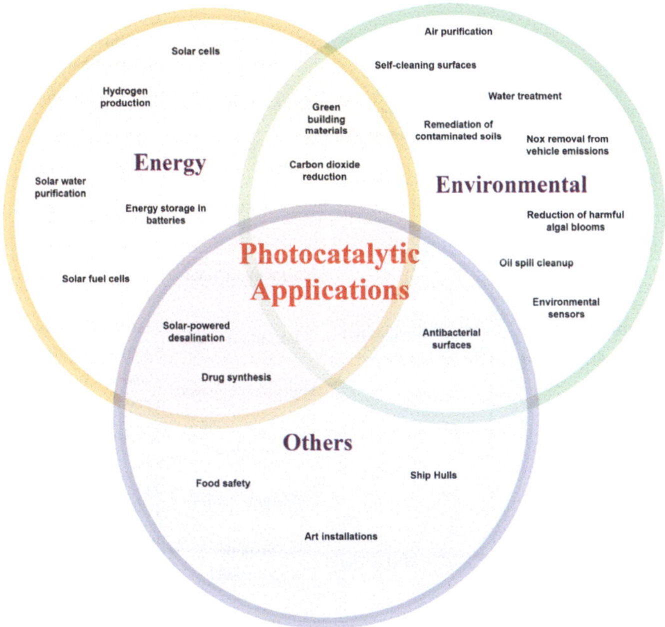

Fig. 3 Versatile applications of photocatalysts in addressing energy and environmental challenges

(Fig. 3): The nanomaterials revolution has transcended photocatalytic applications beyond water splitting and pollutant degradation. Nanomaterial-based photocatalysts are now employed in diverse areas, such as CO_2 reduction, hydrogen generation, and the synthesis of high-value chemicals. In the realm of energy storage, nanomaterials have shown promise in the development of advanced photoelectrochemical cells and photovoltaic devices.

2.1 General Classification of Nanomaterials

Nanostructured materials are classified based on the shape, dimensionality and phase composition according to their physical and chemical properties. According to their dimensionality nanostructured materials are classified into four categories by Seigel [88]. According to the composition, morphology and distribution of the nanocrystalline component, the nanostructured materials were further classified quite broadly by Gleiter, including several permutations of materials[36]. Schematic representation of the classification of nanomaterials based on shape and dimension, phase composition and nature of material is represented as a figure (Fig. 4) [48].

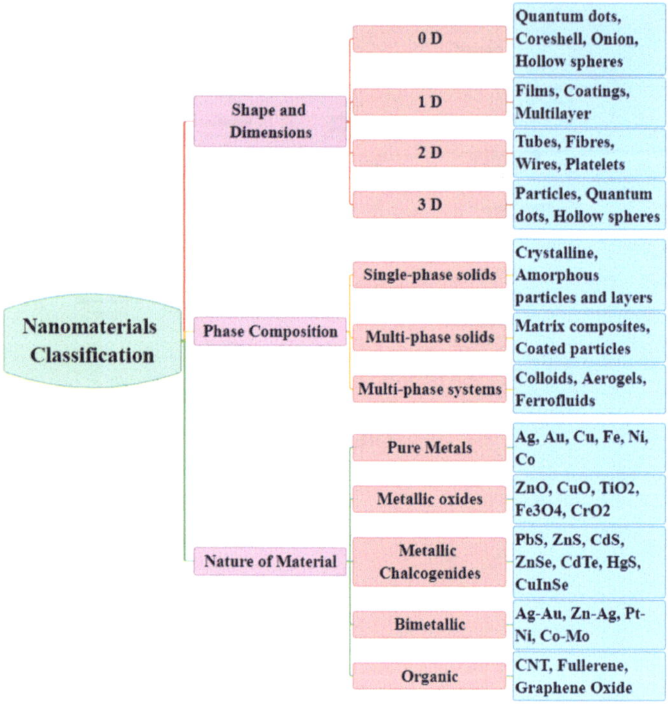

Fig. 4 Nanomaterials broad classification based on the shape and dimension, phase composition and nature of material [48]

Exploring the critical aspects of nanomaterials, including their nature, composition, and classification, it covers various criteria for categorizing nanomaterials, such as morphology, dimensionality, size, agglomeration state, and composition. Carbon-based nanomaterials, including fullerene, graphene, nanodiamond, and carbon nanotubes (CNTs), constitute a distinct category. Morphology plays a crucial role in distinguishing nanomaterials, and their properties often exhibit significant improvements at the nanoscale compared to bulk materials. Optical properties, influenced by size and shape, make nanomaterials suitable for various applications, including catalysis [89].

2.2 *Characterization of Nanomaterials*

Nanoparticles have opened new horizons in various scientific domains due to their unique properties and potential applications. To unlock their full potential, accurate characterization is essential. The following description provides a comprehensive overview of spectroscopy and microscopy techniques frequently employed for the

Table 1 Spectroscopy and microscopy techniques for photocatalytic nanoparticle characterization: insights and applications

Characterization techniques		Principle	Application
Name of the technique	Acronym		
UV-visible spectroscopy	UV-Vis	Absorption and electronic transitions	Quantification of nanoparticle concentration, optical properties
Dynamic light scattering	DLS	Particle size from Brownian motion	Determining hydrodynamic size, size distribution
Atomic force microscopy	AFM	Probe-based imaging and force measurement	Surface topography, mechanical properties of nanoparticles
Transmission electron microscopy	TEM	Electron beam interaction for high-resolution imaging	Morphology, size distribution, crystallography
Scanning electron microscopy	SEM	High-energy electron beam for surface imaging	Morphology, elemental composition, particle agglomeration
Energy dispersive spectroscopy	EDS	Elemental analysis by X-ray detection	Chemical composition mapping, identifying impurities
Powder x-ray diffraction	XRD	X-ray diffraction from crystal lattice planes	Phase identification, crystal structure analysis
Fourier transform infrared spectroscopy	FT-IR	Molecular vibrations and infrared absorption	Chemical composition, functional groups
Raman spectroscopy	–	Inelastic scattering of light and molecular vibrations	Molecular identification, crystal structure, chemical bonding

characterization of nanoparticles. From UV-Visible spectroscopy (UV-Vis) to atomic force spectroscopy (AFM), scanning electron microscopy (SEM), and transmission electron microscopy (TEM), we delve into the principles, advantages, and applications of each method (Table 1). These techniques not only enable the determination of size and morphology but also offer insights into nanoparticle properties critical for tailoring their performance and applications.

SEM-EDS reveals efficient doping. TEM exhibits varied particle sizes. UV-Vis spectrophotometry highlights enhanced visible light absorption. Raman spectroscopy and XRD confirm anatase phase suitable for waste water treatment. Cathodoluminescence and photoluminescence studies demonstrate improved electron-hole pair recombination [95].

2.3 Synthesis Approaches of Nanomaterials

In the realm of nanotechnology, where science meets the infinitesimally small, the ability to manipulate matter at the nanoscale has opened doors to unprecedented advancements in various fields, from electronics and medicine to energy and materials science. One of the key challenges in nanotechnology is the precise fabrication of nanomaterials, which are materials engineered at the nanoscale to exhibit unique properties and functionalities. Two dominant strategies for crafting nanomaterials with precision are the top-down and bottom-up approaches, each with its own set of advantages and limitations. Though there are several methods and techniques for the synthesis and fabrication of nanoparticles, carbon-based graphene, graphene oxide and reduced graphene oxide nanomaterial synthesis is also grouped under the top-down and bottom-up approach [37]. These approaches involve different techniques, each of which plays vital role in modern industry and nanotechnology, and they possess their advantages and disadvantages [10], briefed in the following sections.

2.3.1 Top-Down Approach

The top-down approach, reminiscent of sculpting, involves starting with a bulk material and gradually reducing its size to the nanoscale through various machining, etching, or lithographic techniques. This strategy draws parallels to carving a statue from a block of stone, where material is removed strategically to achieve the desired form, in nanotechnology, the top-down approach has been as staple method for creating nanomaterials, thanks to its established techniques and ability to produce relatively large quantities of nanomaterials. One of the key advantages of the top-down approach is its compatibility with existing manufacturing infrastructure. Semiconductor industries have extensively employed top-down techniques, such as photolithography, to create integrated circuits with ever-shrinking features. Additionally, the top-down approach is well-suited for producing uniform nanomaterials on a large scale, making it ideal for applications where homogeneity is crucial. However, the top-down approach faces challenges when aiming for the nanoscale precision. As dimensions shrink, traditional top-down methods encounter limitations imposed by diffraction limits and tool wear. Achieving atomic-scale precision becomes increasingly difficult, making it challenging to create nanomaterials with custom-designed properties. Furthermore, the removal of excess material can lead to waste, limiting the cost-effectiveness of the approach.

2.3.2 Bottom-Up Approach

Contrastingly, the bottom-up approach takes inspiration from nature's self-assembly processes and involves building nanomaterials atom by atom or molecule by molecule. It is akin to arranging individual building blocks to form a complex structure. This method holds great promise for achieving unparalleled precision and tailoring properties with atomic-scale control. One of the key advantages of the bottom-up approach is its potential to design and synthesize nanomaterials with properties that cannot be achieved through traditional materials. By controlling the interactions between individual components, scientists can create materials with unique optical, electronic, and mechanical characteristics. This approach is particularly exciting for applications such as drug delivery, where nanomaterials can be designed to target specific cells or tissues. However, the bottom-up approach faces its own set of challenges. The synthesis of nanomaterials at the atomic or molecular scale requires intricate control over reaction conditions, which can be time-consuming and expensive. Additionally, scaling up production while maintaining the same level of precision presents a significant hurdle. Quality control and reproducibility are also vital concerns in bottom-up approaches, as slight variations in the synthesis process can lead to inconsistent properties.

The choice between top-down and bottom-up approaches often depends on the specific goals of nanomaterial fabrication. In many cases, a combination of both strategies, often referred to as a "hybrid approach" can offer a balanced solution. This approach leverages the strengths of each method to create complex nanomaterials that would be challenging to achieve using a singly approach. The crafting of nanomaterials with precision is a complex endeavor that requires a deep understanding of both top-down and bottom-up approaches. While each method offers its own advantages and challenges, the convergence of these strategies holds the key to unlocking the vast potential of nanotechnology across diverse applications. As scientists and engineers refine their techniques, the nanomaterials of the future will undoubtedly become integral to our daily lives, ushering in a new era of innovation and discovery.

In summary, photocatalytic nanomaterials hold immense promise in addressing global challenges, from pollution mitigation to renewable energy production. The upcoming sections would provide an overview on recent developments in the field, shedding light on the synthesis strategies, properties, and applications of these materials. As ongoing research continue to innovate and overcome existing limitations, photocatalytic nanomaterials are poised to play an increasingly vital role in building a sustainable and cleaner future.

3 Constructive Routes of Photocatalysts

This section discusses synthesis methods such as physical, chemical, and green methods, along with the parameters involved in their construction. Additionally, it explores how these methods enhance the properties of the resulting nano-photocatalysts.

3.1 Nano-Photocatalyst Synthesis

Over the last decade, scientists have made significant progress in developing techniques for synthesizing and characterizing a wide array of new materials with at least one dimension existing on the nanoscale. These materials encompass nanoparticles, nanolayers, and nanotubes [49]. Despite these advancements, the deliberate design and controlled synthesis of fabrication of nanoscale materials with specific properties remain substantial and persisting challenges in the fields of nanoscience and nanotechnology. The actualization of a circular economy on an industrial scale, begun the evolution of competent green routes in tailoring the nanomaterials wherein the conventional physicochemical routes are havoc to the environment.

3.1.1 Physical Methods

Physical synthesis methods for photocatalysts involve the formation of photocatalytic materials through physical processes without the use of chemical reactions. These methods are characterized by their ability to create photocatalysts with controlled size, shape, and structure. Here are some common physical synthesis methods for producing photocatalytic materials explained in brief in Table 2.

Physical synthesis methods offer advantages such as the ability to produce nanoparticles with controlled sizes and shapes and the avoidance of chemical precursors, making them suitable for certain photocatalysts applications. The choice of method depends on the desired properties of the photocatalyst and the specific requirements of the application.

3.1.2 Chemical Methods

Chemical synthesis methods are commonly employed to produce photocatalysts, which are materials that can initiate chemical reactions when exposed to light. These methods allow for precise control over the composition, structure, and properties of the photocatalytic materials. Here are some chemical synthesis methods used for producing photocatalysts explained in brief in Table 3.

3.1.3 Green Synthesis Method

Green nanotechnology has emerged as a groundbreaking technology aimed at overcoming the limitations of physical and chemical methods of synthesizing nano catalysts. The development of new materials and their applications incorporates green chemistry principles due to the unique nature of engineered nanomaterials and their potential uses. Green synthesis methods of photocatalysts are environmentally friendly approaches for producing photocatalytic materials without the use of toxic chemicals or excessive energy [21]. These methods aim to minimize the environmental impact of photocatalyst production while maintaining or improving their performance. Some notable green synthesis methods for photocatalysts are shown in Fig. 5.

Table 2 Nanomaterials synthesis by physical methods

Physical method	Description
Mechanical milling	• Mechanical milling involves the use of mechanical energy, such as ball milling or attrition milling, to reduce bulk materials into nanoscale photocatalytic particles • The high-energy collisions between particles lead to the mechanical fracture and reduction in size
Gas phase synthesis	• In gas phase synthesis methods like chemical vapor deposition (CVD) and physical vapor deposition (PVD), precursor gases or vaporized materials are introduced into a reactor chamber • Under controlled conditions, these gases or vaporized materials react and deposit as thin films or nanoparticles on a substrate, forming photocatalytic materials
Laser ablation	• Laser ablation involves using a high-energy laser to vaporize and ablate a target material • The vaporized material then condenses into nanoparticles, which can be collected and used as photocatalysts
Sputtering	• Sputtering is a physical deposition technique that uses high-energy ions to eject atoms or molecules from a solid target • These ejected species are then deposited onto a substrate to create thin films of photocatalytic materials
Evaporation	• Evaporation techniques, such as thermal evaporation and electron beam evaporation, involving heating a solid material to create vapor • The vaporized material is then deposited onto a substrate to form thin films or coatings of photocatalysts
Electrospinning	• Electrospinning is a method that uses an electric field to stretch and spin a polymer solution or melt into nanofibers • These nanofibers can be loaded with photocatalytic materials, creating composite materials with enhanced photocatalytic properties
Solvent-free synthesis	• Some physical methods, like the solvent-free solid-state synthesis, involve heating and/or milling precursor materials in the absence of solvents • This can lead to the formation of photocatalytic materials through solid-state reactions
Spray pyrolysis	• Spray pyrolysis involves atomizing a precursor solution into fine droplets and then heating these droplets to induce the formation of photocatalytic nanoparticles • The nanoparticles can be collected on a substrate
Plasma synthesis	• Plasma-based methods use high-temperature plasma discharges to create and deposit nanoparticles • The controlled environment of plasma allows fort the production of photocatalytic nanoparticles with specific properties
Ion implantation	• Ion implantation involves accelerating ions to high energies and implanting them into a substrate or host material to introduce desired photocatalytic properties

Table 3 Nanomaterials synthesis by chemical methods

Chemical method	Description
Sol-gel method	• The sol-gel method involves the hydrolysis and condensation of metal alkoxides or metal salts in a solution to form a sol (a colloidal suspension of nanoparticles) • The sol is then allowed to gel, leading to the formation of a three-dimensional network of nanoparticles • The gel is subsequently heated to remove the solvent and organic residues through calcination, resulting in a porous photocatalyst
Chemical precipitation	• In chemical precipitation, soluble precursors are mixed to induce the formation of insoluble photocatalytic materials • By carefully controlling the pH, temperature, and reaction kinetics, researchers can influence the size, shape, and composition of the resulting nanoparticles
Hydrothermal synthesis	• Hydrothermal synthesis is conducted at elevated temperatures and pressures in an aqueous solution • This method allows for the growth of photocatalytic nanoparticles under controlled conditions, promoting crystallinity and controlled particle size
Solvothermal synthesis	• Solvothermal synthesis is similar to hydrothermal synthesis but is conducted in organic solvents rather than water • It combines the sol-gel and hydrothermal methods, using both a solvent and controlled heating conditions to synthesize photocatalytic materials • This method can lead to the production of photocatalysts with unique properties and morphologies
Co-precipitation	• In co-precipitation, multiple metal precursors are simultaneously precipitated from a solution to form mixed-metal oxide photocatalysts • This method allows for the incorporation of multiple metal ions into the photocatalyst's structure
Ion exchange	• Ion exchange involves the replacement of ions within a solid lattice with ions from a solution • This method is often used to introduce or modify dopants in photocatalytic materials to enhance their properties
Combustion synthesis	• Combustion synthesis relies on a rapid exothermic reaction between fuel and oxidizer in the presence of metal precursors • This method can produce highly crystalline photocatalysts with fine particle sizes
Microwave-assisted synthesis	• Microwave irradiation can be used to accelerate chemical reactions, including the synthesis of photocatalysts • This method can reduce reaction times and improve efficiency
Template-assisted synthesis	• Templates, such as mesoporous materials or biological molecules, can be used to guide the synthesis of photocatalysts, resulting in materials with specific structures and morphologies

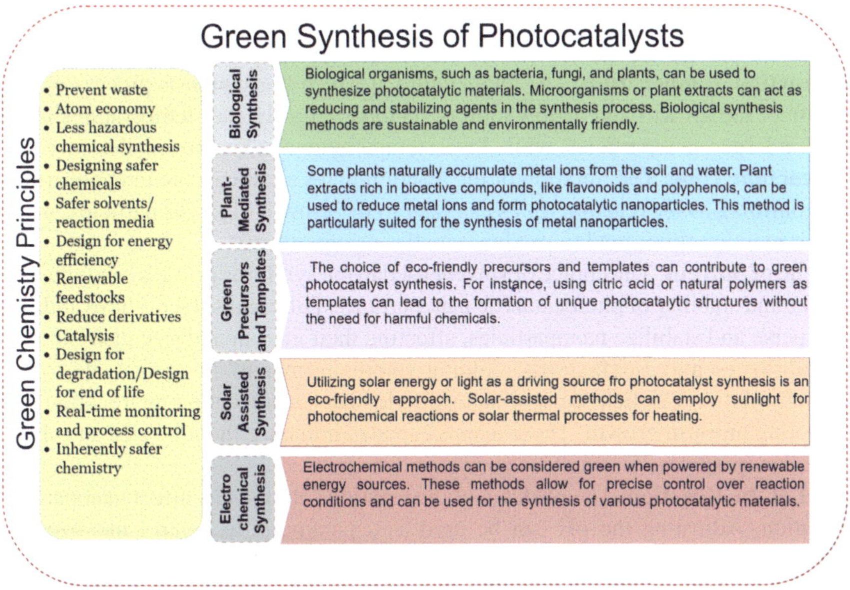

Fig. 5 Green photocatalyst synthesis applying green chemistry principles

Green synthesis methods should also consider the recyclability and reusability of materials and resources, minimizing waste and environmental impact. Green Synthesis methods of photocatalysts are not only sustainable but also often lead to materials with desirable properties of environmental and energy-related applications. By adopting these eco-friendly approaches, researchers and industries can contribute to reducing the environmental footprint of photocatalysts production while promoting sustainable technologies for a cleaner and greener future.

The choice of synthesis method depends on the desired properties, application, and scalability of the photocatalytic material. Researchers select the most appropriate method to tailor the photocatalyst's structure and composition to meet specific requirements, whether it be for environmental remediation, energy conversion, or other photocatalysis-related applications.

3.2 *Parameters Affecting the Size of the Photocatalyst*

The size of a photocatalyst, especially when dealing with nanoparticles, is a crucial factor that can significantly influence its properties and performance in various applications. Several parameters can affect the size of photocatalytic nanoparticles during their synthesis. These parameters are briefed as follows.

Precursor Concentration: The concentration of the chemical precursors used in the synthesis process can impact nanoparticle size. Typically, higher precursor concentrations lead to larger nanoparticles, while lower concentrations tend to produce smaller ones.

Reaction Temperature: The temperature at which the synthesis process occurs can affect the growth and aggregation of nanoparticles. Higher temperatures often result in larger nanoparticles, while lower temperature can lead to smaller ones.

Reaction Time: The duration of the reaction plays a critical role in nanoparticle size control. Longer reaction times can lead to the growth of larger particles, while shorter reaction times tend to produce smaller nanoparticles.

Type of Solvent: The choice of solvent used in the synthesis process can influence the size and stability of photocatalytic nanoparticles. Different solvents have abilities to disperse and stabilize nanoparticles, affecting their growth and size distribution.

Stabilizers and Surfactants: Adding stabilizing agents or surfactants to the reaction mixture can help control nanoparticle size by preventing aggregation and promoting uniform growth. These agents can also influence the final size and shape of the nanoparticles.

pH of the Reaction: The pH level of the reaction solution can affect nanoparticle formation. Adjusting the pH can be used as a parameter to control the size and stability of photocatalytic nanoparticles.

Precursor Ratio: The ratio of different precursors used in the synthesis process can impact the size and composition of nanoparticles. Varying the molar ratios of precursors can lead to changes in size, composition, and properties.

Type of Reducing Agent: In some synthesis methods, a reducing agent is used to control the growth of nanoparticles. The choice of reducing agent and its concentration can influence nanoparticle size.

Seed-Mediated Growth: Seed-mediated growth involves the use of smaller nanoparticles as seeds to control the growth of larger ones. The size and concentration of the seed nanoparticles can be adjusted to regulate the final nanoparticle size.

Micro-reactors and Flow Chemistry: Using micro-reactors and flow chemistry techniques allows for precise control over reaction conditions, including residence time and temperature gradients, which can influence nanoparticle size.

Ultrasonic or Microwave Irradiation: The application of ultrasonic or microwave irradiation during the synthesis process can lead to rapid nucleation and growth, affecting the final nanoparticle size.

Agitation and Stirring: The degree of agitation or stirring during the reaction can impact nanoparticle size by affecting mass transport and collision rates among reacting species.

Seeding Density: In some cases, the initial density of seed particles in the reaction mixture can affect the final nanoparticle size when using seeded growth methods.

It is important to note that the choice of synthesis method also plays a significant role in nanoparticle size control. Different methods, such as sol-gel, chemical reduction, hydrothermal, and sono-chemical methods, offer varying degrees of control over nanoparticle size. Researcher carefully tune these parameters to achieve the desired size and properties for specific applications, as the size of photocatalytic nanoparticles can greatly influence their catalytic activity, light absorption, and stability.

3.3 Size Dependent Photocatalyst and Their Properties

PCs are materials that can initiate chemical reactions when exposed to light and their properties can significantly depend on their size at the nanoscale. As the size of photocatalytic nanoparticles decreases, various properties change, leading to unique characteristics and improved performance in many applications. Here are some size-dependent properties and applications of photocatalytic nanoparticles.

Increased Surface Area: Nanoscale photocatalysts have a higher surface area per unit mass compared to bulk materials. This increased surface area provides more active sites for catalytic reactions, enhancing their efficiency.

Improved Light Absorption: Smaller nanoparticles typically have a higher absorption capacity for light due to quantum size effects. This means they can absorb a broader spectrum of light, including visible and ultraviolet (UV) light, making them more versatile in various light conditions.

Enhanced Photocatalytic Activity: Size-dependent properties, such as a higher surface-to-volume ratio and improved light absorption, lead to increased photocatalytic activity. This is crucial for applications like water purification, air purification, and pollutant degradation.

Faster Reaction Rates: Smaller photocatalytic nanoparticles can facilitate faster reaction rates because they provide shorter diffusion pathways for reactants and products. This results in quicker turnover times for catalytic reactions.

Tunable Band Gap: The band gap of a semiconductor photocatalyst is size-dependent, allowing for tunable electronic properties. Smaller nanoparticles tend to have wider band gaps, making them suitable for different types of photocatalytic reactions.

Higher Charge Carrier Mobility: Nanoscale photocatalysts often exhibit improved charge carrier mobility, allowing electrons and holes generated during the photocatalytic process to move more freely. This property contributes to efficient charge separation and reduced recombination, enhancing overall photocatalytic efficiency.

Selective Catalysis: Size-dependent properties can influence the selectivity of photocatalytic reactions. Tailoring the size of nanoparticles can lead to selectivity towards specific reactants or products, which is valuable in applications like selective pollutant degradation or selective hydrogen production.

Size-Dependent Stability: Smaller nanoparticles may have different stability profiles compared to their bulk counterparts. Understanding and controlling the stability of nanoscale photocatalysts is essential for long-term performance and practical applications.

Photoluminescence Emission: Nanoscale photocatalysts can exhibit size-dependent photoluminescence emission. This property is used in applications such as sensors and bio-imaging, where the emission intensity or wavelength can provide information about the local environment.

Versatile Applications: Nanoscale photocatalysts find applications in a wide range of fields, including water and air purification, energy conversion (such as hydrogen production and solar cells), environmental remediation, and medical diagnostics.

Reduced Agglomeration: Smaller nanoparticles are less prone to agglomeration, which can improve their stability and maintain their active surface area during catalytic reactions.

It is important to note that specific size-dependent properties of photocatalytic nanoparticles can vary depending on the material used and the synthesis method employed. Researchers continue to explore and fine-tune these properties to optimize the performance of nanoscale photocatalysts for various applications, making them a critical area of study in the fields of nanotechnology and materials science.

4 Energy Storage/Conversion Application of Photocatalyst

Photocatalysts, materials that can harness light energy to drive chemical reactions, have gained significant attention in recent years for their potential applications in energy storage and conversion. These materials are particularly promising because they can facilitate environmentally friendly processes by using solar energy to drive chemical transformations. By harnessing the power of sunlight, photocatalysis offers a green and sustainable path to address our energy needs. Some key energy storage and conversion applications of photocatalysts are discussed here.

4.1 Hydrogen Production

Exponential population growth intensifies energy demands, resource depletion, and environmental harm. This feedback loop raises global temperatures, amplifying the carbon footprint. Renewable energy-driven water splitting for clean hydrogen production offers a sustainable solution, curbing energy demands and carbon emissions [93]. Photocatalysts can play a crucial role in the generation of hydrogen gas through water splitting. When exposed to sunlight, certain photocatalysts can absorb photons and use their energy to split water molecules into hydrogen and oxygen. This hydrogen can be stored and used as a clean fuel source, either directly or in fuel cells, for electricity generation and various industrial applications. Many H_2 photogeneration cycles focus on byproducts rather than H–H bond formation. Protons and electrons for H_2 often come from high energy substrates. H_2 production in some systems relies on sacrificial donors, using redox shuttles to facilitate the process.

Efficient photocatalyst design can reduce wasteful recombination reactions. To scale up solar H_2 production, new strategies are needed, targeting low electrochemical over potentials and harnessing sunlight for both oxidative and reductive reactions. H_2O and HX (where X = Cl) are potential sources, but both present challenges, especially H_2O splitting, which requires complex multielectron processes [29]. Platinum nanoparticles prioritize hydrogen over methane production, with terrace sites favoring methane and low-coordinated sites favoring hydrogen [26]. Semiconductor photocatalysis for hydrogen production from sunlight is a promising energy conversion method. Sulphide semiconductor photocatalysts are particularly appealing due to their strong response to the solar spectrum and high catalytic activity [121]. Enhancing H_2 production efficiency involves surface decoration of TiO_2 nanoparticles with metal/non-metal element, which modifies its electronic properties. Metal dopants such as copper (Cu), gold (Au), and platinum (Pt) play a critical role in separating electron-hole pairs during UV/Visible/solar light irradiation [52]. The S-scheme heterojunction photocatalyst with a nanosheets-on-nanosheets structure provide efficient hydrogen production from water splitting under visible light. $ZnIn_2S_4$ nanosheets are grown on $Bi_4Ti_3O_{12}$ nanosheets using a low-temperature water bath method. The optimal $Bi_4Ti_3O_{12}/ZnIn_2S_4$ (BTO/ZIS) proportion demonstrates remarkable hydrogen production efficiency, attributed to the sheet-sheet heterostructure providing more active sites and promoting efficient electron-hole pair transfer through the S-scheme heterojunction between BTO and ZIS [106]. Six β-keto-enamine-based COFs were prepared and studied for structure-property relationships. Backbone impacts band gap, absorption, and carrier separation, while substituents affect band positions, surface area, and hydrophobicity. TpPa-Cl2 demonstrates the best hydrogen production performance, with an apparent quantum efficiency of 17% at 400 nm. This offers insights into designing efficient COF-photocatalysts [115].

Research on carbon-based nanomaterials (CNMs) has largely focused on classification and synthesis, with limited attention to their role in photocatalytic hydrogen production. CNMs possess excellent properties for enhancing hydrogen evolution. Carbon quantum dot (CQD)-sensitized titanium dioxide (TiO_2) demonstrates impressive hydrogen generation activity. Optimized catalysts, like 0.4 CQD/CdS, significantly boost hydrogen production rates, addressing challenges in energy storage and cost. Despite obstacles, CNMs hold promise for sustainable energy solutions amid growing demand and environmental concerns [2]. Utilizing solar energy for clean hydrogen production and biomass refinement offers low-carbon solutions, using biomass as hole scavengers [94]. Catalytic CO_2 conversion into valuable chemicals and fuels via thermal, electrocatalytic, and photo(electro)catalytic methods holds promise for reducing carbon emissions and achieving carbon neutrality.

Recent advancements in heterogenous catalysis for various conversion routes are summarized, highlighting the role of advanced characterizations and density functional theory calculations in understanding catalytic materials (D. [33]. Graphitic carbon nitride (GCN or $g\text{-}C_3N_4$), a metal-free, 2D compound, holds promise for efficient hydrogen production but faces challenges like defects, surface area, and stability [93]. Graphitic carbon nitride ($g\text{-}C_3N_4$), with its low bandgap (2.7 eV), holds potential for H_2 production using solar or simulated light [72]. A cost-effective method is presented to create lightweight porous carbon sponges using melamine-formaldehyde (MF) foams as a template and raw material. Additionally, when coated with the $g\text{-}C_3N_4$ semiconductor, they serve as an efficient platform for photocatalytic hydrogen production [47].

4.2 Carbon Dioxide Reduction

Photocatalysts can help mitigate the effects of climate change by converting carbon dioxide (CO_2) into valuable fuels and chemicals. By utilizing sunlight, photocatalysts can catalyze the reduction of CO_2 into hydrocarbons like methane or carbon monoxide, which can be used as feedstock for the chemical industry or as an energy carrier. Nickel (Ni)-based TiO_2 photocatalysts immobilized on quartz plates were used in a unique reactor setup to enhance the conversion of CO_2 into fuels under visible light [76]. Noble metal nanoparticles, particularly gold (Au), play crucial roles in photocatalysis, such as Surface Plasmon Resonance (SPR) and Z-scheme bridges. A study investigates the size-dependent effects of Au in the Au/g–C_3N_4/BiOBr system for photocatalytic CO_2 reduction and dye degradation. Large Au nanoparticles primarily act as Z-scheme bridges, enhancing photocatalytic activity under 380 nm light. Small Au nanoparticles mainly exhibit the SPR effect, improving activity under 550 nm light [9]. Gold nanoparticles (Au NPs) of various sizes and Au@SiO_2 core-shell NPs were synthesized and loaded onto Pt/TiO_2 for CO_2 conversion. Under dual LED irradiation, enhanced catalytic activity resulted from localized surface plasmon resonance (LSPR) effects, with an optimal Au NP size of 18 nm. Pt/TiO_2/Au@SiO_2 systems exhibited significantly higher catalytic activity, particularly Pt/TiO_2Au18@SiO_2, which was 3.1 times more active than bare Pt/TiO_2 [11].

Platinum nanoparticles (Pt NPs) significantly influence photocatalytic CO_2 reduction. Smaller Pt NPs enhance charge transfer efficiency, boosting CO_2 reduction and hydrogen evolution [26]. Au/g–C_3N_4 (Au/CN) nanocomposites were created, with g-C3N4 nanosheets as a substrate for different-sized Au nanoparticles (Au NPs) using a constant temperature bath-reduction method. Photoreduction of CO_2–CO and CH_4 under UV light, with yields of 77.5 and 38.5 μmol/g, respectively [57]. Dendritic porous silica nanospheres (DPSNs) were used as carriers to create efficient DPSNs@X% TiO_{2-X} (X%: weight ratio of TiO_2/DSPNs) composite photocatalysts for CO_2 reduction. Small TiO_{2-X} nanoparticles produced only CO, while larger NPs generated both CO and CH_4. DPSNs@80% TiO_{2-X} exhibited exceptional performance with a high CH_4 production rate (124.3 μmol g-$TiO_{2-x}{}^{-1}$ h^{-1}), moderate CO production rate (14.7 μmol g-$TiO_{2-x}{}^{-1}$ h^{-1}), and excellent stability [112].

PbS QDs were used to sensitize TiO_2 catalysts, enabling CO_2 photoreduction across a wide spectrum range (420–610 nm). PbS QDs enhanced photoreduction rates by a factor of approximately 5 under broad band illumination (UV-NIR) [105]. Halide perovskite quantum dots (QDs), typically used in LEDs and solar cells, have found a new role in nonaqueous media. $CsPbBr_3$ QDs under simulated sunlight, catalyzes CO_2 reduction at 23.7 μmol/g h with over 99.3% selectivity [111]. Quantum dots (QDs), semiconductor nanocrystals below 20 nm, offer advantages like high surface area, short charge migration, size-dependent energy bands, and unique optical properties, are emerging in CO_2 reduction applications [79].

4.3 Solar Water Purification

Photocatalysts are employed in solar-driven water purification systems. These materials can break down organic pollutants and dis-infect water when exposed to sunlight, making it a sustainable and energy-efficient way to provide clean drinking water in remote or resource-scarce regions.

Titanium dioxide-based photocatalysis for organic contaminant removal using solar radiation. It shows promising results in both lab and field tests, effectively reducing contaminants like γ-lindane, offering an efficient water purification method [102]. Addressing global water scarcity and energy crises, solar-driven water purification using 2D nanomaterials offers a sustainable solution. These materials exhibit efficient photothermal evaporation, antimicrobial treatment, and photocatalytic degradation, driven by abundant solar energy. Strategies to enhance their performance are explored, highlighting the potential for green energy-driven water purification [110]. The vital intersection of clean water and renewable energy through solar-driven photocatalytic processes, highlights TiO_2-based heterojunction composites as a solution for efficient water purification and hydrogen generation, addressing challenges like broad-spectrum light absorption and charge carrier recombination [84]. Solar-driven photocatalysis holds promise for sustainable water purification, but immobilizing nano-sized photocatalysts is a challenge. This study immobilized graphitic carbon nitride on chitosan to create g-C_3N_4/chitosan hydrogel beads (GCHBs) and assessed their effectiveness in degrading pollutants. The 3D-printed compound parabolic collector (CPC) reactor demonstrated comparable performance to slurry reactors, offering potential for small-scale water treatment in various settings, including rural and disaster-affected areas [122].

Research has focused on harnessing solar radiation for simultaneous water evaporation and photocatalysis to address freshwater scarcity and energy needs. Advances in photothermal materials offer promising applications in water purification, energy production, and sterilization, but challenges persist, signaling opportunities for integrated solar-driven systems [66]. Photocatalysts like titanium dioxide nanowires (TiO_2NWs) and TiO_2NWs with carbon nanotubes (TiO_2NWs/CNTs), when exposed to sunlight, efficiently purify water by intercepting pathogens, generating Reactive Oxygen Species (ROS), and reducing chemical contaminants, offering a sustainable solution for clean water in electricity-poor regions [42]. The study focuses on developing magnetically recyclable, visible-light-active Mn-doped cobalt ferrite photocatalysts for efficient water treatment. These materials showed high efficiency in degrading contaminants like crystal violet dye, offering a cost-effective and sustainable approach to water purification [87].

4.4 Solar Cells

Photocatalysts used in the development of next-generation solar cells, particularly dye-sensitized solar cells (DSSCs) and perovskite solar cells. In DSSCs, a photocatalysts is used to generate electrons when it absorbs photons, contributing to the conversion of sunlight into electricity. Similarly, in perovskite solar cells, certain photocatalysts can enhance the efficiency of light absorption and charge separation, leading to more efficient energy conversion. ZnO NPs were used in the photovoltaic devices due to its wide bandgap, high electron mobility, and growth control. It focuses on low-temperature solution growth methods for preparing ZnO nanostructures with controlled properties [81].

Combining $CsPbBr_3$ QDs with graphene oxide ($CsPbBr_3$ QD/GO) boosts electron consumption by 25.5%, showing promise for halide perovskite QDs in photocatalytic applications [111]. MoSO and MoSSe Janus monolayers exhibit favorable optoelectronic properties with bandgap of 1.61 eV (indirect) and 2.00 eV (direct), ideal or efficient light absorption. Calculations reveal strong device absorption efficiency (90%) across the infrared to ultraviolet spectrum, making them excellent candidates for tandem solar cell designs [103]. A method to produce g-C3N4-reinforced CdS-modified anatase $TiO_2$is proposed, broadening TiO_2's visible spectrum absorption and enhancing charge transfer with CdS spheres and g-C_3N_4. This ternary photocatalyst exhibits improved hydrogen production (233 $\mu molg^{-1}\ h^{-1}$) compared to other semiconductors. DSSCs using these electrodes achieve a high overall conversion efficiency of 4.16%, an open-circuit voltage of 0.75 V, and a short circuit current density of 10.17 mA/cm^2. Matched bandgap heterojunction catalysts enhance photocatalysis and solar cell performance [114].

4.5 Solar Fuel Cells

Photocatalysts can also be integrated into solar fuel cells, which combine the principles of solar cells and electrolysis cells. They can directly convert sunlight into electricity and use that electricity to split water into hydrogen and oxygen, a process known as photoelectrochemical water splitting. This hybrid approach holds potential for efficient energy storage and conversion. $CsPbBr_3$ QDs are employed as photocatalysts to convert CO_2 into solar fuels [111]. Udayabhaskar et al. [99] studied the influence of graphene on the crystallinity, morphology, band gap energy and luminescence of the synthesized samples of pure ceria (nanospheres) and ceria: graphene nanocomposites (nanorice like morphology). They observed increase in the band gap (blue shift) of nanoceria and proposed the feasibility of this material as an active component in the solid oxide fuel cell.

A promising Fe_2O_3/NiO heterostructure photocathode material enhances methanol fuel cell efficiency through photogenerated electron accumulation and improved oxygen reduction reaction. Notably improved performance under simulated sunlight suggests a new path for solar-chemical energy conversion in fuel cells [119].

A study investigates an innovative approach for electricity production and storage using a photocatalytic fuel cell and a supercapacitor. The setup involves adding a third electrode with activated carbon to create a supercapacitor electrode, which works alongside the photocatalytic fuel cell. The fuel cell utilizes a CdS-sensitized mesoporous TiO_2 photoanode and an air cathode with nano particulate carbon as an oxygen reduction electro-catalyst [4].

Efficient hydrogen production from sunlight and water is crucial for carbon neutrality. Traditional methods, like photoelectrochemical water splitting, often use corrosive electrolytes. Studies show a high solar-to-hydrogen (STH) efficiency of 9.2% using pure water and concentrated solar light. This approach overcomes the efficiency challenge in solar hydrogen production [123]. SiF/Ni-NQGDs achieve record-high hydrogen productivity and vanillin yield under simulated sunlight, without additives. They can be easily recycled without performance degradation. It offers insights into harnessing solar energy for electro-synthesis and biomass refinement, addressing energy-intensive and low-efficiency challenges [20].

A hierarchical composite, Co_9S_8/N, S-CNTs-$ZnIn_2S_4$ (CSCNs-ZIS), was created by grafting $ZnIn_2S_4$ nanosheets onto Co_9S_8/N, S-doped carbon nanotubes derived from ZIF-67, featuring a core-shell structure with $ZnIn_2S_4$ nanosheets uniformly enveloping CSCNs, effectively reducing charge-carrier recombination. The optimized CSCNs-ZIS composite achieved a high H_2production rate of 2409.2 μmol h^{-1} g^{-1} and maintained stability over three cycles [116]. Exploring M_2CO_2/MoX_2 (M = Hf, Zr; X = S, Se, Te) for clean energy conversion shows that their properties including type-II band structures for enhanced photocatalysis and suitable band gaps for optical harvesting, make them promising for solar cell applications and addressing energy and environmental challenges [107].

4.6 Energy Storage in Batteries

Some photocatalysts have been explored for their potential use in advanced energy storage systems, such as photoelectrochemical batteries. These batteries use light absorbing materials to store solar energy as chemical energy, which can be released later when needed.

A solar-driven rechargeable lithium-sulfur (Li–S) battery has been developed, capturing and storing solar energy by oxidizing S^{2-} ions to polysulfide ions using a Pt-modified CdS photocatalyst. The battery exhibits a specific capacity of 792 $mAhg^{-1}$ over a 2 h photo charging process with a discharge potential of around 2.53 V versus Li^+/Li. Within just 10 min, it reaches a specific capacity of 199 $mAhg^{-1}$, comparable to conventional lithium-ion batteries, and can even charge under natural sunlight [55]. A rhenium coordination complex confined within a metal-organic framework (MOF) exhibits unique electron accumulating properties under visible-light irradiation. It can store about 15 Cg_{MOF}^{-1} of electric charges for over four weeks without loss. The stored energy can be discharged for electrochemical reactions and H_2 evolution catalysis on demand, and the system can be recharged through light-driven processes for more than 10 cycles with 90% of the initial capacity retained [97].

5 Environmental Application of Photocatalyst

Beyond energy applications, photocatalysts are used in environmental remediation processes. They can break down pollutants and toxins in the air and water when exposed to light, contributing to cleaner and healthier environments.

Photocatalysts, materials that can initiate chemical reactions when exposed to light, have found numerous applications in the field of environmental remediation. These applications harness the photocatalytic properties of these materials to address various environmental issues, improve air and water quality, and mitigate pollution. Addressing freshwater affordability, photothermal materials harness solar energy for eco-friendly purification [118]. Some important environmental applications of photocatalysts are discussed here.

5.1 Air Purification

Photocatalysts are used in air purification systems to remove pollutants such as volatile organic compounds (VOCs) and nitrogen oxides (NO_x). When exposed to ultraviolet (UV) or visible light, photocatalysts like titanium dioxide (TiO_2) can catalyze the degradation of these harmful compounds into less toxic substances, contributing to cleaner air in indoor and outdoor environments.

The growing importance of photocatalysts in air pollution treatment, traces its evolution from TiO_2 photocatalysis to current trends and models for scaling up [82]. Air quality in urban areas is a growing concern, with stricter EU directives and increased diesel vehicle traffic. The potential of photocatalytic concrete products to reduce nitrogen oxides (NO_x) by converting them to nitrates has been studied using UV-A induced degradation test setup, following ISO standards [44]. The effectiveness of two photocatalytic materials in a closed-loop reactor for treating gaseous pollutants were compared. The reactor, accommodating lamps and TiO_2-coated materials, allowed air flow rates from 16 to 1800 m^3/h. A thin, non-woven tissue coated with TiO_2 P-25 and SiO_2 binder and an aluminium honeycomb coated with TiO_2 P-25 were tested. U-shaped lamps emitted at 254 nm with 18 or 35 W power. The honeycomb material showed at least four times higher methanol removal rates in the recirculation regime compared to the folded tissue, attributed to differences in shape, TiO_2 support, photons scattering, and reactant accessibility [98].

Optimal results involve incorporating TiO_2 into concrete pavement's top layer. Combined with cement products, it allows rain to wash away reaction products.

Applications in Antwerp and other sites aim to assess real-life effectiveness. Long-term air-cleaning durability in concrete roads remain a concern. Several trial applications, including field sites in the Leopold II tunnel in Brussels and new pavements in industrial zones in Wijnegem and Lier, Belgium, have been initiated to assess real-life behavior [14]. Photocatalysis practical use in building materials is limited due to low solar energy conversion efficiency and the need for higher pollutant concentrations. Exploring PCO's removal mechanisms for pollutants, photocatalysts options, and reactor designs, it aims to provide comprehensive insights to advance gas-phase photo degradation for commercial air purification applications [16].

Metal organic frameworks (MOFs) with photocatalytic bacterial properties, particularly zinc-imidazolate MOF(ZIF-8), demonstrates >99.9999% inactivation of Escherichia coli (E.coli) within 2 h of simulated solar irradiation [56]. In recent decades, numerous viral outbreaks causing epidemics and pandemics with significant human casualties have occurred. To mitigate the virus's spread, effective personal protective equipment (PPE) and disinfected surfaces are essential. Introducing a TiO_2 nanowires (TiO_2 NWs) based filter designed for PPE, air conditioners, and air purifiers, its efficiency relies on the photocatalytic generation of reactive oxygen species (ROS) when exposed to UV light and the high dielectric constant of TiO_2, promoting better wettability by water droplets carrying germs [43].

Visible-light-active TiO_2–Cu^{2+}@perlite and Ag@TiO_2–Cu^{2+}@perlite materials were developed and characterized showing small size and substrate adherence. A photocatalytic air purifier prototype using low-cost-high-energy LEDs was built and evaluated in dentistry clinics. Ag@ TiO_2–Cu^{2+}@perlite exhibited better disinfection activity, inhibiting up to 99% of bacterial growth, attributed to enhanced better photocatalytic and synergistic effects of TiO_2, Ag, and Cu [68]. Recent research addresses urban air pollution by developing environmentally friendly civil engineering materials with ZnO nanostructures for photocatalytic air purification [53].

P25 TiO_2, a commercial exterior paint, and an air-purifying curtain fabric, required accelerated weathering and UV-preconditioning, respectively, before showing significant activity. These films exhibited decreasing NO_x removal activity and photocatalytic activity for NO removal with increasing irradiation time [38]. A 3D-printed reactor purifies air, removing NO_2, NO, acetaldehyde, and ozone more effectively than self-cleaning glass [39]. Photocatalytic oxidation (PCO), inspired by nature, effectively purifies indoor air by degrading VOCs, but faces challenges [40, 45].

5.2 *Self-cleaning Surfaces*

Photocatalytic coatings are applied to surfaces like glass, concrete, and building facades to create self-cleaning properties. These coatings break down organic contaminants and dirt when exposed to sunlight or artificial light, reducing the need for manual cleaning and maintenance.

Self-cleaning surfaces draw inspiration from nine biological objects: lotus leaves, rice leaves, cicada wings, butterfly wings, snail shell, fish scale, shark skin, pitcher plant, and photosynthesis, offering unique properties like low adhesion, anti-fouling, and photocatalysis [73]. Dielectric polymer materials are for self-powered nano systems and electronic skin but can degrade due to surface contamination. To address this, a triboelectric nanogenerator (TENG) with TiO_2 photocatalysis is developed to generate energy and self-clean. The TENG uses a polydimethylsiloxane (PDMS) friction layer with varying pore sizes. Contamination reduces the open circuit voltage from 125 to 88 V. After 35 min of solar illumination, the self-cleaning interface degrades contaminants, increasing to 115 V, fully restoring 92% of the TENG's performance. It enhances applications in e-skin and wearable devices [63]. Paris Agreement boosted photovoltaic power, desert dust issues, self-cleaning coatings vital [86].

The membrane, based on polyvinyl alcohol cross-linked tannic acid (PVA-TA) and doped with magnetic titanium dioxide ($MTiO_2$) and Mxene, demonstrates excellent separation efficiency and durability. Its bionic moth-eye structure, along with sunlight induced photocatalytic self-cleaning, offers innovative oil-water separation materials [60]. Mg-doped anatase TiO_2 was synthesized using a template-free and surfactant-free solvothermal method and the resulted ceramic glaze exhibited remarkable supehydrophilicity, with a water contact angle of approximately 7°, and demonstrated excellent resistance to stains [17]. Nano-TiO_2 combined with recycled concrete powder produces urban self-cleaning surfaces [61]. Photocatalytic self-cleaning in industrial and experimental materials were assessed using ISO 27448:2009 by a method involving UV activation, oleic acid fouling, and UV light irradiation [83].

A one-step, eco-friendly method was employed for large-scale super hydrophobic surface creation using silica and TiO_2 nanoparticles, along with fluorinated ethylene propylene, were combined and applied to aluminium. The surface displayed excellent performance with nano silica (12 wt%) and TiO_2 (20 wt%), remaining stable in various conditions during fouling for 10 h [5]. Nanomaterials in textiles offer new opportunities and challenges. Review covers self-cleaning and conductive textile's progress, applications, limitations, and future directions [78]. The technology of photocatalytic self-cleaning for fabrics, using ZnO or TiO_2 materials, decomposes organic stains into water and carbon dioxide (CO_2) when exposed to UV light. The self-cleaning principle is beneficial in various applications, including everyday textiles, medical textiles, athletic wear, military uniforms, and outdoor fabrics. This technology helps conserve water, improve air quality, and reduce long-term energy, laundry costs, and time [24].

5.3 Water Treatment

Photocatalysts are used for water purification and wastewater treatment. When suspended in water or coated on surfaces like water filters, they can degrade organic pollutants, kill bacteria and viruses, and remove harmful chemicals, ensuring the supply of safe drinking water and reducing the environmental impact of industrial wastewater.

Urbanization and technological progress led to water-intensive industries facing complex wastewater challenges. Conventional methods struggle; nano-photocatalysis emerges as a promising solution. This review explores recent advancements and future potential in wastewater treatment using nano-photocatalysts, emphasizing the need for standardized evaluation and practical application. Researchers should focus on cost-effective, high-efficiency solutions [71].

The size dependent roles of Au/g–C_3N_4/BiOBr system are attributed to the SPR effect of small Au nanoparticles and the electron transfer center function of large Au nanoparticles. This insight can inform the development of highly active noble metal-loaded photocatalysts [9]. Textile wastewater, laden with hazardous dyes, poses significant environmental pollution. Advanced oxidation processes, particularly photocatalysis using ZnO nanoparticles, offer promising solutions for efficient dye degradation. Several works focus on improving ZnO NP synthesis and optimizing photocatalysis conditions for effective textile wastewater treatment [25]. Preserving photocatalyst-performance by immobilization on substrates is crucial. Common methods are time and cost-inefficient. Using direct heating, ZnO nanorods grew on kanthal wires rapidly (15 min) with low energy consumption [8]. N–TiO_2/rGO, a novel photocatalyst, effectively removes hazardous methylene blue (MB) and Cr(VI) can be carried out simultaneously and results in a removal efficiency of 95.96% [100].

Photocatalytic mixed matrix membranes (MMMs) baked on MIL-101(Cr)-modified h-BN and TiO_2 in polyvinylidene fluoride (PVDF) effectively degrade dyes and remove pollutants. The composite photocatalysts exhibit stability and combat photo-corrosion, achieving impressive COD and TSS removal rates. The h-BN@MIL-101(Cr)/PVDF MMM holds promise for future commercial applications [96]. Heterogeneous photocatalysis with Tin oxide (SnO_2) offers sustainable wastewater treatment due to its low cost, stability, and effectiveness [90]. Nanoparticles are typically produced through chemical or physical means, which can be problematic due to toxic materials, high costs, and extreme conditions. As a result, environmentally friendly green methods, like using Grewia asiatica extract to create zinc oxide nanoparticles, have gained traction. These nanoparticles, characterized using various techniques, were spherical with sizes between 28 and 42 nm. They efficiently degraded methylene blue, demonstrating their potential as eco-friendly, cost-effective catalysts for dye removal [7].

Polluted water resources demand eco-friendly solutions. Moringa seed-derived silver nanoparticles combat microbes and purify water efficiently [70]. Addressing global water contamination, nanohybrids (NHs) and bionanohybrids (BNHs) play a crucial role in removing pollutants, including organics and heavy metals, with multifunctional applications in various fields. Researchers have successfully utilized BNHs such as Cellulose, Chitosan, Gelatin, Pectin, and Silk protein for eco-friendly pollutant removal in water environments [28]. CeO_2 was doped with Ni, Cu, and Fe for photocatalytic 2-chlorophenol removal under 70 W indoor LED light [85]. The need for clean drinking water, coupled with environmental concerns, drives the development of low-cost, non-toxic nanotechnology for wastewater purification. Overall review explores recent advancements and challenges in nanomaterial synthesis, emphasizing sustainability and highlighting nano-photocatalysts in water decontamination applications, offering promising prospects in nanotechnology research [6].

5.4 Remediation of Contaminated Soils

Photocatalysts can assist in the remediation of contaminated soils by breaking down organic contaminants and pollutants. They can be mixed with soil or applied as a surface treatment to accelerate the degradation of harmful substances and restore the soil's health. Heterogenous photocatalytic degradation of the pesticide Diuron in soil under solar light were investigated. The photocatalytic treatment with TiO_2 and solar light efficiently degraded Diuron in the top 4 cm of contaminated soil. Degradation rates were dependent on the irradiation intensity, with higher intensity leading to faster degradation [41]. Photocatalytic oxidation (PCO) with nano-sized titanium dioxide (TiO_2) effectively degrades diphenylarsinic acid (DPAA) in contaminated soil. The study identified arsenate as the final degradation product.

An orthogonal array experimental design optimized operational parameters (soil:water ratio, TiO_2 dosage, irradiation time, and light intensity) with a significant effect of soil:water ratio on DPAA removal efficiency. The method was validated in various soils, with removal efficiencies ranging from 57.0 to 78.6% [104]. Crude oil pollution harms soil, water, air, bioremediation, especially adsorption and photocatalysis, helps remove pollutants [75]. A study investigated soil samples from agricultural lands contaminated with 18 different pesticides, classified into organochlorines (OCP), organophosphorus (OPP), carbamates (Carb), and pyrethroids (Pyrth). The five most abundant pesticides (Atrazine, Chlorpyrifos methyl, Dimethoate, Heptachlor, and Methomyl) were subjected to photocatalytic degradation under UV irradiation with and without 1% TiO_2 or ZnO photocatalysts [27].

5.5 NO_x Removal from Vehicle Emissions

Photocatalytic materials are incorporated into road surfaces, sound barriers, and tunnels to reduce nitrogen oxide (NO_x) emissions from vehicles. These materials can help convert NO_x into less harmful compounds, contributing to improved air quality in urban areas. A fluidized bed of ultrafine TiO_2 particles, with diameters of 7, 20, and 200 nm, is used to treat NO_x via photocatalytic oxidation. The 7 nm TiO_2 bed exhibits high NO_x removal efficiency due to its large surface area. NO_x removal is directly proportional to the specific surface area. In fluidization, agglomerates of 7 and 20 nm particles remain intact due to strong particle adhesion, leading to increasing bed height with gas velocity for these systems. However, the bed expansion levels off for the 200 nm particle system. Additionally, the entrainment rate is smaller for the 7 and 20 nm particle systems compared to the 200 nm system [69]. Photocatalysis promises to convert air pollutants, improve indoor and outdoor air quality, and combat climate change [23].

The addition of TiO_2 to cement mortars creates photocatalytic materials capable of capturing and breaking down atmospheric pollutants like NO_x. A specially designed test apparatus evaluated the mortar's performance under 27 different conditions, varying air humidity (30, 50, 70%), UV-A radiation (10, 25, 40 W/m^2), and pollutant flow rate (1, 3, 5 l/min), all with an initial NO concentration of 20 ppmv. Results showed that higher UV-A levels improved NO_x degradation, while increased humidity and flow rate reduced photocatalytic activity [22]. Findings demonstrate that cement materials can be customized to reduce NO_x levels through photocatalysis and that Portland cements inherently possess the ability to bind NO_x, especially NO_2 [54]. Reducing platinum-group metal (PGM) particle size to single atoms enhances catalyst efficiency, notably for Pd/TiO_2 photocatalysis. Isolated Pd atoms, identified via spectroscopy, exhibit superior NO_x removal capabilities due to resistance to nitrate poisoning and high selectivity [32].

A study suggests that UV/chlorine in freshwater outperforms seawater due to radical scavenging effects. Reactive radicals, including ·OH, ·Cl, ·OCL, and $ClOH^{\cdot -}$, contribute to NO_x removal, ˙OH confirmed through experiments [113]. Urban ground-level NO_x poses health risks, but strategies like traffic restrictions, catalytic converters, and photocatalytic coatings on surfaces mitigate NO_x in cities. A study compares the environmental profiles of NO_x removal by Three-Way Catalyst (TWC) car converters and photocatalytic surface coatings (for pavements). Using Life Cycle Assessment (LCA, EF3 impact method), it evaluated the environmental performance for removing 1 kg of NOx with these technologies [13]. Instead of using free nanoparticles, recent work focuses on the direct growth of ZnO nanostructures on paving materials like tiles and bitumen roads using a cost-effective hydrothermal method. The materials demonstrated the ability to remove various pollutants from car exhaust (O_3, CO_x, NO_x, VOCs), evaluated for durability and the influence of weather conditions on pollutant degradation. Studies also suggest that sunlight-exposed bitumen roads contribute to pollutant emissions and affect the Chapman cycle [53].

Photocatalytic TiO_2 reduces urban NO_x, but challenges include by-products, test standards [91]. CuMgFe layered double hydroxides (CuMgFe-LDH) were synthesized with varying Cu/Mg/Fe ratios and used to create layered double oxide catalysts (CuMgFe-LDO) for NO_x removal via selective catalytic reduction with NH_3 (NH_3-SCR). Optimal catalytic performance was achieved with $Cu_{0.5}Mg_{2.5}Fe_1$-LDO, displaying 80% NO_x conversion at 180–250 °C and excellent SO_2 resistance [109]. TiO_2 utilized photocatalytic asphalt pavement to reduce vehicle pollutants, includes key factors like nano-TiO_2 crystal types, doping methods, temperature, humidity, irradiation, and pollutant characteristics. Currently, it achieves 55% NO_x and 20% SO_2 pollutant degradation but faces challenges due to traffic and weather [58]. A chemical method developed TiO_2-based photocatalytic coating for Hong Kong roadside NO_x reduction. It showed significant effectiveness and potential health benefits through a field trial [59].

5.6 Green Building Materials

Photocatalytic coatings and materials are used in green building construction to improve indoor air quality and energy efficiency. They can break down indoor air pollutants and reduce the need for artificial lighting by reflecting and diffusing natural light. Cement paste samples with varying percentages of TiO_2 nanoparticles were tested for NO_x pollutant degradation. At early stages (65 h), the most effective samples had an 85:15 anatase:rutile ratio with 5.0% TiO_2, while at larger stages (28 days), the 100:0 ratio with 5.0% TiO_2 performed best in terms of photocatalytic activity [18].

The impact of TiO_2 content in silicate coatings on the photocatalytic degradation of SO_2 using simulated solar irradiation, showed photolysis reduced 10% of SO_2, sorption accounted for 30%, and photocatalytic oxidation removed up to 40%. Regardless of TiO_2 content, 71–86% of SO_2 was consistently removed, with 15% of TiO_2 being twice as efficient as 5% TiO_2 [50]. Research at Belgian Road Research Centre (BRRC) explores the durability of photocatalytic building materials for long-term air pollutant remediation. Initial findings include accelerated aging testing and validation of aging procedures for photocatalytic NO_x removal along with future recommendations [15]. Three TiO_2 and SiO_2 hybrid sols were synthesized and applied as coatings on standard gray bricks. The coated bricks exhibited both hydrophobic and photocatalytic properties, degrading stains under solar lamp irradiation [3].

A study examines the impact of substrate morphology and pore structure (wool-wool cement board and autoclaved aerated concrete) on photocatalytic nitric oxide removal efficiency. Both substrates effectively degrade NO. Effective area, roughness, and open pore structure enhance efficiency [35]. Glazed building envelopes are introduced with photoluminescent coatings for smart windows, demonstrating their potential for visual comfort. Innovative experiments confirm their ability to provide daylight diffusion and orientation without artificial lighting. Numerical solutions also show glare reduction benefits year-round [67]. Study in Depok City explores daylight integration to enhance classroom lighting uniformity in strategies 3 and 7 out of 7 different strategies, using Light Shelves and blinds [117]. Recent advancements in cement include photocatalytic properties, offering dual construction and pollution degradation functions [19]. Nano-TiO_2 cement-based materials in green construction due to their effective photocatalytic properties [46].

5.7 Oil Spill Cleanup

Photocatalysts can assist in the cleanup of oil spills in water bodies. When applied to sorbent materials, they can enhance the absorption of oil and promote its decomposition, helping to mitigate the environmental impact of oil spills. Superparamagnetic nanoparticles and β-cyclodextrin composites efficiently remove oil and pollutants [51]. The 2010 Deepwater Horizon oil spill released 4.9 million barrels of oil and gases into the Gulf of Mexico. It adversely affected marine life, vegetation, and coastal communities. To mitigate the damage, strategies like chemical dispersant, mechanical booms, and skimmers were used, but concerns about their toxicity arose. Subsequent research has explored alternative, eco-friendly oil spill response methods [74]. Costly and toxic oil spill remedies require alternatives. TiO_2 and Fe–TiO_2 photocatalysis offers effective solutions [1].

Oxygen-rich SnO_2 quantum dots (QDs) effectively degrade oil pollutants, aiding marine remediation and photocatalyst development [64]. Photocatalytic concrete incorporates Na-g-C_3N_4 to break down petroleum films on marine structures, with factors like irradiation time, intensity, angle, wave scouring time, and photocatalytic region height affecting degradation. Sea water pH and temperature play a minor role. Film thickness limits degradation to around 50%, and insufficient thickness halts the process, offering a multidisciplinary solution to petroleum pollution cleanup [124]. The MF foams, a common building material, are repurposed, retaining their porous network structure after carbonization. These carbon sponges exhibit unique hierarchical mesoporous features, partial graphitization, and the presence of functional groups. They show superior Joule-heating performance for oil spill recovery compared to supported graphene [47].

5.8 Reduction of Harmful Algal Blooms

In bodies of water affected by harmful algal blooms, photocatalysts can be used to break down the toxins produced by these blooms. This application helps to protect aquatic ecosystems and ensures the safety of water resources. ZnO nanorods grown on polypropylene substrates effectively combat algal biofouling in membrane filtration systems. Visible light photocatalysis achieved nearly total algal cell inactivation within 2 h, preventing regrowth. This environmentally friendly method, based on reactive oxygen species (ROS) formation, offers a green solution for membrane fouling control in water treatment plants, eliminating the need for harmful chemicals [92].

A novel AGON coating, loaded with Ag_2CO_3–N:GO was developed on a modified polyurethane sponge for photocatalytic inactivation of Microcystis aeruginosa (M. aeruginosa) and degradation of Microcystin-LR (MC-LR) in eutrophic waters. Optimal performance was observed with AGON coating-4 under visible light, effectively removing chlorophyll a in 5 h. Natural organic matter (NOM) and alkalinity negatively affected algae photocatalytic inactivation. The coating exhibited good reusability, making it a promising solution for addressing harmful algae blooms [30]. A perovskite-type catalyst, La_2CuO_4, for peroxymonosulfate (PMS) activation to combat cyanobacterial blooms. Within 20 min, it achieved remarkable Microcystis aeruginosa and chlorophyll-a removal (97% and 100%, respectively) without significant release of dissolved organic carbon and metals. The catalysts destabilized cells through increased zeta potential (La portion) and improved electron transfer (Cu portion) [34].

5.9 Solar-Powered Desalination

PCs can be integrated into solar-driven desalination processes, where they aid in the removal of impurities and the conversion of seawater into fresh water, addressing water scarcity issues in coastal regions. Solar desalination uses solar energy's greenhouse effect to produce fresh water from seawater or brackish water. Semiconducting oxides like CuO, PbO_2 and MnO_2 enhance conventional solar still efficiency, increasing desalination rate and improving water quality analysis [80]. Janus hydrogel, a novel solar evaporator, efficiently purifies water under sunlight, with high durability, desalination capability, and organic dye degradation. Potential in desalination and wastewater treatment [108]. A biomimetic hydrogel evaporator, inspired by mushroom transpiration, offers efficient solar-driven desalination. It maintains high water circulation, reduces heat transfer, and achieves excellent performance, salt resistance, and photocatalysis [120].

5.10 Environmental Sensors

Photocatalytic materials can be incorporated into environmental sensors to detect pollutants and monitor air and water quality. These sensors can provide real-time data for environmental management and decision-making. Growing populations and industrialization worsen water and air pollution. A rapid synthesis of Dy-doped ZnO nanoparticles via a solution combustion method revealed excellent properties for gas sensing and photocatalysis. Varying Dy (Dysprosium) concentrations influenced morphology and band gap properties. Photodegradation of DR-31 dye occurred within 60 min under UV light, following first-order kinetics. Dy-doped ZnO nanoparticles coated on glass substrates exhibited strong ethanol sensing capabilities, even at low concentrations (25 ppm), with peak sensitivity at 150 °C. the nanoparticle's high surface-to-volume ratio played a vital role in gas sensing and photocatalysis. The optimal Dy concentration for both applications was 4.0 at.wt% [12].

6 Other Applications

Antibacterial surfaces: Photocatalytic coatings can be applied to surfaces in healthcare settings to reduce the spread of pathogens by continuously killing bacteria and viruses [101].

Drug Synthesis: Photocatalysis can be employed in the synthesis of pharmaceuticals and chemicals, enabling more sustainable and energy-efficient processes. Deuterium labeling techniques play a crucial role in pharmaceutical research, aiding in the study of drug absorption, distribution, metabolism, and excretion (ADME).

Deuterium-labeled drugs offer extended half-lives, increased efficacy, site-specific action, and reduced side effects [77].

Ship Hulls: Antifouling coatings using photocatalysis can prevent the attachment of marine organisms to ship hulls, reducing fuel consumption and environmental impact [62].

Food Safety: Photocatalysis can be used to sterilize food surfaces and packaging materials, extending shelf life and improving food safety [65].

Art Installations: Photocatalytic Materials can be used in artistic installations that interact with sunlight to create dynamic visual effects.

These are just some of the many applications of photocatalysis in energy, environmental remediation, and various other fields. Research in this area continues to expand, offering new possibilities for sustainable and efficient processes.

7 Challenges and Future Directions

Despite the remarkable progress in nanomaterial-based photocatalysis, several challenges remain. Issues such as scalability, cost-effectiveness, and long-term stability need to be addressed for practical implementation. Additionally, the potential environmental and health risk associated with nanomaterials must be thoroughly evaluated. In the future, research efforts should focus on developing scalable and sustainable synthesis methods for nanomaterials, optimizing their stability under real-world conditions, and gaining a comprehensive understanding of their environmental impact. Collaboration between multidisciplinary teams, including chemists, physicists, materials scientists, and engineers, will be crucial in advancing the field and realizing its full potential.

8 Conclusion

Photocatalysis, driven by the design and application of photocatalysts, holds immense promise in addressing some of the world most pressing challenges. This comprehensive review has offered insights into the mechanisms, materials, and applications of photocatalysts, highlighting their critical role in environmental remediation, energy conversion, and sustainable chemistry. While challenges remain continued research and innovation will undoubtedly unlock the full potential of photocatalysts, contributing to a cleaner, more sustainable future.

It is important to note that while photocatalysts hold immense promise for various energy-related applications, several challenges remain, including improving their efficiency, stability, and scalability. Its ability to harness the power of sunlight of various applications, from solar energy conversion to air purification, offers a path toward a more sustainable future.

While photocatalysts offer promising solutions to various environmental challenges, their practical implementation depends on factors like material efficiency, cost-effectiveness, and scalability. Ongoing research aims to optimize photocatalytic materials and processes to maximize their effectiveness in addressing environmental issues and improving the sustainability of our planet.

Researchers continue to explore new materials and design strategies to optimize photocatalysts for practical use in energy storage and conversion technologies and environmental issues. As advancements in this field progress, the role of photocatalysts in sustainable energy solutions is likely to become increasingly significant, contributing to a cleaner and greener world for generations to come.

References

1. B. Agyei-Tuffour, S. Gbogbo, D. Dodoo-Arhin, L.N.W. Damoah, J.K. Efavi, A. Yaya, E. Nyankson, Photocatalytic degradation of fractionated crude oil: potential application in oil spill remediation. Cogent Eng. **7**(1), 1744944 (2020). https://doi.org/10.1080/23311916.2020.1744944
2. S.F. Ahmed, P.S. Kumar, B. Ahmed, T. Mehnaz, G.M. Shafiullah, V.N. Nguyen, X.Q. Duong, M. Mofijur, I.A. Badruddin, S. Kamangar, Carbon-based nanomaterials: characteristics, dimensions, advances and challenges in enhancing photocatalytic hydrogen production. Int. J. Hydrog. Energy (2023). https://doi.org/10.1016/j.ijhydene.2023.03.185
3. I. Alfieri, A. Lorenzi, L. Ranzenigo, L. Lazzarini, G. Predieri, P.P. Lottici, Synthesis and characterization of photocatalytic hydrophobic hybrid TiO2–SiO2 coatings for building applications. Build. Environ. **111**, 72–79 (2017). https://doi.org/10.1016/j.buildenv.2016.10.019
4. T.S. Andrade, V. Dracopoulos, P. Lianos, Solar energy conversion and storage using a photocatalytic fuel cell combined with a supercapacitor. Electronics (Switzerland) **10**(3), 1–9 (2021). https://doi.org/10.3390/electronics10030273
5. A. Ansari, N.M. Nouri, A one step self-cleaning surface with robust superhydrophobic and photocatalytic properties: Electrostatic sprayed fluorinated ethylene propylene mixed with nano TiO2–SiO2 particles. Ceram. Int. **49**(1), 57–66 (2023). https://doi.org/10.1016/j.ceramint.2022.07.269
6. F. Asghar, A. Mushtaq, The future of nanomaterial in wastewater treatment: a review. IJCBS **23**(1) (Asghar and Mushtaq, 2023). www.iscientific.org/Journal.html
7. A. Aziz, Z. Memon, A. Bhutto, Efficient photocatalytic degradation of industrial wastewater dye by Grewia asiatica mediated zinc oxide nanoparticles. Optik **272** (2023). https://doi.org/10.1016/j.ijleo.2022.170352
8. S.N.Q.A.A. Aziz, K.C. Meng, S.Y. Pung, Z. Lockman, A. Ul-Hamid, W.K. Tan, Rapid growth of zinc oxide nanorods on kanthal wires by direct heating method and its photocatalytic performance in pollutants removal. J. Ind. Eng. Chem. **118**, 226–238 (2023). https://doi.org/10.1016/j.jiec.2022.11.008
9. Y. Bai, T. Chen, P. Wang, L. Wang, L. Ye, X. Shi, W. Bai, Size-dependent role of gold in g-C3N4/BiOBr/Au system for photocatalytic CO2 reduction and dye degradation. Sol. Energy Mater. Sol. Cells **157**, 406–414 (2016). https://doi.org/10.1016/j.solmat.2016.07.001
10. J. Behari, Principles of nanoscience: an overview. Indian J. Exp. Biol. **48** (2010)
11. S. Bera, J.E. Lee, S.B. Rawal, W.I. Lee, Size-dependent plasmonic effects of Au and Au@SiO2 nanoparticles in photocatalytic CO2 conversion reaction of Pt/TiO2. Appl. Catal. B **199**, 55–63 (2016). https://doi.org/10.1016/j.apcatb.2016.06.025

12. S. Bhatia, N. Verma, R. Kumar, Morphologically-dependent photocatalytic and gas sensing application of Dy-doped ZnO nanoparticles. J. Alloy. Compd. **726**, 1274–1285 (2017). https://doi.org/10.1016/j.jallcom.2017.08.048
13. V. Bisinella, L. Dahl, H. Jensen, T.N. Mikkelsen, T.H. Christensen, Environmental profile of NO_x reduction by a photocatalytic surface coating and a vehicle catalytic converter. J. Environ. Prot. **12**(09), 590–623 (2021). https://doi.org/10.4236/jep.2021.129037
14. E. Boonen, A. Beeldens, Recent photocatalytic applications for air purification in Belgium. Coatings **4**(3), 553–573 (2014). https://doi.org/10.3390/coatings4030553
15. E. Boonen, A. Beeldens, I. Dirkx, V. Bams, Durability of cementitious photocatalytic building materials. Catal. Today **287**, 196–202 (2017). https://doi.org/10.1016/j.cattod.2016.10.012
16. Y. Boyjoo, H. Sun, J. Liu, V.K. Pareek, S. Wang, A review on photocatalysis for air treatment: from catalyst development to reactor design. Chem. Eng. J. **310**, 537–559 (2017). https://doi.org/10.1016/j.cej.2016.06.090
17. T. Cao, W. Dong, Y. Liang, Q. Bao, C. Xu, M. Bai, T. Luo, X. Gu, A simple solvothermal preparation of Mg-doped anatase TiO2 and its self-cleaning application. Sol. Energy **249**, 12–20 (2023). https://doi.org/10.1016/j.solener.2022.11.023
18. C. Cárdenas, J.I. Tobón, C. García, J. Vila, Functionalized building materials: photocatalytic abatement of NOx by cement pastes blended with TiO2 nanoparticles. Constr. Build. Mater. **36**, 820–825 (2012). https://doi.org/10.1016/j.conbuildmat.2012.06.017
19. X. Chen, L. Qiao, R. Zhao, J. Wu, J. Gao, L. Li, J. Chen, W. Wang, M.G. Galloni, F.M. Scesa, Z. Chen, E. Falletta, Recent advances in photocatalysis on cement-based materials. J. Environ. Chem. Eng. **11**(2) (Elsevier Ltd., 2023). https://doi.org/10.1016/j.jece.2023.109416
20. Y. Choi, S. Choi, I. Lee, T.V.T. Nguyen, S. Bae, Y.H. Kim, J. Ryu, S. Park, J. Ryu, Solar biomass reforming and hydrogen production with earth-abundant Si-based photocatalysts. Adv. Mater. (2023). https://doi.org/10.1002/adma.202301576
21. J.A. Dahl, B.L.S. Maddux, J.E. Hutchison, Toward greener nanosynthesis. Chem. Rev. **107**(6), 2228–2269 (2007). https://doi.org/10.1021/cr050943k
22. J.V.S. de Melo, G. Trichês, Evaluation of the influence of environmental conditions on the efficiency of photocatalytic coatings in the degradation of nitrogen oxides (NOx). Build. Environ. **49**, 117–123 (2012). https://doi.org/10.1016/j.buildenv.2011.09.016
23. R. de Richter, S. Caillol, Fighting global warming: the potential of photocatalysis against CO2, CH4, N2O, CFCs, tropospheric O3, BC and other major contributors to climate change. J. Photochem. Photobiol. C: Photochem. Rev. **12**(1), 1–19 (2011). https://doi.org/10.1016/j.jphotochemrev.2011.05.002
24. M. Diaa, A.G. Hassabo, Self-cleaning properties of cellulosic fabrics (A review). Biointerface Res. Appl. Chem. **12**(2), 1847–1855 (AMG Transcend Association, 2022). https://doi.org/10.33263/BRIAC122.18471855
25. H.R. Dihom, M.M. Al-Shaibani, R.M.S. Radin Mohamed, A.S. Al-Gheethi, A. Sharma, M.H.B. Khamidun, Photocatalytic degradation of disperse azo dyes in textile wastewater using green zinc oxide nanoparticles synthesized in plant extract: a critical review. J. Water Process Eng. **47**, 102705 (2022). https://doi.org/10.1016/j.jwpe.2022.102705
26. C. Dong, C. Lian, S. Hu, Z. Deng, J. Gong, M. Li, H. Liu, M. Xing, J. Zhang, Size-dependent activity and selectivity of carbon dioxide photocatalytic reduction over platinum nanoparticles. Nat. Commun. **9**(1) (2018). https://doi.org/10.1038/s41467-018-03666-2
27. M.H. El-Saeid, A. Baqais, M. Alshabanat, Study of the photocatalytic degradation of highly abundant pesticides in agricultural soils. Molecules **27**(3) (2022). https://doi.org/10.3390/molecules27030634
28. P. Elumalai, B. Muthukumar, P. Dhandapani, O. Parthiba Karthikeyan, M. Huang, Role of bionanohybrids for pollutant removal in wastewater environment. Curr. Opin. Environ. Sci. Health **100504**(2023). https://doi.org/10.1016/j.coesh.2023.100504
29. A.J. Esswein, D.G. Nocera, Hydrogen production by molecular photocatalysis. Chem. Rev. **107**(10), 4022–4047. https://doi.org/10.1021/cr050193e
30. G. Fan, Z. Chen, L. Hong, B. Du, Z. Yan, J. Zhan, Y. You, R. Ning, H. Xiao, Simultaneous removal of harmful algal cells and toxins by a Ag2CO3-N:GO photocatalyst coating under visible light. Sci. Total Environ. **741**(2020). https://doi.org/10.1016/j.scitotenv.2020.140341

31. A. Fujishima, K. Honda, Electrochemical photolysis of water at a semiconductor electrode. Nature **238**(5358), 37–38 (1972). https://doi.org/10.1038/238037a0
32. K. Fujiwara, S.E. Pratsinis, Single Pd atoms on TiO2 dominate photocatalytic NOx removal. Appl. Catal. B **226**, 127–134 (2018). https://doi.org/10.1016/j.apcatb.2017.12.042
33. D. Gao, W. Li, H. Wang, G. Wang, R. Cai, Heterogeneous catalysis for CO2 conversion into chemicals and fuels. Trans. Tianjin Univ. **28**(4), 245–264 (Tianjin University, 2022). https://doi.org/10.1007/s12209-022-00326-x
34. P. Gao, Y. He, S. Lu, M. He, Z. Liu, Y. Deng, Z. Liu, T. Xu, H. Zhang, Activation of peroxymonosulfate by La2CuO4 perovskite for synergistic removal of microcystis aeruginosa and microcystin-LR in harmful algal bloom impacted water. Appl. Catal. B **328**(2023). https://doi.org/10.1016/j.apcatb.2023.122511
35. F. Gauvin, V. Caprai, Q.L. Yu, H.J.H. Brouwers, Effect of the morphology and pore structure of porous building materials on photocatalytic oxidation of air pollutants. Appl. Catal. B **227**, 123–131 (2018). https://doi.org/10.1016/j.apcatb.2018.01.029
36. H. Gleiter, Nanostructured materials: basic concepts and microstructure. www.elsevier.com/locate/actamat
37. A. Gutiérrez-Cruz, A.R. Ruiz-Hernández, J.F. Vega-Clemente, D.G. Luna-Gazcón, J. Campos-Delgado, A review of top-down and bottom-up synthesis methods for the production of graphene, graphene oxide and reduced graphene oxide. J. Mater. Sci. **57**(31), 14543–14578) (Springer, 2022). https://doi.org/10.1007/s10853-022-07514-z
38. R. Han, R. Andrews, C. O'Rourke, S. Hodgen, A. Mills, Photocatalytic air purification: effect of HNO3 accumulation on NOx and VOC removal. Catal. Today **380**, 105–113 (2021). https://doi.org/10.1016/j.cattod.2021.04.017
39. R. Han, R. Zhou, C. O'Rourke, A. Mills, 3D printed photocatalytic reactor for air purification. J. Photochem. Photobiol. A **443**, 114843 (2023). https://doi.org/10.1016/j.jphotochem.2023.114843
40. F. He, W. Jeon, W. Choi, Photocatalytic air purification mimicking the self-cleaning process of the atmosphere. Nat. Commun. **12**(1), 2528 (2021). https://doi.org/10.1038/s41467-021-22839-0
41. M.M. Higarashi, W.F. Jardim, Remediation of pesticide contaminated soil using TiO2 mediated by solar light. Catal. Today **76** (2002)
42. E. Horváth, J. Gabathuler, G. Bourdiec, E. Vidal-Revel, M. Benthem Muñiz, M. Gaal, D. Grandjean, F. Breider, L. Rossi, A. Sienkiewicz, L. Forró, Solar water purification with photocatalytic nanocomposite filter based on TiO2 nanowires and carbon nanotubes. NPJ Clean Water **5**(1), 10 (2022). https://doi.org/10.1038/s41545-022-00157-2
43. E. Horváth, L. Rossi, C. Mercier, C. Lehmann, A. Sienkiewicz, L. Forró, Photocatalytic nanowires-based air filter: towards reusable protective masks. Adv. Funct. Mater. **30**(40) (2020). https://doi.org/10.1002/adfm.202004615
44. G. Hüsken, M. Hunger, H.J.H. Brouwers, Experimental study of photocatalytic concrete products for air purification. Build. Environ. **44**(12), 2463–2474 (2009). https://doi.org/10.1016/j.buildenv.2009.04.010
45. A. Jaison, A. Mohan, Y.-C. Lee, Recent developments in photocatalytic nanotechnology for purifying air polluted with volatile organic compounds: effect of operating parameters and catalyst deactivation. Catalysts **13**(2), 407 (2023). https://doi.org/10.3390/catal13020407
46. Z.M. Jia, Y.R. Zhao, J.N. Shi, Adsorption kinetics of the photocatalytic reaction of nano-TiO2 cement-based materials: a review. Construct. Build. Mater. **370** (Elsevier Ltd., 2023). https://doi.org/10.1016/j.conbuildmat.2023.130462
47. S. Kang, M. Chen, Y. Wang, F. Tang, Y. Liu, L. Cui, M. Dong, Engineering a 3D porous carbon sponge as a self-floating solar energy utilization platform for photothermal oil spill recovery and photocatalytic H_2 evolution. Sustain. Energy Fuels **7**(2), 409–419 (2023). https://doi.org/10.1039/D2SE01463G
48. P. Khanna, A. Kaur, D. Goyal, Algae-based metallic nanoparticles: synthesis, characterization and applications. J. Microbiol. Methods **163** (Elsevier B.V., 2019). https://doi.org/10.1016/j.mimet.2019.105656

49. C.C. Koch, *Nanostructured Materials: processing, Properties and Applications* (William Andrew, 2006)
50. P. Krishnan, M.H. Zhang, Y. Cheng, D.T. Riang, L.E. Yu, Photocatalytic degradation of SO2 using TiO2-containing silicate as a building coating material. Constr. Build. Mater. **43**, 197–202 (2013). https://doi.org/10.1016/j.conbuildmat.2013.02.012
51. A. Kumar, G. Sharma, M. Naushad, S. Thakur, SPION/β-cyclodextrin core–shell nanostructures for oil spill remediation and organic pollutant removal from waste water. Chem. Eng. J. **280**, 175–187 (2015). https://doi.org/10.1016/j.cej.2015.05.126
52. V. Kumaravel, S. Mathew, J. Bartlett, S.C. Pillai, Photocatalytic hydrogen production using metal doped TiO2: a review of recent advances. Appl. Catal. B: Environ. **244**, 1021–1064 (Elsevier B.V., 2019). https://doi.org/10.1016/j.apcatb.2018.11.080
53. M. Le Pivert, O. Kerivel, B. Zerelli, Y. Leprince-Wang, ZnO nanostructures based innovative photocatalytic road for air purification. J. Clean. Prod. **318**, 128447 (2021). https://doi.org/10.1016/j.jclepro.2021.128447
54. B.Y. Lee, A.R. Jayapalan, M.H. Bergin, K.E. Kurtis, Photocatalytic cement exposed to nitrogen oxides: effect of oxidation and binding. Cem. Concr. Res. **60**, 30–36 (2014). https://doi.org/10.1016/j.cemconres.2014.03.003
55. N. Li, Y. Wang, D. Tang, H. Zhou, Integrating a photocatalyst into a hybrid Lithium-Sulfur battery for direct storage of solar energy. Angew. Chem. Int. Ed. **54**(32), 9271–9274 (2015). https://doi.org/10.1002/anie.201503425
56. P. Li, J. Li, X. Feng, J. Li, Y. Hao, J. Zhang, H. Wang, A. Yin, J. Zhou, X. Ma, B. Wang, Metal-organic frameworks with photocatalytic bactericidal activity for integrated air cleaning. Nat. Commun. **10**(1), 2177 (2019). https://doi.org/10.1038/s41467-019-10218-9
57. X. Li, C. Liu, D. Wu, J. Li, P. Huo, H. Wang, Improved charge transfer by size-dependent plasmonic Au on C3 N4 for efficient photocatalytic oxidation of RhB and CO2 reduction. Chin. J. Catal. **40** (2019). http://www.sciencedirect.com/science/journal/18722067
58. X. Li, F. Wang, L. You, S. Wu, C. Yang, L. Zhang, D. Maria Barbieri, A review on photocatalytic asphalt pavement designed for degradation of vehicle exhausts. Transp. Res. Part D: Transp. Environ. **115**, 103605 (2023). https://doi.org/10.1016/j.trd.2023.103605
59. X. Li, P. Wang, S. Han, Y. Huang, W. Ho, S.S.H. Ho, S. Lee, M. Wang, Development and application of photocatalytic coating for roadside NOx mitigation in Hong Kong. Chin. Chem. Lett. **108709**(2023). https://doi.org/10.1016/j.cclet.2023.108709
60. Z. Li, L. Yin, S. Jiang, L. Chen, S. Sang, H. Zhang, A photocatalytic degradation self-cleaning composite membrane for oil-water separation inspired by light-trapping effect of moth-eye. J. Membr. Sci. **669**, 121337 (2023). https://doi.org/10.1016/j.memsci.2022.121337
61. G. Liao, W. Yao, A. She, C. Shi, J. Zuo, D. Wu, Interfacial design of nano-TiO2 modified recycled concrete powder for building self-cleaning. Colloids Surf. A **661**, 130925 (2023). https://doi.org/10.1016/j.colsurfa.2023.130925
62. D. Liu, H. Shu, J. Zhou, X. Bai, P. Cao, Research progress on new environmentally friendly antifouling coatings in marine settings: a review. Biomimetics **8**(2), 200 (2023). https://doi.org/10.3390/biomimetics8020200
63. H. Liu, Y. Feng, J. Shao, Y. Chen, Z.L. Wang, H. Li, X. Chen, Z. Bian, Self-cleaning triboelectric nanogenerator based on TiO2 photocatalysis. Nano Energy **70**, 104499 (2020). https://doi.org/10.1016/j.nanoen.2020.104499
64. J. Liu, Q. Zhang, X. Tian, Y. Hong, Y. Nie, N. Su, G. Jin, Z. Zhai, C. Fu, Highly efficient photocatalytic degradation of oil pollutants by oxygen deficient SnO2 quantum dots for water remediation. Chem. Eng. J. **404**, 127146 (2021). https://doi.org/10.1016/j.cej.2020.127146
65. S. Liu, X. Gao, H. Fan, M. Zhang, G.I.N. Waterhouse, S. Zhu, Green and recyclable graphitic carbon nitride/chitosan/polyvinyl alcohol photocatalytic films with efficient antibacterial activity for fruit packaging. Int. J. Biol. Macromol. **236**, 123974 (2023). https://doi.org/10.1016/j.ijbiomac.2023.123974
66. Y. Lu, H. Zhang, D. Fan, Z. Chen, X. Yang, Coupling solar-driven photothermal effect into photocatalysis for sustainable water treatment. J. Hazard. Mater. **423**, 127128 (2022). https://doi.org/10.1016/j.jhazmat.2021.127128

67. F. Marchini, C. Chiatti, C. Fabiani, A.L. Pisello, Development of an innovative translucent–photoluminescent coating for smart windows applications: An experimental and numerical investigation. Renew. Sustain. Energy Rev. **184**(2023). https://doi.org/10.1016/j.rser.2023.113530
68. J.H. Martínez-Montelongo, I.E. Medina-Ramírez, Y. Romo-Lozano, J.A. Zapien, Development of a sustainable photocatalytic process for air purification. Chemosphere **257**, 127236 (2020). https://doi.org/10.1016/j.chemosphere.2020.127236
69. S. Matsuda, H. Hatano, A. Tsutsumi, Ultrafine particle fluidization and its application to photocatalytic NOx treatment. Chem. Eng. J. **82** (2001)
70. H.M. Mehwish, M.S.R. Rajoka, Y. Xiong, H. Cai, R.M. Aadil, Q. Mahmood, Z. He, Q. Zhu, Green synthesis of a silver nanoparticle using Moringa oleifera seed and its applications for antimicrobial and sun-light mediated photocatalytic water detoxification. J. Environ. Chem. Eng. **9**(4), 105290 (2021). https://doi.org/10.1016/j.jece.2021.105290
71. S. Mishra, B. Sundaram, A review of the photocatalysis process used for wastewater treatment. Mater. Today: Proc. (2023). https://doi.org/10.1016/j.matpr.2023.07.147
72. S.R. Nagella, R. Vijitha, B. Ramesh Naidu, K.S.V. Krishna Rao, C.S. Ha, K. Venkateswarlu, Benchmarking recent advances in hydrogen production using g-C3N4-based photocatalysts. Nano Energy **111** (Elsevier Ltd., 2023). https://doi.org/10.1016/j.nanoen.2023.108402
73. S. Nishimoto, B. Bhushan, Bioinspired self-cleaning surfaces with superhydrophobicity, superoleophobicity, and superhydrophilicity. RSC Adv. **3**(3), 671–690 (2013). https://doi.org/10.1039/C2RA21260A
74. E. Nyankson, D. Rodene, R.B. Gupta, Advancements in crude oil spill remediation research after the deepwater horizon oil spill. Water Air Soil Pollut. **227**(1), 29 (2016). https://doi.org/10.1007/s11270-015-2727-5
75. E. Okoh, Z.R. Yelebe, B. Oruabena, E.S. Nelson, O.P. Indiamaowei, Clean-up of crude oil-contaminated soils: bioremediation option. Int. J. Environ. Sci. Technol. **17**(2), 1185–1198 (2020). https://doi.org/10.1007/s13762-019-02605-y
76. O. Ola, M. Mercedes Maroto-Valer, S. Mackintosh, Turning CO2 into valuable chemicals. Energy Procedia **37**, 6704–6709 (2013). https://doi.org/10.1016/j.egypro.2013.06.603
77. W. Ou, C. Qiu, C. Su, Photo- and electro-catalytic deuteration of feedstock chemicals and pharmaceuticals: a review. Chin. J. Catal. **43**(4), 956–970 (Science Press, 2022). https://doi.org/10.1016/S1872-2067(21)63928-1
78. E. Pakdel, J. Wang, S. Kashi, L. Sun, X. Wang, Advances in photocatalytic self-cleaning, superhydrophobic and electromagnetic interference shielding textile treatments. Adv. Coll. Interface Sci. **277**, 102116 (2020). https://doi.org/10.1016/j.cis.2020.102116
79. Y.H. Park, G. Murali, J.K.R. Modigunta, I. In, S. In Il, Recent advances in quantum dots for photocatalytic CO2 reduction: a mini-review. Front. Chem. **9** (Frontiers Media S.A., 2021). https://doi.org/10.3389/fchem.2021.734108
80. S.G. Patel, S. Bhatnagar, J. Vardia, S.C. Ameta, Use of photocatalysts in solar desalination. Desalination **189**(1–3), 287–291 (2006). https://doi.org/10.1016/j.desal.2005.07.010
81. T. Pauporté, Synthesis of ZnO nanostructures for solar cells—A focus on dye-sensitized and perovskite solar cells, in *The Future of Semiconductor Oxides in Next-Generation Solar Cells* (Elsevier, 2018), pp. 3–43. https://doi.org/10.1016/B978-0-12-811165-9.00001-6
82. Y. Paz, *Photocatalytic Treatment of Air* (2009), pp. 289–336. https://doi.org/10.1016/S0065-2377(09)00408-6
83. H. Peeters, S. Lenaerts, S.W. Verbruggen, Benchmarking the photocatalytic self-cleaning activity of industrial and experimental materials with ISO 27448:2009. Materials **16**(3), 1119 (2023). https://doi.org/10.3390/ma16031119
84. K. Perović, H.M. dela Rosa, M. Kovačić, H. Kušić, U. Štangar, L. F. Fresno, D.D. Dionysiou, A. Loncaric Bozic, Recent achievements in development of TiO2-based composite photocatalytic materials for solar driven water purification and water splitting. Materials **13**(6), 1338 (2020). https://doi.org/10.3390/ma13061338
85. M.T. Qamar, S. Iqbal, M. Aslam, A. Alhujaily, A. Bilal, K. Rizwan, H.M.U. Farooq, T.A. Sheikh, A. Bahadur, N.S. Awwad, H.A. Ibrahium, R.S. Almufarij, E.B. Elkaeed, Transition

metal doped CeO2 for photocatalytic removal of 2-chlorophenol in the exposure of indoor white light and antifungal activity. Front. Chem. **11**(2023). https://doi.org/10.3389/fchem.2023.1126171
86. J. Qin, H. Lu, A review of self-cleaning coatings for solar photovoltaic systems: theory, materials, preparation, and applications. Environ. Sci. Pollut. Res. **30**(40), 91591–91616 (2023). https://doi.org/10.1007/s11356-023-28550-5
87. M. Rashid, W. Hassan, M. Aadil, H.H. Somaily, N.M. Mahdi, R. Lataef, A.G. Taki, K. Srithilat, D.F. Baamer, S.M. Albukhari, M.A. Salam, A. Ilyas, Solar-light-driven and magnetically recoverable doped nano-ferrite: an ideal photocatalyst for water purification applications. Opt. Mater. **135**, 113192 (2023). https://doi.org/10.1016/j.optmat.2022.113192
88. R.W. Siegel, *Synthesis and Processing of Nanostructured Materials* (1993)
89. M. Rizwan, A. Shoukat, A. Ayub, B. Razzaq, M.B. Tahir, Types and classification of nanomaterials, in *Nanomaterials: synthesis, Characterization, Hazards and Safety* (Elsevier, 2021), pp. 31–54. https://doi.org/10.1016/B978-0-12-823823-3.00001-X
90. H. Roy, T.U. Rahman, M.A.R. Khan, M.R. Al-Mamun, S.Z. Islam, M.A. Khaleque, M.I. Hossain, M.Z.H. Khan, M.S. Islam, H.M. Marwani, A. Islam, M.M. Hasan, M.R. Awual, Toxic dye removal, remediation, and mechanism with doped SnO2-based nanocomposite photocatalysts: a critical review. J. Water Process Eng. **54** (Elsevier Ltd., 2023). https://doi.org/10.1016/j.jwpe.2023.104069
91. H.S. Russell, L.B. Frederickson, O. Hertel, T. Ellermann, S.S. Jensen, A Review of photocatalytic materials for urban NOx remediation. Catalysts **11**(6), 675 (2021). https://doi.org/10.3390/catal11060675
92. P. Sathe, M.T.Z. Myint, S. Dobretsov, J. Dutta, Removal and regrowth inhibition of microalgae using visible light photocatalysis with ZnO nanorods: a green technology. Sep. Purif. Technol. **162**, 61–67 (2016). https://doi.org/10.1016/j.seppur.2016.02.007
93. R. Sharma, M. Almáši, S.P. Nehra, V.S. Rao, P. Panchal, D.R. Paul, I.P. Jain, A. Sharma, Photocatalytic hydrogen production using graphitic carbon nitride (GCN): a precise review. Renew. Sustain. Energy Rev. **168**(2022). https://doi.org/10.1016/j.rser.2022.112776
94. C. Shi, F. Kang, Y. Zhu, M. Teng, J. Shi, H. Qi, Z. Huang, C. Si, F. Jiang, J. Hu, Photoreforming lignocellulosic biomass for hydrogen production: optimized design of photocatalyst and photocatalytic system. Chem. Eng. J. **452** (Elsevier B.V., 2023). https://doi.org/10.1016/j.cej.2022.138980
95. R.A. Solano, A.P. Herrera, D. Maestre, A. Cremades, Fe-TiO2 nanoparticles synthesized by green chemistry for potential application in waste water photocatalytic treatment. J. Nanotechnol. **2019**(2019). https://doi.org/10.1155/2019/4571848
96. A.V. Sonawane, Z.V.P. Murthy, A metal organic framework decorated 2-dimensional nanomaterial based nanocomposite photocatalyst for photocatalytic degradation of dyes from textile industry wastewater. Environ. Sci.: Water Res. Technol. (2023). https://doi.org/10.1039/D3EW00386H
97. P.M. Stanley, F. Sixt, J. Warnan, Decoupled solar energy storage and dark photocatalysis in a 3D metal-organic framework. Adv. Mater. **35**(1). https://doi.org/10.1002/adma.202207280
98. J. Taranto, D. Frochot, P. Pichat, Photocatalytic air purification: Comparative efficacy and pressure drop of a TiO2-coated thin mesh and a honeycomb monolith at high air velocities using a 0.4 m^3 close-loop reactor. Sep. Purif. Technol. **67**(2), 187–193 (2009). https://doi.org/10.1016/j.seppur.2009.03.017
99. R. Udayabhaskar, R.V. Mangalaraja, S.F. Sahlevani, V.T. Perarasu, B. Karthikeyan, D. Contreras, M.A. Gracia-Pinilla, Graphene induced band gap widening and luminescence quenching in ceria:graphene nanocomposites. J. Alloy. Compd. **770**, 1221–1228 (2019). https://doi.org/10.1016/j.jallcom.2018.08.226
100. M. Utami, S. Wang, M.M. Musawwa, T.E. Purbaningtias, M. Fitri, I. Yuspita, O.H. Abd-Elkader, K.K. Yadav, G. Munusamy-Ramanujam, D. Bang, S.W. Chang, B. Ravindran, Simultaneous photocatalytic removal of organic dye and heavy metal from textile wastewater over N-doped TiO2 on reduced graphene oxide. Chemosphere **332**(2023). https://doi.org/10.1016/j.chemosphere.2023.138882

101. L. Valenzuela, A. Iglesias, M. Faraldos, A. Bahamonde, R. Rosal, Antimicrobial surfaces with self-cleaning properties functionalized by photocatalytic ZnO electrosprayed coatings. J. Hazard. Mater. **369**, 665–673 (2019). https://doi.org/10.1016/j.jhazmat.2019.02.073
102. A. Vidal, Developments in solar photocatalysis for water purification. Chemosphere **36**(12), 2593–2606 (1998). https://doi.org/10.1016/S0045-6535(97)10221-1
103. H.S. Waheed, M. Asghar, H.S. Ahmad, T. Abbas, H. Ullah, R. Ali, M.J.I. Khan, M.W. Iqbal, Y.-H. Shin, M.S. Khan, R. Neffati, Janus MoSO and MoSSe monolayers: a promising material for solar cells and photocatalytic applications. Phys. Status Solidi (b) **260**(1). https://doi.org/10.1002/pssb.202200267
104. A.N. Wang, Y. Teng, X.F. Hu, L.H. Wu, Y.J. Huang, Y.M. Luo, P. Christie, Diphenylarsinic acid contaminated soil remediation by titanium dioxide (P25) photocatalysis: degradation pathway, optimization of operating parameters and effects of soil properties. Sci. Total Environ. **541**, 348–355. https://doi.org/10.1016/j.scitotenv.2015.09.023
105. C. Wang, R.L. Thompson, P. Ohodnicki, J. Baltrus, C. Matranga, Size-dependent photocatalytic reduction of CO2 with PbS quantum dot sensitized TiO2 heterostructured photocatalysts. J. Mater. Chem. **21**(35), 13452–13457 (2011). https://doi.org/10.1039/c1jm12367j
106. J. Wang, Y. Shi, H. Sun, W. Shi, F. Guo, Fabrication of Bi4Ti3O12/ZnIn2S4 S-scheme heterojunction for achieving efficient photocatalytic hydrogen production. J. Alloy. Compd. **930**(2023). https://doi.org/10.1016/j.jallcom.2022.167450
107. J. Wen, Q. Cai, R. Xiong, Z. Cui, Y. Zhang, Z. He, J. Liu, M. Lin, C. Wen, B. Wu, B. Sa, Promising M2CO2/MoX2 (M = Hf, Zr; X = S, Se, Te) heterostructures for multifunctional solar energy applications. Molecules **28**(8) (2023). https://doi.org/10.3390/molecules28083525
108. J. Wen, X. Li, H. Zhang, S. Zheng, C. Yi, L. Yang, J. Shi, Architecting Janus hydrogel evaporator with polydopamine-TiO2 photocatalyst for high-efficient solar desalination and purification. Sep. Purif. Technol. **304**(2023). https://doi.org/10.1016/j.seppur.2022.122403
109. X. Wu, J. Liu, L. Liu, Y. Du, X. Wu, Superior CuMgFe mixed oxide catalysts engineered by tuning the redox cycle for enhancing NOx removal performance. J. Environ. Chem. Eng. **10**(6) (2022). https://doi.org/10.1016/j.jece.2022.108824
110. Z. Xie, Y.-P. Peng, L. Yu, C. Xing, M. Qiu, J. Hu, H. Zhang, Solar-inspired water purification based on emerging 2D materials: status and challenges. Solar RRL **4**(3). https://doi.org/10.1002/solr.201900400
111. Y.-F. Xu, M.-Z. Yang, B.-X. Chen, X.-D. Wang, H.-Y. Chen, D.-B. Kuang, C.-Y. Su, A $CsPbBr_3$ perovskite quantum dot/graphene oxide composite for photocatalytic CO2 reduction. J. Am. Chem. Soc. **139**(16), 5660–5663 (2017). https://doi.org/10.1021/jacs.7b00489
112. X. Xuan, S. Tu, H. Yu, X. Du, Y. Zhao, J. He, H. Dong, X. Zhang, H. Huang, Size-dependent selectivity and activity of CO2 photoreduction over black nano-titaniasgrown on dendritic porous silica particles. Appl. Catal. B **255**(2019). https://doi.org/10.1016/j.apcatb.2019.117768
113. S. Yang, Z. Han, X. Pan, B. Liu, D. Zhao, Nitrogen oxide removal from simulated flue gas by UV-irradiated electrolyzed seawater: efficiency optimization and pH-dependent mechanisms. Chem. Eng. J. **354**, 653–662 (2018). https://doi.org/10.1016/j.cej.2018.07.191
114. C. Yavuz, S.E. Ela, Fabrication of g-C3N4-reinforced CdS nanosphere-decorated TiO2 nanotablet composite material for photocatalytic hydrogen production and dye-sensitized solar cell application. J. Alloy. Compd. **936**(2023). https://doi.org/10.1016/j.jallcom.2022.168209
115. L. Yin, Y. Zhao, Y. Xing, H. Tan, Z. Lang, W. Ho, Y. Wang, Y. Li, Structure-Property relationship in β-keto-enamine-based covalent organic frameworks for highly efficient photocatalytic hydrogen production. Chem. Eng. J. **419**(2021). https://doi.org/10.1016/j.cej.2021.129984
116. M. Yu, N. Zhang, X. Xue, X. Zhang, X. Ren, R. Feng, Y. Zhao, M. Sun, T. Yan, Highly efficient visible-light photocatalytic hydrogen production using ZIF-derived Co9S8/N, S-CNTs-ZnIn2S4 composite. Chem. Phys. Lett. **821**(2023). https://doi.org/10.1016/j.cplett.2023.140470

117. H.U.A. Zaky, O.C. Dewi, W. Widyarko, C.D. Sulistiani, Daylight and artificial lighting integration in achieving lighting uniformity in educational building. J. Arch. Res. Des. Stud. **7**(1) (2023). https://doi.org/10.20885/jars.vol7.iss1.art5
118. C. Zhang, H. Liang, Z. Xu, Z. Wang, Harnessing solar-driven photothermal effect toward the water-energy nexus. Adv. Sci. **6**(18), 1900883 (2019). https://doi.org/10.1002/advs.201900883
119. D. Zhang, Y. Wang, Y. Wang, Y. Zhang, X.M. Song, Fe2O3/NiO photocathode for photocatalytic methanol fuel cell: an insight on solar energy conversion. J. Alloy. Compd. **815**(2020). https://doi.org/10.1016/j.jallcom.2019.152377
120. H. Zhang, X. Li, X. Liu, Y. Du, W. Xie, S. Zheng, L. Yang, J. Shi, D. Jing, Biomimetic hydrogel with directional heat regulation for efficient solar desalination. Chem. Eng. J. **473**(2023). https://doi.org/10.1016/j.cej.2023.145484
121. K. Zhang, L. Guo, Metal sulphide semiconductors for photocatalytic hydrogen production. Catal. Sci. Technol. **3**(7), 1672–1690 (2013). https://doi.org/10.1039/c3cy00018d
122. Q. Zheng, A. Aiello, Y.S. Choi, K. Tarr, H. Shen, D.P. Durkin, D. Shuai, 3D printed photoreactor with immobilized graphitic carbon nitride: a sustainable platform for solar water purification. J. Hazard. Mater. **399**, 123097 (2020). https://doi.org/10.1016/j.jhazmat.2020.123097
123. P. Zhou, I.A. Navid, Y. Ma, Y. Xiao, P. Wang, Z. Ye, B. Zhou, K. Sun, Z. Mi, Solar-to-hydrogen efficiency of more than 9% in photocatalytic water splitting. Nature **613**(7942), 66–70 (2023). https://doi.org/10.1038/s41586-022-05399-1
124. Y. Zhou, M. Elchalakani, P. Du, C. Sun, Cleaning up oil pollution in the ocean with photocatalytic concrete marine structures. J. Clean. Prod. **329**, 129636 (2021). https://doi.org/10.1016/j.jclepro.2021.129636

MXene as an Uprising Nanosheet for Environmental Remediation

S. M. Abisha, S. Anandan, R. Vijayaraghavan, and Panneerselvam Sathishkumar

Abstract Photocatalysis is a greener process used over the years in environmental applications, in which the sunlight is utilized as energy to produce holes and electrons in semiconducting material that eliminate pollutants by producing reactive oxygen species. Photodegradation using photocatalysis represents an environmentally favorable, economically viable, and sustainable approach to the elimination of contaminants. Organic contaminants were extensively present in the environment through the realms of agriculture, animal husbandry, and medicine. Due to their excessive utilization as well as untreated wastewater has emerged as a primary factor contributing to the worldwide concern of antimicrobial resistance. Mitigation of CO_2 emission and heavy metal adsorption is also an environmental issue to be addressed through eco-friendly methods. There were so many semiconducting materials explored and used in this field. Materials like boron nitride, carbon nitride, black phosphorous, transition metal dichalcogenides, metal oxides, and metal sulfides have been studied widely. These materials hold good properties required for photocatalyst yet some drawbacks like rapid recombination of charge carriers, hydrophobic nature, and limited surface area. MXene is a novel 2D material that exhibits new and enhanced properties such as hydrophilicity, excellent metallic conductivity, and large surface area. The photocatalytic potential of this material remains unexplored, despite its extensive investigation in the fields of electrochemistry and energy storage. The subject under consideration has several obstacles pertaining to its stability and activity, yet it also exhibits favorable qualities that warrant further exploration. MXene has been documented as an effective photocatalyst under solar light illumination due to its ability to adjust its band gap. Photocatalysis using MXene may potentially contribute to the mitigation of environmental pollution. In this chapter, studies on MXene and its evolving synthesis methods, as well as classifications of MXene, are discussed. The main aim

S. M. Abisha · R. Vijayaraghavan · P. Sathishkumar (✉)
Department of Chemistry, School of Advanced Sciences, Vellore Institute of Technology (VIT), Vellore, India
e-mail: sathishkumar.p@vit.ac.in

S. Anandan
Nanomaterials and Solar Energy Conversion Lab, Department of Chemistry, National Institute of Technology, Trichy 620015, India

P. Sathishkumar (ed.), *Photocatalysis for Energy and Environmental Applications*, Green Energy and Technology, https://doi.org/10.1007/978-981-97-1939-6_5

is to elaborate more on the MXene-based composite's efficiency and applications. Hence, MXene as a blooming material in environmental remediation is discussed.

1 Introduction

1.1 What are MXenes?

MXene is a fast-evolving 2D material in various fields including energy storage, electrocatalysis, photocatalysis, biomedical, and so on. MXene was discovered in 2011 by a research group of Yuri Gogotsi, Babak Anasori, and Michael Naguib at Drexel University [1, 2]. MXenes are 2D transition metal carbides and/or nitrides. The empirical formula of MXene is $M_{n+1}X_nT_x$, where M stands for one or more early transition elements (such as Sc, Ti, Zr, Hf, V, Nb, Ta, Cr, and Mo), $(n + 1)$ comprises $n = 1–4$ layers, X stands for carbide or nitride or carbonitride, oxycarbides, T_x represents surface functional groups like −O, −OH, −F, −Cl, −Br, −I, −S, −Se, −Te, imido (−NH), or termination free surfaces on the outermost exposed M atoms [3–5]. MXenes are usually synthesized through the chemical etching process of MAX phases which are layered ternary carbides or nitrides. The general formula for the MAX phase is $M_{n+1}AX_n$. where A stands different from MXene which mostly includes group 11–16 elements of the periodic table or non-MAX phase. More than one "A" atomic layer (e.g., $M_{n+1}A_2X_n$) or carbide layers of A elements (e.g.: $M_nA_3X_{n+2}$) would be present in MAX phases. MAX phases were first discovered in the 1960s [5, 6]. More than 100 MXenes were predicted through theoretical calculations of which 70 MXenes were synthesized experimentally. In theoretical calculations as well as through experiments more than 155 MAX phases have been discovered so far. Hence, MXenes etched from the MAX phase inherit parent properties with distinctive properties. Structurally, in MAX phases, the octahedral blocks are of transition metal carbide, nitride, or carbonitride sheets; hence, it's a layered hexagonal material. Octahedron centers are occupied by X atoms which bond together with pure A layer, wherein MXenes are M_6X octahedra which results in edge-sharing subunits with hexagonal crystal sublattice of M where the interstitial sites are occupied by X atoms. Based on T_x and M interactions, different sites on M are occupied by surface groups [7, 8]. MXenes exist in various forms like flakes, thin films, ink, nanosheet, clay, and so on. When MXene exposed to environment (air, heat, light and water), it is oxidized to carbon-dioxide, carbon monoxide and respective metal oxides of transition metals because of the presence of carbon layer in MXene which undergoes oxidation easily [9]. The stability of MXene flakes can be improved by using capping agent or any modifications during the synthesis process. Colloidal solution of MXene is found to be degraded after 15 days of synthesis. Based on the Acute Toxicity Rating Scale developed by the Fish and Wildlife Service (FWS), Ti_3C_2Tx nanosheets are rated as practically harmless [10]. MXene's acute toxicity to humans and animals has been found to be negligible in tests for $Mo_2Ti_2C_3T_x$,

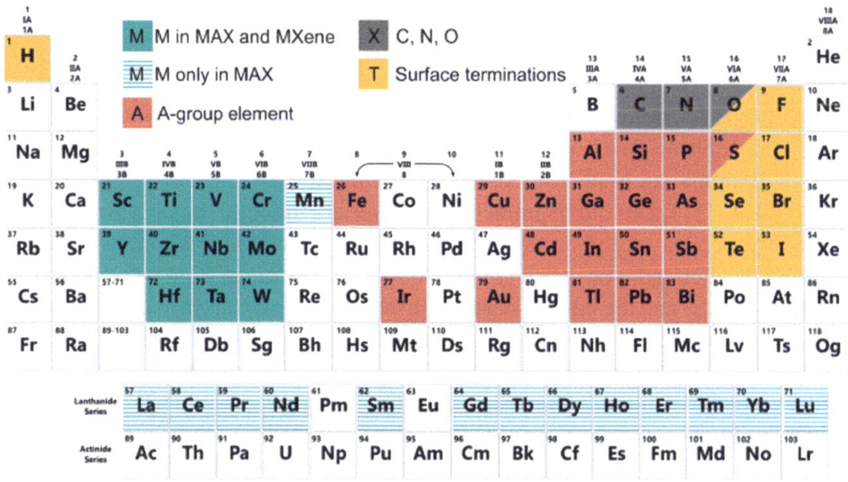

Fig. 1 Pictorial representation of periodic table which illustrates the MAX phase and MXene's compositional elements along with possible surface terminal groups. Adapted with permission from Ref. [5]

Mo_2CT_x, Nb_2CT_x, $Nb_4C_3T_x$, $Ta_4C_3T_x$, and $Ti_3C_2T_x$ [11] . The possible MAX phase and MXene compositions were illustrated in Fig. 1.

1.2 Classification of MXene

MXenes are classified based on the composition of M and the number of different transition metals present as M in single MXene composition. The MXene empirical types are M_2CT_x, $M_3C_2T_x$, $M_4C_3T_x$, and $M_5C_4T_x$. Types of MXenes are framed based on the transition elements present in the composition: Mono transition metal MXenes, double ordered transition metal MXenes, solid solution MXenes, divacancy MXenes, and high entropy MXenes. From the name of mono transition metal MXene itself it can be understood that one transition metal plays a role as M, for example $Ti_3C_2T_x$, Nb_2CT_x, $Hf_3C_2T_x$, $Zr_3C_2T_x$, Mo_2CT_x, $Ti_4N_3T_x$, V_2CT_x, Ti_3CNT_x, $Nb_4C_3T_x$, $Mo_3C_2T_x$, $V_4C_3T_x$, Ti_2CT_x, $Ta_4C_3T_x$, and $Mo_4C_3T_x$, whereas double transition metal MXenes comprises two different transition metals in their composition like $Cr_2TiC_2T_x$, $Mo_2Ti_2C_3T_x$, $Cr_2VC_2T_x$, $Mo_2TiC_2T_x$, and so on. This type is again classified into two: one is out-plane ordered double transition metal MXenes, and the second is in-plane ordered double transition metal MXenes. The first one was discovered in 2014 which forms a sandwich structure with two transition metals. Within each M-X block two of M' layers and either two or more layers of M'' were sandwiched. It has $P6_3/mmc$ symmetry. Out-plane ordered MXenes are Mo_2CT_x, Mo_2TiCT_x, $Mo_2Ti_2C_3T_x$, and $Mo_2ScC_2T_x$. The second one was discovered in 2017, where two transition metal atoms (M', M'') are arranged to form a hexagonal structure whereas the centers of hexagons are occupied by M''. Here the order is within the M layers. It has C2/c monoclinic symmetry. Vacancy of MXenes also falls under in-plane ordered MXene; here one transition metals play a role as M. While etching the A layer of MAX phase, the minor amount of M is etched away which results in Vacancy MXenes. Examples of in-plane MXenes are $V_{1.5}Cr_{1.5}C_2$, $W_{4/3}C$, $Mo_{4/3}C$, $Mo_{1.33}CT_x$, and $W_{1.33}CT_x$. Solid solutions influence all other classified MXenes which consist of two types based on the arrangements of transition metal with varied M on its respective M sites and types of elements and its composition of X on its respective X sites. One is order and the other is random. In-plane and out-plane MXene falls under ordered solid solutions, whereas in random it has two categories one is full-range mixing of two to four transition metals and the other is mixed X elements. A few example for solid solution MXene are (Ti, Nb)$_4$C$_3$, (Nb, Zr)$_4$C$_3$, (Ti, V)$_2$C, and (Ti, Nb)$_2$C. The high entropy MXene was discovered in 2021 by relating configurational entropy and stability of the material. Investigation on more classification is under study and in 2022 the oxycarbide MXene was discovered by confirming the presence of oxygen in the sublattices of X sites. The overall classification in MXene family were reprented in Fig. 2.

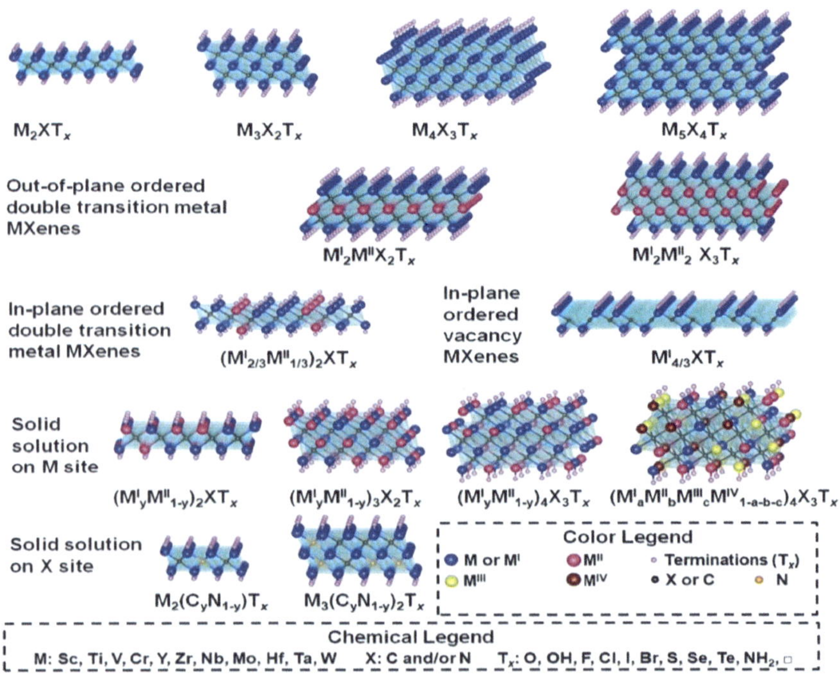

Fig. 2 Schematic representation of the classification of MXene family. Adapted with permission from Ref. [7]

1.3 Synthesis of MXene

Since its discovery, the MXene family has grown faster over the years because of the untiring research on its synthesis to make the large scale feasible. There were two approaches employed in the synthesis of MXene. Based on the choice of application of the material, either of these approaches is chosen for the synthesis.

1.3.1 Bottom-Up Approach

The bottom-up approach method was used in the synthesis of MXene with different stoichiometries. This method includes chemical vapor deposition (CVD), template method, and plasma-enhanced pulsed laser deposition (PEPLD) synthesis methods. Material obtained from this method is defect-free. Synthesis of single-layered MXene cannot be synthesized through this method.

Chemical vapor deposition method

Xu et al. [12] developed this method by synthesizing 2D ultrathin α-Mo_2C crystals in 2015 by using Cu and Mo foil as substrate and methane as carbide source. They tried with 1085 °C for the chemical decomposition on the substrate surface which results in deposition. Liquid Cu/Mo interface is formed through the melting of Cu and Mo foil at high temperature. In the beginning, Cu foil was set above the Mo foil after the interface formation Mo atoms diffused to the liquid Cu surface and the methane decomposed to give C atoms. The diffused Mo atoms ended up in the formation of Mo_2C crystals by reacting with C atoms. The finally formed product has a larger surface area and size can be controlled. The thickness of the material can be controlled by the concentration of methane. Through this method, MXene can be synthesized free of defects, and most of the material prepared through these methods was used for energy storage applications.

Template method

Transition metal carbides (TMCs) and nitrides (TMNs) were synthesized by using transition metal oxides (TMOs) as templates by carbonizing or ammoniating them, and the structural properties depend on TMO used as template. Other than the CVD method, the template method can be employed to synthesize TMNs and TMCs with higher yields. Using salt-template synthesis methods, Xiao et al. [13] synthesized 2D-MoN nanosheets. NaCl coating on MoO_3 was done under an Ar atmosphere at 280 °C by annealing precursor of Mo and NaCl crystals, and then at 650 °C it was ammoniated slowly by NH_4, and the final wash with de-ionized water MoN is obtained. The obtained one has a similar hexagonal structure as MoO_3 which is confirmed through XRD. Two-step synthesis method for layered MoN is reported by Joshi et al. [14]; in the first step, layered MoO_3 is prepared by heating Mo wire in oxygen atmosphere closer to its melting point then on a closer substrate and was let to be deposited. In the second step, it was ammoniated at 700 °C for 8 h by using

NH_4/Ar in the ratio of 1:4. The synthesized MoN has a layered structure like its metal oxide precursor. Jia et al. [15] used Mo_2O nanosheets as a template and for carbon and nitrogen sources, they used dicyandiamide. First, they prepared Mo_2O by chemical vapor reduction of Mo_3O. In the second step, synthesized Mo_2O and $C_2H_4N_4$ were taken in the ratio of 10:1 for calcination at 450 °C followed by 700 °C for 2 h and 4 h in total. They took SEM images as confirmation of morphological similarity of Mo_2O and MoN nanosheets. Through this method nanosheets forms of TMCs and TMNs were synthesized with less defect and high yields.

Plasma-enhanced pulsed laser deposition method

Zhang et al. in 2017 [16, 17] have devised a process known as PEPLD (plasma-enhanced physical vapor deposition) for the synthesis of extensive, ultrathin films of face-centered cubic (FCC) Mo_2C on sapphire (0001) substrates. The PEPLD technique integrates the benefits of plasma-enhanced chemical vapor deposition (PECVD) and pulsed laser deposition (PLD) methodologies. During the synthesis procedure, the sapphire substrate was subjected to a temperature of 700 °C to facilitate the deposition of a high-quality Mo_2C film. The CH_4 plasma was employed as a carbon source, undergoing a reaction with the molybdenum vapor produced by the pulsed laser. The resulting layer of FCC Mo_2C exhibits a high degree of uniformity and smoothness, with the ability to precisely adjust its thickness ranging from 2 to 25 nm through modulation of the laser pulse rate. Nevertheless, the films acquired exhibit significantly reduced crystalline quality compared to the samples generated using chemical vapor deposition (CVD), as evidenced by the presence of numerous stacking flaws visible in TEM pictures. They claimed the emergence of the FCC structure can be attributed to an orientation relationship and minimal lattice misfit between $\{111\}Mo_2C$ and $\{0003\}$ sapphire. The PEPLD method is a relatively new technique, and it is still under development which can be used to synthesize films with different thicknesses and crystalline qualities.

1.3.2 Top-Down Approach

The top-down approach is extensively used because of its selective etching synthesis methods. In this approach, the general synthesis process falls into two essential steps: one is the exfoliating of A layer from the MAX phase through processes like etching and the second is the delamination of multilayered MXene. Delamination of MXene can be achieved through the intercalation process. In the intercalation process, intercalants were used such as dimethyl sulfoxide, tetramethylammonium hydroxide or tetrabutylammonium hydroxide, dimethylformamide, N-Methyl-2-pyrrolidone, and lithium chloride. Delaminated MXenes shows enhanced efficiency of the material because of the larger surface area exposure than the non-delaminated ones. These two processes were achieved through various synthesis methods. Some of them are explained below.

HF etching

HF etching was chosen as the initial approach to synthesize MXene due to its ability to selectively erode the Al layer in the MAX phase [1]. To etch HF and produce MXene, the following stages are usually involved: First, preparation of MAX phase material, which consists of layers of "M" (transition metal) atoms interleaved with layers of "A" (typically aluminum) atoms and "X" (carbon and/or nitrogen) atoms. In a controlled environment, the MAX phase precursor and concentrated hydrofluoric acid (HF) are mixed. The MAX phase precursor's "A" layer is selectively removed by the HF. The "M" and "X" layers, which form the framework of the MXene structure, are left behind by this. After centrifuging the suspension to precipitate the powder, any remaining HF and byproducts are washed out using de-ionized water. To ensure the stability and safety of the MXene product, this step is essential. To get rid of any leftover water, the washed MXene powder is dried, frequently with the use of drying processes. To acquire the desired qualities, additional treatments like delamination, functionalization, or layer stacking may be applied to the synthesized multilayered MXene, depending on the application [18]:

$$\begin{aligned}
&Ti_3AlC_2(s) + 3H^+(aq) \rightarrow Ti_3C_2(s) + Al^{3+}(aq) + \tfrac{3}{2}H_2(g)\\
&Al^{3+}(aq) + 3F^-(aq) \rightarrow AlF_3(aq)\\
&Al^{3+}(aq) + 6F^-(aq) \rightarrow AlF_6^{3-}(aq)\\
&Ti_3AlC_2(s) \rightarrow Ti_3C_2(s) + 3e^- + Al^{3+}(aq)
\end{aligned}$$

Molten salt etching

The molten salt synthesis of MXene is a method for producing these materials using molten salts as the etchant. Urbankowski et al. [19] used the molten salt etching method first in 2016 to synthesize MXenes. The powder of Ti_4AlN_3 mixed with a 1:1 mixture ratio of fluoride salt is heated for 30 min at 550 °C. The selective etching was confirmed through the absence of Ti-containing fluorides but with the presence of five different fluoride phases containing Al in it. The dilute sulfuric acid is used to dissolve the Al-containing fluorides which is washed with de-ionized water through centrifugation and decanting. XRD confirms the presence of Ti_4NT_x (T = −OH, −F) and Ti_4AlN_3. Further tetrabutylammonium hydroxide (TBAOH) is used for delamination. The resulting product is washed and decanted with de-ionized water. It was then probe-sonicated and centrifuged after adding de-ionized water to the powder; finally it was filtered then flakes of unlayered $Ti_4N_3T_x$ (T = −OH, −F) were collected. The breaking strength of material is reduced compared with MXene obtained from HF etching. The molten salt synthesis of MXene has several advantages over other methods. It is less hazardous because it does not use HF. It is also more efficient because the etching process is faster and more complete. Additionally, it is more scalable because the molten salt can be used to etch large quantities of MAX phase powder. The molten salt etching process was more efficient than the traditional solution etching process. The molten salt synthesis of MXene has been further developed by other researchers, who used the Lewis-acid molten salt method of synthesize. For example, Li et al. [20] used a redox-controlled A-site etching method in molten salts to synthesize MXenes from several MAX phase precursors with Si, Zn, and Ga as the A element. This method allowed them to synthesize MXenes with a wider range of compositions and properties.

Bifluoride synthesis method

Halim et al. [21] in 2014 found a method which replaces HF etching using ammonium bifluoride (NH_4HF_2). Later, Feng et al. [22] produced Ti_3C_2 using bifluorides ($NaHF_2$, KHF_2, NH_4HF_2) and it resulted with larger interplanar spacing MXenes. 1 M bifluoride solution was prepared in 100 ml de-ionized water; in that, one gram of the precursor, i.e., Ti_3AlC_2 powder, is added at 60 °C and the mixture was kept stirring for different times and the product was washed until it reached pH ≈ 7, and then centrifuged and dried at 80 °C for 12 h in vacuum oven. They optimized the reaction condition with time. For $NaHF_2$ it took 8 h as its etching time and for NH_4HF_2 12 h of etching time; then for KHF_2 it took 24 h of time to get fine MXenes. In the HF etching method, intercalants were used to increase d-spacing, where in the bifluoride synthesis method Na^+, K^+, NH^{4+} and their hydrated cations act as intercalants which are the reason for larger interplanar spacing because of their size. It's a single-stage process with a safe and effective method for the synthesis of MXenes:

$$\begin{aligned}
&Ti_3AlC_2 + XHF_2 \rightarrow X_aAlF_b + AlF_3 + H_2 + Ti_3C_2\\
&AlF_3 + cH_2O \rightarrow AlF_3 \cdot cH_2O\\
&Ti_3C_2 + XHF_2 + H_2O \rightarrow Ti_3C_2F_x(OH)_yX_z
\end{aligned}$$

where X stands for K^+, Na^+, NH^{4+} and the varied composition numbers are represented by a, b, c, x, y, and z.

Urea glass route

The urea glass route synthesis of MXene is a relatively new method that was first reported in 2015 by Ma et al. [23] which is a green and sustainable method that does not use any hazardous chemicals. MXenes like Mo_2C and Mo_2N were synthesized from $MoCl_5$ as metal precursor through this method. Ethanol was added into the precursor, and as a result Mo-orthoesters were formed. The mixture was then supplemented with solid urea, which was stirred into the solution until it was completely dissolved. The resulting gel-like product is treated under N_2 gas flow along with 800 °C of heating. The product obtained is silvery black in color. In the case of MoO_x, the powder of MoO_x exhibits a high degree of purity and no other crystalline side phases were detected using XRD. It is also a mild method that produces MXene nanosheets with high crystallinity and uniformity.

Alkaline-assisted hydrothermal method

In 2018, Li et al. [24] reported this method. MXene $Ti_3C_2T_x$ is synthesized from MAX phase Ti_3AlC_2. In this method, the Al atoms in the MAX phase are oxidized by hydroxide anions (OH^-) resulting in dissolved aluminum hydroxides in alkali and

OH or O formed into terminal groups on the exposed Ti atoms. They followed the Bayer process to dissolve Al hydroxides which formed during the alkali treatment, and it removes the jamming effect in MXene formation. The reaction was carried under an argon atmosphere. By applying a series of hydrothermal temperatures and NaOH solution concentrations, the reaction condition is optimized. At 270 °C and 275.5 M concentration of NaOH treatment gives good results as the result of their work. MXene synthesized through this method has more OH and O terminations than HF-etched MXenes.

Electrochemical synthesis method

In 2018, Yang et al. [25] employed binary aqueous electrolyte instead of fluorine in this electrochemical approach to delaminate Ti_3C_2. The anode alone underwent an etching process to yield $Ti_3C_2T_x$, and even both the anode and cathode of the two electrodes constructed in this system are bulk Ti_3AlC_2. They took a mixture of 1 M NH_4Cl and 0.2 M tetramethylammonium hydroxide (TMAOH) with a pH of >9 to prevent the etching that occurred on the surface and allow the electrolyte ions to get into the deeper layers of the anode. The bulk anode gradually delaminated when a low voltage of +5 V was applied. To create individual sheets of $Ti_3C_2T_x$, the sediment and suspended powders of $Ti_3C_2T_x$ were then pounded and added to 25% w/w TMAOH.

Intercalation and Delamination

Synthesizing a few or single-layer MXene exhibits different and enhanced properties than their bulk material. Physical, electrochemical, and chemical properties of the material change because of an increase in the surface area to the volume ratio. Single-layer MXene can be achieved through various methods like chemical intercalation, mechanical cleavage, liquid exfoliation, and so on. In the delamination process, the intercalants are used to intercalate in the layers of MXene thereby the interlayer spacing increases and through sonication the interaction between the layers weakens; finally, the delamination of layers occurs. Ghidiu et al. [26] in 2014 tried the molten salt method in which LiF + HCl is used for etching the Al layer simultaneously hydrated Li ion intercalate between the M–X bonds resulting in delamination of multilayer MXene to single layer. The selection of intercalant depends on the etching method. Sonication plays a role in the delamination of layered MXene. Mashtalir et al. [27] first reported delamination of MXene using intercalants and sonication process. Some of the chemical intercalants are ethanol, acetone, thiophene, formaldehyde, tetrahydrofuran, chloroform, hexane, toluene, hydrazine, dimethylformamide, dimethyl sulfoxide, tetrabutylammonium hydroxide, and so on. Among these intercalants, they tried with DMSO and urea. Multilayered MXene synthesized from HF etching was taken and stirred in DMSO for 18 h then the residual DMSO was removed by several washes; it was subjected to sonication for 4 h after dispersing it in water. Finally, the delamination of multilayer MXene occurred and single-layer MXene flakes were obtained. Wang et al. [28] used aqueous solution of 5 wt% of NaOH to disperse MXene prepared through HF etching stirred for 2 h; Na^+ ions intercalate into MX layers. In the process of Na^+ ions intercalation, sulfanilic acid diazonium

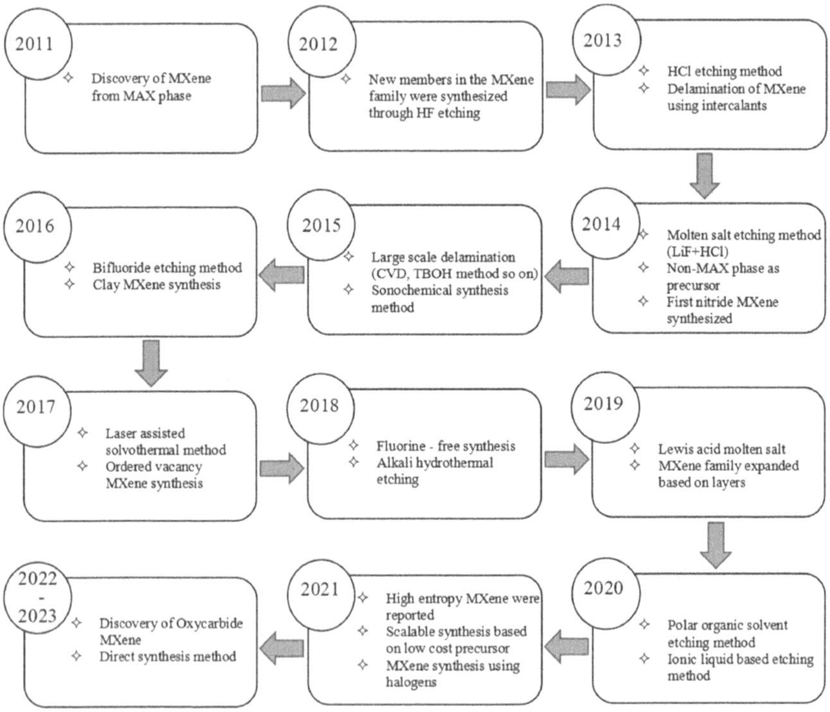

Fig. 3 Illustration of remarkable developments in MXene over the years

salt is added then stirring continued for 4 h under 5 °C. Finally, the stable MXene dispersion is formed because of aryl group intercalation and phenylsulfonic group interaction with MXene's negative surface charge weakens the interaction between the layers resulting in delamination. The dimension and stability of MXene can be controlled in this process, and the delamination of MXene is still under research. The following illustration shows the notable developments and discoveries that happened to MXene as years go from its discovery Fig. 3.

1.4 MXene Composite Synthesis

The potential of MXene in diverse fields is widely recognized; yet it's prone to oxidation in moisture, air, and at high temperatures. Hence, it is important to frame the synthesis method with mild conditions, especially for less layered MXenes. However, the material's limitations can be effectively addressed through the implementation of composite fabrication strategies, thereby enhancing the overall efficiency of the composite material. Novel materials can be synthesized by combining MXene with various other materials like metal oxides, metal sulfides, reduced graphene, carbon nitride, bismuth compounds, and so on. Composites can be synthesized through various methods like sonication, hydrothermal treatment, solvothermal process, calcination, mechanical stirring, and many more. As mentioned earlier, MXene has potential in various fields from energy storage to cleaning the environment. Some of the MXene-based composites andtheirs synthesis with diverse applications are listed in the following Table 1 as well as the contaminants used for the study are addressed in the reactant column. As for the hydrogen evolution study, sacrificial agents like methanol, lactic acid, Na_2S/Na_2SO_3, triethanolamine (TEOA), and so on are used as hole scavengers which are also mentioned in the table along with the composite performance rate. Interestingly, in nitrogen reduction reaction, the MXene nanocomposite acts as robust catalyst to produce NH_3. Ammonium formate is used as a hole scavenger in photocatalytic reduction of 4-nitroaniline (4-NA) to p-phenylenediamine (PPD) at the rate of 180 mg/L/h. $KHCO_3$ aqueous solution is the electrolyte used in M-TCCN || $BiVO_4$ electrode system for the photoelectrocatalytic reduction of CO_2 to hydrocarbon at the rate of 25.1 mM/h/g. The possibility of photogenerated electron transfer is higher for the composite because of the increased interfacial area when the photocatalyst is formed through a chemical reaction. Most of the MXenes used for the study mentioned here were synthesized through either HF or the molten salt etching method, and a few processes like stirring, sonication, and hydrothermal were employed more on the fabrication steps of composite preparation.

Referring to the above table, synthesis of MXene nanocomposites as well as the enhanced efficiency of the synthesized composite are also evident. Some picks from the list were as follows: $TiO_2@C/g\text{-}C_3N_4$ reduces nitrogen to ammonia at a high-performance rate of 250.6 μ mol h^{-1} g^{-1} than $TiO_2@C$ and $g\text{-}C_3N_4$. $TiO_2@C/g\text{-}C_3N_4$ is prepared by the calcination of melamine where $TiO_2@C$ is formed from the oxidation of Ti_3C_2 [19]. In this, Xu et al. prepared $Ti_3C_2/g\text{-}C_3N_4$ with rich Ti^{3+} species which suppresses charge recombination. $CdS/2.5\%Ti_3C_2$ is prepared through hydrothermal reaction. This composite performance on hydrogen production is 136.6 times higher than its pristine CdS, and the total rate of production is 14,342 μ mol h^{-1} g^{-1}. Wang et al. used a hydrothermal method to synthesize $In_2S_3/Ti_3C_2T_x$ hybrid for the degradation of methyl orange. The performance rate is 18 mg/L/h. Ti_3C_2 acts as a $BiVO_4$ nanoparticle growth supporter as well as a precursor for oxidized 2D carbon formation which facilitates the interfacial charge transfer; thus, it efficiently degrades Rhodamine B. Ultrathin MX/Bi_2WO_6, has 2D/2D heterojunction which is synthesized by a hydrothermal method and gives yield of CH_4 and CH_3OH to 4.6 times than its pure Bi_2WO_6. CO_2 absorption is increased because of its improved specific area and pore structure [58].

MXene materials have been employed in several uses, such as doping, as supporting materials, and as co-catalysts, with the aim of enhancing the overall efficiency of the material. The implementation of a composite strategy proved to be effective in optimizing the performance of the materials under investigation.

Table 1 Table for list of MXene-based composites synthesized through various synthetic methods for photocatalytic application

Sample	MXene (synthetic method)	Sample synthesis	Reactant	Sacrificial agent	Rate	Precursor
g-C_3N_4/3%Ti_3C_2/2%Pt	Ti3C2 flakes (HF 48%, 20 h, 60 °C and H2O delamination, 12 h, ultrasonication)	(1) Ti_3C_2 stirring dispersions (2) Pt UV deposition	H_2O	10 vol% triethanolamine (TEOA)	5100 μmol/h/gcat	–
g-C_3N_4/$Ti_3C_2T_x$ (1: 0.3)	Multilayer Ti_3C_2 (HF 49%, 24 h)	Grinding in a mortar	H_2O	10 vol% TEOA	88 μmol/h/gcat	–
CdS/0.5%$Ti_3C_2T_x$	Ti_3C_2 flakes (LiF 1 g/ HCl 9 M, 24 h, 35 °C)	(1) Ultrasonication (2) Stirring in water	4-NA	40 mg ammonium formate in 30 mL solution	180 mg/L/h	–
P25/5%Ti_3C_2–OH	Multilayer Ti_3C_2 (HF 49%, 24 h and KOH 2 M, 4 h)	Stirring in water	CO_2	–	28.35 μmol/h/gcat	–
a-Fe_2O_3/Ti_3C_2 (1: 2)	Multilayer Ti_3C_2	(1) Stirring in ethanol (2) Ultrasonication	Rhodamine B (RhB)	–	5 mg/L/h	–
g-C_3N_4/Ti_3C_2 (100: 3)	Ti_3C_2 flakes (HF 40%, 24 h and H_2O intercalation, 5 h, ultrasonication)	(1) Ultrasonication (2) Stirring in water at 60 °C	Ciprofloxacin	–	18 mg/L/h	–
TiO_2/5%Ti_3C_2	Ti_3C_2 flakes (LiF 1 g/ HCl 6 M, 24 h, 35 °C)	Sonication	H_2O	25% Methanol	2650 μmol/h/gcat	–
			H_2O			–
Ag_3PO_4/2%Ti_3C_2	Ti_3C_2 flakes (NaF 3.35 g/HCl 36–38 wt%, 12 h, 60 °C)	(1) Stirring in water with $AgNO_3$ (2) Adding Na_2HPO_4	Tetracycline hydrochloride (TC-H), etc	–	192 mg/L/h	–

(continued)

Table 1 (continued)

Sample	MXene (synthetic method)	Sample synthesis	Reactant	Sacrificial agent	Rate	Precursor
3%Ti_3C_2/g-C_3N_4	Ti_3C_2 flakes (LiF 1.5 g/HCl 6 M, 24 h, 35 °C)	(1) Sonication in HCl (2) Stirring	H_2O	10 vol% TEOA	73.3 μmol/h/gcat	–
TiO_2/0.5%Ti_3C_2/1%CoS_x	Multilayer Ti_3C_2 (HF 49%, 4 h)	(1) Stirring in 2-methylimidazole (2) Hydrothermal 140 °C for 12 h with thioacetamide	H_2O	20 vol% methanol	950 μmol/h/gcat	$Co(NO_3)_2$, 2-methylimidazole and thioacetamide
CdS/MoS_2/2%$Ti_3C_2T_x$	Ti_3C_2 flakes (HF 49%, 72 h, ultrasonication in H_2O, 2 h)	(1) MoS_2 synthesis (2) Stirring with Ti_3C_2 (3) Add CH_4N_2S and $Cd(CH_3COO)_2$ (4) Hydrothermal 160 °C for 24 h	H_2O	0.25 M Na_2S and 0.35 M Na_2SO_3	9679 μmol/h/gcat	$Cd(CH_3COO)_2$, CH_4N_2S, MoS_2
0.4%Ti_2C/g-C_3N_4	Ti_2C flakes (NH_4F 16 g/HCl 9 M, 24 h)	(1) Stirring ethanol (2) 550 °C, 4 h in muffle	H_2O	10 vol% TEOA	950 μmol/h/gcat	Melamine
10%TiO_2@C/g-C_3N_4	Multilayer Ti_3C_2 (HF 49%, 4 h)	(1) Stirring in water (2) 550 °C, 2 h in muffle	N_2	20 vol% methanol	250 μmol/h/gcat	Melamine
Pd–Ti_3C_2/g-C_3N_4(1:10)	Multilayer Ti_3C_2 (HF 40%, 24 h)	(1) Grinding (2) 500 °C, 2 h in muffle (3) Pd electrodeposition	CO_2	0.1 M $KHCO_3$	25,100 μmol/h/gcat	Thiourea

(continued)

Table 1 (continued)

Sample	MXene (synthetic method)	Sample synthesis	Reactant	Sacrificial agent	Rate	Precursor
0.001 mol TiO_2/Ti_3C_2	Multilayer Ti_3C_2 (HF 49%, 24 h, 60 °C)	(1) Stirring (2) Hydrothermal 180 °C, 18 h	Methyl orange (MO)	–	40 mg/L/h	Ti_SO_4
TiO_2/5%Ti_3C_2	Ti_3C_2 flakes (HF 48%, 15 h and DMSO delamination, 15 h)	(1) Stirring in ice-water bath (2) Heated 95 °C, 4 h	H_2O	25% methanol	43 μmol/h/gcat	$TiCl_4$
TiO_2/5%Ti_2C	Ti_2C flakes (HF 10%, 10 h and DMSO delamination)					
TiO_2/5%Nb_2C	Nb_2C flakes (HF 48%, 90 h and 20% isopropyl alcohol delamination)					
CdS/2.5%Ti_3C_2	Ti_3C_2 nanoparticles (HF 49%, 20 h, 60 °C and H_2O delamination, ultrasonication, 5 h)	(1) Stirring in water (2) Hydrothermal 180 °C, 12 h	H_2O	Lactic acid (88 vol%)	14,342 μmol/h/g	$Cd(Ac)_2$
						Thiourea
$TiO_2/C/BiVO_4$ (1:1079)	Ti_3C_2 flakes (LiF 1.5 g/ HCl 6 M, 48 h, 50 °C)	(1) Stirring in water (2) Hydrothermal 100 °C, 6 h	RhB	–	3.1 mg/L/h	$Bi(NO_3)_3$
						NH_4VO_3
TiO_2/Ti_3C_2 (1:1)	Multilayer Ti_3C_2 (HF 40%, 26 h, 60 °C)	(1) Stirring in 10 M NaOH (2) Hydrothermal 180 °C, 10 h	Methylene blue (MB)	–	8.5 mg/L/h	P25

(continued)

Table 1 (continued)

Sample	MXene (synthetic method)	Sample synthesis	Reactant	Sacrificial agent	Rate	Precursor
BiOBr/Ti_3C_2 (250: 1)	Ti_3C_2 flakes (LiF 3 g/ HCl 9 M, 24 h, 35 °C)	(1) Stirring (2) Refluxed 80 °C, 2 h	RhB	–	24 mg/L/h	$Bi(NO_3)_3$ and KBr
2%Ti_3C_2/Bi_2WO_6	Ti_3C_2 flakes (HF 40%, 72 h and DMSO delamination, ultrasonication, 1 h)	(1) Stirring (2) Hydrothermal 120 °C, 24 h	CO_2	–	2.22 μmol/h/gcat	$Bi(NO_3)_3$
						Na_2WO_4
$Bi_{0.9}Gd_{0.1}Fe_{0.8}Sn_{0.2}O_3$/ Ti_3C_2	Multilayer Ti_3C_2 (HF 39%, 36 h)	(1) Stirring in 0.01 M acetic acid and ethylene glycol (2) Sonicated, 2 h, 60 °C (3) stirring 1 h, 80 °C	Congo red	–	–	$Bi_{1-x}Gd_xFe_{1-y}Sn_y$
In_2S_3/TiO_2@ $Ti_3C_2T_x$ (1:0.123)	Multilayer Ti_3C_2 (HF 50%, 20 h)	(1) Stirring (2) Hydrothermal 180 °C, 24 h	MO	–	18 mg/L/h	$In(NO_3)_3$
						CH_3CSNH_2
ZnS/0.75 wt%Ti_3C_2	Ti_3C_2 flakes (HF, 24 h, 25 °C)	(1) Stirring in ethanol–glycerol (2) Hydrothermal 180 °C, 10 h	H_2O	20 vol% lactic acid	502.6 μmol/h/gcat	$ZnCl_2$

(continued)

Table 1 (continued)

Sample	MXene (synthetic method)	Sample synthesis	Reactant	Sacrificial agent	Rate	Precursor
$Ti_2C/3\%TiO_2/1\%Ag$	Multilayer Ti_2C (HF 48%)	(1) Stirring for volatiles evaporation (2) Annealing in H_2 at 400 °C	Salicylic acid	–	32.4 μmol/h	Titanium isopropylate
TiO_2/Ti_3C_2 (12 h)	Multilayer Ti_3C_2 (HF 49%, 12 h, 60 °C)	Hydrothermal 160 °C for different time, $NaBF_4$ and HCl	MO	–	24 mg/L/h	–
TiO_2/Ti_3C_2 (20 h)	Multilayer Ti_3C_2 (HF 49%, 12 h, 60 °C)	Hydrothermal 200 °C for different time, NH_4F	MB	–	6 mg/L/h	–
HC–TiO_2	Ti_3C_2 flakes (tetramethylammonium hydroxide 25%, 24 h)	Hydrothermal 160 °C, 9 h	H_2O	10 vol% TEOA	33.04 μmol/h/gcat	–
$4\%Cu_4/TiO_2@Ti_3C_2T_x$-12 h	Multilayer Ti_3C_2 (HF 49%, 12 h, 60 °C)	(1) Hydrothermal 160 °C for different time, $NaBF_4$ and HCl (2) Photodepositing copper nanodots	H_2O	6.7 vol% methanol	764 μmol/h/gcat	–
$Ti_3C_2/TiO_2/CuO$ (100:1)	Multilayer Ti_3C_2 (HF 49%, 24 h, 60 °C)	(1) Dissolved in water (2) Annealing in argon, 500 °C, 30 min	MO	–	15 mg/L/h	–

(continued)

Table 1 (continued)

Sample	MXene (synthetic method)	Sample synthesis	Reactant	Sacrificial agent	Rate	Precursor
C/TiO_2-700 °C–150 sccm	Multilayer Ti_3C_2 (HF 40%, 90 h, 55 °C)	Heated in CO_2 at different temperature and different rate, 1 h	H_2O	10 vol% TEOA	480 μmol/h/gcat	–
TiO_2/Ti_3C_2 (TT550 °C)	Multilayer Ti_3C_2 (HF 50%, 48 h)	Calcination at different temperature	CO_2	–	4.4 μmol/h/gcat	–
Nb_2O_5/C/Nb_2C-1 h	Multilayer Nb_2C (HF 50%, 90 h)	Annealing in CO_2, 850 °C for different time	H_2O	25% methanol	7.81 μmol/h/gcat	–
Microporous-MXene/TiO_{2-x} nanodots	Multilayer Ti_3C_2 (HF 50%, 90 h)	30% H_2O_2, 10 min	RhB, etc.	–	–	–
C/TiO_2	Multilayer Ti_2C (HF 40%, 2.5 h)	High-energy ball milling in air, 1.5 h, 200 rpm	MB	–	2.13 mg/L/h	–
TiO_2/Ti_3C_2@AC-48 h	Multilayer Ti_3C_2 (HF 49%, 24 h)	Heated in H_2O for different time at 60 °C	H_2O	29 g/L ascorbic acid (AA)	33.4 μmol/h/gcat	–
Ti_3C_2/TiO_2-500/Pt	Multilayer Ti_3C_2 (HF 40%, 72 h)	(1) Hydrothermal in 1 M NaOH and 30% H_2O_2, 140 °C, 12 h (2) Immersed in 0.1 M HCl, 24 h (3) Annealing in muffle for different time	H_2O	20 vol% methanol	H_2 1596.35 μmol/h/gcat	–
				0.01 M $AgNO_3$	O2 500 μmol/h/gcat	

(continued)

Table 1 (continued)

Sample	MXene (synthetic method)	Sample synthesis	Reactant	Sacrificial agent	Rate	Precursor
				–	H_2 526 μmol/h/gcat and O_2 315 μmol/h/gcat	
LDC-S-TiO_2/C	Multilayer Ti_3C_2 (HF 40%, 48 h, 45 °C)	(1) Ball mixing with sulfur (2) Hydrothermal 155 °C,12 h (3) Annealing in CO_2 at 700 °C for 2 h (4) Annealing in air at 450 °C, 2 h	H_2O	10% methanol	333 μmol/h/gcat	–
TiO_2/Ti_3C_2	Multilayer Ti_3C_2 (HF 30%, 10 h, 40 °C)	Hydrothermal 160 °C for 12 h, $NaBF_4$ and HCl	Carbamazepine	–	1.48 mg/L/h	–
Ti_3C_2/TiO_2/15%MoS_2	Multilayer Ti_3C_2 (HF 40%, 72 h)	(1) Hydrothermal 160 °C for 12 h with $NaBF_4$ and HCl (2) Hydrothermal 200 °C for 24 h with Na_2MoO_4 and CN_2H_4S	H_2O	TEOA	6425 μmol/h/gcat	$NaBF_4$, HCl, Na_2MoO_4 and CN_2H_4S

Adopted from Ref. [29] and further references cited therein

1.5 MXene in Photocatalysis

Photocatalysis is one of the cleanest cutting-edge technologies for environmental cleanup such as the degradation of organic contaminants, the reduction of CO_2, and heavy metal adsorption. Photocatalysis is a process where a certain amount of light energy falls on the semiconductor which is equal to or greater than the band gap energy; then the electrons in the valence band are excited to the conduction band. Holes were created in the valence band where the oxidation reaction occurs with the pollutant and generated electrons to react with the dissolved oxygen subsequently to produce hydroxyl radicals followed by the degradation of environmental pollutants [30]. In photocatalytic reaction are employed three stages: (1) The generation of photoexcited carriers was facilitated through the absorption of incident light by the photocatalyst, (2) it facilitates the isolation of the charge carrier and transports it to the photocatalyst surface, and (3) surface redox reactions occurred by the generated charge carriers. Those three processes determine the photocatalytic activities. The rate determining step is the separation and migration of the photogenerated charge carrier to the photocatalyst surface.

For a great photocatalysis process, an ideal photocatalyst is needed. A photocatalyst should possess following conditions: Band gap should be tunable, facilitate a good amount of interfacial charge transfer, should generate high photogenerated electron and hole pairs, and must inhibit recombination of charge carriers resulting in higher hydroxyl radical production which is the primary oxidant in the photocatalytic system. The photocatalyst material's chemical element can alter its valence state reversibly, accommodating a hole in the semiconductor without decomposing, for example Ti^{3+}–Ti^{4+} in TiO_2 [31].

MXene is well known among 2D materials because of its properties like hydrophilicity, biocompatibility, high surface area, activated metallic hydroxide site, adjustable band gap, ease of functionalization, and high metallic conductivity. The band gap of MXene varies in the range of 0.05–2.87 eV and the surface terminations of MXene have –OH and –F; then it has 0.05–0.2 eV, respectively. The band gap of 2D Ti_3CO_2 is about 0.24 eV [32, 33]. Hence, MXene is under investigation in the field of photocatalysis because it possesses excellent photocatalytic properties.

MXene has potential co-catalyst properties such as electron reservoir, regulated band gap, good electrical and conducting properties, and a large number of adsorption sites on the surface. Heterojunctions like Schottky junction, R-, Z-, S-schemes were formed between MXene and other photocatalysts like GO, rGO, MoS_2, Bi_2WO_6, $MoSe_2$, etc. As a co-catalyst, the MXene helps to trap the electrons which effectively reduce the recombination rate as well the higher band gap can be modulated for the composites. Nanosheets of MXene have a large surface area and massive exposed active sites. MXene exhibits a conjugated structure and the positive sites in metals and negative sites of carbide result in stronger interaction of the active site with different reactants which favor the transfer of exciton between them. The semiconducting property of MXene is improved by the presence of surface functionalized groups. conducting property of MXene is transformed to semiconducting through surface functionalization groups. Ti_3C_2 MXene withstands wider pH range from acidity to alkaline medium. (i) The terminal functional groups like –OH, –O, and –F on the surface of MXene make good contact with water molecules and organic pollutants and strong connection with the other semiconductors in the composite. The electronegativity of MXene acts as an anchor site to construct heterojunctioned composites as well as light absorbance is improved further, for example in BOC/Ti_3C_3 composite the charge carrier separation efficiency is enhanced, in $CuFe_2O_4$/Ti_3C_2 the photocatalytic activity is enhanced, in the case of g-C_3N_4/Ti_3C_2, MXene suppress the recombination rate. (ii) MXene acts as an electron mediator because of the band bending effect which enhances the electron–hole pair isolation because of high-density electronic states around the Fermi level. The photogenerated electron is transferred from the semiconductor to MXene. MXene is known for good electron transfer because of its excellent electrical conductivity due to its metallic-like character which helps to transfer the photogenerated electron to the surface of MXene which traps it further, resulting in the enhancement of charge carrier separation. (iii) It has the photothermal conversion feature in it which enhances the photocatalytic activity of the material. In the surface of Ti_3C_2, a large number of atoms of Ti were exposed which brings stronger redox reactivity than other carbon-based 2D materials. Morphological control plays an important role in strengthening the photocatalytic activity. The heterojunction with the same dimension of 2D material, i.e., face-to-face (2D/2D) configuration, has a larger interfacial area with a shorter distance for charge carrier transportation, whereas the heterojunctions formed between different dimensions like point-to-face (0D/2D) and line-to-face (1D/2D) are not efficient like 2D/2D configuration. MXene is thermodynamically metastable in nature when it is treated in high temperatures, hence the transport of charge carriers was enhanced along with its electron conductivity. In 2D/2D heterojunction, composite material of Ti_3C_2 MXene/Mo_2S synthesized through hydrothermal method formed R-scheme which showed 97.4% methyl orange degradation rate after 30 min irradiation and 380.2 μmol/h/g rate of H_2 evolution under visible light irradiation. Among the heterojunctions, most of the MXene-based nanocomposites preferably form a Schottky and Z-scheme; from the above discussed heterojunction mechanisms Schottky and Z-scheme are effective and efficient one for photocatalytic reactions because of its recombination inhibition possibility. Double heterojunction formation also plays an efficient role in the suppression of exciton recombination. Those heterojunction mechanisms were illustrated as follows in Fig. 4.

1.6 MXene-Based Photocatalyst

1.6.1 Organic Contaminant Degradation

The demand for the development of noble metal-free co-catalysts has been growing significantly in recent times. After the evolution of MXene, those catalysts became replaceable because of MXene's unique properties. The key features of MXene as a

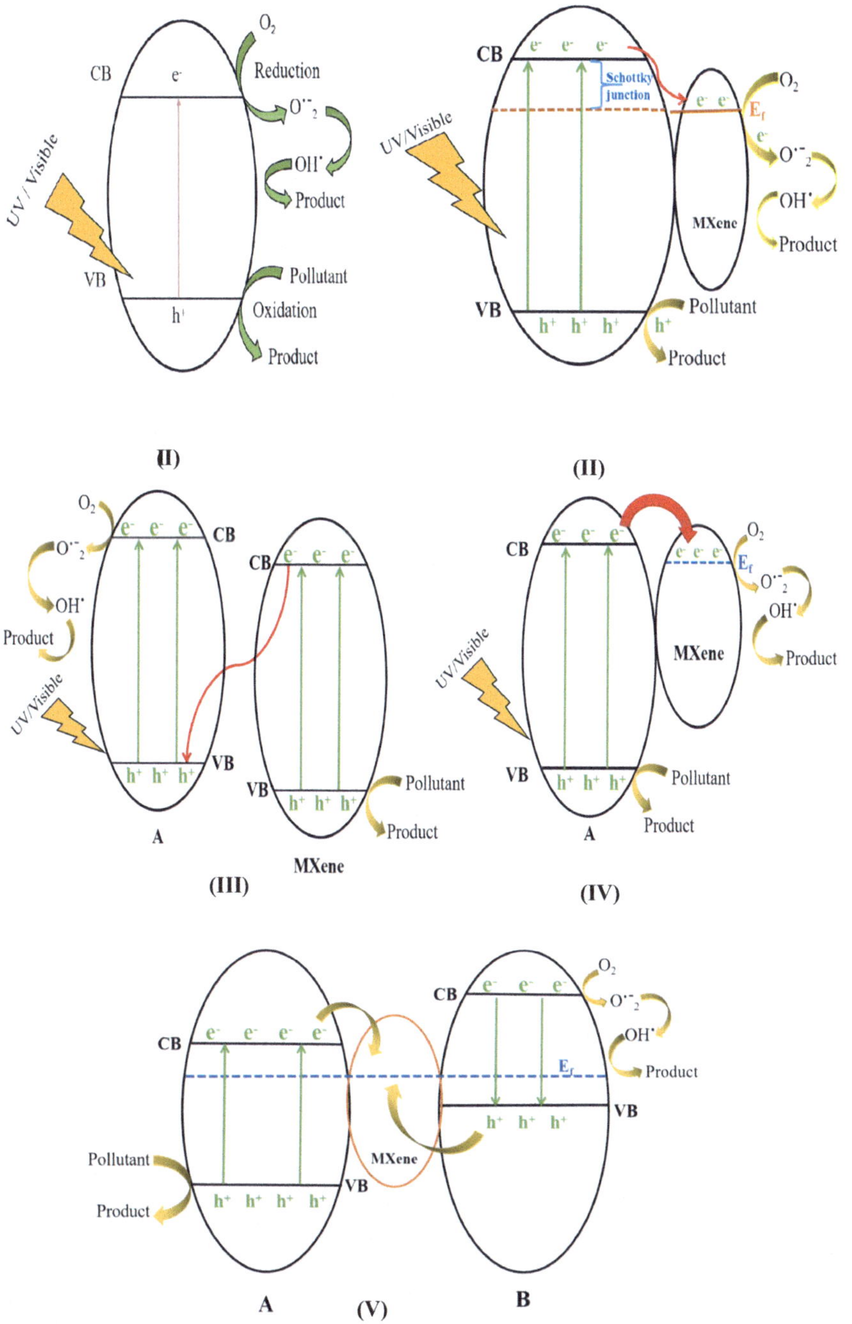

Fig. 4 Schematic representation of (I) photocatalysis, (II) Schottky, (III) S-, (IV) R-, and (V) Z-schemes mechanisms

co-catalyst are its high surface area, abundant surface functional groups, semiconductor nature with an adjustable band gap, and fast photogenerated charge carrier separation ability. MXene-based photocatalysts have emerged as a competitive candidate for other 2D materials such as graphitic carbon nitride (g-C_3N_4) and transition metal dichalcogenides (TMDs). In the Ag_2WO_4/Ti_3C_2 case, MXene acts as a co-catalyst, i.e., electron sink, and forms a Schottky heterojunction which prevents electrons from getting back to their respective photocatalyst causing the inhibition in the recombination rate. The hydrophilicity and water dispersibility of MXene improves the composite binding rate through a large number of positive ion absorption. Electrons from the Ag-based semiconductor are transferred to the surface Ti_3C_2 of the Ag_2WO_4/Ti_3C_2 composite because of MXene's electrical conductivity; in the mean time electron transfer rate increases and degradation occurs on the surfaces of MXene [34]. The heterojunction of $CdS/Ti_3C_2T_x$ composite was constructed by Xie et al., in which MXene promotes the photoreduction reaction caused by electrons and inhibits photocorrosion caused by holes. The presence of surface and interface oxygen vacancies in the nanocomposite of $Bi_2O_2CO_3/Ti_3C_2T_x$ showed enhancement in the photocatalytic degradation [35]. The photoredox ability and separation of charge carriers were enhanced in the binary composite of Bi_2WO_6/C_3N_4 due to the presence of Ti_3C_2. In the ternary system of $Bi_2WO_6/C_3N_4/Ti_3C_2$, nanocomposite interfacial contact plays a role in the separation and transfer of charge carriers [36]. Among the studied MXenes, niobium carbide (Nb_2CT_x) has a lower Fermi level compared to titanium carbide ($Ti_3C_2T_x$) hence it can serve as a better co-catalyst. Cao et al. [37] synthesized carrier-assisted $CuFe_2O_4/Ti_3C_2T_x$ composite where they used $Ti_3C_2T_x$ because of its potential properties as a co-catalyst to overcome the agglomeration, limited surface area, and recombination of $CuFe_2O_4$ nanoparticle due to its strong magnetic and intermolecular interaction. This composite efficiently degraded sulfamethazine than the monomers. g-$C_3N_4/Ti_3C_2/TiO_2$ nanotube array's in situ growth on Ti meshes synthesized through the calcination process showed an efficient degradation rate of Rhodamine dye and Tetracycline compared with the conventional g-C_3N_4/TNTAs under visible light irradiation. In this composite, TiO_2 present on the Ti meshes possesses more active sites and contact areas, whereas the recombination inhibition and charge carrier transport enhancement occur because of close interface adsorption between Ti mesh, g-C_3N_4, and Ti_3C_2 compounds [38]. Xu et al. [39] constructed 2D/2D heterostructure through in situ coupling of 1% wt of Ti_3C_2 with Bi_2WO_6 and showed good degradation efficiency by complete removal of amoxicillin. They found improvement in the properties of light absorption and photothermal conversion. Inside this composite, the transport of the charge carrier's distance and time is reduced which enhances the charge carrier separation and results in complete degradation within 40 min. Liu et al. [40] prepared sandwiched Co_3O_4/MXene composite which activates PMS to degrade 95% of bisphenol A at 20 mg/L composite within 7 min.

Light sources were important for the photocatalysis; hence the choice of light source must complement the band gap of the respective material used in the photocatalysis reaction. White LEDs are cost-effective as well as Xe lamp is used as a source of a wide range for the photocatalysis study (Table 2).

Table 2 List of MXene and MXene-based composites employed in degradation of various pollutants and the reaction condition with efficiency is discussed in the table

MXene co-catalyst	Main catalyst	Final catalyst (with the inclusion of other co-catalysts)	Pollutant	Light source	Catalyst content	Contaminant	Degradation efficiency (%)
$Ti_3C_2T_x$	$Ti_3C_2T_x$	Blender made $Ti_3C_2T_x$	Methylene blue	300 W visible light	0.02 mg/mL	2 ppm/mL	98.00
$Ti_3C_2T_x$	BiOI/(001)TiO_2	BiOI/(001)TiO_2/$Ti_3C_2T_x$	Rhodamine B	500 W Xe lamp ($\lambda >$ 420 nm)	20 mg/L	50 mg/L	–
$Ti_3C_2T_x$	Graphene/TiO_2/g-C_3N_4	$Ti_3C_2T_x$/graphene/TiO_2/g-C_3N_4	Tetracycline	300 W Xe lamp	0.6 g/L	15 mg/L	76.70
$Ti_3C_2T_x$	Graphene/TiO_2/g-C_3N_4	$Ti_3C_2T_x$/graphene/TiO_2/g-C_3N_4	Ciprofloxacin	300 W Xe lamp	0.6 g/L	3 mg/L	55.20
$Ti_3C_2T_x$	Graphene/TiO_2/g-C_3N_4	$Ti_3C_2T_x$/graphene/TiO_2/g-C_3N_4	Bisphenol A	300 W Xe lamp	0.6 g/L	5 mg/L	50.00
$Ti_3C_2T_x$	Graphene/TiO_2/g-C_3N_4	$Ti_3C_2T_x$/graphene/TiO_2/g-C_3N_4	Rhodamine B	300 W Xe lamp	0.6 g/L	20 mg/L	72.40
$Ti_3C_2T_x$	TiO_2	TiO_2/$Ti_3C_2T_x$	Bisphenol A	300 W high-pressure Hg lamp	0.1 mg/mL	15 mg/L	90.50
$Ti_3C_2T_x$	TiO_2	TiO_2/$Ti_3C_2T_x$	Rhodamine B	300 W high-pressure Hg lamp	0.1 mg/mL	15 mg/L	80.00
$Ti_3C_2T_x$	TiO_2	TiO_2/$Ti_3C_2T_x$	Phenol	300 W high-pressure Hg lamp	0.1 mg/mL	15 mg/L	90.00
$Ti_3C_2T_x$	g-C_3N_4	g-C_3N_4/$Ti_3C_2T_x$–Ag NPs	Aniline	300 W Xe lamp	100 mg/mL	20 mg/mL	81.80
Nb_2CT_x	Bi_2WO_6	Bi_2WO_6/Nb_2CT_x	Rhodamine B	500 W Xe lamp	15 mg/L	0.5 g/L	99.80
Nb_2CT_x	Bi_2WO_6	Bi_2WO_6/Nb_2CT_x	Methylene blue	500 W Xe lamp	15 mg/L	0.5 g/L	92.70

(continued)

Table 2 (continued)

MXene co-catalyst	Main catalyst	Final catalyst (with the inclusion of other co-catalysts)	Pollutant	Light source	Catalyst content	Contaminant	Degradation efficiency (%)
Nb_2CT_x	Bi_2WO_6	Bi_2WO_6/Nb_2CT_x	Tetracycline hydrochloride	500 W Xe lamp	15 mg/L	0.5 g/L	83.10
$Ti_3C_2T_x$	CeO_2	$CeO_2/Ti_3C_2T_x$	Tetracycline	350 W Xe lamp	1 mg/mL	10 mL/L	80.20
$Ti_3C_2T_x$	Ag_2WO_4	$Ag_2WO_4/Ti_3C_2T_x$	Tetracycline hydrochloride	300 W Xe lamp	20 mg/mL	20 mg/mL	62.90
$Ti_3C_2T_x$	Ag_2WO_4	$Ag_2WO_4/Ti_3C_2T_x$	Sulfadimidine	300 W Xe lamp	20 mg/mL	20 mg/mL	88.60
$Ti_3C_2T_x$	SiC	$SiC/Ti_3C_2T_x$ QDs	Nitrous oxide	300 W Xe lamp ($\lambda >$ 420 nm)	0.5 g/L	280 ppb	74.60
$Ti_3C_2T_x$	MoS_2	$MoS_2/Ti_3C_2T_x$	Ranitidine	UV-free 25 W LED lamp	1 mg/mL	10 mg/L	88.40
$Ti_3C_2T_x$	Ag/g-C_3N_4	Ag/g-$C_3N_4/Ti_3C_2T_x$	Tetracycline hydrochloride	300 W Xe lamp ($\lambda >$ 420 nm)	1 mg/mL	20 mg/L	92.10
$Ti_3C_2T_x$	Ag/g-C_3N_4	Ag/g-$C_3N_4/Ti_3C_2T_x$	Rhodamine B	300 W Xe lamp ($\lambda >$ 420 nm)	1 mg/mL	10 mg/L	99.20
$Ti_3C_2T_x$	ZnO	$ZnO/Ti_3C_2T_x$	Rhodamine B	UV light ($\lambda =$ 365 nm)	0.4 mg/mL	10–5 mol/L	97.50
$Ti_3C_2T_x$	Ag_2S	$Ag_2S/Ti_3C_2T_x$	Methylene blue	300 W Xe lamp ($\lambda >$ 420 nm)	0.5 g/mL	20 mg/L	98.00
$Ti_3C_2T_x$	TiO_2	$TiO_2@Ti_3C_2T_x$	Rhodamine B	300 W Xe lamp	0.4 mg/mL	10 mg/L	97.00
$Ti_3C_2T_x$	α-$Fe_2O_3/ZnFe_2O_4$	α-Fe_2O_3/ $ZnFe_2O_4@Ti_3C_2T_x$	Rhodamine B	300 W Xe lamp ($\lambda >$ 400 nm)	5 mg/mL	10 mg/L	95.00

(continued)

Table 2 (continued)

MXene co-catalyst	Main catalyst	Final catalyst (with the inclusion of other co-catalysts)	Pollutant	Light source	Catalyst content	Contaminant	Degradation efficiency (%)
$Ti_3C_2T_x$	TiO_2@Ti_3C_2/g-C_3N_4	TiO_2@$Ti_3C_2T_x$/g-C_3N_4	Rhodamine B	300 W Xe lamp (λ > 400 nm)	1 mg/mL	10 mg/L	99.90
$Ti_3C_2T_x$	TiO_2@Ti_3C_2/g-C_3N_4	TiO_2@$Ti_3C_2T_x$/g-C_3N_4	Aniline	300 W Xe lamp (λ > 400 nm)	1 mg/mL	20 mg/L	76.40
$Ti_3C_2T_x$	Ti^{3+}–TiO_2	$Ti_3C_2T_x$/Ti^{3+}–TiO_2	Acetaldehyde	500 W Xe lamp	0.1 g	500 ppm	27.00
$Ti_3C_2T_x$	Ag/Ag_3PO_4	$Ti_3C_2T_x$–Ag/Ag_3PO_4	Methyl orange	300 W Xe lamp (λ > 420 nm)	2 mg/mL	20 mg/L	93.00
$Ti_3C_2T_x$	$BiPO_4$	$BiPO_4$/$Ti_3C_2T_x$	Rhodamine B	300 W Xe lamp	2 mg/mL	10 mg/L	100.00
$Ti_3C_2T_x$	a-Fe_2O_3	a-Fe_2O_3/$Ti_3C_2T_x$	Rhodamine B	500 W Xe lamp (λ > 420 nm)	2 mg/mL	10 mg/L	98.00
$Ti_3C_2T_x$	$CuFe_2O_4$	$CuFe_2O_4$/$Ti_3C_2T_x$	Sulfamethazine	300 W Xe lamp (λ > 420 nm)	25 mg	40 mg/L	59.40
$Ti_3C_2T_x$	TiO_2/BiOCl	$Ti_3C_2T_x$/TiO_2/BiOCl	Rhodamine B	500 W Xe lamp	1 mg/mL	10 mg/L	78.36
$Ti_3C_2T_x$	TiO_2/BiOCl	$Ti_3C_2T_x$/TiO_2/BiOCl	Tetracycline	500 W Xe lamp	1 mg/mL	20 mg/L	<20.00
$Ti_3C_2T_x$	La- and Mn-codoped $BiFeO_3$	La- and Mn-codoped $BiFeO_3$/$Ti_3C_2T_x$	Congo red	300 W Xe lamp (λ > 420 nm)	1 mg/mL	100 mg/L	100.00
$Ti_3C_2T_x$	TiO_2-black phosphorus	$Ti_3C_2T_x$/TiO_2-black phosphorus	Tetracycline hydrochloride	300 W Xe lamp (λ > 420 nm)	1 mg/mL	10 mg/L	92.70
$Ti_3C_2T_x$	TiO_2-black phosphorus	$Ti_3C_2T_x$/TiO_2-black phosphorus	Rhodamine B	300 W Xe lamp (λ > 420 nm)	1 mg/mL	20 mg/L	99.09
$Ti_3C_2T_x$	Bi_2MoO_6	Bi_2MoO_6/$Ti_3C_2T_x$	Tetracycline	50 W White LED lamp (λ > 400 nm)	2 mg/mL	15 mg/L	99.00

(continued)

Table 2 (continued)

MXene co-catalyst	Main catalyst	Final catalyst (with the inclusion of other co-catalysts)	Pollutant	Light source	Catalyst content	Contaminant	Degradation efficiency (%)
$Ti_3C_2T_x$	Bi_2MoO_6	$Bi_2MoO_6/Ti_3C_2T_x$	Cr(VI)	50 W White LED lamp (λ > 400 nm)	2 mg/mL	15 mg/L	100.00
$Ti_3C_2T_x$	BiOBr	$BiOBr/Ti_3C_2T_x$	Rhodamine B	300 W Xe lamp (λ > 420 nm)	0.5 mg/mL	20 mg/L	89.30
$Ti_3C_2T_x$	MoS_2	$MoS_2/Ti_3C_2T_x$	Methyl orange	400 W metal halide lamp	1 mg/mL	50 mg/L	97.40
$Ti_3C_2T_x$	$BiFeO_3$	$BiFeO_3/Ti_3C_2T_x$	Acetophenone	300 W Xe lamp	100 mg/L	30 ppm	100.00
$Ti_3C_2T_x$	$BiFeO_3$	$BiFeO_3/Ti_3C_2T_x$	Congo red	300 W Xe lamp	100 mg/L	–	100.00
$Ti_3C_2T_x$	BiOBr	$BiOBr/Ti_3C_2T_x$	NO_x gas	450 W Hg lamp (λ < 290 nm)	–	1 ppm	61.90
$Ti_3C_2T_x$	BiOBr	$BiOBr/Ti_3C_2T_x$	Rhodamine B	300 W Xe lamp (λ > 420 nm)	1 mg/mL	10 mg/L	100.00
$Ti_3C_2T_x$	$NiCo_2S_4$	$NiCo_2S_4/Ti_3C_2T_x$	Rhodamine B	250 W Xe lamp (cutoff 400–800)	0.1 mg/mL	50 mg/L	99.00
$Ti_3C_2T_x$	$ZnIn_2S_4$	$ZnIn_2S_4/Ti_3C_2T_x$	Tetracycline	300 W Xe lamp (λ > 420 nm)	0.2 mg/mL	20 mg/L	–
$Ti_3C_2T_x$	$ZnIn_2S_4$	$ZnIn_2S_4/Ti_3C_2T_x$	Cr(VI)	300 W Xe lamp (λ > 420 nm)	0.4 mg/mL	20 mg/L	93.80
$Ti_3C_2T_x$	$Fe–TiO_2$	$Fe-TiO_2/Ti_3C_2T_x$	Carbamazepine	300 W Xe lamp	2 g dm^{-3}	14 mg/dm^3	99.00
Nb_2CT_x	Nb_2O_5	Nb_2O_5/Nb_2CT_x	Tetracycline hydrochloride	500 W Xe lamp	1 mg/mL	10 mg/L	91.20
Nb_2CT_x	Nb_2O_5	Nb_2O_5/Nb_2CT_x	Rhodamine B	500 W Xe lamp	1 mg/mL	10 mg/L	98.50

(continued)

Table 2 (continued)

MXene co-catalyst	Main catalyst	Final catalyst (with the inclusion of other co-catalysts)	Pollutant	Light source	Catalyst content	Contaminant	Degradation efficiency (%)
$Ti_3C_2T_x$	$CoO@TiO_2$	$CoO@TiO_2/Ti_3C_2T_x$	Phenol	300 W Xe lamp ($\lambda >$ 420 nm)	0.2 mg/mL	20 mg/L	96.00
$Ti_3C_2T_x$	Perylene diimide supramolecular	Perylene diimide supramolecular/ $Ti_3C_2T_x$	Rhodamine B	300 W Xe lamp ($\lambda >$ 420 nm)	0.2 mg/mL	20 mg/L	~75.00
$Ti_3C_2T_x$	$NiFe_2O_4$	$NiFe_2O_4/Ti_3C_2T_x$	Methylene blue	Visible-light lamp	–	–	74.00
$Ti_3C_2T_x$	$SnNb_2O_6$	$SnNb_2O_6/Ti_3C_2T_x$	Tetracycline hydrochloride	300 W Xe lamp ($\lambda >$ 420 nm)	0.25 mg/ mL	10 mg/L	70.00
$Ti_3C_2T_x$	$SnNb_2O_6$	$SnNb_2O_6/Ti_3C_2T_x$	Rhodamine B	300 W Xe lamp ($\lambda >$ 420 nm)	0.25 mg/ mL	10 mg/L	98.00
$Ti_3C_2T_x$	Bi_2WO_6	$Bi_2WO_6/Ti_3C_2T_x$	Amoxicillin	300 W Xe lamp (AM 1.5 cutoff filter)	0.4 mg/mL	20 ppm	100.00
$Ti_3C_2T_x$	MIL-125-$NH_2(TiO_2)$	MIL-125-$NH_2(TiO_2)$/ $Ti_3C_2T_x$	Tetracycline hydrochloride	300 W Xe lamp ($\lambda >$ 420 nm)	0.2 mg/mL	20 mg/L	83.00
$Ti_3C_2T_x$	Ag_3PO_4	$Ag_3PO_4/Ti_3C_2T_x$	2,4-Dinitrophenol	300 W Xe lamp ($\lambda >$ 420 nm)	0.4 mg/mL	20 mg/L	68.40
$Ti_3C_2T_x$	CaIn2S_4	CaIn2S_4-$Ti_3C_2T_x$	Tetracycline hydrochloride	400 W metal halogen lamp ($\lambda > 420$ nm)	1 mg/mL	20 mg/L	96.00
$Ti_3C_2T_x$	g-C_3N_4/TiO_2	g-$C_3N_4/TiO_2/Ti_3C_2T_x$	NO_x gas	300 W Xe lamp ($\lambda >$ 420 nm)	0.125 mg/ mL	500 ppb	66.30
$Ti_3C_2T_x$	$Bi_2O_2CO_3$	$Bi_2O_2CO_3/Ti_3C_2T_x$	Tetracycline	300 W Xe lamp	55 mg	20 mg/L	81.00

(continued)

Table 2 (continued)

MXene co-catalyst	Main catalyst	Final catalyst (with the inclusion of other co-catalysts)	Pollutant	Light source	Catalyst content	Contaminant	Degradation efficiency (%)
$Ti_3C_2T_x$	Carbon nanotubes	Carbon nanotubes-$Ti_3C_2T_x$	Rhodamine B	Visible-light lamp	2 mg/mL	0.01 M in 100 mL	75.00
$Ti_3C_2T_x$	MoS_2	MoS_2@$Ti_3C_2T_x$	Waste oil (liquid paraffin)	1000 W high-pressure Hg lamp	1 mg/mL	30 ppm	80.50
$Ti_3C_2T_x$	MoS_2	MoS_2@$Ti_3C_2T_x$	Methyl orange	1000 W high-pressure Hg lamp	1 mg/mL	30 ppm	98.00
$Ti_3C_2T_x$	In_2S_3/TiO_2	In_2S_3/TiO_2@$Ti_3C_2T_x$	Methyl orange	300 W Xe lamp ($\lambda >$ 420 nm)	0.6 mg/mL	20 mg/L	92.10
$Ti_3C_2T_x$	–	$Ti_3C_2T_x$	Methylene blue	UV-C lamp	0.025 mg/mL	0.05 mg/mL	81.00
$Ti_3C_2T_x$	–	$Ti_3C_2T_x$	Acid blue 80	UV-C lamp	0.025 mg/mL	0.06 mg/mL	62.00
$Ti_3C_2T_x$	TiO_2	$TiO_2/Ti_3C_2T_x$	Carbamazepine	UV light	1 mg/mL	5 mg/L	98.67
$Ti_3C_2T_x$	$SrTiO_3$	$SrTiO_3/Ti_3C_2T_x$	UO^{22+}	300 W Xe lamp	0.33 mg/mL	50 ppm	77.00
$Ti_3C_2T_x$	g-C_3N_4	g-$C_3N_4/Ti_3C_2T_x$	Ciprofloxacin	500 W Xe lamp	0.2 mg/mL	20 mg/L	40.00
Ti_2CT_x	3%TiO_2/1%Ag	Ti_2CT_x/3%TiO_2/1%Ag	Salicylic acid	150 W medium-pressure Hg lamp	0.625 mg/mL	0.25 μm/mL	97.10
Ti_2CT_x	3%TiO_2/1%Ag_2O	Ti_2CT_x/3%TiO_2/1%Ag_2O	Salicylic acid	150 W medium-pressure Hg lamp	0.625 mg/mL	0.25 μm/mL	95.80

(continued)

Table 2 (continued)

MXene co-catalyst	Main catalyst	Final catalyst (with the inclusion of other co-catalysts)	Pollutant	Light source	Catalyst content	Contaminant	Degradation efficiency (%)
Ti_2CT_x	3%TiO_2/1%Pd	Ti_2CT_x/3%TiO_2/1%Pd	Salicylic acid	150 W medium-pressure Hg lamp	0.625 mg/mL	0.25 μm/mL	88.70
Ti_2CT_x	3%TiO_2/1%PdO	Ti_2CT_x/3%TiO_2/1%PdO	Salicylic acid	150 W medium-pressure Hg lamp	0.625 mg/mL	0.25 μm/mL	86.10
Ti_2CT_x		Ti_2CT_x	Salicylic acid	150 W medium-pressure Hg lamp	0.625 mg/mL	0.25 μm/mL	91.70
$Ti_3C_2T_x$	5%Fe/ZSM-5 zeolite/potassium peroxymonosulfate	5%Fe/5% $Ti_3C_2T_x$/ZSM-5 zeolite/potassium peroxymonosulfate	Phenol	300 W Xe lamp	3 mg/mL	20 mg/L	98.00
$Ti_3C_2T_x$	5%Fe/ZSM-5 zeolite/potassium peroxymonosulfate	5%Co/5% $Ti_3C_2T_x$/ZSM-5 zeolite/potassium peroxymonosulfate	Phenol	300 W Xe lamp	3 mg/mL	20 mg/L	91.00
$Ti_3C_2T_x$	Ultrathin nickel metal–organic framework	Ultrathin nickel metal–organic framework/$Ti_3C_2T_x$	Methylene blue	300 W Xe lamp	0.5 mg/mL	20 mg/L	–
$Ti_3C_2T_x$	TiO_2	TiO_2/$Ti_3C_2T_x$	Methylene blue	300 W Xe lamp ($\lambda >$ 420 nm)	0.5 mg/mL	20 mg/L	94.47

(continued)

Table 2 (continued)

MXene co-catalyst	Main catalyst	Final catalyst (with the inclusion of other co-catalysts)	Pollutant	Light source	Catalyst content	Contaminant	Degradation efficiency (%)
$Ti_3C_2T_x$	TiO_2	$TiO_2/Ti_3C_2T_x$	4-chlorophenol	300 W Xe lamp ($\lambda >$ 420 nm)	0.5 mg/mL	20 mg/L	83.65
$Ti_3C_2T_x$	TiO_2/g-C_3N_4	$TiO_2/Ti_3C_2T_x$/g-C_3N_4	Rhodamine B	500 W high-pressure Hg lamp	1 mg/mL	100 mg/L	–
Mo_2CT_x	MoS_2/In_2S_3	$Mo_2C/MoS_2/In_2S_3$	Cr(VI)	350 W Xe lamp	1 mg/mL	40 mg/L	99.10
$Ti_3C_2T_x$	BiOBr	BiOBr/$Ti_3C_2T_x$	Rhodamine B	350 W Xe lamp	0.5 mg/mL	20 mg/L	93.30
$Ti_3C_2T_x$	Bi_2WO_6	$Bi_2WO_6/Ti_3C_2T_x$	Rhodamine B	15 W white LED	0.4 mg/mL	10 mg/L	99.00
$Ti_3C_2T_x$	Bi_2WO_6	$Bi_2WO_6/Ti_3C_2T_x$	Methylene blue	15 W white LED	0.4 mg/mL	10 mg/L	80.00
$Ti_3C_2T_x$	Bi_2WO_6	$Bi_2WO_6/Ti_3C_2T_x$	Methyl orange	15 W white LED	0.4 mg/mL	10 mg/L	86.00
$Ti_3C_2T_x$	Bi_2WO_6	$Bi_2WO_6/Ti_3C_2T_x$	Tetracycline	15 W white LED	0.4 mg/mL	20 mg/L	97.00
$Ti_3C_2T_x$	CeO_2	$CeO_2/Ti_3C_2T_x$	Rhodamine B	500 W Hg lamp	1 mg/mL	20 mg/L	75.00
$Ti_3C_2T_x$	Co_3O_4	$Co_3O_4/Ti_3C_2T_x$	Rhodamine B	–	1 mg/mL	5 mg/L	–
$Ti_3C_2T_x$	Co_3O_4	$Co_3O_4/Ti_3C_2T_x$	Methylene blue	–	1 mg/mL	12.5 mg/L	92.37

Adopted from Ref. [41] and further references cited therein

1.6.2 Degradation Mechanism of Contaminants

Increasing industrialization and development of technologies causes environmental crisis such as global warming, clean water scarcity, and ecosystem imbalance due to pollutants like harmful azo dyes, pharmaceutical compounds, endocrine disrupting chemicals, insecticides, pesticides, and so on which affect the well-being of humans and other living organisms. Dissolved oxygen concentration levels in the aquatic system would collapse, and it destroys the aquatic animals and plants. These contaminations were treated before getting into the environment through adsorption, ion exchange, sonodegradation, precipitation, and chlorination. Among these, photocatalysis is considered to be a greener, cost-effective method and degrades pollutants without secondary contaminants. As discussed earlier, MXene-based nanocomposites have excellent performance in photocatalysis. In this session, we will discuss the proposed mechanism of contaminant degradation. Global research works on finding ways to tap the whole potential of MXene in the photocatalysis field.

Tu et al. [42] synthesized the g-C_3N_4/amorphous Ti-peroxo complex and studied Rhodamine B (RhB) and Tetracycline (TC) degradation mechanism. Degradation of Rhodamine B undergoes three pathways: first pathway is direct and gradual way of intermediate formation then in the second by transformation through a deamination reaction. In the third pathway, degradation via indirect deamination and dihydroxylation were carried out. Finally, the mineralized inorganic product is formed by chromophore cleavage and ring opening. In the case of Tetracycline degradation: in first pathway fragmention of TC occurs via dehydration and the ring is formed via deamination; then in second pathway, superoxide anion radical presence causes a demethylation effect subsequently the intermediates formed through dehydroxylation, oxidation, and ring-opening reaction. Finally, in pathway three hydroxylation and ring-opening reaction occur to degrade TC into CO_2, H_2O, and small molecules.

Liu et al. [43] studied Levofloxacin (LEV) degradation using $BiOBr/Bi_2MoO_6$/ oxygen-functionalized Ti_3C_2 MXene/exfoliated montmorillonite. and oxygen vacancies co-embed $Bi_2O_2CO_3/Ti_3C_2Tx$, respectively. They proposed three pathways for LEV degradation. In the first pathway, piperazine ring is opened by the superoxide anion radical attack on the N–C bond of the aromatic ring, and the intermediates in this pathway were formed through demethylation and decarboxylation. At the end of this pathway, the piperazine ring is cleaved. In pathway two, the piperazine ring is destroyed by the removal of certain intermediate and single-bonded CH_3 through defluorination and decarboxylation. The third pathway was the destruction of the quinolone moiety structure via demethylation, and replacement of the hydroxyl group in a few intermediates occurs. At the end, the degraded products were CO_2, H_2O and NO_3^-, F^-, and $CO_3{}^{2-}$ minerals (small molecules). Precisely when the Schottky heterojunctioned $Bi_2O_2CO_3/Ti_3C_2T_x$ exposed to visible light the photogenerated electrons alone transfer from $Bi_2O_2CO_3$ conduction band via surface oxygen vacancies (OVs) to $Ti_2C_3T_x$ via interface OVs because of its lower Fermi level. Because of the strong oxidizing property of holes, it directly degrades the LEV, whereas the accumulated electrons in the MXene cause superoxide anion radical formation by reducing dissolved oxygen present in the aqueous solution. Further the superoxide anion radical degrades LEV. An example of the systematic illustration of $Bi_2O_2CO_3$/$Ti_3C_2T_x$ on LEV degradation is mentioned below

$$
\begin{aligned}
&Bi_2O_2CO_3 + h\upsilon \rightarrow e^- + h^+ \\
&e^- + \text{dissolved oxygen}\,(O_2) \rightarrow {}^{\cdot}O_2^- \\
&h^+ + LEV \rightarrow \text{degradation products} \\
&{}^{\cdot}O_2^- + LEV \rightarrow \text{degradation products}
\end{aligned}
$$

Liu [40] prepared a Co_3O_4/MXene composite to study one of the endocrine-disrupting chemicals, i.e., Bisphenol A using a peroxymonosulfate system. Production of hydroxyl radical from composite and sulfate radicals from PMS was responsible for the degradation of contaminants. They proposed the mechanism of contaminant degradation in two pathways: first BPA decomposed into monohydroxylated BPA by either the attack of hydroxyl radical or hydroquinone/isopropenylphenol reaction, then it undergoes hydroxylation to form quinone of dihydroxylated BPA. In other the sulfate radical attacked the BPA and converted it into biphenyldiol. Further attacks of hydroxyl and sulfate radicals degrade the intermediates into lower molecules and acids. As for the material mechanism, the PMS is activated by the cobalt ion's active radical in which the peroxide bond in PMS is cleaved by an electron from Co^{2+} to produce hydroxyl radical. Co^{2+} on MXene transfer electron to PMS where the surface sulfate radicals are generated and degrade the BPA because of its strong oxidizing potential. $SO_5{}^{\cdot -}$ and Co^{2+} were produced by the reaction between PMS and Co^{3+}. Reaction mechanisms are illustrated below

$$
\begin{aligned}
&Co^{2+} + HSO_5^- \rightarrow Co^{3+} + SO_4^- + {}^{\cdot}OH \\
&Co^{2+} + HSO_5^- \rightarrow Co^{3+} + SO_4^{\cdot -} + OH^- \\
&Co^{3+} + HSO_5^- \rightarrow Co^{2+} + SO_5^{\cdot -} + H^+ \\
&SO_4^{\cdot -} + OH^- \rightarrow {}^{\cdot}OH + SO_4^{2-} \\
&SO_4^{\cdot -} + H_2O \rightarrow {}^{\cdot}OH + HSO_4^- \rightarrow SO_4^{2-} + {}^{\cdot}OH + H^+ \\
&SO_4^{\cdot -}/{}^{\cdot}OH + [BPA] \rightarrow [BPA] - SO_4^{\cdot -}/{}^{\cdot}OH \rightarrow [BPA]+ \rightarrow (\ldots\ldots) \rightarrow CO_2 + H_2O
\end{aligned}
$$

Organic contaminant degradation mechanisms and pathways were proposed by various research groups. For example, the following illustration explains about the starting material structure of a pollutant followed by its possible final intermediate before mineralization, and at the end release of CO_2, H_2O, and smaller molecules or lower chain molecules was represented Fig. 5.

1.6.3 Water Purification

Pathogenic contamination, water pollution, and desalination are considered as the bigger challenge as well as the reason for crisis in the demand of clean drinking water. There were various technologies developed to solve this issue, but the membrane technology stands out in recent days because of its affordability, great selectivity,

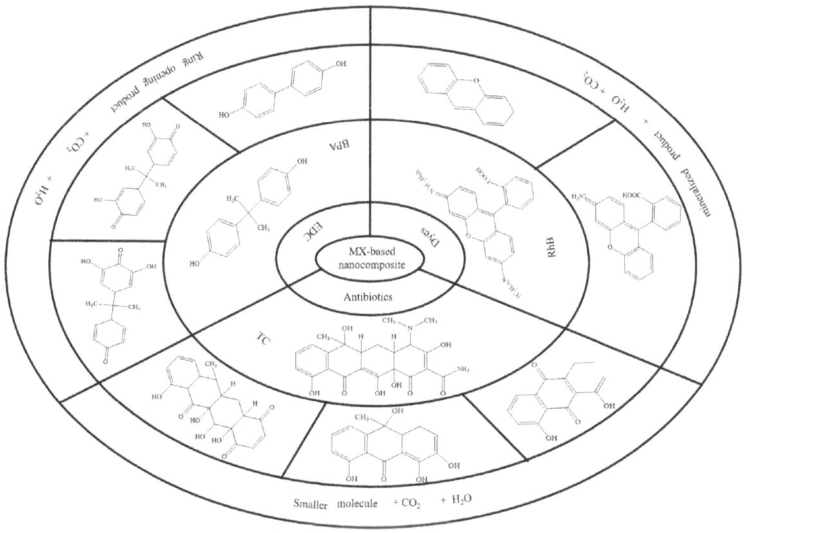

Fig. 5 Schematic representation of proposed possible degradation pathway of RhB, TC, and BPA

and environmental compatibility. Its selectivity in adsorption, absorption, filtration, and distillation is better than conventional separation methods. Polymer material are preferably used in membrane technology because it has apt features like packing density, cost effective so on yet it has problems in organic solvents and aqueous medium as swelling and fouling properties respectively and also has stability issues with high temperature [44]. 2D material such as MoS_2, graphite, graphene oxide, reduced graphene oxide, and WS_2 exhibit antibacterial activity. Rasool et al. [45] experimented the antibacterial property of MXene which is helpful in antifouling behavior needed in the separation membrane. MXene's hydrophilicity because of its surface terminations, adsorption capacity, and its ability in molecular filtration makes it a suitable material for various applications in water purification. The pure MXene has a disadvantage over its merits like easy delamination due to electrostatic and hydrogen bonding, and the swelling property affects the efficiency in separation. Hence the research groups like Gogotsi et al. [46] prepared membrane of PVDF-supported $Ti_3C_2T_x$ through vacuum filtration method which showed potential separation as membranes; the permeability is ~37.4 $Lm^{-2}\ h^{-1}\ bar^{-1}$ for Mg^{2+}, Ca^{2+}, Li^{2+}, Al^{3+}, Ni^{2+}, Na^{2+}, and K^+ ions. MXene interlayer spacing is lesser hence the permeability is slower. The pure MXene membrane has a swelling rate of 16.6 Å but after self-crosslinking the swelling decreases to 15.4 Å and the stability is up to 70 h for monovalent ion separation. The permeation rates of a few monovalent ions such as Li^+, Na^+, and K^+ were 0.0283, 0.222, and 0.232 mol $h^{-1}\ m^{-2}$, respectively (Lu-85). The easy delamination of multilayered MXene affect the separation efficiency of the membrane. Ding et al. [47] synthesized MXene with non-swelling property by Al^{3+} ion insertion to the laminated nanosheets of MXene. Electrostatic interaction is strengthened by these ions with MXene's functional group which results in less swelling and the stability is enhanced to 400 h with the rejection rate of NaCl being 89.5–99.5%, whereas Wang et al. [48] built a pillar structured lamellar membrane by mixing sodium alginate (SA) and MXene $Ti_3C_2T_x$. Hydrogen bonding and Van der Waals forces attached both the material in this composite SA-$Ti_3C_2T_x$ which is then submerged in solutions containing cations like Ca^{2+}, Ba^{2+}, Mn^{2+}, and Al^{3+}. In this, pristine MXene has 13.8 Å interlayer spacing then it dilated to 15.2 Å by the Mn^{2+} and Ca^{2+} cations which act as pillar between the MXene nanosheets. It shows 100% $NaSO_4$ salt rejection as well as lesser swelling; the selectivity is enhanced because of the improvement in d-spacing between the MXene nanosheets [48]. Ag has good antibacterial property; Mahmou et al. [49] tried MXene/Ag composite which showed enhanced water permeability, rate of rejection, and antifouling property. 21% of this composite is comprised of Ag nanoparticles and in synthetic sea water, it shows high stability. Wang [50] experimented with the nanosheet membrane comprised of g-C_3N_4 and MXene for the purification of water, and they used it for the removal of various molecules such as Evan blue, rhodamine, gold nanoparticles, and cytochrome C with the rate of 87.2%, 75.5%, 99.5%, and 93.1%, respectively. Kang [51] and Liu and coworkers [52] fabricated GO/MXene membrane through the vacuum filtration method. Lamellar membrane mentioned previously showed better molecular separation efficiency than the GO/MXene membrane synthesized later. In the previous membrane composite, it reject methyl blue, methylene blue, rose bengal and brilliant blue with 68%, 99.5%, 93.5%, and 100% respectively based on the hydrated radii because the interlayer space is around 5 Å for the prepared $Ti_3C_2T_x$/GO membrane, whereas in the later membrane small organic dyes were separated such as neutral red, methylene blue, CV, brilliant blue with >99% rate of rejection. Membranes like this composite also reject molecules like bovine serum albumin and humic acid. Factors such as pH, temperature, pressure, and water uptake can considerably affect the molecule separation efficiency of the membrane. The proper use of MXene-based membranes enables the passage of cations bearing hydration radii that exceed the distance between layers within MXene interlayers. Membranes based on MXene can be enhanced using appropriate intercalation or crosslinking techniques, which effectively regulate the structure of pores.

As for its antibacterial activity, $Ti_3C_2T_x$-based membrane worked against both gram positive and negative bacteria such as E.coli and B.subtilis which was studied by Rasool [53] on agar medium through optical density and colony growth rate. 200 μg/mL of $Ti_3C_2T_x$ causes a loss of 98% bacterial cell viability at the exposure time of 4 h. The efficiency of this membrane is better than that of graphene oxide membrane. The composite of $Ti_3C_2T_x$/PVDF membrane exhibits antibacterial activity rate of ~73% and ~63% which is compared with the control PVDF toward the previously discussed bacteria respectively, wherein the $Ti_3C_2T_x$ membrane exhibit >99% under same condition. Jastrzebsa and his fellow researchers [54] worked on various MXene based membrane such as $Ti_3C_2/Al_2O_3/Ag$, $Ti_3C_2/SiO_2/Pd$ and $Ti_3C_2/SiO_2/Ag$ membranes, and they confirmed the bioactive property on gram negative bacteria strain of MXenes. On the enhancement of antibacterial activity, Zhu et al. [55] worked on near-infrared (NIR) light on the MXene-based membrane, i.e., Ag/$Ti_3C_2T_x$. In this composite membrane, the Ag has antibacterial properties; after irradiation the 200 μg/mL of composite membrane inhibits the growth of E.coli in 0–6 h. Bacteria present in the surrounding area were destroyed by the photothermal heat produced by NIR which strengthen the antibacterial activity of $Ag/Ti_3C_2T_x$ membrane. The versatile properties of MXene make it work in fields like membrane technology. Table 3 shows some of the MXene-based membranes in the separation of ions, molecules, and pathogens present in feed solution and the efficiency rate of the membrane along with its permeability.

1.6.4 CO_2 Reduction

In the process of urbanization, the consumption of fossil fuel is increased due to anthropogenic activity causing an energy crisis and greenhouse effect. There should be a sustainable way to mitigate CO_2 emissions. Photocatalysis plays a role in the reduction of CO_2 to value-added chemical fuels such as formic acid, carbon monoxide, formaldehyde, methanol, methane, and ethanol with ambient temperature and pressure. The following Fig. 6 illustrates the possible value-added fuel production from photocatalytic reduction of CO_2 and the required holes, and protons for the process are represented.

The linear configuration of CO_2 orientation is more stable and difficult for conversion but when photoexcitation of electron happened, CO_2^- anions formed then its configuration changed to bent which has lower LUMO making it easy for the conversion process [56]. In this process, holes work for the hydrogenated product formation and electrons for reduction. In aqueous media, the formation of CH_3OH as a product is more compared with others and which can be used as fuel directly. pH of the system affects the solubility and CO_2's chemical form [30]. Conversion of CO_2 into a single value-added fuel is more preferable to the mixture formation hence the research continues on constructing an apt semiconductor toward selectivity and should be a good reductant. Some of the constructed semiconductor based on bandgap energy level is represented in Fig. 7.

MXene's co-catalytic activity makes it into the application of CO_2 reduction but the oxygen vacancy MXene, i.e., $Ti_3C_2O_2$ makes it selective towards photocatalytic reduction of CO_2. Ultrathin Ti_3C_2/Bi_2WO_6 nanosheet was synthesized by Cao et al. [58], and it was found that 5 wt% of this material showed higher photoreduction of CO_2 in which the methane and methanol formation is 4.6 times higher compared to Bi_2WO_6 pure nanosheet. It confirms that MXene isa good co-catalyst for CO_2 photoreduction reaction. Quantum dots of MXene with Cu_2O form 0D/1D heterojunction producing methanol 8.25 and 2.15 times higher than other combinations of these materials [59]. $Ti_3C_2(OH)_2/g\text{-}C_3N_4$ nanocomposite showed a better yield of CO than graphitic carbon nitride which is >90% [60]. MXene-based nanocomposites play an excellent role in photocatalytic reduction of CO_2 which helps to mitigate CO_2 emission. Some of the photocatalysts based on MXene are listed below along with the synthesis method, its reaction condition, mechanism, and efficiency in Table 4.

Table 3 MXene-based membranes for the separation of ions, molecules, and pathogens from water

Type of membranes	Fabrication method	Feed solution/ concentration	Rejection (%)	Permeability ($Lm^{-2}\ h^{-1}\ bar^{-1}$)
$Ti_3C_2T_x$	Vacuum filtration	RB	85	1084
		EB	90	
		CC (Each 10–20 mg/L)	97	
$Ti_3C_2T_x$	Vacuum filtration	NaCl (10,000 mg/ L)	56–64	10
		BSA (2000 mg/L)		
$Ti_3C_2T_x$	Vacuum filtration	CR	92	405
		GN	80	
		$MgCl_2$	2.3	
		Na_2SO_4	13.2	
		NaCl (Each 100–1000 mg/L)	13.8%	
$Ti_3C_2T_x$	Vacuum filtration	E. coli	> 99	37.4
		B. subtilis	> 99	
$Ti_3C_2T_x$	Vacuum filtration	Na_2SO_4	50–99	5–15.25
		Mg_2SO_4		
		$MgCl_2$		
		NaCl		
		$VOSO_4$		
$Ti_3C_2T_x$–Ag	Vacuum-assisted filtration	RB	79.9	~420
		MG	92.3	
		BSA (50–100 mg/L)	>99%	
$Ti_3C_2T_x$–GO	Vacuum filtration	BB	95.4	~ 25 L
		Rose Bengal	94.6	
		MLB	40	
		MLR	5	
		$MgSO_4$	<1	
		NaCl (Each 10 mg/L)		
$Ti_3C_2T_x$–GO	Vacuum filtration	RB	>97 (dyes)	89.6
		MB		
		CV		
		NR (Each 10 mg/ L)		
		Na_2SO_4	61	

(continued)

Table 3 (continued)

Type of membranes	Fabrication method	Feed solution/ concentration	Rejection (%)	Permeability (Lm^{-2} h^{-1} bar^{-1})
		NaCl (Each 5 mM)	23	
$Ti_3C_2T_x$–GO	Vacuum filtration	Chrysoidine G	>99% (dyes)	71.9
		MLB		
		NR		
		CV		
		BB		
		HA		
		BSA		
		Na_2SO_4	61	
		NaCl (Each 10 mg/L)	23	
$Ti_3C_2T_x$–GO	Vacuum filtration	MO	>95	~8.5–11
		MLB		
		Acid yellow 14		
		IC		
		Eosin (Each 10 mg/L)		
$Ti_3C_2T_x$–TiO_{ssss}	Spin coating	Dextran (3000 mg/L)	>95	~90

CC: cytochrome C; MLB: methylene blue; RB: rhodamine B; EB: Evan blue; MO: methyl orange; IC: indigo carmine; HA: humic acid; BB: brilliant blue; NR: neutral red; CV: crystal violet; CR: Congo red; GN: gentian violet; MG: methyl green; MLR: methylene red
Adopted from Ref. [44] and further references cited therein

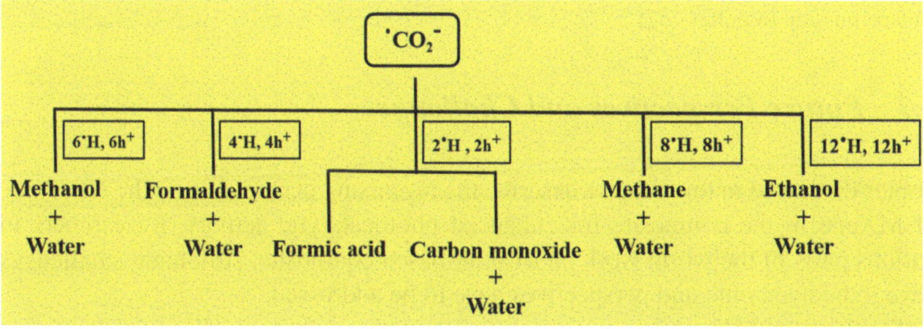

Fig. 6 Pathways to the production of value-added fuel from photocatalytic reduction of CO_2

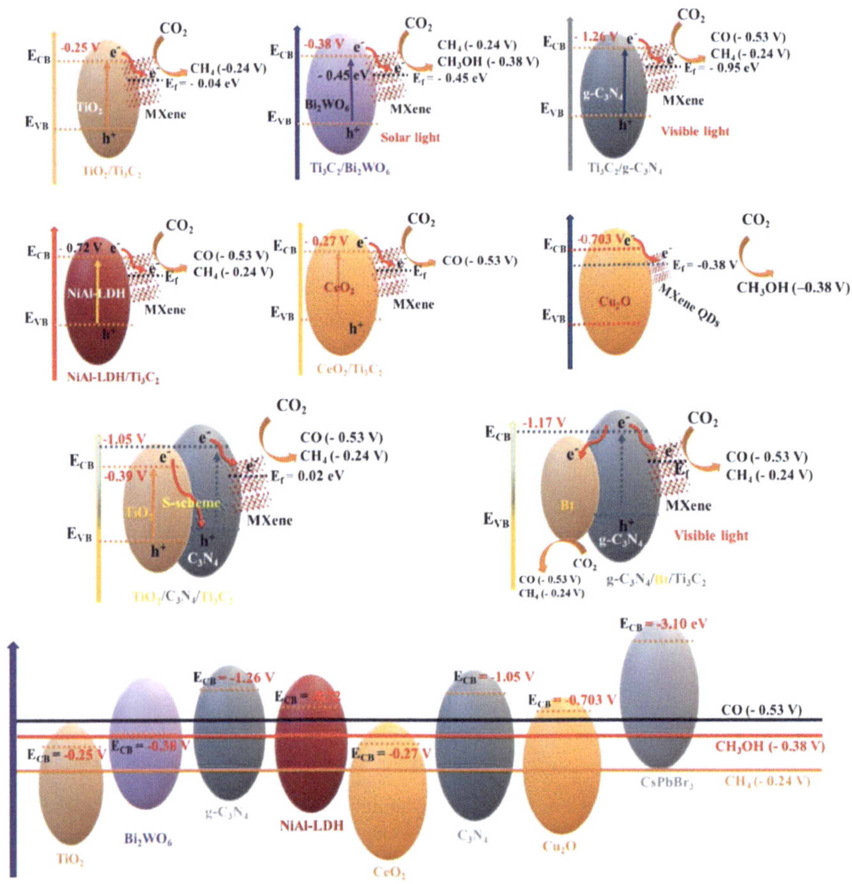

Fig. 7 Overall representation of some semiconductor energy level on CO_2 reduction. Reprinted with permission from Ref. [57]

1.7 Future Perspectives and Challenges

Topics discussed in this chapter describe the promising perspective on the efficiency of MXene in the composite for enhanced photocatalytic activity. Researchers in various parts of the world work on MXene-based composite; still, many challenges have to be overcome and perspectives have to be addressed.

Synthesis of MXene depends on an etching process through HF or an alternative based on different compositions yet it has toxicity which is harmful to the environment. Even though MXene is a potential 2D material because of its synthesis methods, the commercialization of MXene is constrained. Hence framing a new greener method for MXene synthesis is important and needed.

MXene has very low oxidation resistance when it is exposed to moisture, air, or high temperatures because of its negatively charged surface and positively charged

Table 4 Studies on the catalytic performance of MXene for CO_2 reduction

Materials	Synthesis	Production	Improvement	Production rate (μ mol·L^{-1}·h^{-1})	Reaction condition	Mechanism
Ti_3C_2/C_3N_4	Sonochemical method	CO/CH_4	3.64 times	CO: 10.67 CH_4: 4.19	UV–vis irradiation for 4 h	Schottky heterojunction
Ti_3C_2/Bi_2WO_6	In situ grown on Ti_3C_2	CH_4/CH_3OH	4.6 times	CH_4: 1.78 CH_3OH: 0.44	Anaerobic/solar irradiation	2D/2D heterojunction
Ti_3C_2/g-C_3N_4	Calcining the mixture of multilayered Ti_3C_2 particles and urea	CO/CH_4	8.1 times	CO: 5.19 CH_4: 0.044	Visible light irradiation ($\lambda \geq 420$ nm)	Interface contact between 2D g-C_3N_4 and Ti_3C_2
Ti_3C_2/Cu_2O		CH_3OH	8.25 times 2.15 times	CH_3OH: 78.50	300 W Xe lamp irradiation	Ti_3C_2 QDs promote charge transfer
$TiO_2/C_3N_4/Ti_3C_2$	Interfacial self-assembly	CO/CH_4	3 times 8 times	CO: 4.39 CH_4: 1.20	Anaerobic/xenon lamp irradiation	Heterojunction (lower carrier recombination rate)

Adapted from Ref. [61] and further references cited therein

edges, hence, working on its anti-oxidant property is important for its stability. Chemical modifications on MXene can improve its drawbacks. Working on stability and reusability of the material is important, so effective investigation of MXene is vital for its practical application.

Preserving MXene in the dark with an inert gas environment is suggested and preferred to store it in low temperatures. MXene's rapid oxidation results in a diminished lifespan of the material. The lifetime of MXene is very limited, so future studies should address this trial by researching mitigation methods.

Scalable synthesis on a large scale is not practical so far; it is difficult to synthesize MXene in large quantities with high quality, good purity level, and consistency in size. Commercializing the material is in the stack because its synthesis methods are limited, to laboratory scale as well as the stability and storage of MXene are challenging.

The primary studies confirmed that MXenes are practically non-toxic yet very few studies claim they might have notable toxicity. Studies on the toxicity of MXene are also limited. Hence long-term toxicity study on MXene must be done.

The potential of functional groups on the MXene surface studies was initiated and identified by theoretical studies. MXene's surface functional terminations make it a semiconductor but experimental evidence was needed to prove it further and exploration of surface functionalities on MXene is necessary to research their characteristics and additional properties which might give a new perspective on the material.

Since, MXene has potential photoreactivity similar to precious metal based co-catalysts. In future MXene can be able to replace the noble metal-based reaction and its application through cost effective manner.

Because of MXene's unique characteristics and excellent properties as a co-catalyst, it extended its application in environmental remediation such as water purification, CO_2 reduction, organic pollutant degradation, and so on but exploration on air purification is limited and needs more focus on this area of research.

New members in the MXene family have to be discovered and explored like oxygen vacancy which has distinctive properties than other MXenes to overcome its problems like non-homogenous surfaces, performance efficiency, and so on.

Photocatalytic activity of semiconductors is enhanced after making composites with MXene is evident; computational studies also confirm it, yet the effect of MXene on composites, structure-activity relationships as well as understanding of heterojunction formations has to be figured out in depth through mechanisms with evidence.

Engineering MXene from morphology control, band gap tuning, defect creation, and interlayer spacing to structural changes cause significant impact on normal performance. Designing MXene with a particular property might increase its scope in various fields.

The studied application at the laboratory level has to be implemented in real time by applying it in real water systems because it has varied variables like pH, combination of compounds which interfere with the efficiency, and performance of the material. Constructing a material with good performance in a real-time wastewater system is quite challenging yet necessary to study.

The great potential of MXene in a diverse field holds scope for various applications and hence the basic research on this material is in the early stage, so more study on this material's versatile properties and characteristics has numerous challenges on its way for future research to untap the whole potential of MXene in the photocatalysis field.

1.8 Conclusion

This chapter emphasizes MXene as one among the 2D materials which is rising in material science field due to its characteristics and its applications in various fields. It has been demonstrated that MXene is a promising material in environmental remediation. MXene is getting near to graphene in material research because of its stand-out properties. Year-by-year synthesis methods from HF etching are evolving along with the new MXene family member discoveries like high entropy MXene. In the photocatalysis field, MXene is playing a prominent role along with photocatalysts like GO, rGO, g-C_3N_4, MO_2S_2, metal oxides, etc. Laboratory-level works showed good results, whereas applying them to the industrial level there are challenges that have been discussed. In the contaminant removal, various parameters like pH, catalyst dosage, contact time, temperature, and stable intermediate formations can influence the performance of the materials. MXenes are potential materials in photocatalysis though challenges have to be overcome. However, this material is in the early stage of research hence more focus and in-depth investigation are needed to make this material a practically viable one.

References

1. M. Naguib et al., Two-dimensional nanocrystals produced by exfoliation of Ti3AlC2. Adv. Mater. **23**(37), 4248–4253 (2011). https://doi.org/10.1002/adma.201102306
2. V. Natu, R. Pai, M. Sokol, M. Carey, V. Kalra, M.W. Barsoum, 2D Ti3C2Tz MXene synthesized by water-free etching of Ti3AlC2 in polar organic solvents. Chem **6**(3), 616–630 (2020). https://doi.org/10.1016/j.chempr.2020.01.019
3. Y. Wei, P. Zhang, R.A. Soomro, Q. Zhu, B. Xu, Advances in the synthesis of 2D MXenes. Adv. Mater. **33**(39) (2021). https://doi.org/10.1002/adma.202103148
4. V. Kamysbayevet al., Covalent surface modifications and superconductivity of two-dimensional metal carbide MXenes. Science (1979) **369**(6506), 979–983 (2020). https://doi.org/10.1126/science.aba8311
5. B. Anasori, M. Naguib, G. Editors, Two-dimensional MXenes. MRS Bull. **48**(3), 238–244 (2023). https://doi.org/10.1557/s43577-023-00500-z
6. M.W. Barsoum, The MN+1AXN phases: a new class of solids. Progr. Solid State Chem. **28**(1–4), 201–281 (2000). https://doi.org/10.1016/S0079-6786(00)00006-6
7. M. Naguib, M.W. Barsoum, Y. Gogotsi, Ten years of progress in the synthesis and development of MXenes. Adv. Mater. **33**(39) (2021). https://doi.org/10.1002/adma.202103393

8. K.R.G. Lim, M. Shekhirev, B.C. Wyatt, B. Anasori, Y. Gogotsi, Z.W. Seh, Fundamentals of MXene synthesis. Nat. Synth. **1**(8), 601–614 (2022). https://doi.org/10.1038/s44160-022-00104-6
9. A. Iqbal, J. Hong, T.Y. Ko, C.M. Koo, Improving oxidation stability of 2D MXenes: synthesis, storage media, and conditions. Nano Converg. **8**(1), 9 (2021). https://doi.org/10.1186/s40580-021-00259-6
10. G.K. Nasrallah, M. Al-Asmakh, K. Rasool, K.A. Mahmoud, Ecotoxicological assessment of $Ti_3C_2T_x$ (MXene) using a zebrafish embryo model. Environ. Sci. Nano **5**(4), 1002–1011 (2018). https://doi.org/10.1039/C7EN01239J
11. G. Murali et al., A review on MXene synthesis, stability, and photocatalytic applications. ACS Nano **16**(9), 13370–13429 (2022). https://doi.org/10.1021/acsnano.2c04750
12. C. Xu et al., Large-area high-quality 2D ultrathin Mo2C superconducting crystals. Nat. Mater. **14**(11), 1135–1141 (2015). https://doi.org/10.1038/nmat4374
13. X. Xiao et al., Salt-templated synthesis of 2D metallic MoN and other nitrides. ACS Nano **11**(2), 2180–2186 (2017). https://doi.org/10.1021/acsnano.6b08534
14. S. Joshi, Q. Wang, A. Puntambekar, V. Chakrapani, Facile synthesis of large area two-dimensional layers of transition-metal nitride and their use as insertion electrodes. ACS Energy Lett. **2**(6), 1257–1262 (2017). https://doi.org/10.1021/acsenergylett.7b00240
15. J. Jia et al., Ultrathin N-doped Mo2C nanosheets with exposed active sites as efficient electrocatalyst for hydrogen evolution reactions. ACS Nano **11**(12), 12509–12518 (2017). https://doi.org/10.1021/acsnano.7b06607
16. Z. Zhang, F. Zhang, H. Wang, C. Ho Chan, W. Lu, J. Dai, Substrate orientation-induced epitaxial growth of face centered cubic Mo_2C superconductive thin film. J. Mater. Chem. C **5**(41), 10822–10827 (2017). https://doi.org/10.1039/C7TC03652C
17. F. Zhang et al., Plasma-enhanced pulsed-laser deposition of single-crystalline Mo_2C ultrathin superconducting films. Phys. Rev. Mater. **1**(3), 034002 (2017). https://doi.org/10.1103/PhysRevMaterials.1.034002
18. M. Alhabeb et al., Guidelines for synthesis and processing of two-dimensional titanium carbide (Ti3C2Tx MXene). Chem. Mater. **29**(18), 7633–7644 (2017). https://doi.org/10.1021/acs.chemmater.7b02847
19. P. Urbankowski et al., Synthesis of two-dimensional titanium nitride Ti_4N_3 (MXene). Nanoscale **8**(22), 11385–11391 (2016). https://doi.org/10.1039/C6NR02253G
20. Y. Li et al., A general Lewis acidic etching route for preparing MXenes with enhanced electrochemical performance in non-aqueous electrolyte. Nat. Mater. **19**(8), 894–899 (2020). https://doi.org/10.1038/s41563-020-0657-0
21. J. Halim et al., Transparent conductive two-dimensional titanium carbide epitaxial thin films. Chem. Mater. **26**(7), 2374–2381 (2014). https://doi.org/10.1021/cm500641a
22. A. Feng et al., Fabrication and thermal stability of NH4HF2-etched Ti3C2 MXene. Ceram Int. **43**(8), 6322–6328 (2017) https://doi.org/10.1016/j.ceramint.2017.02.039
23. L. Ma, L.R.L. Ting, V. Molinari, C. Giordano, B.S. Yeo, Efficient hydrogen evolution reaction catalyzed by molybdenum carbide and molybdenum nitride nanocatalysts synthesized via the urea glass route. J. Mater. Chem. A Mater. **3**(16), 8361–8368 (2015). https://doi.org/10.1039/C5TA00139K
24. T. Li et al., Fluorine-Free Synthesis of High-Purity $Ti_3C_2T_x$ (T = OH, O) via Alkali treatment. Angew. Chem. Int. Ed. **57**(21), 6115–6119 (2018). https://doi.org/10.1002/anie.201800887
25. S. Yang et al., Fluoride-free synthesis of two-dimensional titanium carbide (MXene) using a binary aqueous system. Angewandte Chem. **130**(47), 15717–15721 (2018). https://doi.org/10.1002/ange.201809662
26. M. Ghidiu, M. R. Lukatskaya, M.-Q. Zhao, Y. Gogotsi, M.W. Barsoum, Conductive two-dimensional titanium carbide 'clay' with high volumetric capacitance. Nature **516**(7529), 78–81 (2014). https://doi.org/10.1038/nature13970
27. O. Mashtalir et al., Intercalation and delamination of layered carbides and carbonitrides. Nat. Commun. **4**(1), 1716 (2013). https://doi.org/10.1038/ncomms2664

28. H. Wang et al., Surface modified MXene Ti3C2 multilayers by aryl diazonium salts leading to large-scale delamination. Appl. Surf. Sci. **384**, 287–293 (2016). https://doi.org/10.1016/j.apsusc.2016.05.060
29. Y. Sun et al., 2D MXenes as co-catalysts in photocatalysis: synthetic methods. Nanomicro Lett. **11**(1), 79 (2019). https://doi.org/10.1007/s40820-019-0309-6
30. E. Karamian, S. Sharifnia, On the general mechanism of photocatalytic reduction of CO2. J. CO2 Utiliz. **16**, 194–203 (2016). https://doi.org/10.1016/j.jcou.2016.07.004
31. K. Kabra, R. Chaudhary, R.L. Sawhney, Treatment of hazardous organic and inorganic compounds through aqueous-phase photocatalysis: a review. Ind. Eng. Chem. Res. **43**(24), 7683–7696 (2004). https://doi.org/10.1021/ie0498551
32. K. Zhang et al., Insights into different dimensional MXenes for photocatalysis. Chem. Eng. J. **424**, 130340 (2021). https://doi.org/10.1016/j.cej.2021.130340
33. N.H. Solangi et al., MXene as emerging material for photocatalytic degradation of environmental pollutants. Coord. Chem. Rev. **477**, 214965 (2023). https://doi.org/10.1016/j.ccr.2022.214965
34. Y. Fang, Y. Cao, Q. Chen, Synthesis of an Ag2WO4/Ti3C2 Schottky composite by electrostatic traction and its photocatalytic activity. Ceram. Int. **45**(17), 22298–22307 (2019). https://doi.org/10.1016/j.ceramint.2019.07.256
35. B. Tan, Y. Fang, Q. Chen, X. Ao, Y. Cao, Construction of Bi2O2CO3/Ti3C2 heterojunctions for enhancing the visible-light photocatalytic activity of tetracycline degradation. J. Colloid Interface Sci. **601**, 581–593 (2021). https://doi.org/10.1016/j.jcis.2021.05.155
36. K. Wu, S. Song, H. Wu, J. Guo, L. Zhang, Facile synthesis of Bi2WO6/C3N4/Ti3C2 composite as Z-scheme photocatalyst for efficient ciprofloxacin degradation and H2 production. Appl. Catal. A Gen. **608** (2020). https://doi.org/10.1016/j.apcata.2020.117869
37. Y. Cao et al., Fabrication of novel CuFe2O4/MXene hierarchical heterostructures for enhanced photocatalytic degradation of sulfonamides under visible light. J. Hazard. Mater. **387** (2020). https://doi.org/10.1016/j.jhazmat.2020.122021
38. Y. Diao et al., In-situ grown of g-C3N4/Ti3C2/TiO2 nanotube arrays on Ti meshes for efficient degradation of organic pollutants under visible light irradiation. Colloids Surf. A Physicochem. Eng. Asp. **594** (2020). https://doi.org/10.1016/j.colsurfa.2020.124511
39. D. Xu et al., Interfacial engineering of 2D/2D MXene heterostructures: face-to-face contact for augmented photodegradation of amoxicillin. Chem. Eng. J. **426**, 31246 (2021). https://doi.org/10.1016/j.cej.2021.131246
40. Y. Liu et al., Sandwich-like Co3O4/MXene composite with enhanced catalytic performance for Bisphenol A degradation. Chem. Eng. J. **347**, 731–740 (2018). https://doi.org/10.1016/j.cej.2018.04.155
41. G. Murali et al., A Review on MXene synthesis, stability, and photocatalytic applications. ACS Nano **16**(9), 13370–13429 (American Chemical Society, 2022). https://doi.org/10.1021/acsnano.2c04750
42. W. Tu et al., Carbon nitride coupled with Ti3C2-Mxene derived amorphous Ti-peroxo heterojunction for photocatalytic degradation of rhodamine B and tetracycline. Colloids Surf. A Physicochem. Eng. Asp. **640**, 128448 (2022). https://doi.org/10.1016/j.colsurfa.2022.128448
43. K. Liu et al., Oxygen-functionalized Ti3C2 MXene/exfoliated montmorillonite supported S-scheme BiOBr/Bi2MoO6 heterostructures for efficient photocatalytic quinolone antibiotics degradation. Chem. Eng. J. **457** (2023). https://doi.org/10.1016/j.cej.2023.141271
44. I. Mahar et al., Two-dimensional transition metal carbides and nitrides (MXenes) for water purification and antibacterial applications. Membranes (Basel) **11**(11), 869 (2021). https://doi.org/10.3390/membranes11110869
45. K. Rasool, M. Helal, A. Ali, C.E. Ren, Y. Gogotsi, K.A. Mahmoud, Antibacterial activity of Ti3C2Tx MXene. ACS Nano **10**(3), 3674–3684 (2016). https://doi.org/10.1021/acsnano.6b00181
46. C.E. Ren, K.B. Hatzell, M. Alhabeb, Z. Ling, K.A. Mahmoud, Y. Gogotsi, Charge- and size-selective ion sieving through $Ti_3C_2T_x$ MXene membranes. J. Phys. Chem. Lett. **6**(20), 4026–4031 (2015). https://doi.org/10.1021/acs.jpclett.5b01895

47. L. Ding, Y. Wei, Y. Wang, H. Chen, J. Caro, H. Wang, A two-dimensional lamellar membrane: MXene nanosheet stacks. Angew. Chem. Int. Ed. **56**(7), 1825–1829 (2017). https://doi.org/10.1002/anie.201609306
48. J. Wang et al., Ion sieving by a two-dimensional Ti3C2Tx alginate lamellar membrane with stable interlayer spacing. Nat. Commun. **11**(1), 3540 (2020). https://doi.org/10.1038/s41467-020-17373-4.
49. R.P. Pandey, K. Rasool, V.E. Madhavan, B. Aïssa, Y. Gogotsi, K.A. Mahmoud, Ultrahigh-flux and fouling-resistant membranes based on layered silver/MXene ($Ti_3C_2T_x$) nanosheets. J. Mater. Chem. A Mater. **6**(8), 3522–3533 (2018). https://doi.org/10.1039/C7TA10888E
50. Y. Wang et al., Water transport with ultralow friction through partially exfoliated g-C_3N_4 nanosheet membranes with self-supporting spacers. Angew. Chem. Int. Ed. **56**(31), 8974–8980 (2017). https://doi.org/10.1002/anie.201701288
51. K.M. Kang et al., Selective molecular separation on $Ti_3C_2T_x$—Graphene oxide membranes during pressure-driven filtration: comparison with graphene oxide and MXenes. ACS Appl. Mater. Interfaces **9**(51), 44687–44694 (2017). https://doi.org/10.1021/acsami.7b10932
52. T. Liu, X. Liu, N. Graham, W. Yu, K. Sun, Two-dimensional MXene incorporated graphene oxide composite membrane with enhanced water purification performance. J. Memb. Sci. **593**, 117431 (2020). https://doi.org/10.1016/j.memsci.2019.117431
53. K. Rasool, M. Helal, A. Ali, C. E. Ren, Y. Gogotsi, K.A. Mahmoud, Antibacterial activity of $Ti_3C_2T_x$ MXene. ACS Nano **10**(3), 3674–3684 (2016). https://doi.org/10.1021/acsnano.6b00181
54. A. Rozmysłowska-Wojciechowska et al., Influence of modification of Ti_3C_2 MXene with ceramic oxide and noble metal nanoparticles on its antimicrobial properties and ecotoxicity towards selected algae and higher plants. RSC Adv. **9**(8), 4092–4105 (2019). https://doi.org/10.1039/C8RA07633B
55. X. Zhu et al., A near-infrared light-mediated antimicrobial based on Ag/$Ti_3C_2T_x$ for effective synergetic antibacterial applications. Nanoscale **12**(37), 19129–19141 (2020). https://doi.org/10.1039/D0NR04925E
56. Ž. Kovačič, B. Likozar, M. Huš, Photocatalytic CO2 reduction: a review of ab initio mechanism, kinetics, and multiscale modeling simulations. ACS Catal. **10**(24), 14984–15007 (2020). https://doi.org/10.1021/acscatal.0c02557
57. A. Sreedhar, I.N. Reddy, J.-S. Noh, Photocatalytic and electrocatalytic reduction of CO2 and N2 by Ti3C2 MXene supported composites for a cleaner environment: a review. J. Clean. Prod. **328**, 129647 (2021). https://doi.org/10.1016/j.jclepro.2021.129647
58. S. Cao, B. Shen, T. Tong, J. Fu, J. Yu, 2D/2D heterojunction of ultrathin MXene/Bi2WO6 nanosheets for improved photocatalytic CO2 reduction. Adv. Funct. Mater. **28** (2018)
59. Z. Zeng, Y. Yan, J. Chen, P. Zan, Q. Tian, P. Chen, Boosting the photocatalytic ability of Cu_2O nanowires for CO_2 conversion by MXene quantum dots. Adv. Funct. Mater. **29**(2) (2019). https://doi.org/10.1002/adfm.201806500
60. Q. Tang, Z. Sun, S. Deng, H. Wang, Z. Wu, Decorating g-C3N4 with alkalinized Ti3C2 MXene for promoted photocatalytic CO2 reduction performance. J. Colloid Interface Sci. **564**, 406–417 (2020). https://doi.org/10.1016/j.jcis.2019.12.091
61. L. Zhang, W. Song, H. Liu, H. Ding, Y. Yan, R. Chen, Influencing factors on synthesis and properties of MXene: a review. Processes **10**(9), 1744 (2022). https://doi.org/10.3390/pr10091744

Green Synthesis of Metal/Metal Oxide Nanoparticles and Their Photocatalytic Degradation of Dyes

M. Jeevarathinam, M. Sivagami, Nethra Kuruthukulangara, S. P. Ragavi, D. Thirumalai, and I. V. Asharani

Abstract Adopting an environmentally friendly approach to nanoparticle production is vital for striking a balance between industrial growth and pollution control. The spotlight has recently been on green-synthesized nanomaterials, particularly for their role in environmental cleanup. Numerous natural resources, including certain plants and microorganisms, are being studied for their efficacy as potent reducing agents in nanoparticle synthesis. These naturally derived extracts demonstrate the ability to efficiently introduce a wide range of functional groups onto surfaces, resulting in synergistic benefits when used together. The textile industry utilizes organic dyes, with the subsequent release of these dyes in effluent. Due to its limited capacity for biodegradation, it exerts a substantial ecological impact. Photocatalytic degradation represents a viable approach for the efficient treatment of these dyes. A promising solution for managing these dyes lies in photocatalytic degradation. In this process, pollutants are subjected to irradiation by employing visible or UV light, in the presence of either homogeneous or heterogeneous catalysts. A significant quantity of research has delved into the potential of nanoscale metal oxide nanoparticles as heterogeneous photocatalysts for breaking down toxic dyes. In this section, we delve into the eco-conscious techniques utilized for the synthesis of metal and metal oxide nanoparticles, with specific emphasis on cobalt/cobalt oxide, iron/iron oxide, and copper/copper oxide nanoparticles. Recognized for their photocatalytic properties, these nanoparticles' roles in breaking down harmful dyes are discussed. The object of this chapter is to shed light on the green synthesis processes, mechanisms, and dye degradation.

Keywords Nanoparticles · Eco-friendly synthesis · Material characterization · Photocatalysis · Product evaluation

---

M. Jeevarathinam · M. Sivagami · N. Kuruthukulangara · S. P. Ragavi · I. V. Asharani (✉)
Department of Chemistry, School of Advanced Sciences, Vellore Institute of Technology, Vellore 632014, Tamil Nadu, India
e-mail: asharani.iv@vit.ac.in

D. Thirumalai
Department of Chemistry, Thiruvalluvar University, Vellore 632 115, Tamil Nadu, India

P. Sathishkumar (ed.), *Photocatalysis for Energy and Environmental Applications*, Green Energy and Technology, https://doi.org/10.1007/978-981-97-1939-6_6

1 Introduction

Recently, there has been significant growth in the utilization of nanotechnology across numerous scientific domains [1]. Nanotechnology involves the design and fabrication of materials and structures with dimensions and configurations at the nanoscale [2], offering a fresh perspective to the scientific field. It has garnered substantial interest in areas such as energy, electronics, cosmetics, healthcare, drug delivery systems, biomedical devices, electron transistors, and nonlinear optical devices. Particles within the size range of 1–100 nm are often referred to as "nanoparticles" [3]. Lately, there has been a growing interest in applying nanotechnology to address environmental issues. Nanoparticles exhibit advantageous properties and high efficacy, making them well-suited for mitigating environmental pollution. This is primarily because of remarkable catalytic activity and great surface-to-volume ratio [4–6]. Effluents containing textile dyes are known to have significant adverse effects on both terrestrial and aquatic ecosystems. These effluents pose a particular challenge due to their elevated thermal and optical intensity, which allows them to persist in the environment for extended periods without significant degradation. These dyes can effectively absorb and scatter incident light in waterbodies like rivers and lakes. By limiting the penetration of light, they decrease the photosynthetic activity of algae, thereby impacting the aquatic food chain and biodiversity. Moreover, the deposition of dyes in the soil and sediments, particularly in areas with sewage discharges, can lead to various ecological imbalances within the ecosystem [7, 8].

This chapter aims to provide comprehensive information on nanoparticles, encompassing their fundamental principles, advancements, and current trends. It serves as a valuable resource by covering synthesis techniques, characteristics, and applications related to nanoparticles within a single platform. While it is not possible to encompass the entirety of nanoparticle literature, this chapter highlights key research from both the past and present. By condensing the achievements in metal/metal oxide nanoparticles, such as cobalt/cobalt oxide nanoparticles (Co/CoO NPs), iron/iron oxide nanoparticles (Fe/FeO NPs), and copper/copper oxide nanoparticles (Cu/CuO NPs), into one location, this book chapter offers essential knowledge for researchers in the field.

1.1 Properties of Nanoparticles

Nanotechnology has garnered significant interest owing to its versatile attributes, opening a myriad of potential applications [9, 10]. Notably, metal and metal oxide nanoparticles have recently gained substantial attention across fields such as chemistry, energy, environmental science, agriculture, electronics, and more as illustrated in Fig. 1 [11–14].

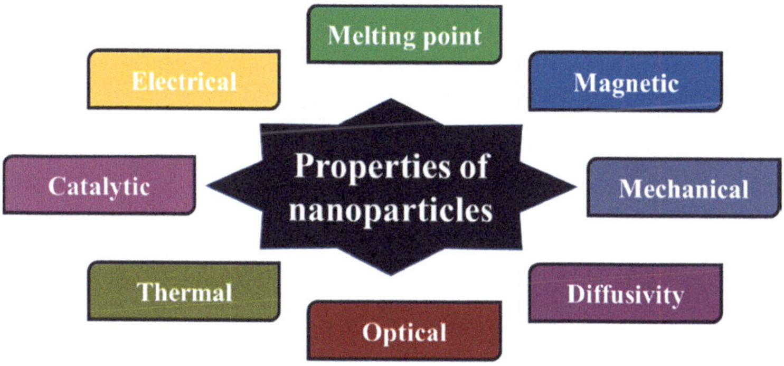

Fig. 1 Properties of nanoparticles

1.1.1 Physical Properties

The physical characteristics of nanoparticles are impacted by various factors, including their color, shape, size, and morphology. A combination of particle size, shape, and configuration contributes to the size effect, crystal structure, and lattice parameters of these nanoparticles [15].

1.1.2 Chemical Properties

The chemical configuration and content of nanoparticles play pivotal roles in determining their intended functionality. Surface energy, chemical potential, oxidation processes, and catalytic activity collectively exert a substantial impact on the nanoparticle composition [16].

1.1.3 Optical Properties

The optical properties of nanoparticles, including absorption, transmission, reflection, and light emission, are governed by their electronic structure, which, in turn, is determined by the surface atoms. When particle dimensions are reduced to 10 nm (nm), optical features are utilized to observe the size-related effects [17, 18].

1.1.4 Mechanical Properties

Nanostructured materials exhibit a higher thermal expansion coefficient, which measures how material size changes with temperature, primarily because of the increased presence of defects compared to bulk materials [19, 20].

1.1.5 Catalytic Properties

The catalytic performance of a material is notably influenced by its surface area. Studies suggest that diminishing the size of the material results in an exponential increase in the proportion of surface atoms. Consequently, the surface area with a greater concentration of surface atoms exhibits greater catalytic activity compared to the bulk material.

1.1.6 Diffusivity

Materials with nanostructures are small and possess a larger surface area when compared with bulk materials. These interfaces facilitate easier diffusion pathways. Therefore, self-diffusion is more effective along the defect sites in nanostructured materials than in their bulk counterparts.

1.1.7 Magnetic Properties

Nanostructured materials demonstrate higher saturation magnetization and magnetic coercivity values. Saturation magnetization signifies the point at which an external magnetic field can no longer alter a material's magnetization, rendering it magnetically saturated. Magnetic coercivity, conversely, represents the amount of magnetic field required to reduce the saturation magnetization of a ferromagnetic material to zero. Both coercivity and saturation magnetization serve as crucial indicators of a magnet's strength. The reduced size of nanostructured materials enhances these properties. Small particles within these materials exhibit a single magnetic domain behavior, making them easily attracted by an external magnetic field, ultimately leading to improved magnetic characteristics.

1.1.8 Melting Point

In contrast to bulk materials, nanostructured materials exhibit a greater concentration of atoms at their surface and within grain boundaries. As the size decreases, the quantity of surface atoms increases. Melting occurs when an atom's vibrational amplitude exceeds its bonding distance.

1.2 Classification of Nanoparticles

Nanoparticles are categorized based on their origin, size, and material composition, as shown in Fig. 2.

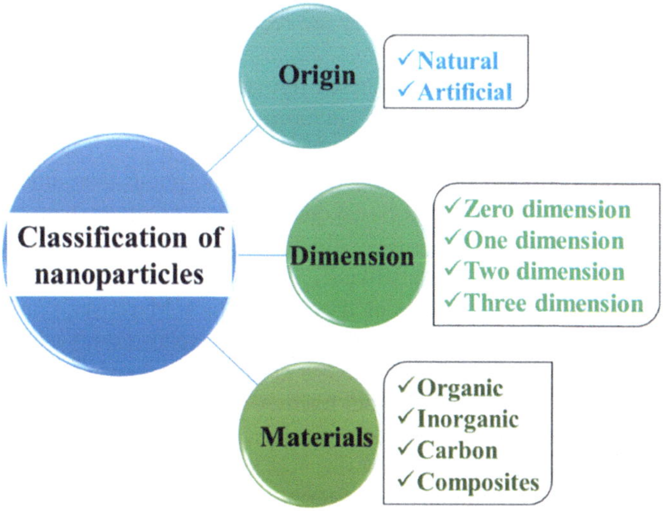

Fig. 2 Classification of nanoparticles based on origin, dimension, and materials

Origin-based nanoparticles are further divided into two types artificial and natural. Numerous natural nanoparticles exist in diverse forms. Examples include protein molecules, viruses, gels like gelatine, minerals such as clay, aerosols like fog, liquid colloids like milk and blood, and mineralized structures like shells, bones, and corals. Additional instances include insect wings, opals, lotus leaves, spider silk, ocean sprays, gecko feet, and volcanic ash. On the other hand, semiconductor nanoparticles and carbon nanotubes are prime examples of synthetic nanomaterials, which are deliberately produced through mechanical and manufacturing processes [21–23].

Based on their dimensions, nanomaterials can be divided into four groups: one dimensional (1D), two dimensional (2D), three dimensional (3D), and zero dimensional (0D). Zero-dimensional nanomaterials have all their dimensions confined to the nanoscale. Examples of 0D materials encompass quantum dots, fullerenes, and nanoparticles. One-dimensional nanomaterials, which include nanotubes, nanofibers, nanorods, and nanowires, go beyond the nanoscale. Two-dimensional nanomaterials are those having more than the nanoscale, in two dimensions, with structures like nanosheets, nanofilms, and nanolayers. Finally, in all dimensions, three-dimensional nanomaterials are not limited to nanoscale. This category comprises bulk powders, nanoparticle dispersions, and arrays of nanowires and nanotubes [24, 25]. Nanoparticles can be classified into four primary categories based on their structural composition: organic, inorganic, carbon-based, and composite nanoparticles.

Organic-based Nanoparticles: These are formed from organic compounds. Organic molecules transform the nanoscale, leading to the creation of nanomaterials. Such materials can be produced through self-assembly or by molding organic compounds into specific structures. Notable examples include micelles, liposomes, ferritin, and dendrimers. They are recognized for their non-toxicity and biodegradability [22, 26].

Inorganic-based Nanoparticles: These originate from non-carbon elements like metal and metal oxides. During their formation, atoms can be packed into various shapes like stars, ellipsoids, spheres, oblates, cubes, and cylinders, often retaining the crystalline features of metal compounds. However, amorphous inorganic nanoparticles also exist. Among these, metal-based quantum dots (1–10 nm) are particularly noteworthy for their unique properties, which bridge the gap between bulk materials and atom-sized entities. Metals like Fe, Ag, Ni, Au and metal oxides such as ZnO, TiO_2, and MnO_2 are crafted using these techniques. Moreover, semiconductor nanoparticles from materials like silicon and ceramics are also synthesized [26, 27].

Carbon-based Nanoparticles: These are primarily made of carbon, with sp2-bonded carbon atoms, and come in various forms. Carbon nanofibers, spheres, hollow tubes, graphene, and fullerenes serve as typical examples of such nanomaterials.

Composite Nanoparticles: Composite materials are formed by combining nanoparticles with another nanoparticle, nanoparticles with bulk-type materials, and nanoparticles with large-scale materials. Composites can incorporate a diverse range of materials, including organic/inorganic substances, metals, carbon-based, ceramics, and bulk polymers. The morphologies of these materials vary depending on the synthesis method employed and the characteristics needed for their desired uses [22, 26].

1.3 Synthesis of Nanoparticles Using Various Techniques

The synthesis of nanoparticles primarily involves two main methods: the bottom-up and top-down approaches. Figure 3 displays diagrams illustrating the different nanoparticle synthesis techniques [28].

1.3.1 Top-Down Approach

The top-down method is employed to transform large bulk materials into minuscule nano-sized particles. When elucidating the top-down approach, it is essential to highlight that, while it is straightforward to use, it falls short in producing irregularly shaped and exceedingly small particles. The principal limitation of this method lies in the difficulty of attaining precise control over particle size and shape.

1.3.2 Bottom-Up Approach

The bottom-up method is frequently termed the "constructive technique." In contrast to the top-down approach, the bottom-up method involves the creation of nanoparticles by building them up through the development and self-assembly of their

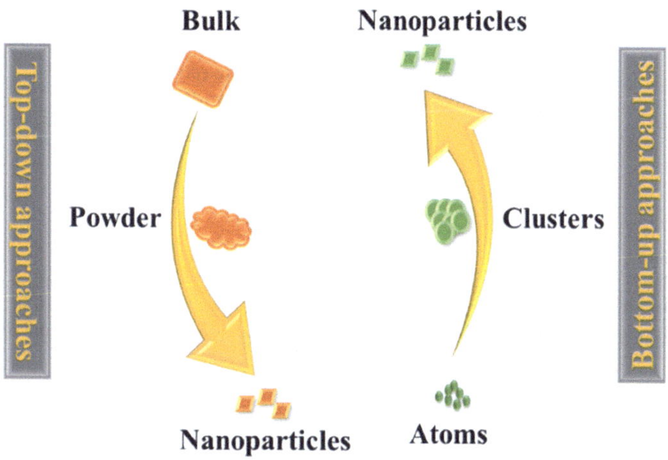

Fig. 3 Schematic representation of various techniques for nanoparticle synthesis

constituent units, which consist of atoms and molecules. This approach enables the production of nanoparticles with meticulous control over their shapes, sizes, and chemical compositions [29, 30].

1.4 Synthesis Methods for Nanoparticles

Nanoparticles can be produced using chemical physical, biological, and physical methods, as illustrated in Fig. 4. Table 1 outlines the pros and cons associated with each of these synthesis methods: physical, chemical, and biological.

1.4.1 Physical Method

Various physical techniques, such as spray pyrolysis, laser ablation, ultrasonication, and radiolysis, are utilized in the synthesis of nanoparticles. The physical synthesis process involves evaporating metal atoms that subsequently condense on different substrates. During this condensation phase, metal atoms reorganize and aggregate, forming small metallic nanoparticle clusters. These techniques, often following a top-down strategy, offer the advantage of being solvent-free and yielding uniformly sized nanoparticles. However, the downside of physical methods is their lower cost-effectiveness due to the substantial waste they produce. Additionally, these methods typically require specialized equipment, chemicals, radiative heating, and significant energy consumption, contributing to elevated operational costs [31, 32].

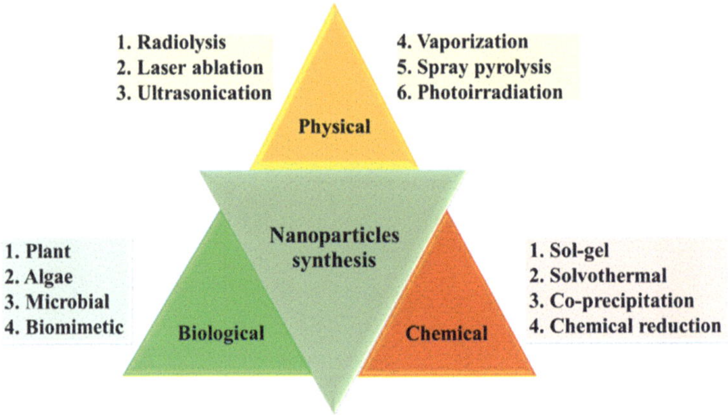

Fig. 4 Synthesis methods for nanoparticles

1.4.2 Chemical Method

Many chemical techniques utilize potentially hazardous reducing agents, like ethylene glycol, hydrazine, sodium citrate, and sodium borohydride ($NaBH_4$). However, these techniques frequently produce unsatisfactory particle sizes and size distributions, which require the inclusion of capping agents. These additional agents can include polymers like polyethylene glycol and polyvinyl alcohol, compounds like ascorbic acid, polyvinylpyrrolidone (PVP), and dendrimers, as well as surfactants such as cetyltrimethylammonium bromide (CTAB), sodium dodecyl sulfate, and phospholipids. These agents are essential for stabilization, and preventing the agglomeration of nanoparticles [33, 34].

1.4.3 Biological Method

Biological synthesis of nanoparticles employs plant extracts and microorganisms such as bacteria, algae, and fungi. Utilizing microbes for nanoparticle synthesis is eco-friendly, cultivating these microorganisms can be more expensive than using plant products, which don't require maintaining microbial cultures [35, 36]. Additionally, nanoparticle synthesis using plants is often more efficient than with microorganisms, making it a low-cost and scalable method for producing nanoparticles in large quantities [37, 38].

1.4.4 Green Synthesis of Nanoparticles

By utilizing plant extracts, metallic nanoparticle synthesis is relatively easy, straightforward, and rapid compared to synthesis involving microorganisms, which demands

Table 1 The advantages and disadvantages of different methods for the synthesis of nanoparticles

Synthesis methods	Implementation plans	Advantages	Disadvantages
Physical approach	Laser ablation Sputtering Ball milling/ mechanical milling Physical vapor deposition (PVD)	Ensuring the absence of toxic chemicals Achieving high speeds Maintaining uniformity in the size and shape of nanoparticles	High energy consumption Significant waste generation Requirement of substantial costs Presence of imperfections on nanoparticle surfaces Changes in nanoparticle surface chemistry and physicochemical properties Reduced stability Extensive dilution is needed Time-intensive process for achieving thermal stability Need for expansive working space
Chemical approach	Sol-gel Chemical vapor deposition (CVD) Chemical reduction Hydrothermal Spray pyrolysis Thermal decomposition	Precise size distribution Enhanced yield Excellent thermal stability Manageable particle size, shape, and morphology Superior purity Straightforward and replicable Extensive versatility in surface chemistry	Challenging process control Utilization of harmful organic solvents Significant costs involved Dangerous to humans and the environment Use of hazardous chemicals Production of toxic by-products Tendency to form aggregates Extended reaction times needed
Biological approach	Bacteria Fungus Algae	Easily achievable and environmentally friendly Non-hazardous and safe Biologically compatible Straightforward and uncomplicated Suitable for mass production Stable over the long term	Challenging to regulate size, shape, and crystal growth Unclear mechanism of action Possible risk of bacterial contamination and transfection Time-consuming processing procedures

Fig. 5 Advantages of environmentally friendly synthesis methods

highly sterile conditions [39]. Plants contain a variety of phytochemicals that play a crucial role in reducing and stabilizing nanoparticles. Peptides and terpenoids found in plants assist in stabilizing nanoparticles during photo-fabrication. Additionally, proteins present in plants prevent nanoparticle agglomeration, enhancing their stability. Among various inorganic nanoparticles, metal nanoparticles hold particular significance due to their cost-effectiveness, ease of synthesis, and promising applications [40, 41]. Figures 5 and 6 illustrate the advantages of green nanoparticle synthesis and the general mechanism of nanoparticle formation using phytoconstituents [42, 43].

1.4.5 Mechanism of Formation of Plant-Mediated Nanoparticles

1.5 Different Metal/Metal Oxide Nanoparticles

1.5.1 Cobalt and Cobalt Oxide Nanoparticles

Cobalt nanoparticles and their composites have proven to be potent catalysts for the oxidation of water and efficient absorbers of light with a wide bandgap [44]. Cobalt can exist in various oxidation states, including Co, CoO, and Co_3O_4. However, because of its structural and redox properties, cobalt oxide, a non-stoichiometric polymorph has attracted a lot of interest [45]. Cobalt oxide nanoparticles (Co_3O_4 NPs), known for their variable oxidation states, have a wide array of applications across various products. These applications span from serving as catalysts, pigments, and magnetic materials to being employed as anode materials in rechargeable batteries,

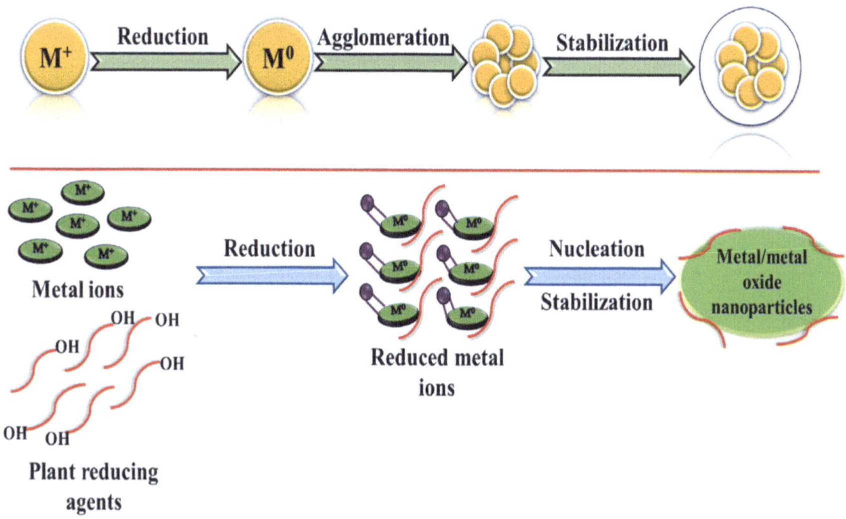

Fig. 6 Schematic representation of mechanism for the formation of nanoparticles by phytoconstituents

high-temperature solar selective absorbers, gas sensors, electronic devices, magnetic systems for solar energy utilization, photocatalysts for water purification, and diverse biological applications, including cancer treatment [46–50]. The high activity and selectivity of cobalt oxide catalysts are attributed to adsorbed oxygen, differences in oxygen vacancies, and oxygen defects in different cobalt states within cobalt oxide with mixed valences [51, 52]. Co_3O_4 nanoparticles exhibit a characteristic cubic spinel crystal structure. Approximately one-third of the cobalt atoms within this structure exist in the Co(II) oxidation state, occupying tetrahedral sites, while the remaining two-thirds are in the Co(III) oxidation state, residing in octahedral sites [53–55] (Fig. 7 and Table 2).

1.5.2 Iron and Iron Oxide Nanoparticles

Since the 1990s, Fe NPs have been employed in water treatment, owing to their distinctive properties like a large specific surface area, remarkable adsorption behavior, and extensive distribution of reactive surface sites. Iron oxide (FeO), which predominantly manifests as magnetite (Fe_3O_4) and hematite/maghemite (Fe_2O_3), stands out as one of the prominent metal nanoparticles. FeO NPs have garnered significant attention owing to their exceptional characteristics and versatility in numerous applications. These applications span from catalysis, antimicrobial agents,

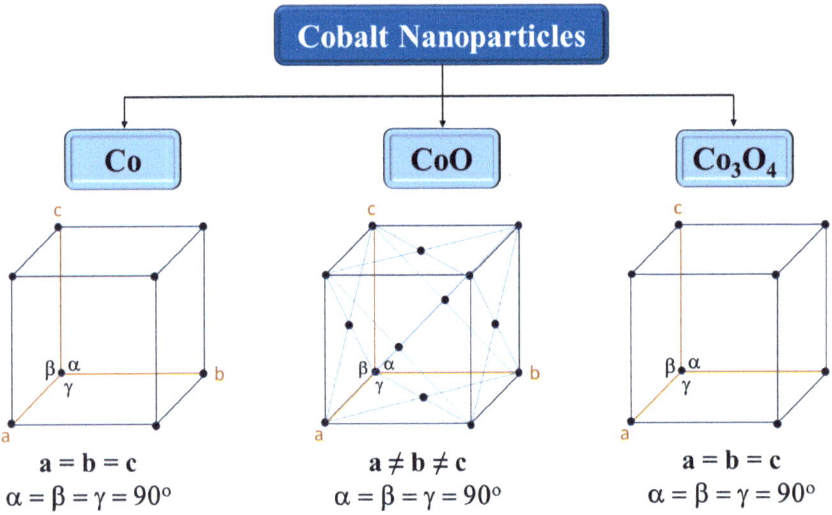

Fig. 7 Structure of cobalt/cobalt oxide nanoparticles

Table 2 Green synthesis and catalytic applications of Co/CoO/Co_3O_4 NPs

Plant name	NPs	Morphology	Particle size (nm)	Application	Refs.
Celosia argentea	Co	Spherical	27	Degradation of methylene blue	[56]
Vitis rotundifolia	CoO	Rhombus		Degradation of acid blue-74	[57]
Taraxacum officinale	Co_3O_4	Spherical	50–100	Direct yellow-142 and methyl orange	[58]
Catharanthus roseus	Co		27	Degradation of toxic alizarin red S dye	[59]
Punica granatum	CoO	Spherical	40–80	Remazol brilliant orange 3R	[60]
Coriandrum sativum	Co_3O_4	Honeycomb		Rhodamine B	[61]
Curcuma longa	Co_3O_4	Spherical	22	Degradation of methylene blue	[62]
Orange peel	Co_3O_4	Spherical	25–28	Degradation of methyl orange and malachite green	[63]
Phoenix dactylifera L	CoO	Spherical	80	Degradation of Congo red	[64]

cosmetics, food coatings, and antibacterial agents [65]. Their wide range of applications is attributed to several features: their heightened magnetic permeability, nanometric size, cost-effectiveness, straightforward synthesis, colloidal stability, adaptable surface chemistry, strong chemical stability, and ease of dispersion in aqueous solutions [66–68].

In recent years, environmentally friendly iron nanoparticle materials produced through green synthesis methods have gained popularity. Conventional synthetic techniques can be employed to create iron nanoparticles; however, these methods involve the use of flammable chemicals, toxic, corrosive, specialized equipment, and higher costs. To address the limitations of existing synthesis methods, the biosynthesis of nanoparticles using natural product extracts has emerged as an environmentally friendly alternative approach, simple and cost-effective. A brief review of the literature reveals that there are only a limited number of articles reporting the synthesis of FeO NPs using plant sources. Some noteworthy examples include the utilization of leaf extracts from Camellia sinensis [69], *Eucalyptus Globulus* [70], *Aloe vera* [71], *Sageretia thea* [72], *Ailanthus excels* [73], *Rhus punjabensis* [74], root extracts from *Arisaema amurense* [75], and fruit extracts from *Cynometra ramiflora* [76] for the recent synthesis of FeO NPs.

Numerous studies have explored the potential applications of various nanoparticles, including metals like silver (Ag) and gold (Au), as well as metal oxides like TiO_2 and ZnO, as photocatalysts for the degradation of textile dyes [77, 78]. Iron oxide nanoparticles (FeO NPs) stand out due to their exceptional catalytic and reductive properties, making them well-suited for the treatment of wastewater. Additionally, they offer the advantage of easier separation compared to other nanomaterials that need centrifugation for separation purposes [79]. FeO NPs find extensive use in diverse applications, including the heavy metals removal, antibiotics, and dyes from water sources. They also play a vital part in biological applications, facilitating site-specific medicine delivery and the destruction of tumor cells [80, 81].

Iron nanomaterials are gaining increasing attention from researchers across various fields due to enhanced reactivity of surface sites, high catalytic activities, and their versatile properties. It has a wide range of applications in different fields [82–89]. Iron oxide nanoparticles are also utilized as materials for the removal of dyes to promote environmental health [90]. An alternative approach involves using iron nanoparticles to catalyze the breakdown of pollutants and remediate polluted areas. Research has shown successful degradation of a diverse range of dye categories using this method [91]. While iron millimetric particles have been much explored and applied, the next frontier is transitioning to nanoscale particles. Due to their larger surface areas, nanoparticles can degrade contaminants orders of magnitude faster than their millimetric counterparts [92]. Initial studies have demonstrated that iron nanoparticles can achieve over 98% dye degradation within just one hour at 30 °C. Furthermore, employing iron nanoparticles in conjunction with temperature increases has been exhibited to accelerate the dye degradation rate [93, 94] (Fig. 8 and Table 3).

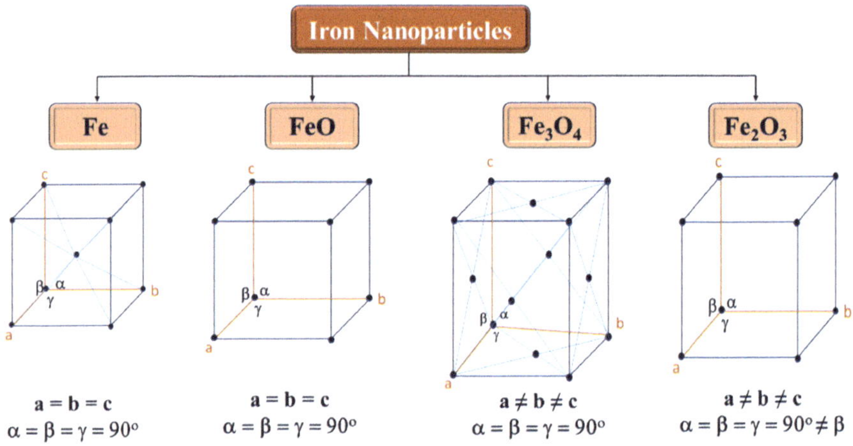

Fig. 8 Structure of iron/iron oxide nanoparticles

1.5.3 Copper Nanoparticles

Copper nanoparticles (Cu NPs) have recently gained significant interest owing to their unique physical and chemical. They have emerged as promising candidates for a wide range of domestic and commercial applications, encompassing catalysis, energy-based research, medical and pharmaceutical uses, cosmetics, imaging, and environmental applications [105]. Cu NPs can be synthesized from different plant parts, like bark, seeds, stems, flowers, leaves, roots, and fruits resulting in different sizes and forms [106, 107]. The size and shape of Cu NPs can be adjusted by different concentrations of metal and the plant extract in the reaction mixture. The plant leaf extracts contain several bioactive metabolites, such as phenols, terpenoids, flavonoids, tannins, and proteins which serve as reducing and capping agents, facilitating the conversion of metallic salts into nanoparticles [108–110]. Concerning their crystalline structures, copper (Cu) exhibits a face-centered cubic (fcc) phase, while CuO adopts a monoclinic phase, and Cu_2O displays a cubic phase, as indicated by their respective JCPDS card numbers. The synthesis temperature plays a key role in determining the formation of these phases, with CuO, Cu_2O, and Cu forming at temperatures of 100, 180, and 220 °C, respectively [111] (Fig. 9 and Table 4)

Table 3 Green synthesis and catalytic applications of Fe/FeO nanoparticles

Botanical name	NPs	Morphology	Particle size (nm)	Applications	Refs.
Green tea leaves	Fe	Agglomerated and circular	20–50	Degradation of anthraquinone (RBB-R) & azo (DR80) dye mixture	[95]
Peltophorum pterocarpum leaves	Fe_2O_3	Irregular rod-like shape	–	Catalytic degradation of methylene blue dye	[96]
Punica granatum seed	Fe_2O_3	Agglomerated with semi-spherical	25–55	Photocatalytic degradation of reactive blue 4 dye	[97]
Psidium Guajava-Moringa Oleifera leaves	FeO	Spherical rod	–	Photocatalytic degradation of methylene blue dye	[98]
Daphne mezereum leaves	FeO	Spherical	9.2	Degradation of methyl orange dye	[99]
Carica papaya leaves	α-Fe_2O_3	Agglomerated	21.59	Photocatalytic degradation of remazol yellow RR dye	[100]
Cynometra ramiflora fruit	Fe_2O_3	Spherical	58.50–78.13	Photocatalytic degradation of methylene blue dye	[76]
Withania coagulants	FeO	Spindle-shaped nanoarrays	15–20	Photocatalytic degradation of Safranin dye	[101]
Teucrium polium leaves	FeO	Spherical	5.68–30.29	Photocatalytic degradation of methyl orange dye	[102]
Delonix regia flower	Fe_3O_4, α-Fe_2O_3	Agglomerated	60–95	Photocatalytic degradation of Congo red and methylene blue dyes	[103]
Terminalia chebula Seed	FeO	Crystalline	–	Photocatalytic degradation of methylene blue dye	[104]

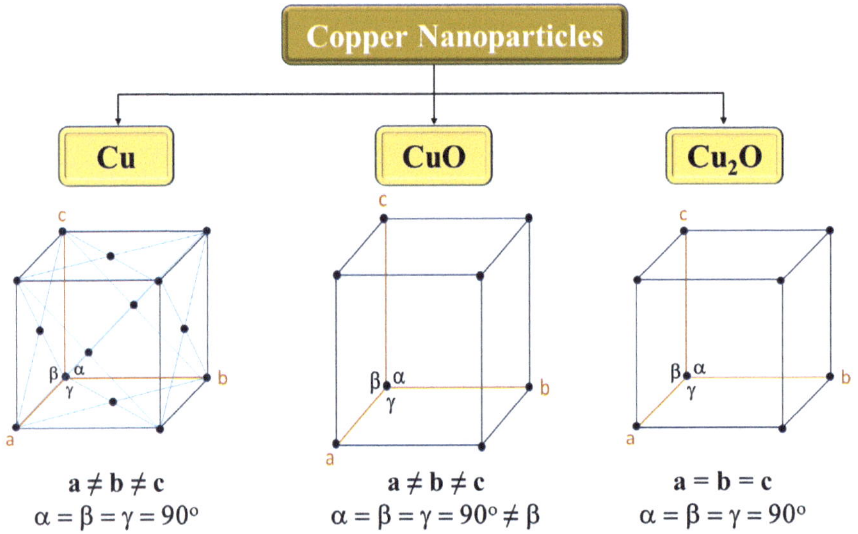

Fig. 9 Structure of copper/copper oxide nanoparticles

Table 4 Green synthesis and catalytic activity of copper/copper oxide nanoparticles

Plant name	NPs	Morphology	Particle size (nm)	Application	Refs.
Annona muricata leaves	CuO	Spherical	30–40	Degradation of reactive red 120, methyl orange	[112]
Celastrus paniculatus leaves	Cu	Spherical	2–10	Degradation of methylene blue dye	[113]
Cinnamon bark	Cu	Spherical	–	Degradation of methylene blue (MB) and methyl orange (MO)	[105]
Jatropha curcas leaves	Cu	Spherical	10 ± 1	Photocatalytic activity against methylene blue (MB)	[114]
Ficus carica extract	Cu	Spherical	61	Degradation performance of Alizarin Yellow R dye	[115]
Camelia sinensis leaves	Cu	–	60 ± 6	Bromophenol blue	[116]
Ruellia tuberosa	CuO	Nanorods	83.23	Crystal violet	[117]
Terminalia arjuna bark	Cu_2O	–	–	Degradation of methylene blue	[118]
Musa acuminate fruits	Cu_2O	Octahedron shaped	–	Degradation of methylene blue	[119]
Guava fruits juice	Cu_2O	Octahedron morphology	Microspheres 10, 4 μm	Degradation of methylene blue	[120]
Aloe vera	Cu_2O	Octahedral and spherical	24–61 nm	Degradation of methylene blue	[121]

1.6 Characterization Techniques for Nanoparticles Formation

The preparation of nanoparticles can be verified through a range of different techniques, like UV-visible spectroscopy (UV-vis), Fourier-transform infrared spectroscopy (FTIR), X-ray diffraction analysis (XRD), X-ray photoelectron spectroscopy (XPS), Energy-dispersive spectroscopy (EDS), Dynamic light scattering (DLS), Scanning electron microscopy (SEM), Transmission electron microscopy (TEM), Atomic force microscopy (AFM), Differential thermal analysis (DTA), and Thermogravimetric analysis (TGA) (Fig. 10).

1.6.1 XRD

XRD is a non-destructive method used to identify a material's crystallographic structure, size, and identification of phase.

Its operation is grounded in the fundamental concept of constructive interference occurring between monochromatic X-rays and a crystalline sample. In the XRD process, these X-rays are collimated and aimed at a nanomaterial sample. Upon interacting with the sample, incident rays produce a diffracted ray that is subsequently detected, analyzed, and recorded. The resulting diffraction pattern is displayed as an intensity distribution of diffracted photons at varying angles and provides information [122]. Sharp peaks show a very high crystalline nature and broadening of peaks shows the amorphous nature of the sample. From the obtained results the average particle size will be calculated by using the Debye Scherrer equation in Eq. 1.

$$D = K\lambda/\beta \cos\theta \tag{1}$$

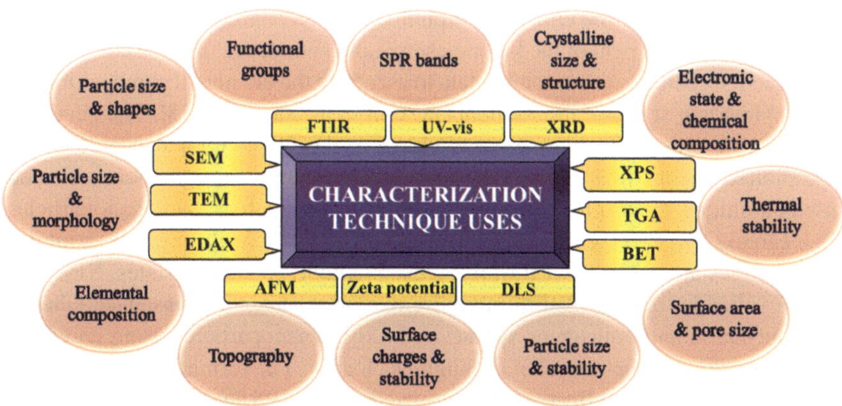

Fig. 10 Various characterization techniques adopted for nanoparticles

The method of synthesis plays a key role in determining the size and crystallinity of particles. In a comparative research study led by Murtaza Hasan and colleagues, they synthesized iron oxide nanorods using both a chemical approach involving precipitation reduction and a green method that utilized *Withania coagulans* extract. Subsequent characterization revealed that the average crystallite size achieved through the green approach was smaller, measuring 26 nm, compared to the chemical method's 34 nm. Furthermore, the iron oxide nanoparticles produced via the green method exhibited a standard face-centered cubic (FCC) crystalline structure. Notably, the catalytic activity of these green-synthesized iron oxide nanorods surpassed that of their chemically prepared counterparts [123]. In a distinct investigation conducted by Kuldeep Kumar and the team, they synthesized monoclinic phase CuO nanoparticles (NPs) by varying the precursor's concentrations (5, 10, 50 mmol kg^{-1}). Their findings indicated that as the precursor concentration increased, the average crystallite size decreased from 9.240 to 8.424 nm. Interestingly, the interplanar distance of CuO, measuring at 0.242 nm, remained consistent across all concentrations. This occurrence can be ascribed to the observation that with the augmentation of precursor concentration, there was a marginal decrease in the dimensions of CuO nano-spheres or nanorods [124]. In summary, the green synthesis method, which incorporates plant components, has proven to be an effective approach without adversely impacting the interplanar distance while promoting good crystallinity in the sample. This characteristic is particularly valuable for enhancing catalytic activity.

1.6.2 XPS

X-ray Photoelectron Spectroscopy (XPS), sometimes referred to as Electron Spectroscopy for Chemical Analysis (ESCA), is an extremely valuable quantitative technique used to probe an element's electronic structure, elemental composition, and oxidation states within a material. During XPS analysis, a sample is exposed to X-ray irradiation, resulting in the emission of photoelectrons from the material's surface. Through the measurement of the kinetic energy and abundance of these emitted electrons, scientists can acquire essential data to ascertain the elemental composition and chemical bonding states of the elements present in the sample [125]. The generated XPS spectrum represents the number of electrons detected within a specific energy range plotted against their kinetic energy. It's important to note that each element exhibits a unique spectrum. To illustrate, let's consider cobalt oxide nanoparticles, which can exist in two forms: CoO and Co_3O_4. In CoO, cobalt typically exhibits a +2 oxidation state, while in Co_3O_4, it can exist in both +3 and +2 states. XPS analysis is a valuable tool for confirming these oxidation states. For instance, Ghasem Najafpour-Darzi and colleagues synthesized Co_3O_4 nanoparticles using an extract from *Trigonella foenum-graceum* (Fenugreek) leaves. The formation of Co_3O_4 and the oxidation states were effectively confirmed through XPS analysis, which revealed characteristic spin-orbit splitting. The observed difference between the Co $2p_{3/2}$ and

Co $2p_{1/2}$ peaks was measured at 15.3 eV, providing solid confirmation of the oxidation of approximately 2/3 of the Co^{2+} ions to Co^{3+} [126]. Similarly, Kalarical Janardhanan Sreeram et al. synthesized CoO nanoparticles using starch and characterized them through XPS analysis. The observed splitting of 2p orbital peaks, occurring at around 795.2 and 780 eV, indicated the presence of a +2 oxidation state for cobalt atoms within the CoO nanoparticles [127]. In summary, due to the various forms and phases that metal-based nanoparticles can assume, it is of paramount importance to determine their oxidation states accurately. The use of green synthesis methods allows researchers to fine-tune reaction conditions to achieve the desired nanoparticle forms. Consequently, XPS emerges as an indispensable tool for the comprehensive characterization of metal nanoparticles.

1.6.3 SEM and TEM

Transmission Electron Microscopy (TEM) is a highly valuable technique for the detailed examination of nanoparticle characteristics such as shape, dispersion, and precise size measurement. TEM enables atomic-level visualization with an impressive resolution of approximately 1 nm [128]. Whereas, Scanning Electron Microscopy (SEM) serves as a versatile tool that offers qualitative insights into various aspects of samples, encompassing topography, morphology, composition, and crystallographic patterns, providing high-resolution images. The operational principles of SEM differ significantly from TEM. While TEM involves the transmission of electrons through the sample, SEM directs a focused electron beam onto the specimen surface. Through the capture of interactions between the electron beam and the sample, SEM generates a three-dimensional black-and-white image pixel by pixel [129, 130]. In an experiment performed by Harunar Rashid and colleagues, CuO nanoparticles were prepared with an extract from *Lantana camara flowers*. Microscopic analysis of these CuO nanoparticles revealed 13–28 nm sizes. Intriguingly, the examination also uncovered the existence of rod-shaped CuO nanoparticles alongside spherical particles. These findings explain the capability of the green synthesis method to yield materials with extremely small nanometer-sized particles and spherical shapes [131]. It is essential to note that both the size and shape of nanoparticles play a pivotal role in influencing their catalytic activity.

1.6.4 FTIR

Fourier Transform Infrared (FTIR) analysis is a versatile technique applicable to both liquid and solid samples. It plays a pivotal role in the identification of functional groups and metabolites present on nanoparticle surfaces, making it especially valuable for understanding the processes involved in stabilizing and reducing nanoparticles. Additionally, FTIR is a valuable tool for investigating adsorption phenomena concerning metal oxide nanoparticles, providing insights into the interactions between molecules and the nanoparticle surfaces. When molecules adhere to nanoparticle surfaces, it can lead to shifts in absorption bands, the appearance of new bands, or alterations in intensity spectra. These changes offer valuable insights into the nature of adsorption and the formation of adsorbate-nanoparticle complexes. Typically, peaks observed in the 400–600 cm^{-1} range indicate the formation of metal oxide bonds [132]. For example, in a study conducted by Z Alhalili, CuO nanoparticles were synthesized using an extract from *Eucalyptus Globulus* leaves. Vibrational peaks at 530 and 580 cm^{-1} were detected, corresponding to the CuO bond, thus confirming the successful synthesis of CuO [133, 134]. Similarly, P Chelliah and their research team synthesized Co_3O_4 nanoparticles using *Curcuma longa* plant extract and suggested that Co^{2+} ions were reduced to produce Co_3O_4 NPs through electron donation from curcumin. They observed CoO stretching vibrations at 555 cm^{-1} and Co(III)–O bonds at 659 cm^{-1}. Moreover, the plant extract generated carboxylic acid and amides on the cobalt particles, contributing to the capping, reduction, and stabilization of Co_3O_4 NPs. Minor peaks at 1639 and 1382 cm^{-1} were also evident, showing the presence of carboxylic acid and amide originating from the plant's biomolecules [135]. Key components of FTIR spectra include absorption bands, which correspond to the vibrational modes of chemical bonds within the sample.

1.6.5 BET

The nanoparticle's surface is a key factor in their characterization, directly influencing their performance and properties due to the substantial surface area-to-volume ratio they possess. The preferred method for identifying this surface area is the BET (Brunauer-Emmett-Teller) analysis, despite its somewhat time-consuming nature [136]. The principle of BET analysis is gas physical adsorption on a solid surface and is widely utilized for the rapid and precise determination of nanostructure surface areas [137]. Additionally, during the drying of nanoparticle samples for such measurements, there is a risk of significant hydrogen bonding occurring among nanoparticle surfaces, which can introduce inaccuracies [138]. The green synthesis method can be effectively employed to prevent the agglomeration of particles, ensuring accurate surface area measurements [139]. For instance, in a study conducted by P. K. Chauhan and colleagues, CuO nanoparticles were synthesized using the *rhizome* extract *of Bergenia ciliata.* They reported a surface area of 24 m^2 g^{-1} and demonstrated the efficient removal of methylene blue and methyl red dyes. A larger surface area enhances the adsorption properties of the material [140, 141]. Hence, the green synthesis method for metal nanoparticles enables the design of nanoparticles with significant surface areas, which is advantageous for their adsorption and photocatalytic properties.

1.6.6 TGA

Thermogravimetric Analysis (TGA) serves as a valuable technique for investigating the thermal properties of nanoparticles, with applications spanning various scientific disciplines. Its primary purpose is to elucidate how materials undergo thermal decomposition and experience changes in weight as a function of temperature. TGA involves the continuous monitoring of a sample's mass while subjecting it to controlled heating at a predetermined rate within a controlled environment. The sample is positioned on a balance housed within a furnace and subjected to heating according to a specified temperature program. This environment can be tailored to be reactive, like in the presence of O_2 or air, inert, utilizing gases like nitrogen or argon, or even under vacuum conditions. The primary data obtained from TGA are presented as thermograms, which can be expressed either as the percentage of weight loss over time or about temperature. For data analysis, researchers often rely on the first and second derivatives of the thermogram, illustrating the percentage weight loss as a function of temperature. The application of heat triggers changes in the material, resulting in either an increase or decrease in sample mass. Different systems exhibit various transitions, encompassing processes such as the desorption of adsorbed species (e.g., H_2O), oxidation of metal species, decomposition through pyrolysis, and combustion, notably carbonaceous materials [142].

1.6.7 UV-Vis Spectroscopy

UV-vis spectroscopy is a widely utilized method for quantifying the light absorbed and scattered by a sample. In this method, the sample is positioned between the UV-Vis light source and the detector, allowing light to pass through it. The intensity of the light beam is measured both before and after it traverses the sample, and the combined measurement of scattered and absorbed light is referred to as extinction [143, 144]. UV-visible spectroscopy plays a pivotal role in characterizing the structure of NPs. It is firmly established that the optical absorption spectra of metal NPs are predominantly influenced by surface plasmon resonances (SPR) [145]. Additionally, metal NPs incorporation into semiconductor catalysts proves to be an effective methodology for improving the separation of photogenerated charge carriers and increasing visible light absorption [146–149]. Due to the low Fermi levels of metal NPs, electrons may migrate from semiconductors with higher Fermi levels to the surfaces of metal nanoparticles [150, 151]. Remarkably, noble metal NPs such as Ag, Au, and Cu exhibit strong visible light absorption due to their surface plasmon resonance (SPR) [152, 153]. The optical band gap of nanoparticles plays a key role in predicting their photocatalytic activity. A band gap value below 5 eV is advantageous for nanoparticles' photocatalytic activity. To calculate the optical band gap from a UV-Vis spectrum, Tauc's formula is commonly used (Eq. 2). This formula relates the absorption coefficient (α) to photon energy ($h\nu$) and the band gap energy. By plotting α against $h\nu$ and extrapolating, one can determine the band gap, with values below 5 eV favoring photocatalytic activity [154].

$$(\alpha h\upsilon)^n = A\,(h\upsilon - Eg) \tag{2}$$

In this equation, h is Planck's constant, υ signifies frequency, A is a constant, Eg denotes the band gap, and n represents the exponent value.

1.6.8 Zeta Potential

Zeta potential (ζ) is a valuable parameter for evaluating the surface charges and stability of nanoparticles. Phytoconstituents derived from plant extracts are noted for their significant contribution to the stabilization and reduction of NPs and NCs. The determination of zeta potential involves particle dispersion in a water medium. Zeta potential values have been found to directly correlate with the stability of nanoparticles. Specifically, zeta potential values in the range of 0 to ± 5 are associated with fast coagulation, ± 10 to ± 30 indicate initial stability, ± 30 to ± 40 denote average stability, ± 40 to ± 60 suggest better stability, and values exceeding ± 61 indicate excellent stability [155, 156].

1.7 Dye Degradation

Water pollution is a serious global environmental and human health concern, with organic dyes representing a major source of contamination in natural resources due to their discharge by various industrial processes such as textile, plastic, leather, paper, and food production. Organic dyes are generally characterized as large molecules that possess a range of undesirable traits, including toxicity, non-biodegradability, chemical stability, carcinogenicity, and the ability to hinder the penetration of light in aqueous systems. These attributes contribute to environmental pollution and associated concerns. Numerous chemical and physical techniques were employed to address the degradation of the organic dye. These approaches include adsorption, ion exchange, flotation, membrane filtration, UV radiation, chemical coagulation, and hydrogen peroxide oxidation [157]. Nonetheless, these approaches often face challenges in effectively degrading intricate organic dye molecules, and certain methods have limitations despite their merits. Interestingly, the photocatalytic method sets itself apart owing to its remarkable efficacy in breaking dyes (Fig. 11).

1.7.1 Ion Exchange Method

The ion-exchange method is a reversible chemical process where ions in wastewater, which are charged particles due to electron gain or loss, are exchanged for ions of the same charge attached to a stationary solid. Currently, synthetic organic resins represent the primary type of material employed for this purpose, as their characteristics

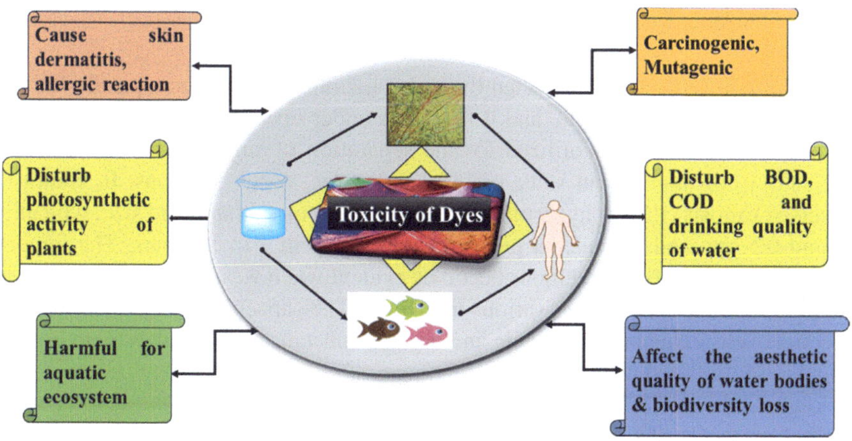

Fig. 11 Toxicities of various dyes

can be customized to suit specific applications. Organic ion-exchange resins are large-molecule polyelectrolytes with the capacity to exchange their mobile ions for ions of the same charge found in wastewater. When it comes to removing dyestuffs from wastewater, ion exchange emerges as a favorable option due to its significant capacity and selectivity toward various dye components. Ion exchange can effectively address reactive dyes, which belong to a highly complex group of dyes extensively employed in the textile industry due to their strong dye-binding characteristics [158, 159].

1.7.2 Chemical Coagulation and Flocculation

Chemical coagulation and flocculation are utilized to aid in the removal of finely dispersed solids by creating flocs that settle more readily, often preceding sedimentation or filtration processes. The conventional sequence involving coagulation, flocculation, and sedimentation is vital in the initial phases of pretreatment for various water purification and wastewater treatment procedures. Various chemicals are utilized in these processes, depending on the specific application. For drinking water treatment, common chemicals such as soda ash (Na_2CO_3), lime ($Ca(OH)_2$), and organic polyelectrolytes like polysaccharide gums, starch, and synthetic polyelectrolytes are employed. These chemicals facilitate the rapid formation of dense and easily settling flocs, thereby improving the efficiency of suspended solids removal. In the case of wastewater treatment, coagulants such as alum ($Al_2(SO_4)_3)14H_2O$, lime ($Ca(OH)_2$), iron salts like ferric sulfate ($Fe_2(SO_4)_3$), ferric chloride ($FeCl_3 \cdot 6H_2O$), and ferrous sulfate ($FeSO_4 \cdot 7H_2O$), as well as organic polyelectrolytes, are used as flocculation aids to assist in the process [160]

1.7.3 Nanofiltration (NF) and Reverse Osmosis (RO)

Nanofiltration (NF) is distinguished through a membrane pore size typically falling within the range of 0.5–2 nm, and it functions under operational pressures ranging from 5 to 40 bars. NF is utilized for the separation of sugars, different multivalent salts from monovalent salts, organic molecules, water, and ions. In contrast, Reverse Osmosis (RO) or hyperfiltration has a tighter membrane pore size, typically around 0.5 nm, and operates at pressures ranging from 7 to 100 bars. The importance of these membrane methods becomes evident when we consider the extensive use of membrane areas across various industries. The capacity of RO membranes to effectively eliminate both inorganic and organic substances also makes them highly valuable for polluted drinking water treatment.

Regarding dye removal from wastewater, there is a positive correlation with applied pressure, pH, total dissolved solids (TDS), and dye concentration in the feed solution. However, a negative relationship exists with the feed temperature. Both solution temperature and applied pressure positively influence the permeate flux in NF and RO membranes. Conversely, the permeate flux decreases with rising dye concentration and pH. It's worth noting that wastewater containing acid dyes, when treated with NF or RO membranes, generally yields a lower rejection rate and a higher permeate flux compared to wastewater containing reactive dyes [161].

1.7.4 Ultraviolet and Hydrogen Peroxide

Utilizing ultraviolet (UV) and solar radiation in conjunction with hydrogen peroxide as a bleaching agent, employing either glass or quartz tubes has proven to be highly effective in achieving complete decolorization of dyes. The combination of hydrogen peroxide (H_2O_2) and UV oxidation represents the advanced oxidation processes (AOPs) that have been effectively employed in treating various water pollutants. These processes involve relatively straightforward reactions, such as the UV photolysis of H_2O_2. This method is distinguished by the production of exceptionally potent oxidizing species, known as [.OH] radicals, in relatively high, stable concentrations. These reactive radicals possess an impressive ability to efficiently degrade and sometimes even mineralize organic contaminants. A key benefit of this method is its avoidance of generating additional disposal concerns post-treatment, and it displays non-selectivity toward a wide array of chemicals [162] (Fig. 12 and Table 5).

1.7.5 Photocatalytic Reaction

Photocatalytic degradation of organic dyes stands out as an efficient, cost-effective method that allows for the full mineralization of organic content. Moreover, this technique doesn't yield any toxic by-products, marking it as environmentally friendly. Photocatalytic reactions occur at the interface of a catalyst and the dye, whether in solution or as a gas. When the catalyst is exposed to energy of photon equivalent to

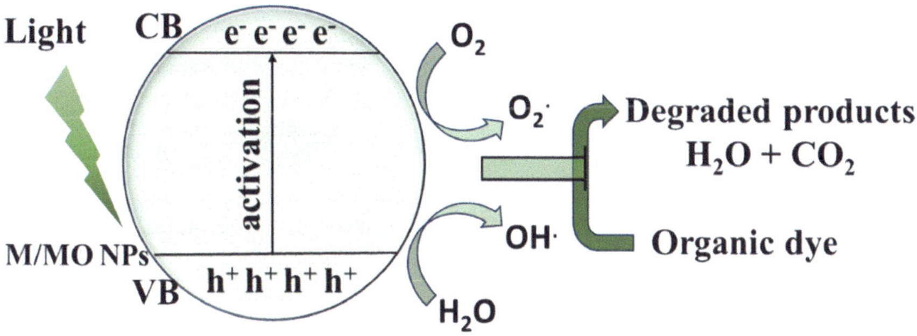

Fig. 12 Illustration of the photocatalytic dye degradation process

Table 5 Deciphering the steps in a typical photocatalytic reaction process

Reaction type	Photocatalytic reaction mechanism
1. Photoexcitation	M/MO NPs + hυ → h^+ (HOMO) + e^- (LUMO)
2. Recombination	$h^+ + e^- \rightarrow$ heat (Δ)
3. Reduction	$O_2 + e^- \rightarrow \cdot O_2^-$ $\cdot O_2^- + H^+ \rightarrow HOO\cdot$ $HOO\cdot + e^- \rightarrow HOO^-$ $HOO^- + H^+ \rightarrow H_2O_2 \rightarrow \cdot OH$ $H_2O_2 + e^- \rightarrow {}^-OH + \cdot OH$
4. Oxidation	$H_2O \rightarrow H^+ + {}^-OH$ ${}^-OH + h^+ \rightarrow \cdot OH$ Dye effluent pollutant + h^+ → ·Pollutant → Intermediates & final products
5. Photodegradation	·OH + Pollutant → H_2O + ·Pollutant → Intermediates & final products

or higher than its band gap, it behaves as a photocatalyst by producing electron-hole pairs. This results in the development of reactive oxygen species (ROS). Specifically, the generated electrons and holes interact with water and molecular oxygen to produce highly reactive radicals, notably the OH and O_2 radicals. The dye molecules are then chemically attacked by these active radicals, which convert them into smaller chemicals, mainly CO_2 and H_2O. Between the semiconductor and dye contact, an array of photocatalytic processes take place simultaneously. Additionally, dye molecules can be directly excited to a high-energy triplet state. Studies suggest that this excited triplet state of dye molecules can donate an electron to the conduction band (CB) of semiconductors like TiO_2, transforming itself into a semi-oxidized radical cation dye (+). This electron, residing in the semiconductor's CB, reacts with O_2 to generate hydroxide radicals, resulting in the direct photooxidation of dye [163]. The general photocatalytic dye degradation process is illustrated in Fig. 13.

Parvathiraja Chelliah et al. performed photocatalytic activity on 10 ppm of MB dye by using Co_3O_4 NPs synthesized by green methods and showed 90% degradation in 40 min of visible light irradiation [62]. Mali et al. performed photocatalytic

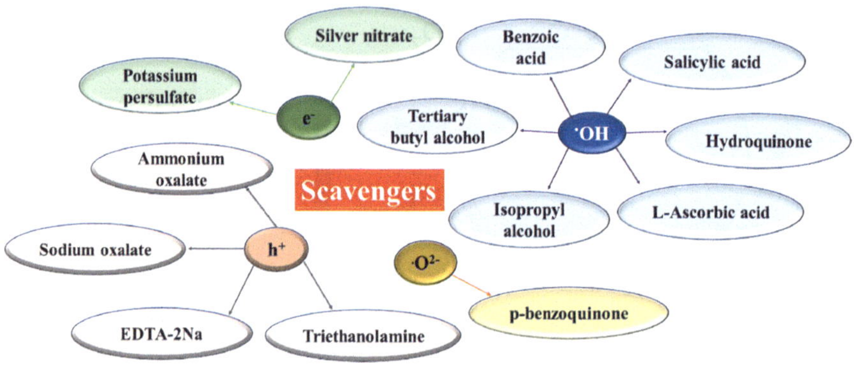

Fig. 13 Illustration of the scavengers for the reactive species

degradation studies for MB dye under sunlight and got 90% degradation efficiency for 120 min by using Cu_2O NPs synthesized by *Celastrus paniculatus* leaves extract [113]. Bishnoi et al. synthesized Fe_2O_3 NPs by using *Cynometra ramiflora* extract and performed photocatalytic degradation of MB dye. They report that 20 ppm of dye was removed for 110 min under sunlight irradiation [76].

1.8 Product Analysis

Analyzing the degraded products after the reaction represents a critical step in the degradation process. Various analysis techniques were employed to examine these degraded products.

1.8.1 Total Organic Carbon (TOC)

Total organic carbon (TOC) serves as the primary parameter for assessing the increasing quantity of organic compounds within a sample. TOC analysis enables the determination of the total organic content in a sample and can detect variations in the organic composition. The levels of TOC can vary depending on the type of solution and can be measured at various concentrations using different systems. However, TOC alone may not provide a comprehensive picture, as carbon-containing compounds can exist in diverse forms and proportions. Typically, TOC analysis involves three main stages: sampling, oxidation, and detection. Initially, samples are introduced into an analyzer, which may employ a flow-injection technique. Oxidation techniques employed encompass chemical oxidation, high-temperature combustion, and photocatalytic oxidation. TOC analyzers utilize conductivity detectors and non-dispersive infrared (NDIR) technology for detection [164].

The demineralization of degraded dye is determined by using the following relationship in Eq. (3).

$$\alpha = \frac{\text{TOC(initial)} - TOC(\text{final})}{\text{TOC(initial)}} \times 100 \tag{3}$$

where the percentage of the dye's mineralization is represented by α. $\text{TOC}_{\text{(initial)}}$ and $\text{TOC}_{\text{(final)}}$ describe the initial and final organic carbon contents, respectively

1.8.2 HPLC-MS/UPLC-MS

High-performance liquid chromatography-mass spectrometry (HPLC-MS) stands as a potent analytical technique employed for both the separation and characterization of compounds, whether known or unknown. It serves the purpose of unraveling the structure of various molecules. This method proves particularly valuable for the analysis of small molecules and delivers enhanced sensitivity and selectivity when conducting trace analysis of substances with multiple components. In the process of HPLC analysis, the solvent employed to segregate components within a liquid sample is referred to as the mobile phase. The mobile phase is transported to a separation column, which is also referred to as the stationary phase, and then proceeds to the detector. Throughout this process, a consistent flow rate is maintained under the control of the solvent transfer pump. A defined amount of the sample is introduced into the column, where the constituents of the sample undergo separation. Downstream of the column, a detector identifies and quantifies each separated compound. Depending on the properties of the dye sample, adjustments are made to the solvent mobile phase and the flow rate to ensure optimal results. The structural formulas of the degradation intermediates were hypothesized based on potential degradation pathways. These were determined using mass spectra (m/z) values and their associated fragmentation patterns [165]. By understanding these degradation pathways, one can readily identify the toxicity of the degraded products.

1.8.3 GC-MS

The gas chromatography (GC) method is a widely used method for confirmation testing because it efficiently splits all the components within a sample and gives a representative spectral output. In the case of the degraded dye sample, it is introduced into the GC device's port, where the sample undergoes vaporization and subsequent separation to initiate the analysis. The sample inlet is sustained at a high temperature, often reaching up to 752°F (400 °C), to ensure that the sample remains in a gaseous state. Then, the specimen is allowed to enter the ionization chamber, where at high voltage, the beam of electrons is accelerated. This acceleration process ensures that only one mass reaches the detector for a given voltage value. To cover a range of masses, the accelerating voltage is adjusted, allowing all fragments to reach the detector [165].

1.8.4 Scavenger-Assisted Analyses of Photocatalytic Degradation

The generation of reactive species and their mechanisms throughout the organic pollutant's degradation have been indirectly studied using scavengers. The investigative use of scavengers has harnessed their high reactivity and non-selective characteristics to explore degradation mechanisms. Catalytic oxidation of organic contaminants is a result of the production of active radicals. Various types of scavengers are employed to identify different reactive species, as illustrated in the diagram below [165].

1.9 Conclusion

The green synthesis approach is gaining significant attention as a promising method for manufacturing metal/metal oxide NPs. It offers numerous advantages compared to physical synthesis routes or chemical wet processing. These advantages encompass biocompatibility, low energy requirements, non-toxic reagents, ease of processing, rapid synthesis, sustainable stability, and scalability of the produced nanoparticles. This chapter provides an overview of recent research on the production of metal/metal oxide NPs through the reaction of phytochemicals that exist in various plant extracts and metal precursors. In recent years, numerous studies have been published in the literature regarding the synthesis of these nanoparticles using green synthesis techniques based on natural plant extracts. Based on the research findings, it is evident that the existence of essential phytochemical components in plant extracts eliminates the need for stabilizing or surfactant agents during nanoparticle synthesis. Nanomaterials produced through green synthesis play a crucial role in protecting aquatic ecosystems from a multitude of harmful pollutants, including heavy metals, fertilizers, industrial effluents, organic compounds, and more. These nanoparticles possess elevated surface-to-volume ratios, increased surface energy, enhanced biocompatibility, and heightened chemical reactivity, rendering them well-suited for a diverse array of applications. Biosynthesis greatly simplifies the creation of sustainable materials.

1.10 Future Aspects

However, certain research gaps still exist, demanding further focus. Economic studies are essential to pinpoint vital areas for reducing the production expenses of nanomaterials. Moreover, upcoming research should delve into the potential advantages of this method in mitigating the discharge of hazardous substances into the environment.

Additionally, it is crucial to explore the suitability of these techniques in crafting materials and structures activated by visible light, facilitating efficient separation of photogenerated electrons and holes. Standardizing protocols for the environmentally friendly production of photocatalytic materials and elucidating the precise mechanisms involved in the eco-friendly synthesis of nanoparticles are equally imperative aspects warranting the attention of the scientific community.

References

1. R.G. Saratale et al., A comprehensive review on green nanomaterials using biological systems: recent perception and their future applications. Colloids Surf. B **170**, 20–35 (2018). https://doi.org/10.1016/j.colsurfb.2018.05.045
2. E.P. Giannelis, Polymer-layered silicate nanocomposites: synthesis, properties and applications. Appl. Organomet. Chem. **12**, 675–680 (1998). https://doi.org/10.1002/(SICI)1099-0739(199810/11)12
3. S.S. Hashmi et al., Potentials of phyto-fabricated nanoparticles as ecofriendly agents for photocatalytic degradation of toxic dyes and waste water treatment, risk assessment and probable mechanism. J. Indian Chem. Soc. **98**, 100019 (2021). https://doi.org/10.1016/j.jics.2021.100019
4. S.C. Kumari, V. Dhand, P.N. Padma, Green synthesis of metallic nanoparticles: a review. Nanomaterials, 259–281 (2021). https://doi.org/10.1016/B978-0-12-822401-4.00022-2
5. O.P. Bolade, A.B. Williams, N.U. Benson, Green synthesis of iron-based nanomaterials for environmental remediation: a review. Environ. Nanotechnol. Monit. Manag. **13**, 100279 (2020). https://doi.org/10.1016/j.enmm.2019.100279
6. C. Surya et al., Costus speciosus leaf extract assisted CS-Pt-TiO_2 composites: synthesis, characterization and their bio and photocatalytic applications. J. Mol. Struct. **1195**, 787–795 (2019). https://doi.org/10.1016/j.molstruc.2019.06.030
7. A. Ahmadi, M. Hajilou, S. Zavari, S. Yaghmaei, A comparative review on adsorption and photocatalytic degradation of classified dyes with metal/non-metal-based modification of graphitic carbon nitride nanocomposites: synthesis, mechanism, and affecting parameters. J. Clean. Prod. **382**, 134967 (2023). https://doi.org/10.1016/j.jclepro.2022.134967
8. Y. Cheng, L. He, G. Xia, C. Ren, Z. Wang, Nanostructured g-C_3N_4/AgI composites assembled by AgI nanoparticles-decorated g-C_3N_4 nanosheets for effective and mild photooxidation reaction. NJC **43**, 14841–14852 (2019). https://doi.org/10.1039/C9NJ02725D
9. X. Wu et al., Synthesis and characterization of Fe and Ni co-doped ZnO nanorods synthesized by a hydrothermal method. Ceram. Int. **40**, 14635–14640 (2014). https://doi.org/10.1016/j.ceramint.2014.06.050
10. S.K. Chaudhuri, L. Malodia, Biosynthesis of zinc oxide nanoparticles using leaf extract of Calotropis gigantea: characterization and its evaluation on tree seedling growth in nursery stage. Appl. Nanosci. **7**, 501–512 (2017). https://doi.org/10.1007/s13204-017-0586-7
11. P.P.N. Vijay Kumar, S.V.N. Pammi, P. Kollu, K.V.V. Satyanarayana, U. Shameem, Green synthesis and characterization of silver nanoparticles using *Boerhaavia diffusa* plant extract and their antibacterial activity. Ind. Crops. Prod. **52**, 562–566 (2014). https://doi.org/10.1016/j.indcrop.2013.10.050
12. S. Panigrahi, S. Kundu, S. Ghosh, S. Nath, T. Pal, General method of synthesis for metal nanoparticles. J. Nanopart. Res. **6**, 411–414 (2004). https://doi.org/10.1007/s11051-004-6575-2
13. A.K. Gupta et al., ZnO-Nanorod processed PC-SET as the light-harvesting model for plasmontronic fluorescence sensor. Sens. Actuat. B Chem. **307**, 127597 (2020). https://doi.org/10.1016/j.snb.2019.127597

14. A.K. Gupta, C.-H. Hsu, C.-S. Lai, Enhancement of the Au/ZnO-NA plasmonic SERS signal using principal component analysis as a machine learning approach. IEEE Photonics J. **12**, 1–11 (2020). https://doi.org/10.1109/JPHOT.2020.3015740
15. S.P. Patil, V.V. Burungale, Physical and chemical properties of nanomaterials, in *Nanomedicines for Breast Cancer Theranostics* (2020), pp. 17–31. https://doi.org/10.1016/B978-0-12-820016-2.00002-1
16. J. Gopal et al., Sustainable ecofriendly phytoextract mediated one pot green recovery of chitosan. Sci. Rep. **9**, 13832 (2019). https://doi.org/10.1038/s41598-019-50133-z
17. A.K. Mittal, U.C. Banerjee, Current status and future prospects of nanobiomaterials in drug delivery, in *Nanobiomaterials in Drug Delivery* (2016), pp. 147–170. https://doi.org/10.1016/B978-0-323-42866-8.00005-8
18. A.B. Asha, R. Narain, Nanomaterials properties, in *Polymer Science and Nanotechnology* (2020), pp. 343–359. https://doi.org/10.1016/B978-0-12-816806-6.00015-7
19. A. Reghunadhan, N. Kalarikkal, S. Thomas, Mechanical property analysis of nanomaterials. *Characterization of Nanomaterials* (2018), pp. 191–212. https://doi.org/10.1016/B978-0-08-101973-3.00007-9
20. D. Guo, G. Xie, J. Luo, Mechanical properties of nanoparticles: basics and applications. J. Phys. D Appl. Phys. **47**, 013001 (2014). https://doi.org/10.1088/0022-3727/47/1/013001
21. G. Cho, Y. Park, Y.-K. Hong, D.-H. Ha, Ion exchange: an advanced synthetic method for complex nanoparticles. Nano Converg. **6**, 17 (2019). https://doi.org/10.1186/s40580-019-0187-0
22. B. Mekuye, B. Abera, Nanomaterials: an overview of synthesis, classification, characterization, and applications. Nano Sel. **4**, 486–501 (2023). https://doi.org/10.1002/nano.202300038
23. M.C. Biswas, A. Chowdhury, M. Md. Hossain, M.K. Hossain, Applications, drawbacks, and future scope of nanoparticle-based polymer composites, in *Nanoparticle-Based Polymer Composites* (2022), pp. 243–275. https://doi.org/10.1016/B978-0-12-824272-8.00002-6
24. L.A. Kolahalam et al., Review on nanomaterials: synthesis and applications. Mater. Today Proc. **18**, 2182–2190 (2019). https://doi.org/10.1016/j.matpr.2019.07.371
25. N. Joudeh, D. Linke, Nanoparticle classification, physicochemical properties, characterization, and applications: a comprehensive review for biologists. J. Nanobiotechnol. **20**, 262 (2022). https://doi.org/10.1186/s12951-022-01477-8
26. V. Singh, P. Yadav, V. Mishra, Recent advances on classification, properties, synthesis, and characterization of nanomaterials, in *Green Synthesis of Nanomaterials for Bioenergy Applications* (Wiley, 2020), pp. 83–97. https://doi.org/10.1002/9781119576785.ch3
27. A. Barhoum et al., Review on natural, incidental, bioinspired, and engineered nanomaterials: history, definitions, classifications, synthesis, properties, market, toxicities, risks, and regulations. Nanomaterials **12**, 177 (2022). https://doi.org/10.3390/nano12020177
28. P. Christian, F. Von der Kammer, M. Baalousha, T. Hofmann, Nanoparticles: structure, properties, preparation and behaviour in environmental media. Ecotoxicology **17**, 326–343 (2008). https://doi.org/10.1007/s10646-008-0213-1
29. S. Drummer, T. Madzimbamuto, M. Chowdhury, Green synthesis of transition-metal nanoparticles and their oxides: a review. Materials **14**, 2700 (2021). https://doi.org/10.3390/ma14112700
30. K. Bloch, K. Pardesi, C. Satriano, S. Ghosh, Bacteriogenic Platinum Nanoparticles for Application in Nanomedicine. Front. Chem. **9** (2021). https://doi.org/10.3389/fchem.2021.624344
31. P. Khandel, R.K. Yadaw, D.K. Soni, L. Kanwar, S.K. Shahi, Biogenesis of metal nanoparticles and their pharmacological applications: present status and application prospects. J. Nanostr. Chem. **8**, 217–254 (2018). https://doi.org/10.1007/s40097-018-0267-4
32. C. Dhand et al., Methods and strategies for the synthesis of diverse nanoparticles and their applications: a comprehensive overview. RSC Adv. **5**, 105003–105037 (2015). https://doi.org/10.1039/C5RA19388E

33. S. Mukherji, S. Bharti, G. Shukla, S. Mukherji, Synthesis and characterization of size- and shape-controlled silver nanoparticles. Phys. Sci. Rev. **4** (2019). https://doi.org/10.1515/psr-2017-0082
34. A. Talib, Hui-Fen-Wu, Biomimetic synthesis of lotus leaf extract-assisted silver nanoparticles and shape-directing role of cetyltrimethylammonium bromide. J. Mol. Liq. **220**, 795–801 (2016). https://doi.org/10.1016/j.molliq.2016.04.100
35. Y. Park, Y.N. Hong, A. Weyers, Y.S. Kim, R.J. Linhardt, Polysaccharides and phytochemicals: a natural reservoir for the green synthesis of gold and silver nanoparticles. IET Nanobiotechnol. **5**, 69 (2011). https://doi.org/10.1049/iet-nbt.2010.0033
36. S.S. Shankar et al., Biological synthesis of triangular gold nanoprisms. Nat. Mater. **3**, 482–488 (2004). https://doi.org/10.1038/nmat1152
37. A.K. Jha, K. Prasad, K. Prasad, A.R. Kulkarni, Plant system: nature's nanofactory. Colloids Surf. B Biointerfaces **73**, 219–223 (2009). https://doi.org/10.1016/j.colsurfb.2009.05.018
38. H. Bar et al., Green synthesis of silver nanoparticles using latex of Jatropha curcas. Colloids Surf. A Physicochem. Eng. Asp. **339**, 134–139 (2009). https://doi.org/10.1016/j.colsurfa.2009.02.008
39. M. Behravan et al., Facile green synthesis of silver nanoparticles using *Berberis vulgaris* leaf and root aqueous extract and its antibacterial activity. Int. J. Biol. Macromol. **124**, 148–154 (2019). https://doi.org/10.1016/j.ijbiomac.2018.11.101
40. G. Jayakumarai, C. Gokulpriya, R. Sudhapriya, G. Sharmila, C. Muthukumaran, Phytofabrication and characterization of monodisperse copper oxide nanoparticles using *Albizia lebbeck* leaf extract. Appl. Nanosci. **5**, 1017–1021 (2015). https://doi.org/10.1007/s13204-015-0402-1
41. M. Naseer, R. Ramadan, J. Xing, N.A. Samak, Facile green synthesis of copper oxide nanoparticles for the eradication of multidrug resistant Klebsiella pneumonia and Helicobacter pylori biofilms. Int. Biodeterior. Biodegrad. **159**, 105201 (2021). https://doi.org/10.1016/j.ibiod.2021.105201
42. A. Pandey, P. Shukla, P.K. Srivastava, Remediation of dyes in water using green synthesized nanoparticles (NPs). IJPE**6**, 68–84 (2020). https://doi.org/10.18811/ijpen.v6i01.08
43. C.L. Keat, A. Aziz, A.M. Eid, N.A. Elmarzugi, Biosynthesis of nanoparticles and silver nanoparticles. Bioresour. Bioprocess. **2**, 47 (2015). https://doi.org/10.1186/s40643-015-0076-2
44. K. Maeda, K. Ishimaki, Y. Tokunaga, D. Lu, M. Eguchi, Modification of wide-band-gap oxide semiconductors with cobalt hydroxide nanoclusters for visible-light water oxidation. Angew. Chem. Int. Ed. **55**, 8309–8313 (2016). https://doi.org/10.1002/anie.201602764
45. B.M. Abu-Zied, K.A. Alamry, Green synthesis of 3D hierarchical nanostructured Co_3O_4/carbon catalysts for the application in sodium borohydride hydrolysis. J. Alloys Compd. **798**, 820–831 (2019). https://doi.org/10.1016/j.jallcom.2019.05.249
46. A. Ledo-Suárez, L. Rodríguez-Sánchez, M.C. Blanco, M.A. López-Quintela, Electrochemical synthesis and stabilization of cobalt nanoparticles. Phys. Status Solidi (A) **203**, 1234–1240 (2006). https://doi.org/10.1002/pssa.200566196
47. P. Alex, S. Majumdar, J. Kishor, I.G. Sharma, Synthesis of cobalt nano crystals in aqueous media and its characterization. Mater. sci. appl. **02**, 1307–1312 (2011). https://doi.org/10.4236/msa.2011.29177
48. H. Tachikawa, Y. Izumi, T. Iyama, S. Abe, I. Watanabe, Aluminum-doping effects on the electronic states of graphene nanoflake: diffusion and hydrogen storage mechanism. Nanomaterials **13**, 2046 (2023). https://doi.org/10.3390/nano13142046
49. I. Mindru et al., A new approach: synthesis of cobalt aluminate nanoparticles using tamarind fruit extract. MSEB **246**, 42–48 (2019). https://doi.org/10.1016/j.mseb.2019.05.031
50. J. Ma, A. Manthiram, Precursor-directed formation of hollow CO_3O_4 nanospheres exhibiting superior lithium storage properties. RSC Adv. **2**, 3187 (2012). https://doi.org/10.1039/C2RA20092A
51. M. Saeed et al., Ag–Co_3O_4: synthesis, characterization and evaluation of its photo-catalytic activity towards degradation of rhodamine B dye in aqueous medium. Chin. J. Chem. Eng. **26**, 1264–1269 (2018). https://doi.org/10.1016/j.cjche.2018.02.024

52. N.G. Sahoo, S. Rana, J.W. Cho, L. Li, S.H. Chan, Polymer nanocomposites based on functionalized carbon nanotubes. Prog. Polym. Sci. **35**, 837–867 (2010). https://doi.org/10.1016/j.progpolymsci.2010.03.002
53. V.R. Shinde, S.B. Mahadik, T.P. Gujar, C.D. Lokhande, Supercapacitive cobalt oxide (Co_3O_4) thin films by spray pyrolysis. Appl. Surf. Sci. **252**, 7487–7492 (2006). https://doi.org/10.1016/j.apsusc.2005.09.00
54. G.B. Smith, A. Ignatiev, G. Zajac, Solar selective black cobalt: preparation, structure, and thermal stability. J. Appl. Phys. **51**, 4186–4196 (1980). https://doi.org/10.1063/1.328276
55. A. Diallo, A.C. Beye, T.B. Doyle, E. Park, M. Maaza, Green synthesis of Co_3O_4 nanoparticles via *Aspalathus linearis*: physical properties. Green Chem. Lett. Rev. **8**, 30–36 (2015). https://doi.org/10.1080/17518253.2015.1082646
56. T. Shahzadi et al., Synthesis of eco-friendly cobalt nanoparticles using celosia argentea plant extract and their efficacy studies as antioxidant, antibacterial, hemolytic and catalytical agent. Arab. J. Sci. Eng. **44**, 6435–6444 (2019). https://doi.org/10.1007/s13369-019-03937-0
57. M.S. Samuel et al., Green synthesis of cobalt-oxide nanoparticle using jumbo Muscadine (Vitis rotundifolia): characterization and photo-catalytic activity of acid Blue-74. J. Photochem. Photobiol. B. **211**, 112011 (2020). https://doi.org/10.1016/j.jphotobiol.2020.112011
58. T. Rasheed, F. Nabeel, M. Bilal, H.M.N. Iqbal, Biogenic synthesis and characterization of cobalt oxide nanoparticles for catalytic reduction of direct yellow-142 and methyl orange dyes. Biocatal. Agric. Biotechnol. **19**, 101154 (2019). https://doi.org/10.1016/j.bcab.2019.101154
59. M. Zaib, T. Shahzadi, I. Muzammal, U. Farooq, *Catharanthus roseus* extract mediated synthesis of cobalt nanoparticles: evaluation of antioxidant, antibacterial, hemolytic and catalytic activities. Inorg. Nano-Met. **50**, 1171–1180 (2020). https://doi.org/10.1080/24701556.2020.1737819
60. I. Bibi et al., Green and eco-friendly synthesis of cobalt-oxide nanoparticle: characterization and photo-catalytic activity. Adv. Powder Technol. **28**, 2035–2043 (2017). https://doi.org/10.1016/j.apt.2017.05.008
61. G. Asha, V. Rajeshwari, G. Stephen, S. Gurusamy, D. Carolin Jeniba Rachel, Eco-friendly synthesis and characterization of cobalt oxide nanoparticles by sativum species and its photocatalytic activity. Mater. Today Proc. **48**, 486–493 (2022). https://doi.org/10.1016/j.matpr.2021.02.338
62. P. Chelliah et al., Green synthesis and characterizations of cobalt oxide nanoparticles and their coherent photocatalytic and antibacterial investigations. Water (Basel) **15**, 910 (2023). https://doi.org/10.3390/w15050910
63. S. Akshatha et al., Microwave assisted green synthesis of p-type Co_3O_4@Mesoporous carbon spheres for simultaneous degradation of dyes and photocatalytic hydrogen evolution reaction. Mater. Sci. Semicond. Process. **121**, 105432 (2021). https://doi.org/10.1016/j.mssp.2020.105432
64. V.D. Rajeswari et al., Green and ecofriendly synthesis of cobalt oxide nanoparticles using Phoenix dactylifera L: antimicrobial and photocatalytic activity. Appl. Nanosci. **13**, 1367–1375 (2023). https://doi.org/10.1007/s13204-021-02038-5
65. R. Prucek et al., The targeted antibacterial and antifungal properties of magnetic nanocomposite of iron oxide and silver nanoparticles. Biomaterials **32**, 4704–4713 (2011). https://doi.org/10.1016/j.biomaterials.2011.03.039
66. C. Prasad, G. Yuvaraja, P. Venkateswarlu, Biogenic synthesis of Fe_3O_4 magnetic nanoparticles using Pisum sativum peels extract and its effect on magnetic and Methyl orange dye degradation studies. J. Magn. Magn. Mater. **424**, 376–381 (2017). https://doi.org/10.1016/j.jmmm.2016.10.084
67. J. Dulińska-Litewka et al., Superparamagnetic iron oxide nanoparticles—Current and prospective medical applications. Materials **12**, 617 (2019). https://doi.org/10.3390/ma12040617
68. A. Syed, S. Saraswati, G.C. Kundu, A. Ahmad, Biological synthesis of silver nanoparticles using the fungus Humicola sp. and evaluation of their cytotoxicity using normal and cancer

cell lines. Spectrochim. Acta A Mol. Biomol. Spectrosc. **114**, 144–147 (2013). https://doi.org/10.1016/j.saa.2013.05.030

69. B. Ahmmad et al., Green synthesis of mesoporous hematite (α-Fe_2O_3) nanoparticles and their photocatalytic activity. Adv. Powder Technol. **24**, 160–167 (2013). https://doi.org/10.1016/j.apt.2012.04.005
70. M. Balamurugan, S. Saravanan, T. Soga, Synthesis of iron oxide nanoparticles by using eucalyptus globulus plant extract. e-J. Surf. Sci. Nanotechnol. **12**, 363–367 (2014). https://doi.org/10.1380/ejssnt.2014.363
71. D. Mukherjee, S. Ghosh, S. Majumdar, K. Annapurna, Green synthesis of α-Fe_2O_3 nanoparticles for arsenic(V) remediation with a novel aspect for sludge management. J. Environ. Chem. Eng. **4**, 639–650 (2016). https://doi.org/10.1016/j.jece.2015.12.010
72. A.T. Khalil et al., Biosynthesis of iron oxide (Fe_2O_3) nanoparticles via aqueous extracts of *Sageretia thea* (Osbeck.) and their pharmacognostic properties. Green Chem. Lett. Rev. **10**, 186–201 (2017). https://doi.org/10.1080/17518253.2017.1339831
73. H.M. Asoufi, T.M. Al-Antary, A.M. Awwad, Green route for synthesis hematite (α-Fe_2O_3) nanoparticles: toxicity effect on the green peach aphid, Myzus persicae (Sulzer). Environ. Nanotechnol. Monit. Manag. **9**, 107–111 (2018). https://doi.org/10.1016/j.enmm.2018.01.004
74. S. Naz et al., Green synthesis of hematite (α-Fe_2O_3) nanoparticles using *Rhus punjabensis* extract and their biomedical prospect in pathogenic diseases and cancer. J. Mol. Struct. **1185**, 1–7 (2019). https://doi.org/10.1016/j.molstruc.2019.02.088
75. K.B. Narayanan, S.S. Han, One-pot green synthesis of hematite (α-Fe_2O_3) nanoparticles by ultrasonic irradiation and their in vitro cytotoxicity on human keratinocytes CRL-2310. J. Clust. Sci. **27**, 1763–1775 (2016). https://doi.org/10.1007/s10876-016-1040-9
76. S. Bishnoi, A. Kumar, R. Selvaraj, Facile synthesis of magnetic iron oxide nanoparticles using inedible *Cynometra ramiflora* fruit extract waste and their photocatalytic degradation of methylene blue dye. Mater. Res. Bull. **97**, 121–127 (2018). https://doi.org/10.1016/j.materresbull.2017.08.040
77. K.Ö. Hamaloğlu, E. Sağ, A. Tuncel, Bare, gold and silver nanoparticle decorated, monodisperse-porous titania microbeads for photocatalytic dye degradation in a newly constructed microfluidic, photocatalytic packed-bed reactor. J. Photochem. Photobiol. A Chem. **332**, 60–65 (2017). https://doi.org/10.1016/j.jphotochem.2016.08.015
78. K.R. Ahammed et al., Microwave assisted synthesis of zinc oxide (ZnO) nanoparticles in a noble approach: utilization for antibacterial and photocatalytic activity. SN Appl. Sci. **2**, 955 (2020). https://doi.org/10.1007/s42452-020-2762-8
79. L.C.A. Oliveira et al., Activated carbon/iron oxide magnetic composites for the adsorption of contaminants in water. Carbon N Y. **40**, 2177–2183 (2002). https://doi.org/10.1016/S0008-6223(02)00076-3
80. S. Vasantharaj, S. Sathiyavimal, P. Senthilkumar, F. Lewis Oscar, A. Pugazhendhi, Biosynthesis of iron oxide nanoparticles using leaf extract of *Ruellia tuberosa*: antimicrobial properties and their applications in photocatalytic degradation. J. Photochem. Photobiol. B **192**, 74–82 (2019). https://doi.org/10.1016/j.jphotobiol.2018.12.025
81. H.S. Devi, M.A. Boda, M.A. Shah, S. Parveen, A.H. Wani, Green synthesis of iron oxide nanoparticles using *Platanus orientalis* leaf extract for antifungal activity. Green Process. Synth. **8**, 38–45 (2019). https://doi.org/10.1515/gps-2017-0145
82. M. Cao et al., Food related applications of magnetic iron oxide nanoparticles: Enzyme immobilization, protein purification, and food analysis. Trends Food Sci. Technol. **27**, 47–56 (2012). https://doi.org/10.1016/j.tifs.2012.04.003
83. A.K.A. Silva, A. Espinosa, J. Kolosnjaj-Tabi, C. Wilhelm, F. Gazeau, Medical applications of iron oxide nanoparticles. Iron Oxides, 425–472 (Wiley, 2016). https://doi.org/10.1002/9783527691395.ch18
84. M. Hasanzadeh, N. Shadjou, M. de la Guardia, Iron and iron-oxide magnetic nanoparticles as signal-amplification elements in electrochemical biosensing. TrAC **72**, 1–9 (2015). https://doi.org/10.1016/j.trac.2015.03.016

85. Y. Wang, H. Zhao, G. Zhao, Iron-copper bimetallic nanoparticles embedded within ordered mesoporous carbon as effective and stable heterogeneous Fenton catalyst for the degradation of organic contaminants. Appl. Catal. B **164**, 396–406 (2015). https://doi.org/10.1016/j.apcatb.2014.09.047
86. J.M. Walker, J.M. Zaleski, A simple route to diverse noble metal-decorated iron oxide nanoparticles for catalysis. Nanoscale **8**, 1535–1544 (2016). https://doi.org/10.1039/C5NR06700F
87. W. Yu et al., Synthesis and electrochemical lithium storage behavior of carbon nanotubes filled with iron sulfide nanoparticles. Adv. Sci.**3** (2016). https://doi.org/10.1002/advs.201600113
88. A. Massoud, H.H. Mahmoud, Evaluation of hybrid polymeric resin containing nanoparticles of iron oxide for selective separation of In(III) from Ga(III). J. Inorg. Organomet. Polym. Mater. **27**, 1806–1815 (2017). https://doi.org/10.1007/s10904-017-0645-2
89. Y. Bentahir et al., Dispersed synthesis of uniform Fe_3O_4 magnetic nanoparticles via in situ decomposition of iron precursor along cotton fibre for Sudan dyes analysis in food samples. Food Addit. Contam. Part A **34**, 1853–1862 (2017). https://doi.org/10.1080/19440049.2017.1357840
90. A. Ebrahiminezhad et al., Plant-mediated synthesis and applications of iron nanoparticles. Mol. Biotechnol. **60**, 154–168 (2018). https://doi.org/10.1007/s12033-017-0053-4
91. D. O'Carroll, B. Sleep, M. Krol, H. Boparai, C. Kocur, Nanoscale zero valent iron and bimetallic particles for contaminated site remediation. Adv. Water Resour. **51**, 104–122 (2013). https://doi.org/10.1016/j.advwatres.2012.02.005
92. C. Zhang, Z. Zhu, H. Zhang, Z. Hu, Rapid decolorization of acid orange II aqueous solution by amorphous zero-valent iron. J. Environ. Sci. **24**, 1021–1026 (2012). https://doi.org/10.1016/S1001-0742(11)60894-2
93. J. Fan, Y. Guo, J. Wang, M. Fan, Rapid decolorization of azo dye methyl orange in aqueous solution by nanoscale zerovalent iron particles. J. Hazard. Mater. **166**, 904–910 (2009). https://doi.org/10.1016/j.jhazmat.2008.11.091
94. X. Fei, X. Xu, G. Jia, Q. Li, Notice of Retraction: Preparation and photocatalytic activity of magnetic Fe_3O_4/TiO_2 photocatalyst, in *2010 The 2nd Conference on Environmental Science and Information Application Technology* (IEEE, 2010), pp. 227–229. https://doi.org/10.1109/ESIAT.2010.5568526
95. A. Truskewycz, R. Shukla, A.S. Ball, Iron nanoparticles synthesized using green tea extracts for the Fenton-like degradation of concentrated dye mixtures at elevated temperatures. J. Environ. Chem. Eng. **4**, 4409–4417 (2016). https://doi.org/10.1016/j.jece.2016.10.008
96. S. Anchan et al., Biogenic synthesis of ferric oxide nanoparticles using the leaf extract of *Peltophorum pterocarpum* and their catalytic dye degradation potential. Biocatal. Agric. Biotechnol. **20**, 101251 (2019). https://doi.org/10.1016/j.bcab.2019.101251
97. I. Bibi et al., Green synthesis of iron oxide nanoparticles using pomegranate seeds extract and photocatalytic activity evaluation for the degradation of textile dye. J. Mater. Res. Technol. **8**, 6115–6124 (2019). https://doi.org/10.1016/j.jmrt.2019.10.006
98. N. Madubuonu et al., Biosynthesis of iron oxide nanoparticles via a composite of *Psidium guavaja-Moringa oleifera* and their antibacterial and photocatalytic study. J. Photochem. Photobiol. B. **199**, 111601 (2019). https://doi.org/10.1016/j.jphotobiol.2019.111601
99. N. Beheshtkhoo, M.A.J. Kouhbanani, A. Savardashtaki, A.M. Amani, S. Taghizadeh, Green synthesis of iron oxide nanoparticles by aqueous leaf extract of *Daphne mezereum* as a novel dye removing material. Appl. Phys. A **124**, 363 (2018). https://doi.org/10.1007/s00339-018-1782-3
100. Md.S.H. Bhuiyan et al., Green synthesis of iron oxide nanoparticle using *Carica papaya* leaf extract: application for photocatalytic degradation of remazol yellow RR dye and antibacterial activity. Heliyon **6**, e04603 (2020). https://doi.org/10.1016/j.heliyon.2020.e04603
101. S. Qasim et al., Green synthesis of iron oxide nanorods using *Withania coagulans* extract improved photocatalytic degradation and antimicrobial activity. J. Photochem. Photobiol. B. **204**, 111784 (2020). https://doi.org/10.1016/j.jphotobiol.2020.111784

102. M.A.J. Kouhbanani, N. Beheshtkhoo, S. Taghizadeh, A.M. Amani, V. Alimardani, One-step green synthesis and characterization of iron oxide nanoparticles using aqueous leaf extract of *Teucrium polium* and their catalytic application in dye degradation. ANSN **10**, 015007 (2019). https://doi.org/10.1088/2043-6254/aafe74
103. A. Chakraborty, S. Sarangapany, U. Mishra, K. Mohanty, Green synthesized magnetically separable iron oxide nanoparticles for efficient heterogeneous photo-Fenton degradation of dye pollutants. J. Clust. Sci. **33**, 675–685 (2022). https://doi.org/10.1007/s10876-021-020 10-x
104. P. Singh, K.R. Singh, R. Verma, J. Singh, R.P. Singh, Efficient electro-optical characteristics of bioinspired iron oxide nanoparticles synthesized by T*erminalia chebula* dried seed extract. Mater. Lett. **307**, 131053 (2022). https://doi.org/10.1016/j.matlet.2021.131053
105. N. Sarwar et al., Citric acid mediated green synthesis of copper nanoparticles using cinnamon bark extract and its multifaceted applications. J. Clean. Prod. **292**, 125974 (2021). https://doi.org/10.1016/j.jclepro.2021.125974s
106. S.K. Chandraker et al., Green synthesis of copper nanoparticles using leaf extract of *Ageratum houstonianum* Mill. and study of their photocatalytic and antibacterial activities. Nano Express. **1**, 010033 (2020). https://doi.org/10.1088/2632-959X/ab8e99
107. J.K. Sharma, M.S. Akhtar, S. Ameen, P. Srivastava, G. Singh, Green synthesis of CuO nanoparticles with leaf extract of *Calotropis gigantea* and its dye-sensitized solar cells applications. J. Alloys Compd. **632**, 321–325 (2015). https://doi.org/10.1016/j.jallcom.2015.01.172
108. N. Sreeju, A. Rufus, D. Philip, Microwave-assisted rapid synthesis of copper nanoparticles with exceptional stability and their multifaceted applications. J. Mol. Liq. **221**, 1008–1021 (2016). https://doi.org/10.1016/j.molliq.2016.06.080
109. M. Ovais et al., Role of plant phytochemicals and microbial enzymes in biosynthesis of metallic nanoparticles. Appl. Microbiol. Biotechnol. **102**, 6799–6814 (2018). https://doi.org/10.1007/s00253-018-9146-7
110. V. Soni et al., Sustainable and green trends in using plant extracts for the synthesis of biogenic metal nanoparticles toward environmental and pharmaceutical advances: a review. Environ. Res. **202**, 111622 (2021). https://doi.org/10.1016/j.envres.2021.111622
111. X. Liu, B. Geng, Q. Du, J. Ma, X. Liu, Temperature-controlled self-assembled synthesis of CuO, Cu_2O and Cu nanoparticles through a single-precursor route. Mater. Sci. Eng. A **448**, 7–14 (2007). https://doi.org/10.1016/j.msea.2006.08.10
112. S. Kayalvizhi et al., Eco-friendly cost-effective approach for synthesis of copper oxide nanoparticles for enhanced photocatalytic performance. Optik (Stuttg). **202**, 163507 (2020). https://doi.org/10.1016/j.ijleo.2019.163507
113. S.C. Mali, A. Dhaka, C.K. Githala, R. Trivedi, Green synthesis of copper nanoparticles using *Celastrus paniculatus* Willd. leaf extract and their photocatalytic and antifungal properties. Biotechnol. Rep. **27**, e00518 (2020). https://doi.org/10.1016/j.btre.2020.e00518
114. M.K. Ghosh, S. Sahu, I. Gupta, T.K. Ghorai, Green synthesis of copper nanoparticles from an extract of *Jatropha curcas* leaves: characterization, optical properties, CT-DNA binding and photocatalytic activity. RSC Adv. **10**, 22027–22035 (2020). https://doi.org/10.1039/D0RA03186K
115. M. Usman et al., Photocatalytic potential of bio-engineered copper nanoparticles synthesized from *Ficus carica* extract for the degradation of toxic organic dye from waste water: growth mechanism and study of parameter affecting the degradation performance. Mater. Res. Bull. **120**, 110583 (2019). https://doi.org/10.1016/j.materresbull.2019.110583
116. A. Ahmed et al., Plant mediated synthesis of copper nanoparticles by using *Camelia sinensis* leaves extract and their applications in dye degradation. Ferroelectrics **549**, 61–69 (2019). https://doi.org/10.1080/00150193.2019.1592544
117. S. Vasantharaj et al., Synthesis of ecofriendly copper oxide nanoparticles for fabrication over textile fabrics: characterization of antibacterial activity and dye degradation potential. J. Photochem. Photobiol. B. **191**, 143–149 (2019). https://doi.org/10.1016/j.jphotobiol.2018.12.02

118. V.K. Mrunal, A.K. Vishnu, N. Momin, J. Manjanna, Cu_2O nanoparticles for adsorption and photocatalytic degradation of methylene blue dye from aqueous medium. Environ. Nanotechnol. Monit. Manag. **12**, 100265 (2019). https://doi.org/10.1016/j.enmm.2019.100265
119. M. Muthukumaran et al., Green synthesis of cuprous oxide nanoparticles for environmental remediation and enhanced visible-light photocatalytic activity. Optik (Stuttg) **214**, 164849 (2020). https://doi.org/10.1016/j.ijleo.2020.164849
120. M. Muthukumaran et al., Green route synthesis and characterization of cuprous oxide (Cu_2O): visible light irradiation photocatalytic activity of MB dye. Mater. Today Proc. **14**, 563–568 (2019). https://doi.org/10.1016/j.matpr.2019.04.179
121. A. Kerour, S. Boudjadar, R. Bourzami, B. Allouche, Eco-friendly synthesis of cuprous oxide (Cu_2O) nanoparticles and improvement of their solar photocatalytic activities. J. Solid State Chem. **263**, 79–83 (2018). https://doi.org/10.1016/j.jssc.2018.04.010
122. S. Upadhyay, K. Parekh, B. Pandey, Influence of crystallite size on the magnetic properties of Fe_3O_4 nanoparticles. J. Alloys Compd. **678**, 478–485 (2016). https://doi.org/10.1016/j.jallcom.2016.03.279
123. S. Sharma, K. Kumar, N. Thakur, S. Chauhan, M.S. Chauhan, Eco-friendly *Ocimum tenuiflorum* green route synthesis of CuO nanoparticles: characterizations on photocatalytic and antibacterial activities. J. Environ. Chem. Eng. **9**, 105395 (2021). https://doi.org/10.1016/j.jece.2021.105395
124. D.R. Baer, M.H. Engelhard, XPS analysis of nanostructured materials and biological surfaces. J. Electron. Spectros. Relat. Phenomena. **178–179**, 415–432 (2010). https://doi.org/10.1016/j.elspec.2009.09.003
125. N. Akhlaghi, G. Najafpour-Darzi, H. Younesi, Facile and green synthesis of cobalt oxide nanoparticles using ethanolic extract of *Trigonella foenumgraceum* (Fenugreek) leaves. Adv. Powder Technol. **31**, 3562–3569 (2020). https://doi.org/10.1016/j.apt.2020.07.004
126. M. Nidhin, K.J. Sreeram, B.U. Nair, Green synthesis of rock salt CoO nanoparticles for coating applications by complexation and surface passivation with starch. J. Chem. Eng. **185–186**, 352–357 (2012). https://doi.org/10.1016/j.cej.2011.11.104
127. M. Rafique et al., A review on synthesis, characterization and applications of copper nanoparticles using green method. NANO **12**, 1750043 (2017). https://doi.org/10.1142/S1793292017500436
128. D. Gnanasangeetha, M. Suresh, A review on green synthesis of metal and metal oxide nanoparticles. Nat. Environ. Pollut. Technol.**19**, 1789–1800 (2020). https://doi.org/10.46488/NEPT.2020.v19i05.002
129. J. Campbell, S. Burkitt, N. Dong, C. Zavaleta, Nanoparticle characterization techniques. in *Nanoparticles for Biomedical Applications* (2020), pp. 129–144. https://doi.org/10.1016/B978-0-12-816662-8.00009-6
130. R. Chowdhury, A. Khan, Md.H. Rashid, Green synthesis of CuO nanoparticles using *Lantana camara* flower extract and their potential catalytic activity towards the aza-Michael reaction. RSC Adv. **10**, 14374–14385 (2020). https://doi.org/10.1039/D0RA01479F
131. S. Rades et al., High-resolution imaging with SEM/T-SEM, EDX and SAM as a combined methodical approach for morphological and elemental analyses of single engineered nanoparticles. RSC Adv. **4**, 49577–49587 (2014). https://doi.org/10.1039/C4RA05092D
132. A. Lassoued, B. Dkhil, A. Gadri, S. Ammar, Control of the shape and size of iron oxide (α-Fe_2O_3) nanoparticles synthesized through the chemical precipitation method. Results Phys. **7**, 3007–3015 (2017). https://doi.org/10.1016/j.rinp.2017.07.066
133. Z. Alhalili, Green synthesis of copper oxide nanoparticles CuO NPs from *Eucalyptus Globoulus* leaf extract: adsorption and design of experiments. Arab. J. Chem. **15**, 103739 (2022). https://doi.org/10.1016/j.arabjc.2022.103739
134. R. Bardestani, G.S. Patience, S. Kaliaguine, Experimental methods in chemical engineering: specific surface area and pore size distribution measurements—BET, BJH, and DFT. Can. J. Chem. Eng. **97**, 2781–2791 (2019). https://doi.org/10.1002/cjce.23632

135. Y. Sahoo et al., Aqueous ferrofluid of magnetite nanoparticles: fluorescence labeling and magnetophoretic control. J. Phys. Chem. B **109**, 3879–3885 (2005). https://doi.org/10.1021/jp045402y
136. O. Nonappa and Ikkala, Hydrogen bonding directed colloidal self-assembly of nanoparticles into 2D crystals, capsids, and supracolloidal assemblies. Adv. Funct. Mater. **28**, (2018). https://doi.org/10.1002/adfm.201704328
137. P. Yugandhar, T. Vasavi, P. Uma Maheswari Devi, N. Savithramma, Bioinspired green synthesis of copper oxide nanoparticles from *Syzygium alternifolium* (Wt.) Walp: characterization and evaluation of its synergistic antimicrobial and anticancer activity. Appl. Nanosci. **7**, 417–427 (2017). https://doi.org/10.1007/s13204-017-0584-9
138. K. Dulta et al., Multifunctional CuO nanoparticles with enhanced photocatalytic dye degradation and antibacterial activity. Sustain. Environ. Res. **32**, 2 (2022). https://doi.org/10.1186/s42834-021-00111-w
139. V. Kamath, P. Chandra, G.P. Jeppu, Comparative study of using five different leaf extracts in the green synthesis of iron oxide nanoparticles for removal of arsenic from water. Int. J. Phytoremediation **22**, 1278–1294 (2020). https://doi.org/10.1080/15226514.2020.1765139
140. E. Mansfield, M. Banash, Thermal analysis of nanoparticles: methods, kinetics, and recent advances, in *Modeling, Characterization, and Production of Nanomaterials* (2023), pp. 535–547. https://doi.org/10.1016/B978-0-12-819905-3.00019-1
141. I. Ahmed, F.A. Mir, J.A. Banday, Synthesis of metal and metal oxide nanoparticles using plant extracts—Characterization and applications. Bionanoscience (2023). https://doi.org/10.1007/s12668-023-01194-y
142. A. Woźniak et al., Size and shape-dependent cytotoxicity profile of gold nanoparticles for biomedical applications. J. Mater. Sci. Mater. Med. **28**, 92 (2017). https://doi.org/10.1007/s10856-017-5902-y
143. J.A.A. Abdullah et al., Green synthesis and characterization of iron oxide nanoparticles by *pheonix dactylifera* leaf extract and evaluation of their antioxidant activity. Sustain. Chem. Pharm. **17**, 100280 (2020). https://doi.org/10.1016/j.scp.2020.100280
144. S. Yin et al., Enhanced electron–hole separation in $SnS_2/Au/g\text{-}C_3N_4$ embedded structure for efficient CO_2 photoreduction. J. Chem. Eng. **406**, 126776 (2021). https://doi.org/10.1016/j.cej.2020.126776
145. G. Li et al., Ag quantum dots modified hierarchically porous and defective TiO_2 nanoparticles for improved photocatalytic CO_2 reduction. J. Chem. Eng. **410**, 128397 (2021). https://doi.org/10.1016/j.cej.2020.128397
146. X. Li et al., Multichannel electron transmission and fluorescence resonance energy transfer in In_2S_3 /Au/rGO composite for CO_2 photoreduction. ACS Appl. Mater. Interfaces **13**, 11755–11764 (2021). https://doi.org/10.1021/acsami.0c18809
147. U. Caudillo-Flores, M.J. Muñoz-Batista, M. Fernández-García, A. Kubacka, Bimetallic Pt-Pd co-catalyst Nb-doped TiO_2 materials for H_2 photo-production under UV and visible light illumination. Appl. Catal. B **238**, 533–545 (2018). https://doi.org/10.1016/j.apcatb.2018.07.047
148. F. Zhang, Y.-H. Li, M.-Y. Qi, Z.-R. Tang, Y.-J. Xu, Boosting the activity and stability of Ag–Cu_2O/ZnO nanorods for photocatalytic CO_2 reduction. Appl. Catal. B **268**, 118380 (2020). https://doi.org/10.1016/j.apcatb.2019.118380
149. Z. Yang et al., Visible light-assisted thermal catalytic reverse water gas reaction over Cu–CeO_2: the synergistic of hot electrons and oxygen vacancies induced by LSPR effect. Fuel **315**, 123186 (2022). https://doi.org/10.1016/j.fuel.2022.123186
150. X. Yue et al., Photo-assisted thermal catalytic Fischer-Tropsch synthesis over Co–Cu/CeO_2. Inorg. Chem. Front. **9**, 1258–1269 (2022). https://doi.org/10.1039/d2qi00004k
151. P.G. Krishna et al., Photocatalytic activity induced by metal nanoparticles synthesized by sustainable approaches: a comprehensive review. Front. Chem. **10** (2022). https://doi.org/10.3389/fchem.2022.917831
152. R. Beranek, H. Kisch, Tuning the optical and photoelectrochemical properties of surface-modified TiO_2. Photochem. Photobiol. Sci. **7**, 40–48 (2008). https://doi.org/10.1039/b711658f

153. A.A. Barzinjy, H.H. Azeez, Green synthesis and characterization of zinc oxide nanoparticles using *Eucalyptus globulus* Labill. leaf extract and zinc nitrate hexahydrate salt. S. N. Appl. Sci. **2**, 991 (2020). https://doi.org/10.1007/s42452-020-2813-1
154. A.A. Sharwani, K.B. Narayanan, M.E. Khan, S.S. Han, Photocatalytic degradation activity of goji berry extract synthesized silver-loaded mesoporous zinc oxide (Ag@ZnO) nanocomposites under simulated solar light irradiation. Sci. Rep. **12**, 10017 (2022). https://doi.org/10.1038/s41598-022-14117-w
155. S. Prakash et al., Green synthesis of copper oxide nanoparticles and its effective applications in Biginelli reaction, BTB photodegradation and antibacterial activity. Adv. Powder Technol. **29**, 3315–3326 (2018). https://doi.org/10.1016/j.apt.2018.09.009
156. Nanotechnology and functional materials for engineers by yaser dahman. MRS Bull. **43**, 638–638 (2018). https://doi.org/10.1557/mrs.2018.196
157. A.A. Khan, M. Mondal, Effective materials in the photocatalytic treatment of dyestuffs and stained wastewater, in *Photocatalytic Degradation of Dyes* (2021), pp. 91–111. https://doi.org/10.1016/B978-0-12-823876-9.00017-2
158. V. Yargeau, Water and wastewater treatment: chemical processes, in *Metropolitan Sustainability* (2012), pp. 390–405. https://doi.org/10.1533/9780857096463.3.390
159. M.F. Abid, M.A. Zablouk, A.M. Abid-Alameer, Experimental study of dye removal from industrial wastewater by membrane technologies of reverse osmosis and nanofiltration. Iranian. J. Environ. Health. Sci. Eng. **9**, 17 (2012). https://doi.org/10.1186/1735-2746-9-171-9
160. I.K. Konstantinou, T.A. Albanis, TiO_2-assisted photocatalytic degradation of azo dyes in aqueous solution: kinetic and mechanistic investigations. Appl. Catal. B **49**, 1–14 (2004). https://doi.org/10.1016/j.apcatb.2003.11.010
161. D. Bidwai, N.K. Sahu, G. Swati, Rare earth doped nanomaterials for visible-light-driven photocatalytic degradation of organic dyes, in *Nanostructured Materials for Visible Light Photocatalysis* (2022), pp. 251–275. https://doi.org/10.1016/B978-0-12-823018-3.00011-7
162. S. Das, H. Mahalingam, Dye degradation studies using immobilized pristine and waste polystyrene-TiO_2/rGO/g-C_3N_4 nanocomposite photocatalytic film in a novel airlift reactor under solar light. J. Environ. Chem. Eng. **7**, 103289 (2019). https://doi.org/10.1016/j.jece.2019.103289
163. J. Ćirković et al., Visible-light photocatalytic degradation of mordant blue 9 by single-phase $BiFeO_3$ nanoparticles. J. Environ. Chem. Eng. **9**, 104587 (2021). https://doi.org/10.1016/j.jece.2020.104587
164. D.C. Khandekar, A.R. Bhattacharyya, R. Bandyopadhyaya, Role of impregnated nano-photocatalyst (SnxTi(1–x)O_2) inside mesoporous silica (SBA-15) for degradation of organic pollutant (Rhodamine B) under UV light. J. Environ. Chem. Eng. **7**, 103433 (2019). https://doi.org/10.1016/j.jece.2019.103433
165. A. Das, M.K. Adak, N. Mahata, B. Biswas, Wastewater treatment with the advent of TiO_2 endowed photocatalysts and their reaction kinetics with scavenger effect. J. Mol. Liq. **338**, 116479 (2021). https://doi.org/10.1016/j.molliq.2021.116479

Photocatalytic Degradation of Organic Pollutants Using Heterojunction Nanocomposite Based on Bismuth Compound

Ancy Kurian and S. Sumathi

Abstract In the pursuit of sustainable and effective photocatalytic materials, bismuth-based nanocomposites with their distinctive electronic and structural properties have arisen as promising materials. This book chapter investigates the photocatalytic activity of various bismuth-based nanocomposites which includes bismuth tungstate (Bi_2WO_6), bismuth vanadate ($BiVO_4$), bismuth oxide (Bi_2O_3), bismuth molybdate (Bi_2MoO_6), bismuth sulphide (Bi_2S_3), bismuth subcarbonates ($(BiO)_2CO_3$) and shedding light on their potential applications in environmental remediation. Through an exhaustive review of recent research, the synthesis techniques, structural characteristics and photocatalytic mechanisms are explored. In addition, we emphasise the role of bismuth-based nanocomposites in addressing contemporary challenges, such as water purification, pollutant degradation, comprehensive understanding of the photocatalytic capabilities of bismuth-based nanocomposites and their future prospects in the advancement of sustainable and green technologies.

Keywords Bismuth based compounds · Photocatalyst · Nanocomposite · Heterojunction

1 Introduction

In this era of industrialization and urbanization, population growth and water pollution has become a global issue. Water sources are being contaminated by substances like metals, organic compounds, pathogens and emerging pollutants. This leads to health problems and also disrupts the delicate balance of the aquatic ecosystem. The search, for the ways to treat water has resulted in the development of different approaches which includes conventional methods like coagulation, flocculation and

A. Kurian · S. Sumathi (✉)
School of Advanced Sciences, Department of Chemistry, Vellore Institute of Technology, Vellore, Tamil Nadu 632014, India
e-mail: sumathishanmugam2003@gmail.com

P. Sathishkumar (ed.), *Photocatalysis for Energy and Environmental Applications*, Green Energy and Technology, https://doi.org/10.1007/978-981-97-1939-6_7

membrane filtration have shown some success, but they may not fully address emerging pollutants. Moreover, these methods often produce an amount of sludge and waste which was a challenge for disposal. As a result, advanced oxidation processes (AOPs) are being explored as a solution. AOPs involve the generation of oxygen species to break down pollutants that are resistant to traditional methods.

Photocatalysis offers a sustainable means of converting light energy into chemical energy without limitations. Its eco-friendly nature, requiring minimal energy input and leaving a negligible carbon footprint, renders it highly promising. Two prominent examples of photocatalytic processes include the reduction of CO_2 to hydrocarbons and the splitting of water into hydrogen gas. Photocatalysis finds significant environmental applications, including contaminant degradation [1–4], antibacterial treatment [5–8] and sterilization [9, 10]. In the realm of photocatalytic reactions, photocatalysts play pivotal roles in governing both solar light absorption efficiency and photocatalytic capacity. When exposed to light, semiconductors generate electrons and holes, which can be harnessed for the decomposition of organic and inorganic contaminants.

Prominent semiconducting materials employed in photocatalysis include titanium dioxide (TiO_2), zinc oxide (ZnO), and various metal oxides [11]. These materials, upon exposure to visible light, produce active species like, superoxide radicals ($O_2^{\cdot-}$), singlet oxygen ($^1O_2$) and hydroxyl radicals (·OH) with potent oxidation properties. Consequently, these species facilitate the breakdown of pollutants into harmless molecules. One of the primary challenges faced by photocatalysts is their relatively low photocatalytic efficiency [12]. This limitation largely stems from the recombination of electrons and holes. Upon exposure to light, electrons get excited from the valence band (VB) to the conduction band (CB). Within a short time frame, typically in the picosecond to nanosecond range, electron-hole pairs recombine either within the photocatalyst or on its surface [13–15]. This recombination process occurs swiftly due to the substantial Coulombic forces acting between the photogenerated electrons and holes [16, 17].

To gain perspective on the timescales involved, it is instructive to contrast the calculation of Coulombic forces with gravitational forces. Notably, the Coulombic constant, with a value of 8.99×10^9 N m^2/C^2, exhibits a significant contrast to the gravitational constant, which is measured at 6.67×10^{-11} N m^2/kg^2. Consequently, the process of recombination of electron-hole pairs generated by photons occurs at a significantly faster pace compared to the downward motion of a falling object [18]. Within a single photocatalyst, the formidable force acting upon photogenerated electrons and holes leads to their swift recombination which significantly hampers the ability of these photogenerated charge carriers to engage in specific chemical reactions. Furthermore, when assessing the capabilities of a photocatalyst, its broad absorption of sunlight and its redox potential are two pivotal attributes. A narrow band gap is advantageous for achieving a wide range of light absorption by a single photocatalyst. Conversely, to ensure a high redox capacity, the photocatalyst should possess a conduction band (CB) positioned at a high energy level and a valence band (VB) situated at a lower energy level, corresponding to a larger band gap. Regrettably, these two requirements are inherently incompatible, presenting a challenge.

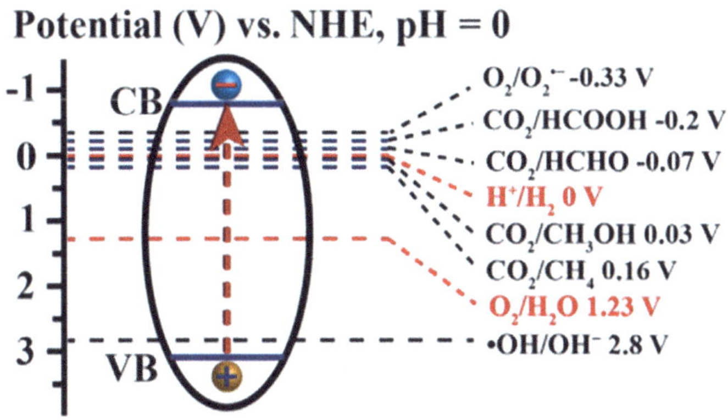

Fig. 1 Redox potentials at the conduction and valence bands, relative to the normal hydrogen electrode (NHE) at pH0

This dilemma prompts the emergence of heterojunctions as a solution. Heterojunction photocatalysts are produced through the amalgamation of two distinct semiconductor photocatalysts. Figure 1 presents a comprehensive exposition of a range of exemplary redox potentials [19–22]. From a thermodynamic perspective, the impetus behind a specific reaction is determined by the differential between the conduction band (CB) and valence band (VB) potentials, coupled with the potential energy associated with the reaction. In essence, the concept of distinguishing between redox and gravitational potential energies is apt for illustration. Consequently, if a monocomponent photocatalyst aims to exhibit robust redox capabilities, it should feature conduction band exhibiting elevated energy levels, while the valence band demonstrates notable energy positions. Conversely, if optimal solar efficiency in harnessing solar radiation is the goal, the CB and VB levels must be shallow. The simultaneous existence of both of these situations is unachievable in a monocomponent photocatalyst, prompting researchers to investigate heterojunction photocatalytic devices [23–25].

Recent investigations have explored the potential of bismuth-containing compounds as alternatives to conventional photocatalysts. These compounds exhibit outstanding photocatalytic performance and antibacterial properties while remaining non-toxic [26, 27]. Additionally, they possess the ability to convert light into heat and demonstrate significant light absorption in the near-infrared region. The distinctive electronic structure of these compounds manifests in a valence band comprising hybrid orbitals derived from oxygen's 2p orbitals and bismuth's 6s orbitals. In contrast, the valence band of titanium dioxide (TiO_2) consists solely of oxygen 2p orbitals. Reports suggest that the well-dispersed Bi 6s orbitals enhance the mobility of photogenerated charge carriers and reduce the band gap. Consequently, trivalent bismuth-containing catalysts exhibit superior photocatalytic activity compared to

TiO_2 photocatalysts, both in visible and ultraviolet (UV) light regions. Most bismuth-based compounds typically possess band gaps less than 3.0 electron volts (eV), rendering them efficient visible-light excited photocatalysts.

A variety of bismuth compounds have been prepared and investigated for their photocatalytic properties, including $BiVO_4$ [28], Bi_2S_3 [29], Bi_2O_3 [30], $BiPO_4$ [31], Bi_2MO_6 where M represents Chromium, Molybdenum, and tungsten [32], $(BiO)_2CO_3$ [33], and BiOX where X denotes Chlorine, Bromine, and Iodine [34]. Many photocatalytic materials that utilize bismuth (Bi) as a component are known for their layered structure, which is associated with optimal optical properties [35–37]. Notably, the positions of the conduction band (CB) and valence band (VB) in Bi-based photocatalysts can be effectively tailored to accommodate a range of photo-redox reactions. The modification aims to establish a conduction band (CB) potential that is more negatively charged and a valence band (VB) potential that is more positively charged. This, in turn, results in an increased bandgap and a limitation in the absorption of solar photons [38, 39]. To enhance the photocatalytic activity of Bi-based photocatalysts, researchers employ various strategies, with the formation of heterojunctions with other nanomaterials proving particularly effective. This chapter primarily focuses on different heterojunctions and nanocomposites involving bismuth-based compounds, elucidating their structural and optical properties in the context of degrading organic pollutants.

2 General Photocatalytic Mechanism

The commencement of the photocatalytic process ensues upon the absorption of photons possessing energies equal to or surpassing the bandgap energy (Eg) by the photocatalyst. This absorption event induces the elevation of electrons from the valence band (VB), corresponding to the lower energy state, to the conduction band (CB), denoting the higher energy state [40, 41]. As a result of this process, positively charged holes in the valence band (h+) are produced (Fig. 2). These photo-excited electrons and holes subsequently migrate to the surface of the photocatalyst, where they initiate reduction and oxidation processes, respectively. The redox potential of the generated holes, determined by pH and the specific semiconductor photocatalyst in use, typically falls within the range of +1.0 to +3.5 V. This redox potential imparts the hole with strong oxidizing capabilities, rendering it an effective oxidant [42]. The hole can undergo direct oxidation reactions on the photocatalyst's surface or participate in indirect oxidation processes by reacting with water to yield hydroxyl radicals. These hydroxyl radicals play a crucial role in promoting the oxidative mineralization of pollutants. Additionally, the photo-generated electrons may engage with an oxygen scavenger, leading to the production of superoxide radical species, which in turn contribute to the degradation of contaminants [43].

While electrons and holes migrate toward the surface of the photocatalyst, the potential for recombination still exist. Recombination events can result in the conversion of the energy carried by charge carriers into either photons or phonons, which

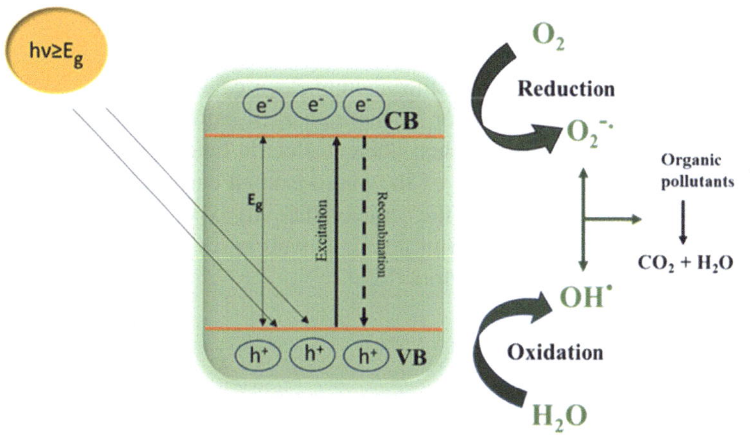

Fig. 2 General mechanism of degradation of pollutants by a photocatalyst

represent the vibrational energy of lattice atoms. It is worth noting that a high rate of recombination diminishes the quantum yield of the photocatalytic process [44]. Typically, defect sites within the photocatalyst serve as the loci for electron and hole recombination [45]. To mitigate the undesirable recombination of electrons and holes, specific approaches can be employed. These include structural alterations applied to nanoparticles based on bismuth and the utilization of diverse heterojunction systems [46]. These strategies are designed with the aim of impeding recombination events and thereby enhancing the photocatalytic efficacy of the catalysts [47]. Through the manipulation of band structures utilizing photocatalysts related to bismuth, studies have indicated a reduction in the recombination between the valence and conduction bands [48].

3 Heterojunction Construction

Heterojunctions, which represent the interfaces created by combining two dissimilar semiconductor materials, have emerged as a pivotal tool in enhancing the effectiveness and specificity of photocatalytic processes applied for the degradation of organic impurities. Through the optimization of charge separation kinetics, the integration of a heterojunction structure comprising two distinct semiconductors is anticipated to yield improvements in photocatalytic performance [49]. In fact, aligning the band gap of these two semiconductors within the heterojunction can substantially enhance the efficiency of carrier separation. In situations where the two dissimilar semiconductors exhibit contrasting Fermi energy levels (EF) or work functions (W), the formation of a built-in electric field (ED) is likely at the heterojunction interface. This occurs due to the spontaneous movement of electrons from the semiconductor with the higher EF towards the semiconductor with the lower EF [50, 51].

When subjected to illumination, electrons and holes that have been excited by light (referred to as non-equilibrium carriers) can be compelled to migrate between the two semiconductor materials due to the inherent electric field present. This migration serves to prevent the recombination of these charged particles. It is worth emphasizing that the establishment of this inherent electric field at the interface of the heterojunction and the behaviour of the photo-excited carriers depend on several factors. These factors encompass the semiconductivity type (whether it's n-type or p-type), the work function (or Fermi level), as well as the conduction and valence band potentials of the semiconductors [52].

3.1 Type I Heterojunction

A Type-I heterojunction forms when the conduction band (CB) and valence band (VB) of the photocatalytic system II (PS II) have lower and higher energy levels than those of photocatalytic system I (PS I), respectively (Fig. 3). Thus, electrons and holes tend to migrate and accumulate within the photocatalytic system PS I. In theory, Type-I heterojunctions do not offer any advantage in terms of separating charge carriers, and consequently, they do not enhance photocatalytic activity. Type-I heterojunctions are frequently encountered when composite materials consist of a photocatalytic system with a narrow band gap and another with a wide band gap. This prevalence of Type-I heterojunctions arises due to the predominance of wide band gap semiconductors and certain constraints related to the positions of valence and conduction bands [53].

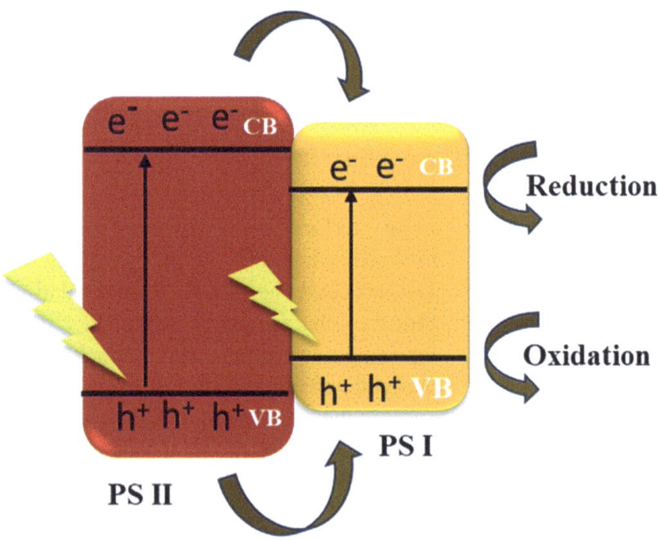

Fig. 3 Type I heterojunction

3.2 Type II Heterojunction

The initial proposition of utilising interparticle electron transfer as a means to mitigate charge recombination was put up by Serpone et al. in 1984 [54]. It was later extended to interparticle hole transfer more than a decade later when two interconnected semiconductors exhibited staggered band structures [55]. During illumination, photogenerated electron-hole pairs were formed, and electrons were transferred from both photocatalysts referred to as PS I and PS II (Fig. 4). Although type-II charge transfer is considered advantageous for promoting charge separation and has led to investigations and developments, it has certain drawbacks upon closer examination. The increased efficacy of charge separation is achieved at the rate of diminished redox ability, hence it is unfavourable from a thermodynamic perspective.

In cases where the transfer of charge carriers follows the type-II, photogenerated electrons tend to accumulate on the conduction band of PS I, characterized by a very low reduction potential. Conversely, photogenerated holes accumulate on the valence band of PS II, which has a weak oxidation potential. In such instances, this configuration may impede the driving force for specific photocatalytic reactions. From a dynamic standpoint, the repulsion between the electrons in PS II and PS I hinders the continuous transfer of electrons from PS II to PS I. Simultaneously, repulsive interactions between holes in PS II and those transferred from PS I impede spatial charge separation. Likewise, electrostatic attraction between electrons and holes in PS II prevents electrons from moving from PS II to PS I. Similar charge interactions occur in PS I, preventing hole transfer between PS I and PS II. Therefore, achieving the anticipated charge transfer in a type-II heterojunction becomes challenging. Electrostatic repulsive forces in this context can be likened to the repulsion between two like magnetic poles. The magnitude of these repulsive forces between electrons or

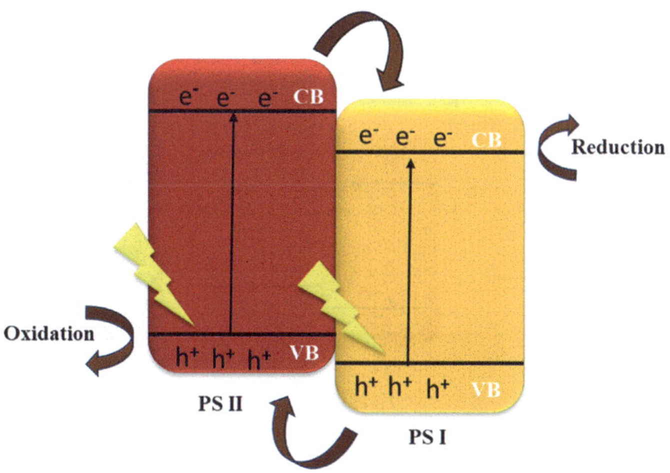

Fig. 4 Type II heterojunction

holes hinders their transfer in a type-II heterojunction. In summary, the proposed charge transfer mode in type-II heterojunctions is contentious because it fails to effectively segregate photogenerated electrons and holes. Additionally, this transfer mode reduces the overall redox capacity of the heterojunction. Conversely, photogenerated holes are transferred from PS I to PS II, resulting in the spatial separation of electrons and holes [56, 57].

3.3 Z-Scheme Heterojunction

Drawing inspiration from the natural photosynthesis process, an artificial Z-scheme photocatalyst, initially proposed by Bard in 1979 [58], holds significant promise for achieving high photocatalytic efficiency. In contrast to type-II heterojunctions, Z-scheme photocatalysts share the same band structure configuration but exhibit notably distinct charge carrier transfer mechanisms. In Z-scheme photocatalysts, the photocatalytic system consists of PS II and PS I, with redox ion pairs functioning as electron intermediaries (Fig. 5). When exposed to light, photo-excited electrons are transferred from the conduction band (CB) of PS I to the valence band (VB) of PS II through the mediation of redox ion pairs. However, as a result of the considerable reduction potential, electrons from the CB of PS II additionally engage with these redox pairs. This interaction interferes with the electron transfer process from the CB of PS I, leading to a limited redox capacity within the system. Furthermore, redox mediator ions, commonly used as electron intermediaries, result in light absorption and reduce the availability of solar photons for utilization. Additionally, these redox mediator ions are sensitive to various pH. In order to confront these issues, progress was achieved through the development of second-generation Z-scheme photocatalysts, which integrated solid-phase materials to facilitate electron mediation [14, 39, 59].

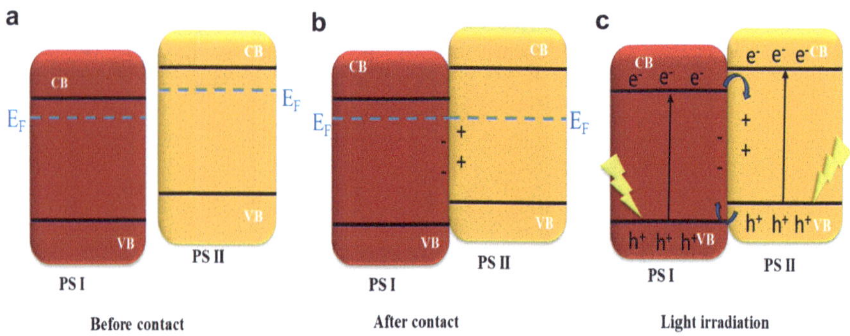

Fig. 5 The route of charge transfer within the Z-scheme

3.4 S-Scheme Heterojunction

An S-scheme, also known as a step-scheme, heterojunction comprises of the photocatalytic system I (PS I) and photocatalytic system II (PS II). These two photocatalysts have staggered band structures, equivalent to a type-II heterojunction, but with a fundamentally distinct charge transfer mechanism. The approach for enhancing bismuth based materials through the creation of S-type heterojunctions has gained significant popularity and widespread adoption in the field of research and development.

The transfer of charge dynamics within a step scheme heterojunction can be elucidated as depicted in Fig. 6. In comparison to photocatalytic system I (PS I), photocatalytic system II (PS II) exhibits a higher conduction band (CB) and valence band (VB) positions and a reduced work function. When the two photocatalytic systems are brought into close proximity, electrons naturally migrate from PS II to PS I, leading to the formation of an electron depletion region in PS II and an electron accumulation region near the interface in PS I. As a result, PS I becomes negatively charged, while PS II carries a positive charge. Simultaneously, an internal electric field emerges, extending from PS II to PS I. This internal electric field facilitates the transfer of photogenerated electrons from PS I to PS II, rather than the reverse.

Upon collision of PS I and PS II (as depicted in Fig. 6b, c), their Fermi energy levels must align. This adjustment causes an elevation in the Fermi levels of PSI and a corresponding reduction in the Fermi levels of PS II. This band bending effect compels photogenerated electrons in the CB of PS I and holes in the VB of PS II to recombine within the interface region, akin to how water flows downstream due to gravitational pull. Under the influence of Coulombic attraction, photoinduced electrons in the CB of PS I and holes in the VB of PS II tend to recombine at the interface. In summary, the recombination of electrons within the conduction band of PS I and holes within the valence band of PS II is regulated by the synergistic influence of the internal electric field, band bending, and Coulombic attraction. Consequently, recombination eliminates non-productive electrons and holes, while retaining photoinitiated electrons in the CB of PS II and holes in the VB of PS I for subsequent photocatalytic reactions [14, 60].

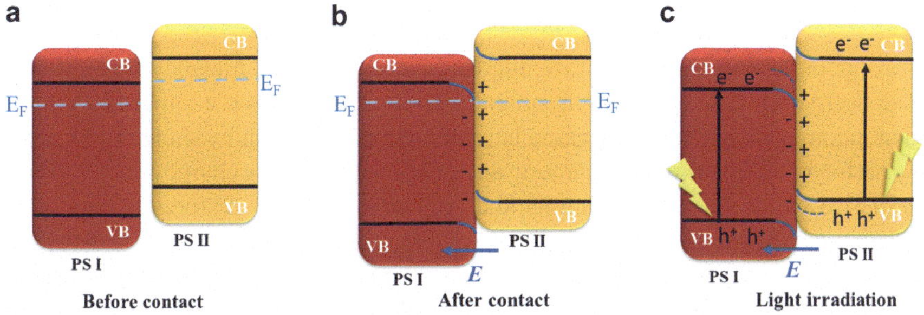

Fig. 6 Mechanism of transfer process of photogenerated charge carrier in S-scheme configuration

4 Organic Contaminants

Organic contaminants are chemically active compounds released from industries. These may reach marine waters by migrating from their point of origin via water currents. The widespread application of organic pollutants, encompassing dyes and antibiotics, within a diverse array of industries like textiles, paper, cosmetics, pharmaceuticals, paint, leather, and plastics has markedly intensified environmental contamination, with a pronounced emphasis on waterborne ecosystems. On a global scale, the annual production of synthetic dyes surpasses 700,000 metric tonnes, with documented records indicating the availability of more than 100,000 discrete commercially obtainable dye compositions. The textile industry is responsible for approximately 15% of the dye found in worldwide wastewater, but the agricultural, chemical, and pharmaceutical sectors collectively contribute to nearly 97% of the overall volume of wastewater generated [61].

Despite the well-documented hazardous, mutagenic, and carcinogenic properties associated with dyes, it remains a common practise to discharge a significant portion of surplus dyes produced during dyeing and finishing procedures into the environment, sometimes without subsequent remedial measures. The release of effluent from these wastewater streams into natural aquatic environments is a big cause for concern, since dye exposure has the potential to cause problems in the kidneys, reproductive organs, liver, brain, and central nervous system, which are all very important parts of the body. Also, tiny amounts of pigments, especially concentrations below 1 ppm, can be seen in aquatic environments. These pigments block the sun's rays and change the photosynthetic processes in water plants. Based on their ionic charge, dyes can be divided into two groups: cationic and anionic dyes. Anionic dyes include Congo red, reactive black, methyl orange, ethyl violet and Orange II. Whereas dyes like methylene blue, rhodamine B, crystal violet, brilliant green are cationic dyes which are commonly used in the textile industry for dyeing [62]. The chemical formula and structure of some of the dyes are given in Table 1.

5 Bismuth Based Heterojunctions as Photocatalyst

In recent scientific investigations, bismuth-based heterojunctions have gained prominence as promising photocatalysts due to their impressive efficiency in capturing solar energy for a range of environmental applications. When compared to traditional photocatalysts, bismuth-based heterojunctions stand out by showcasing superior performance in terms of separating and effectively utilizing charge carriers. Their distinctive electronic configurations and adaptable crystal structures makes them excellent candidates for harnessing both visible and infrared light, thus broadening their potential use in various photocatalytic processes. Some noteworthy bismuth-based compounds shedding light on their potential contributions to the field of advanced photocatalysis are listed in Table 2.

Table 1 Structure and chemical formula of dyes

Dye name	Chemical structure	Chemical formula	λ_{max} (nm)
Methylene blue (cationic)	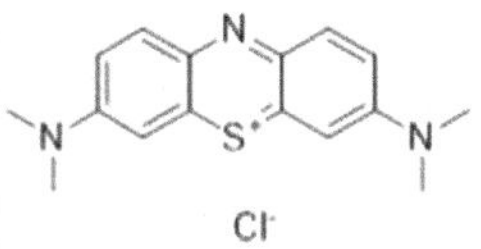	$C_{16}H_{18}ClN_3S$	663
Crystal violet (cationic)	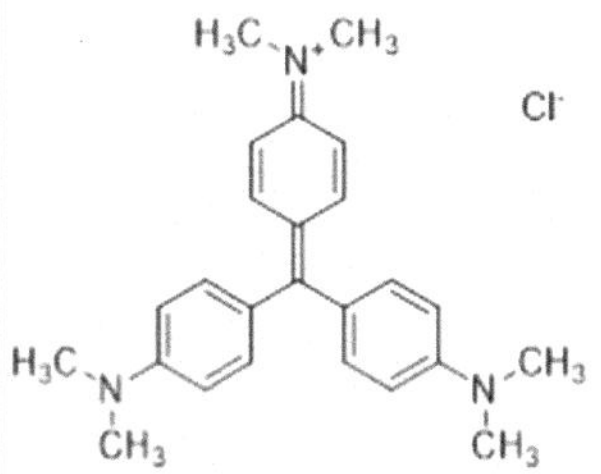	$C_{25}N_3H_{30}Cl$	590
Rhodamine B (cationic)	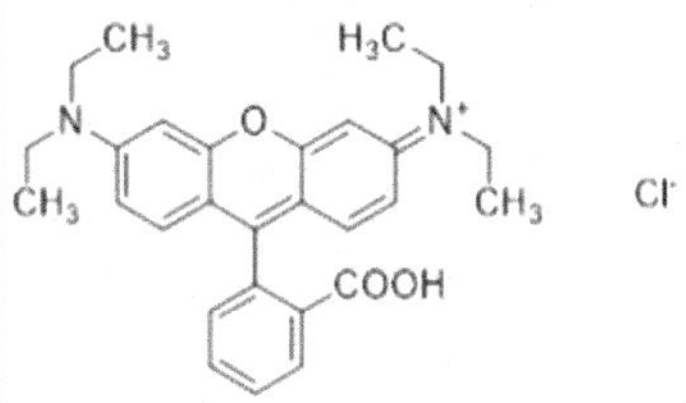	$C_{28}H_{31}ClN_2O_3$	543
Methyl orange (anionic)	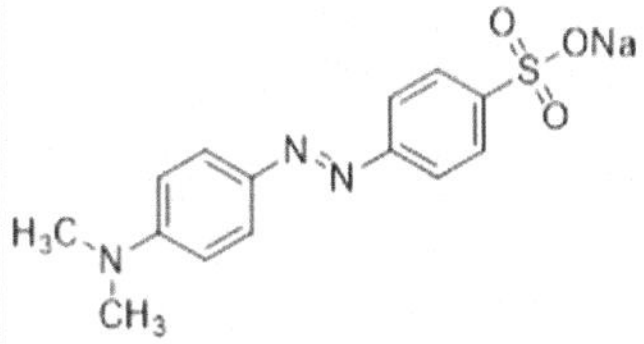	$C_{14}H_{14}N_3NaO_3S$	462
Eriochrome black T (anionic)	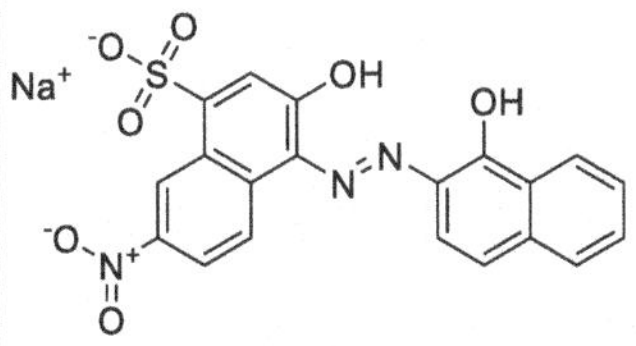	$C_{20}H_{12}N_3O_7SNa$	531
Alizarin yellow R (anionic)	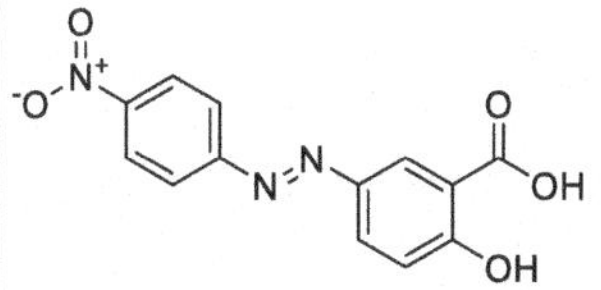	$C_{13}H_9N_3O_5$	367

Table 2 Bismuth based composites reported in the literature

Nanocomposite	Pollutant	K_{app} for single nanocomposite (min^{-1})	K_{app} for Heterojunction (min^{-1})	References
Bi_2S_3/Bi_2WO_6	Phenol	0.0010	0.0062	[63]
Ag_2O/Bi_2WO_6	Methyl orange	0.00	0.016	[64]
ZrO_2/Bi_2WO_6	Rhodamine B	0.0012	0.0083	[65]
g-C_3N_4/Bi_2WO_6	Methylene blue	0.0025	0.0171	[66]
CdS/Bi_2WO_6	Methyl orange	0.00863	0.0103	[67]
$Ag/AgBr/Bi_2WO_6$	Rhodamine B, methyl orange, and phenol	0.01, 0.0 and 0.02	0.03, 0.04 and 0.055	[68]
P-$La_2Ti_2O_7/Bi_2WO_6$	Rhodamine B	0.011	0.062	[69]
g-C_3N_4/Bi_2MoO_6	Rhodamine B	0.00264	0.06484	[70]
Graphene/Bi_2MoO_6	Reactive red dye	0.010	0.022	[71]
$Ag/AgBr/Bi_2MoO_6$	Alizarin red S (ARS) and phenol	0.00512	0.02912	[72]
Polyaniline/Bi_2MoO_6	Rhodamine B	0.0049	0.0335	[73]
$Bi_2MoO_6/Bi_2Ti_2O_7$	Glycolic acid	0.010	0.01354	[74]
$Ag/AgBr/Bi_2MoO_6$	Methylene blue	0.002	0.01014	[75]
Bi_2O_3/Bi_2S_3	RhB oxidation and Cr (VI) reduction	0.01143	0.05953	[76]
$Bi_2O_3/BiOCl$	Photocalytic mineralization of oxytetracycline and ampicillin	Not available	0.0068 and 0.007	[77]
$BiVO_4/BiPO_4$	Methylene blue	0.122	0.800	[78]
$BiVO_4/Bi_2WO_6$	Photocatalytic antifouling of microorganism	0.05213	0.0704	[79]
$BiIO_4/BiVO_4$	Rhodamine B	0.108	0.173	[80]
$BiFeO_3/BiVO_4$	Rhodamine B	0.05	1.46	[81]
$BiVO_4/BiPO_4$/GO	Rhodamine B	0.007	0.035	[82]

5.1 Bi_2WO_6

Bismuth tungstate was investigated by Sun et al. [83] for its photocatalytic potential in degrading dyes and facilitating oxygen generation when exposed to sunlight. The research encompassed the synthesis of Bi_2WO_6 in several structural forms, including nanoparticles, nanoplates, and 3D hierarchical designs [84–87]. Karbasi et al. synthesised flower like Bi_2WO_6 as depicted in Fig. 7 and achieved complete bacterial inactivation in two hours under visible light [88].

N containing biochar (N1-B), which is synthesised through the treatment of biochar with urea, assumes a significant role within the framework of coupling with bismuthal nanocomposite BW/N1-B in which Bi_2WO_6 (BW) is commonly utilised as a substrate for the deposition of a specific substance, which is employed for the aim of eliminating chromium and rhodamine B. The photocatalytic performance of Bi_2WO_6 is significantly enhanced by the heterojunction assembly, primarily due to enhanced conductivity, a decreased band gap, and the formation of a two-dimensional sheet-like structure that provides additional surface reaction sites. The prepared composite BW/N1-B with a ratio of urea to biochar 2:1 showed enhanced activity with a degradation efficiency of 99.1% for 10 mg/L rhodamine B within 45 min and attained 96.7% reduction for Cr (VI) within 30 min [89].

Phosphorous-doped $La_2Ti_2O_7$ was layered with Bi_2WO_6 in a well-controlled manner to produce another Bi_2WO_6 heterojunction (P-$La_2Ti_2O_7$/Bi_2WO_6). The introduction of non-metallic phosphorus through doping has resulted in changes to the electronic configuration and valence states within the Bi_2WO_6 structure, leading to an enhancement in its photocatalytic activity by 99.02% within 80 min under visible light [69]. Bi_2WO_6/$BiPO_4$ with Z-scheme heterojunction showed excellent methylene blue degradation with simulated solar-light irradiation, which is nearly

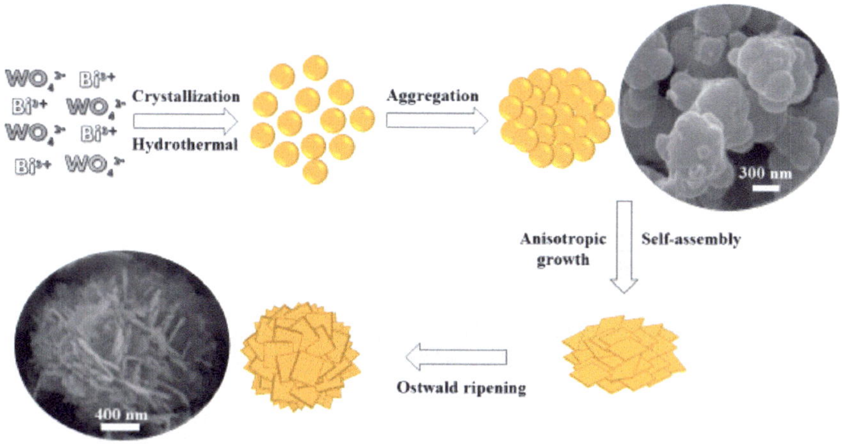

Fig. 7 The growth process of Bi_2WO_6 microspheres resembling flowers. Adapted from [88]. Open access publication by MDPI

1.88 times and 4.29 times higher than the pristine Bi_2WO_6 and $BiPO_4$ respectively [90]. Likewise, the coupling of Bi_2WO_6 and the metal-organic framework NH_2-MIL-125(Ti) leads to creation of an S-scheme heterojunction, which greatly enhances the pigments to degrade upon visible light exposure [91].

5.2 Bi_2MoO_6

A platinum-coated g-C_3N_4/Bi_2MoO_6 heterostructure composite is reported as a photocatalyst with greater photocatalytic activity than purified g-C_3N_4 and Bi_2MoO_6, respectively. In this instance, a g-C_3N_4 matrix resembling a wrinkly flower like bismuth molybdenum oxide microplates with platinum decoration, results in an increase in specific surface area, an increase in visible light absorption, and a promotion of photocatalytic methylene blue degradation [70]. Polyaniline/Bi_2MoO_6 ($PANI_x$/Bi_2MoO_6) nanocomposites synthesised via in situ polymerization of PANI upon Bi_2MoO_6 resembled the band structures of Bi_2MoO_6 and PANI, resulting in outstanding hole transfer with accessible active sites, which ultimately led to nearly 100% photodegradation of rhodamine B in 120 min using visible light [73].

In Bi_2MoO_6/diatomite nanocomposite, nano meter-scale Bi_2MoO_6 crystals were observed and anchored to the diatomite surface. Using rhodamine B, its photocatalytic activity was evaluated. The nano-Bi_2MoO_6/diatomite composite exhibited the greatest photocatalytic activity for degradation of RhB 97.6% in 60 min, while also being highly stable and recyclable. It was observed that holes and oxygen radicals functioned as the most active species and key agents of reactivity [92]. (Bi_2MoO_6/$Bi_2Ti_2O_7$) demonstrated an enhanced photocatalytic activity and chemical stability for glycolic acid removal, reaching up to 99.9% at pH4 in 150 min. Heterojunction interface in the synthesised nanocomposite induced charge transfer, thereby enhancing photocatalytic performance through the active participation of superoxide radicle, holes, and hydroxide radicle [74]. Balasurya et al., synthesized a Bi_2MoO_6–Ag_2MoO_4 nanocomposite and achieved a remarkable degradation (91.8%) of MB dye, surpassing the individual degradation rates of Ag_2MoO_4 (54.66%) and Bi_2MoO_6 (58.6%). The degradation process adhered to a pseudo-first-order kinetic model and exhibited excellent conformity to the Langmuir–Hinshelwood model. Figure 8 depicts different characterisation and photocatalytic results of the composite compared to their pristine photocatalyst. Moreover, this prepared nanocomposite not only functions as a robust solar-driven photocatalyst but also exhibits antimicrobial properties, rendering it a highly promising candidate for the treatment of environmental water [93].

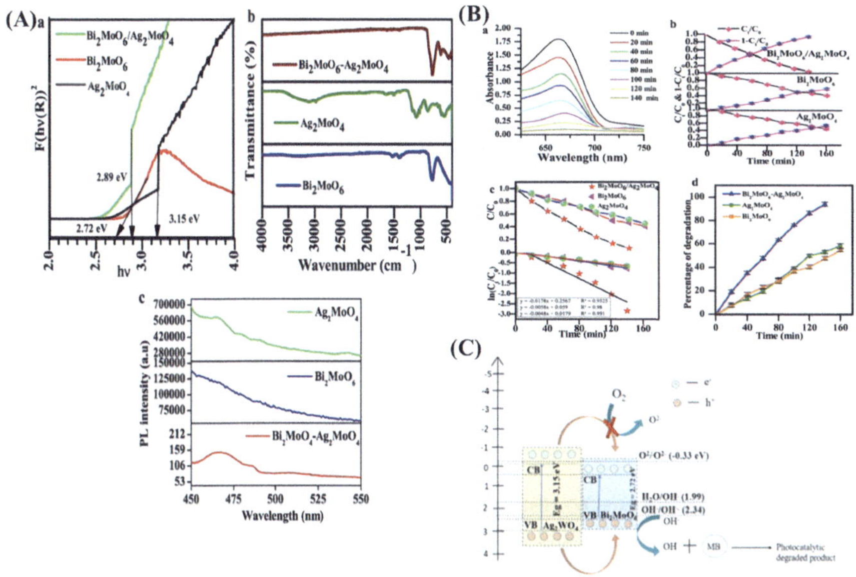

Fig. 8 **A**) UV-DRS, FTIR and PL spectra of the composite, **B**) UV spectrum of degradation of methylene blue, half life period comparison and degradation efficiency comparison of Bi_2MoO_6, Ag_2MoO_4 and Bi_2MoO_6–Ag_2MoO_4, **C**) photocatalytic degradation mechanism of Bi_2MoO_6–Ag_2MoO_4. Adapted from [93] under copyright permission from publisher

5.3 *Bi_2O_3 and Bi_2S_3*

Bi_2O_3 having forbidden energy gap value of 2.5–2.8 eV, which is a promising photocatalyst with a valence band edge that is sufficiently positive to generate oxygen. Furthermore, Bi_2O_3 has gained prominence as the favoured semiconductor owing to its diverse array of crystal structures, each characterized by either n-type or p-type properties. Consequently, it can be utilised in the sensor, catalysis as well as in electrochemical fields [94, 95]. Bi_2O_3 was found to form a bond with bismuth vanadate [96], bismuth sulphide [76], bismuth oxychloride [97], g-C_3N_4 [98], bismuth tungstate [99] and carbon quantum dots [100]. The coupling of bismuth oxide with other nanomaterials mainly by combining with molybdenum disulfide, amplified its properties as MoS_2 exhibited this enhancement as a result of its graphene-like constitution, resulting in a tuning of photo response and charge carrier transport. Thus, MoS_2 was layered onto Bi_2O_3 microplates, which was followed by Bi_2S_3 incorporation, resulting in a three-phase heterojunction $Bi_2O_3/Bi_2S_3/MoS_2$ composite as shown in Fig. 9. This incorporation enhanced photocatalytic water oxidation nearly 529.1 $\mu molh^{-1}\ g^{-1}{}_{cat}$ which is 1.5 and 1.25 times higher than the pristine semiconductor photocatalyst [101].

A novel Bi_2O_3/TiO_2 nanofiber which follows a step (S-scheme) heterojunction is reported. This heterojunction is capable of preventing the recombination of electrons

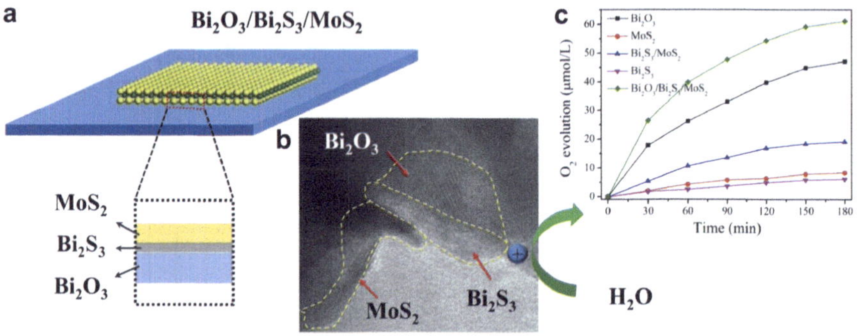

Fig. 9 **a)** Representation of different photocatalysts in $Bi_2O_3/Bi_2S_3/MoS_2$, **b)** TEM image of the composite $Bi_2O_3/Bi_2S_3/MoS_2$, **c)** photocatalytic oxygen evolution of Bi_2O_3, MoS_2, Bi_2S_3/MoS_2, Bi_2S_3 and $Bi_2O_3/Bi_2S_3/MoS_2$. Adapted from [101] under copyright permission from the publisher

on TiO_2 and holes on Bi_2O_3, resulting in a significant improvement in the removal of phenol using visible light and had observed a significant decline in the absorption at 269 nm with the synthesised nanocomposite [102]. Lu et al. synthesized p-n heterojunction photocatalysts, specifically Bi_2O_3/Bi_2SiO_5, using bismuth nitrate and nano silicone dioxide as precursors through a straight forward one-step synthesis method. As the SiO_2 content increased, Bi_2O_3 gradually transformed into Bi_2SiO_5, resulting in the creation of a p-n heterojunction. This improved carrier separation and reduced recombination, ultimately enhancing photocatalytic performance [103].

Bi_2S_3/Bi_2WO_6 had attained a significant phenol degradation with visible light within 2 h, the removal efficiency indicated by TOC reduction, was observed as 4.6% for bare Bi_2WO_6 and 28.7% for the Bi_2S_3/Bi_2WO_6 composite and confirmed the mineralization potential of the composite [63], on coupling with cadmium sulphide for the transformation of CO_2 into CH_3OH under visible light conditions, it was observed that Bi_2S_3 exhibited superior photocatalytic activity and responsiveness to visible light in comparison to CdS. Additionally, the incorporation of Bi_2S_3 led to an enhancement in the performance of CdS, resulting in the most notable methanol yield of 613 μmol/g achieved with a Bi_2S_3/CdS heterojunction photocatalyst featuring a 15% weight ratio of Bi_2S_3 to CdS. This yield was three times greater than that of CdS and twice as large as that of Bi_2S_3 [104].

5.4 *$BiVO_4$*

Bismuth vanadate boasts a band gap of 2.4 eV and exhibits remarkable activity under visible light, attributed to the distinctive valence band (VB) structure, which involves the hybridization of bismuth with its s and p subshells, alongside the contribution of vanadium's 3d orbital to the conduction band (CB). Nonetheless, the challenge of insufficient charge carrier separation persists. Multiple methods for synthesising

$BiVO_4$ have been reported which includes both solid state reaction (SSR) and melting reactions at high temperatures have been utilised to produce monoclinic $BiVO_4$ [105]. The tetragonal variant of $BiVO_4$ was synthesized through precipitation at room temperature [106]. The forbidden energy gap for the two different crystal forms are monoclinic 2.4 eV and tetragonal 2.9 eV, monoclinic $BiVO_4$ holds good for the fabrication of visible light active photocatalysts and is more effective.

Until now, numerous heterojunctions have been synthesised, including gold-coated $BiVO_4$ composite [107]. Among the evaluated composites of $BiPO_4/BiVO_4$, the 10% $BiPO_4/BiVO_4$ composite demonstrated superior photocatalytic efficacy, achieving a remarkable 92.1% degradation of MB following 5 h of visible light illumination. Furthermore, this composite exhibited a rate constant (k value) of 0.503 h^{-1}, which was twice that of pristine $BiVO_4$. This noteworthy enhancement in photocatalytic performance can be primarily attributed to the well-matched heterojunction interface established between $BiPO_4$ and $BiVO_4$, effectively facilitating the separation of photoinduced electron-hole pairs [78], Peng. et al., synthesised $Bi_2WO_6/BiVO_4$ which have a comparison with 600 degree calcined and uncalcined heterojunction showing a complete degradation of rhodamine B within 30 min under visible light [108], $BiIO_4/BiVO_4$ [80], and $BiFeO_3/BiVO_4$ [81]. In addition, a ternary nanocomposite g-C_3N_4/Bi/$BiVO_4$ demonstrates excellent efficiency in aerobic photocatalytic oxidation for degrading organic pollutants like Rhodamine B but also displays remarkable performance in photocatalytic CO_2 reduction under anaerobic conditions. Furthermore, it enables the direct conversion of Rhodamine B into solar fuels within a single-step anaerobic reactor showcasing S-type heterojunction [109].

5.5 *Bismuth Oxychloride, Bismuth Oxyfluoride, Bismuth Oxybromide and Bismuth Oxyiodide*

The band gap of bismuth oxyhalides (BiOX) such as BiOBr, BiOF, BiOI, and BiOCl are 2.64 eV [110], 3.64 eV [111], 1.77 eV [112] and 3.22 eV [113]. The valence band (VB) and conduction band (CB) of BiOX where X denotes fluorine, chlorine, bromine, and iodine, are O 2p, and Bi 6p orbitals. Bismuth oxyhalide family, which includes $Bi_3O_5I_5$, $Bi_7O_9I_3$, Bi_5O_7I, $Bi_4O_5I_2$, and BiOI, exhibits excellent photocatalytic efficacy in the presence of solar light [114]. The scientific literature reports a variety of morphological alterations observed in bismuth oxyiodide. The aforementioned formations encompass three-dimensional microspheres, hollow microspheres, flower-like microspheres, nanosheet structures, square-like arrangements, flake-like two-dimensional structures, and one-dimensional nanofibers [115]. Diverse stages of bismuth oxyiodide were harnessed, along with the incorporation of metallic gold nanoparticles on its surface, to facilitate the degradation of oxytetracycline within wastewater [116].

Kong et al., synthesised AgBr–BiOBr heterojunction photocatalysts, Fig. 10 depicting various characterization results of the composite with various loadings

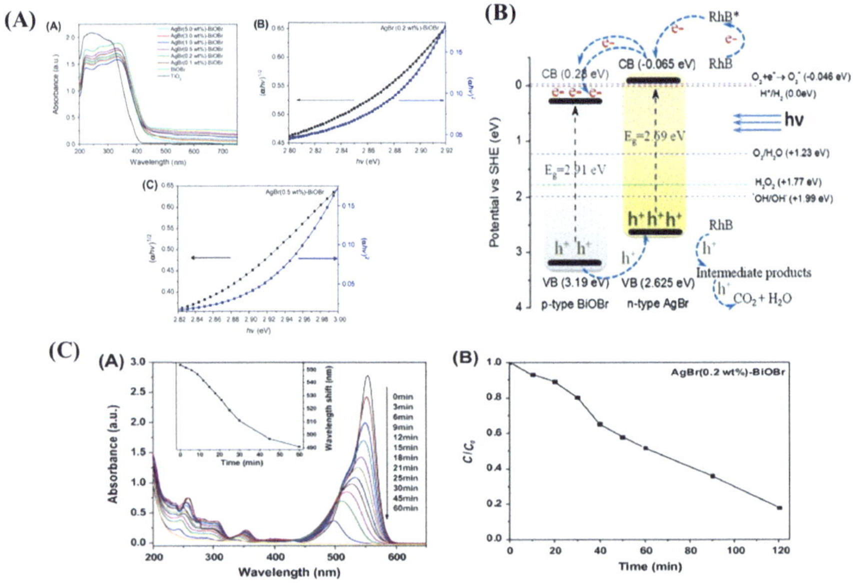

Fig. 10 **A**) UV-Vis diffuse reflectance spectra of BiOBr and AgBr–BiOBr samples, the plots of $(\alpha hv)^{1/2}$ and $(\alpha hv)^2$ versus photo energy of AgBr (0.2 wt%)–BiOBr sample and AgBr (0.5 wt%)–BiOBr. **B**) photocatalytic mechanism of AgBr–BiOBr and the dependence of TOC on time for photodegradation of RhB on AgBr (0.2 wt%) BiOBr, **C**) UV-vis spectral changes of RhB as a function of photo-reaction time; Inset: wavelength shifts as a function of the decrease in absorption maximum. Adapted from [117] under copyright permission from the publisher

of AgBr on BiOBr (up to 0.5 wt%). AgBr (0.2 wt%)–BiOBr showcased highest degradation of rhodamine B within 30 min. However, these AgBr–BiOBr materials gradually lost their photoactivity in the cycling photocatalytic tests due to the as-formed Ag_2O reducing the intimate interface of AgBr–BiOBr catalysts, masking the light excitation, and serves as the recombination center of photogenerated charge carriers, resulting in the deactivation of the catalysts [117].

5.6 *Bismuth Subcarbonate*

Bismuth sub-carbonate ($(BiO)_2CO_3$ or Bi_2CO_5) is the only well known solid carbonate. The band gap is 3.4 eV, so the band gap transition can only be activated by irradiation with wavelengths shorter than 365 nm [118]. The CB of $(BiO)_2CO_3$ comprises of primarily hybridised O 2p and Bi 6p orbitals, whereas the VB is made up of O 2p, Bi 6p, and C 2p orbitals. Photocatalytic heterojunctions, comprised of bismuth sub-carbonate ($Bi_2O_2CO_3$), particularly the $Bi_2O_2CO_3/Bi_2MoO_6$ combination, demonstrated remarkable efficacy when subjected to visible-light illumination. They outperformed than pure $Bi_2O_2CO_3$ in the degradation of ciprofloxacin and two distinct dyes, rhodamine B and methyl orange. This enhanced performance is attributed to the efficient separation as well as exchange of charge carriers, encompassing photogenerated active species such as holes (h^+) and superoxide radical anions ($^{\cdot}O_2$) [119].

6 Conclusion

In summary, this book chapter has provided a valuable foundation for understanding the photocatalytic degradation of organic contaminants using heterojunction nanocomposites based on bismuth compounds. Throughout this chapter, various aspects of bismuth-based heterojunction nanocomposites, including their synthesis methods, structural characterization, and their exceptional photocatalytic properties were discussed. Their capacity to effectively break down a wide range of organic pollutants, their versatility and potential for real-world applications were explored. Furthermore, we addressed key mechanisms and reaction pathways involved in the photocatalytic degradation process, providing insights into the fundamental understanding of these nanocomposites performance.

References

1. P. Kumari, A. Kumar, Advanced oxidation process: a remediation technique for organic and non-biodegradable pollutant. Results Surf. Interfaces **11**, 100122 (2023). https://doi.org/10.1016/j.rsurfi.2023.100122
2. J. Briffa, E. Sinagra, R. Blundell, Heavy metal pollution in the environment and their toxicological effects on humans. Heliyon **6**(9), e04691 (2020). https://doi.org/10.1016/j.heliyon.2020.e04691
3. M.A. Al-Nuaim, A.A. Alwasiti, Z.Y. Shnain, The photocatalytic process in the treatment of polluted water. Chem. Pap. **77**(2), 677–701 (2022). https://doi.org/10.1007/s11696-022-02468-7
4. E.O. Ezugbe, S. Rathilal, Membrane technologies in wastewater treatment: a review. Membranes **10**(5), 89 (2020). https://doi.org/10.3390/membranes10050089
5. S. Topçu, M.G. Şeker, In vitro antimicrobial effects and inactivation mechanisms of 5,8-Dihydroxy-1,4-Napthoquinone. Antibiotics **11**(11), 1537 (2022). https://doi.org/10.3390/antibiotics11111537
6. Z. Zhou, B. Li, X. Liu, Z. Li, S. Zhu, Y. Liang, Z. Cui, S.L. Wu, Recent progress in photocatalytic antibacterial. ACS Appl. Bio Mater. **4**(5), 3909–3936 (2021). https://doi.org/10.1021/acsabm.0c01335
7. Z. Razzaq, A. Khalid, P. Ahmad, M. Farooq, M.U. Khandaker, A. Sulieman, I.U. Rehman, S. Shakeel, A. Khan, Photocatalytic and antibacterial potency of titanium dioxide nanoparticles: a cost-effective and environmentally friendly media for treatment of air and wastewater. Catalysts **11**(6), 709 (2021). https://doi.org/10.3390/catal11060709
8. M. Ghosh, M. Mondal, S. Mandal, A. Roy, S. Chakrabarty, G. Chakrabarti, S. Pradhan, Enhanced photocatalytic and antibacterial activities of mechanosynthesized TiO2–Ag nanocomposite in wastewater treatment. J. Mol. Struct. **1211**, 128076 (2020). https://doi.org/10.1016/j.molstruc.2020.128076

9. A. Saravanan, P.S. Kumar, S. Jeevanantham, S. Karishma, A. Kiruthika, Photocatalytic disinfection of micro-organisms: mechanisms and applications. Environ. Technol. Innov. **24**, 101909 (2021). https://doi.org/10.1016/j.eti.2021.101909
10. M. Gong, S. Xiao, Y. Lei, C. Dong, J. Ji, D. Zhang, M. Xing, Research progress of photocatalytic sterilization over semiconductors. RSC Adv. **9**(34), 19278–19284 (2019). https://doi.org/10.1039/c9ra01826c
11. P. Afanasiev, C. Anastasescu, R. State, A. Vasile, F. Papa, I. Balint, Photocatalytic degradation of organic and inorganic pollutants to harmless end products: assessment of practical application potential for water and air cleaning. Catalysts **13**(2), 380 (2023). https://doi.org/10.3390/catal13020380
12. X. Li, Y. Chen, Y. Tao, L. Shen, Z. Xu, Z. Bian, H. Li, Challenges of photocatalysis and their coping strategies. Chem Catal. **2**(6), 1315–1345 (2022). https://doi.org/10.1016/j.checat.2022.04.007
13. C. Yuan, I. Ie, J. Zheng, C. Hung, Z. Lin, C. Shih, A review of electrical assisted photocatalytic technologies for the treatment of multi-phase pollutants. Catalysts **11**(11), 1332 (2021). https://doi.org/10.3390/catal11111332
14. Q. Xu, L. Zhang, B. Cheng, J. Fan, S-scheme heterojunction photocatalyst. Chem **6**(7), 1543–1559. https://doi.org/10.1016/j.chempr.2020.06.010
15. J. Schneider, M. Curti, Spectroscopic and kinetic characterization of photogenerated charge carriers in photocatalysts. Photochem. Photobiol. Sci. **22**(1), 195–217 (2022). https://doi.org/10.1007/s43630-022-00297-x
16. S. Ghosh, D. Sarkar, S. Bastia, Y.S. Chaudhary, Band-structure tunability via modulation of excitons in semiconductor nanostructures: manifestation in photocatalytic fuel generation. Nanoscale (2023). https://doi.org/10.1039/d3nr02116e
17. T.S. Sherkar, C. Momblona, L. Gil-Escrig, J. Ávila, M. Sessolo, H.J. Bolink, L.J.A. Koster, Recombination in perovskite solar cells: significance of grain boundaries, interface traps, and defect ions. ACS Energy Lett. **2**(5), 1214–1222 (2017). https://doi.org/10.1021/acsenergylett.7b00236
18. H. Wang, X. Li, X. Zhao, C. Li, X. Song, P. Zhang, P. Huo, X. Li, A review on heterogeneous photocatalysis for environmental remediation: from semiconductors to modification strategies. Chin. J. Catal. **43**(2), 178–214 (2022). https://doi.org/10.1016/s1872-2067(21)63910-4
19. N.S. Peighambardoust, E. Sadeghi, U. Aydemir, Lead halide perovskite quantum dots for photovoltaics and photocatalysis: a review. ACS Appl. Nano Mater. **5**(10), 14092–14132 (2022). https://doi.org/10.1021/acsanm.2c02787
20. A.M. Hassan, Z. Wang, Y.H. Ahn, M. Azam, A.A. Khan, U. Farooq, M. Zubair, Y. Cao, Recent defect passivation drifts and role of additive engineering in perovskite photovoltaics. Nano Energy **101**, 107579 (2022). https://doi.org/10.1016/j.nanoen.2022.107579
21. S. Lettieri, M. Pavone, A. Fioravanti, L.S. Amato, P. Maddalena, Charge carrier processes and optical properties in TIO_2 and TIO_2-based heterojunction photocatalysts: a review. Materials **14**(7), 1645 (2021). https://doi.org/10.3390/ma14071645
22. H. Zhang, J. Liu, T. Xu, W. Ji, X. Zong, Recent advances on small band gap semiconductor materials ($\leq$2.1 eV) for solar water splitting. Catalysts **13**(4), 728 (2023). https://doi.org/10.3390/catal13040728
23. G. Lee, Y. Hou, H. Huang, W. Wang, C. Lyu, J.J. Wu, Microwave-Assisted solvothermal synthesis of chalcogenide composite photocatalyst and its photocatalytic CO_2 reduction activity under simulated solar light. Catalysts **10**(7), 789 (2020). https://doi.org/10.3390/catal10070789
24. V. Jeyalakshmi, R. Mahalakshmy, K.R. Krishnamurthy, B. Viswanathan, Photocatalytic reduction of carbon dioxide by water: a step towards sustainable fuels and chemicals. Mater. Sci. Forum **734**, 1–62 (2012). https://doi.org/10.4028/www.scientific.net/msf.734.1
25. Y. Wu, J. Zhang, B. Long, H. Zhang, The thermodynamic stability, electronic and photocatalytic properties of the $ZnWO_4$(100) surface as predicted by screened hybrid density functional theory. ACS Omega **6**(23), 15057–15067 (2021). https://doi.org/10.1021/acsomega.1c01214

26. T.O. Ajiboye, O.A. Oyewo, D.C. Onwudiwe, The performance of bismuth-based compounds in photocatalytic applications. Surf. Interfaces **23**, 100927 (2021). https://doi.org/10.1016/j.surfin.2021.100927
27. T.O. Ajiboye, D.C. Onwudiwe, Bismuth sulfide based compounds: properties, synthesis and applications. Results Chem. **3**, 100151 (2021). https://doi.org/10.1016/j.rechem.2021.100151
28. A.M. La Cruz, U.M. García-Pérez, Photocatalytic properties of BiVO4 prepared by the co-precipitation method: degradation of rhodamine B and possible reaction mechanisms under visible irradiation. Mater. Res. Bull. **45**(2), 135–141 (2010). https://doi.org/10.1016/j.materresbull.2009.09.029
29. G. Zhao, Y. Zheng, Z. He, Z. Lu, L. Wang, C. Li, F. Jiao, C. Deng, Synthesis of Bi_2S_3 microsphere and its efficient photocatalytic activity under visible-light irradiation. Trans. Nonferrous Metals Soc. China **28**(10), 2002–2010 (2018). https://doi.org/10.1016/s1003-6326(18)64844-7
30. Q. Shi, Y. Zhang, D. Sun, S. Zhang, T. Tang, X. Zhang, S. Cao, Bi_2O_3-sensitized TiO_2 hollow photocatalyst drives the efficient removal of tetracyclines under visible light. Inorg. Chem. **59**(24), 18131–18140 (2020). https://doi.org/10.1021/acs.inorgchem.0c02598
31. R. Kumar, P. Raizada, A.A.P. Khan, V. Nguyen, Q. Van Le, S. Ghotekar, S. Rangabhashiyam, V. Gandhi, A. Singh, P. Singh, Recent progress in emerging $BiPO_4$-based photocatalysts: synthesis, properties, modification strategies, and photocatalytic applications. J. Mater. Sci. Technol. **108**, 208–225 (2022). https://doi.org/10.1016/j.jmst.2021.08.053
32. H. Chawla, A. Chandra, P.P. Ingole, S. Garg, Recent advancements in enhancement of photocatalytic activity using bismuth-based metal oxides Bi_2MO_6 ($M = W$, Mo, Cr) for environmental remediation and clean energy production. J. Ind. Eng. Chem. **95**, 1–15 (2021). https://doi.org/10.1016/j.jiec.2020.12.028
33. V. Dutta, S. Sharma, P. Raizada, A.A.P. Khan, A.M. Asiri, A.K. Nadda, P. Singh, Q. Van Le, C. Huang, D. Nguyen, S. Pansambal, V. Nguyen, Recent advances and emerging trends in $(BiO)_2CO_3$ based photocatalysts for environmental remediation: a review. Surf. Interfaces **25**, 101273 (2021). https://doi.org/10.1016/j.surfin.2021.101273
34. X. Lv, F.L.Y. Lam, X. Hu, A review on bismuth oxyhalide (BIOX, $X = Cl$, Br, I) based photocatalysts for wastewater remediation. Front. Catal. **2**. https://doi.org/10.3389/fctls.2022.839072
35. M. Shi, H.Y. Yang, Z. Zhao, G. Ren, X. Meng, Bismuth-based semiconductors applied in photocatalytic reduction processes: fundamentals, advances and future perspectives. Chem. Commun. **59**(29), 4274–4287 (2023). https://doi.org/10.1039/d3cc00580a
36. F. Guo, Z. Chen, X. Huang, L. Cao, X. Cheng, W. Shi, L. Chen, Ternary $Ni_2P/Bi_2MoO_6/g\text{-}C_3N_4$ composite with Z-scheme electron transfer path for enhanced removal broad-spectrum antibiotics by the synergistic effect of adsorption and photocatalysis. Chin. J. Chem. Eng. **44**, 157–168 (2022). https://doi.org/10.1016/j.cjche.2021.08.024
37. A. Walsh, Y. Yan, M.N. Huda, M. Al-Jassim, S. Wei, Band edge electronic structure of $BiVO_4$: elucidating the role of the Bi s and V d orbitals. Chem. Mater. **21**(3), 547–551 (2009). https://doi.org/10.1021/cm802894z
38. Q. Han, Advances in preparation methods of bismuth-based photocatalysts. Chem. Eng. J. **414**, 127877 (2021). https://doi.org/10.1016/j.cej.2020.127877
39. J. Li, H. Yuan, W. Zhang, B. Jin, Q. Feng, J. Huang, Z. Jiao, Advances in Z-scheme semiconductor photocatalysts for the photoelectrochemical applications: a review. Carbon Energy **4**(3), 294–331 (2022). https://doi.org/10.1002/cey2.179
40. H. Yang, B. Yang, W. Chen, J. Yang, Preparation and photocatalytic activities of TiO_2-Based composite catalysts. Catalysts **12**(10), 1263 (2022). https://doi.org/10.3390/catal12101263
41. R.V.H. Hahn, S. Rodríguez-Bolívar, P. Rodosthenous, E.S. Skibinsky-Gitlin, M. Califano, F.M. Gómez-Campos, Optical absorption in N-dimensional colloidal quantum dot arrays: influence of stoichiometry and applications in intermediate band solar cells. Nanomaterials **12**(19), 3387 (2022). https://doi.org/10.3390/nano12193387
42. M.B. Tahir, M. Sohaib, M. Sagir, M. Rafique, Role of Nanotechnology in photocatalysis, In *Elsevier eBooks* (2022), pp. 578–589. https://doi.org/10.1016/b978-0-12-815732-9.00006-1

43. S. Shinde, C. Bhosale, K. Rajpure, Hydroxyl radical's role in the remediation of wastewater. J. Photochem. Photobiol. B-Biol. **116**, 66–74 (2012). https://doi.org/10.1016/j.jphotobiol.2012.08.003
44. D. Mateo, J.L. Cerrillo, S. Durini, J. Gascon, Fundamentals and applications of photo-thermal catalysis. Chem. Soc. Rev. **50**(3), 2173–2210 (2021). https://doi.org/10.1039/d0cs00357c
45. W. Zhou, H. Fu, Defect-mediated electron–hole separation in semiconductor photocatalysis. Inorg. Chem. Front. **5**(6), 1240–1254 (2018). https://doi.org/10.1039/c8qi00122g
46. M.H. Abdurahman, A.Z. Abdullah, W. Da Oh, N.F. Shopware, M.F. Gasim, P.U. Okoye, A. Ul-Hamid, A.R. Mohamed, Tunable band structure of synthesized carbon dots modified graphitic carbon nitride/bismuth oxychlorobromide heterojunction for photocatalytic degradation of tetracycline in water. J. Colloid Interface Sci. **629**, 189–205 (2023). https://doi.org/10.1016/j.jcis.2022.08.172
47. M.Z. Rahman, F. Raziq, H. Zhang, J. Gascon, Key strategies for enhancing H_2 production in transition metal oxide based photocatalysts. Angew. Chem. (2023). https://doi.org/10.1002/anie.202305385
48. Z. Saddique, M. Imran, A. Javaid, S. Latif, N. Hussain, P. Kowal, G. Boczkaj, Band engineering of BiOBr based materials for photocatalytic wastewater treatment via advanced oxidation processes (AOPs)—A review. Water Resour. Ind. **29**, 100211 (2023). https://doi.org/10.1016/j.wri.2023.100211
49. A. Shawky, R.M. Mohamed, S-scheme heterojunctions: emerging designed photocatalysts toward green energy and environmental remediation redox reactions. J. Environ. Chem. Eng. **10**(5), 108249 (2022). https://doi.org/10.1016/j.jece.2022.108249
50. Z. Hu, X. Li, P.L. Hernandez-Martinez, S. Zhang, P. Gu, W. Du, W. Xu, H.V. Demir, H. Liu, Q. Xiong, Interfacial charge and energy transfer in van der Waals heterojunctions. InfoMat**4**(3) (2022). https://doi.org/10.1002/inf2.12290
51. S. Xue, W. Huang, W. Lin, W. Xing, M. Shen, X. Ye, X. Liang, C. Yang, Y. Hou, Z. Yu, X. Wang, Interfacial engineering of lattice coherency at ZnO–ZnS photocatalytic heterojunctions. Chem. Catal. **2**(1), 125–139 (2022). https://doi.org/10.1016/j.checat.2021.11.019
52. C. Tong, L. Li, L.M. Liu, O.V. Prezhdo, Synergy between ion migration and charge carrier recombination in metal-halide perovskites. J. Am. Chem. Soc. **142**(6), 3060–3068 (2020). https://doi.org/10.1021/jacs.9b12391
53. L. Schumacher, R. Marschall, Recent advances in semiconductor heterojunctions and Z-Schemes for photocatalytic hydrogen generation. Top. Curr. Chem. **380**(6) (2022). https://doi.org/10.1007/s41061-022-00406-5
54. N. Serpone, E. Borgarello, M. Grätzel, Visible light induced generation of hydrogen from H2S in mixed semiconductor dispersions; improved efficiency through inter-particle electron transfer. J. Chem. Soc. **6**, 342–344 (1984). https://doi.org/10.1039/c39840000342
55. R. Mishra, S. Bera, R. Chatterjee, S. Banerjee, S. Bhattacharya, A. Biswas, S. Mallick, S. Roy, A review on Z/S-scheme heterojunction for photocatalytic applications based on metal halide perovskite materials. Appl. Surf. Sci. Adv. **9**, 100241 (2022). https://doi.org/10.1016/j.apsadv.2022.100241
56. J. Low, J. Yu, M. Jaroniec, S. Wageh, A.A. Al-Ghamdi, Heterojunction photocatalysts. Adv. Mater. **29**(20), 1601694 (2017). https://doi.org/10.1002/adma.201601694
57. J. Low, C. Jiang, B. Cheng, S. Wageh, A.A. Al-Ghamdi, Y. Wang, A review of direct Z-scheme photocatalysts. Small Methods **1**(5), 1700080 (2017). https://doi.org/10.1002/smtd.201700080
58. A.J. Bard, Photoelectrochemistry and heterogeneous photo-catalysis at semiconductors. J. Photochem. Photobiol. **10**(1), 59–75 (1979). https://doi.org/10.1016/0047-2670(79)80037-4
59. G. Ma, S. Chen, Y. Kuang, S. Akiyama, T. Hisatomi, M. Nakabayashi, N. Shibata, M. Katayama, T. Minegishi, K. Domen, Visible light-driven Z-scheme water splitting using oxysulfide H_2 evolution photocatalysts. J. Phys. Chem. Lett. **7**(19), 3892–3896 (2016). https://doi.org/10.1021/acs.jpclett.6b01802
60. M. Xia, X. Zhao, Y. Zhang, W. Pan, D.Y. Leung, Rational catalyst design for spatial separation of charge carriers in a multi-component photocatalyst for effective hydrogen evolution. J.

Mater. Chem. A, Mater. Energy Sustain. **10**(48), 25380–25405 (2022). https://doi.org/10.1039/d2ta06609b
61. F.M.D. Chequer, G.A.R. De Oliveira, E.R.A. Ferraz, J. Cardoso, M.V.B. Zanoni, D.P. De Oliveira, Textile dyes: dyeing process and environmental impact, In *InTech eBooks* (2013). https://doi.org/10.5772/53659
62. M. Batool, M.F. Nazar, A. Awan, M.B. Tahir, A. Rahdar, A.E. Shalan, S. Lanceros-Méndez, M.N. Zafar, Bismuth-based heterojunction nanocomposites for photocatalysis and heavy metal detection applications. Nano-Struct. Nano-Objects **27**, 100762 (2021). https://doi.org/10.1016/j.nanoso.2021.100762
63. Z. Zhang, W. Wang, L. Wang, S. Sun, Enhancement of visible-light photocatalysis by coupling with narrow-band-gap semiconductor: a case study on Bi_2S_3/Bi_2WO_6. ACS Appl. Mater. Interfaces **4**(2), 593–597 (2012). https://doi.org/10.1021/am2017199
64. H. Yu, R. Liu, X. Wang, P. Wang, J. Yu, Enhanced visible-light photocatalytic activity of Bi_2WO_6 nanoparticles by Ag_2O cocatalyst. Appl. Catal. B-Environ. **111–112**, 326–333 (2012). https://doi.org/10.1016/j.apcatb.2011.10.015
65. Z. Zhang, W. Wang, E. Gao, M. Shang, J. Xu, Enhanced photocatalytic activity of Bi_2WO_6 with oxygen vacancies by zirconium doping. J. Hazard. Mater. **196**, 255–262 (2011). https://doi.org/10.1016/j.jhazmat.2011.09.017
66. Y. Wang, X. Bai, C. Pan, J. He, Y. Zhu, Enhancement of photocatalytic activity of Bi_2WO_6 hybridized with graphite-like C_3N_4. J. Mater. Chem. **22**(23), 11568 (2012). https://doi.org/10.1039/c2jm16873a
67. L. Ge, J. Liu, Efficient visible light-induced photocatalytic degradation of methyl orange by QDs sensitized CdS-Bi_2WO_6. Appl. Catal. B-Environ. **105**(3–4), 289–297 (2011). https://doi.org/10.1016/j.apcatb.2011.04.016
68. M. Tasviri, N. Miri, N. Mohaghegh, Apatite-coated Ag/AgBr/Bi_2WO_6 nanocomposite: synthesis, characterization, and application as an efficient visible-light-driven photocatalyst. Res. Chem. Intermed. **45**(3), 1039–1055 (2018). https://doi.org/10.1007/s11164-018-3660-1
69. J. Li, Y. Zhao, M. Xia, H. An, H. Bai, J. Wei, B. Yang, G. Yang, Highly efficient charge transfer at 2D/2D layered P-$La_2Ti_2O_7/Bi_2WO_6$ contact heterojunctions for upgraded visible-light-driven photocatalysis. Appl. Catal. B-Environ. **261**, 118244 (2020). https://doi.org/10.1016/j.apcatb.2019.118244
70. Z. Jia, F. Lyu, L. Zhang, S. Zeng, S. Liang, Y.Y. Li, Lu, Pt nanoparticles decorated heterostructured g-C_3N_4/Bi_2MoO_6 microplates with highly enhanced photocatalytic activities under visible light. Sci. Rep. **9**(1) (2019). https://doi.org/10.1038/s41598-019-42973-6
71. P. Wang, Y. Ao, C. Wang, J. Hou, J. Qian, A one-pot method for the preparation of graphene–Bi_2MoO_6 hybrid photocatalysts that are responsive to visible-light and have excellent photocatalytic activity in the degradation of organic pollutants. Carbon **50**(14), 5256–5264 (2012). https://doi.org/10.1016/j.carbon.2012.06.063
72. G. Tian, Y. Chen, X. Meng, J. Zhou, W. Zhou, K. Pan, C. Tian, Z. Ren, H. Fu, Hierarchical composite of Ag/AgBr nanoparticles supported on Bi_2MoO_6 hollow spheres for enhanced visible-light photocatalytic performance. Collect. Czech. Chem. Commun. **78**(1), 117–123 (2012). https://doi.org/10.1002/cplu.201200198
73. T. Feng, H. Yin, H. Jiang, X. Chai, X. Li, D. Li, J. Wu, X. Liu, B. Sun, Design and fabrication of polyaniline/Bi_2MoO_6 nanocomposites for enhanced visible-light-driven photocatalysis. New J. Chem. **43**(24), 9606–9613 (2019). https://doi.org/10.1039/c9nj01651a
74. Q. Rong, D. Zhang, Y. Liu, Z. Zha, X. Geng, C. Shen, J. Yang, Synthesis of $Bi_2MoO_6/Bi_2Ti_2O_7$ Z-scheme heterojunction as efficient visible-light photocatalyst for the glycolic acid degradation. J. Nanosci. Nanotechnol. **19**(12), 7635–7644 (2019). https://doi.org/10.1166/jnn.2019.16779
75. Y. Peng, Y. Zhang, F. Tian, J. Zhang, J. Yu, Structure tuning of Bi_2MoO_6 and their enhanced visible light photocatalytic performances. Crit. Rev. Solid State Mater. Sci. **42**(5), 347–372 (2017). https://doi.org/10.1080/10408436.2016.1200009
76. C. Yu, P. Yang, L. Tie, S. Yang, S. Dong, J. Sun, J. Sun, One-pot fabrication of β-Bi_2O_3@Bi_2S_3 hierarchical hollow spheres with advanced sunlight photocatalytic RhB oxidation and Cr(VI)

reduction activities. Appl. Surf. Sci. **455**, 8–17 (2018). https://doi.org/10.1016/j.apsusc.2018.04.201
77. B. Priya, P. Raizada, N. Singh, P. Thakur, P. Singh, Adsorptional photocatalytic mineralization of oxytetracycline and ampicillin antibiotics using Bi_2O_3/BiOCl supported on graphene sand composite and chitosan. J. Colloid Interface Sci. **479**, 271–283 (2016). https://doi.org/10.1016/j.jcis.2016.06.067
78. S. Wu, H. Zheng, Y. Lian, Y. Wu, Preparation, characterization and enhanced visible-light photocatalytic activities of $BiPO_4$/$BiVO_4$ composites. Mater. Res. Bull. **48**(8), 2901–2907 (2013). https://doi.org/10.1016/j.materresbull.2013.04.041
79. X. Zhang, Y. Gong, X. Dong, X. Zhang, M. Chun, F. Shi, Fabrication and efficient visible light-induced photocatalytic activity of Bi_2WO_6/$BiVO_4$ heterojunction. Mater. Chem. Phys. **136**(2–3), 472–476 (2012). https://doi.org/10.1016/j.matchemphys.2012.07.013
80. H. Huang, L. Liu, N. Tian, Novel $BiIO_4$/$BiVO_4$ composite photocatalyst with highly improved visible-light-induced photocatalytic performance for rhodamine B degradation and photocurrent generation. RSC Adv. **5**(2), 1161–1167 (2015). https://doi.org/10.1039/c4ra12916d
81. T. Fan, C. Chen, Z. Tang, Hydrothermal synthesis of novel $BiFeO_3$/$BiVO_4$ heterojunctions with enhanced photocatalytic activities under visible light irradiation. RSC Adv. **6**(12), 9994–10000 (2016). https://doi.org/10.1039/c5ra26500b
82. Y. Wang, K. Ding, R. Xu, D. Yu, W. Wang, P. Gao, B. Liu, Fabrication of $BiVO_4$/$BiPO_4$/GO composite photocatalytic material for the visible light-driven degradation. J. Clean. Prod. **247**, 119108 (2020). https://doi.org/10.1016/j.jclepro.2019.119108
83. H. Sun, C. Zou, W. Tang, Designing double Z-scheme heterojunction of g-C_3N_4/Bi_2MoO_6/Bi_2WO_6 for efficient visible-light photocatalysis of organic pollutants. Colloids Surf. A **654**, 130105 (2022). https://doi.org/10.1016/j.colsurfa.2022.130105
84. N.Đ Phú, L.H. Hoàng, P.V. Hai, T.Q. Huy, X. Chen, W.C. Chou, Photocatalytic activity enhancement of Bi_2WO_6 nanoparticles by Ag doping and Ag nanoparticles modification. J. Alloy. Compd. **824**, 153914 (2020). https://doi.org/10.1016/j.jallcom.2020.153914
85. Z. Jiang, X. Tan, J. Xu, Y. Huang, Piezoelectric-induced internal electric field in Bi_2WO_6 nanoplates for boosting the photocatalytic degradation of organic pollutants. ACS Appl. Nano Mater. **5**(5), 7588–7597 (2022). https://doi.org/10.1021/acsanm.2c01698
86. H. Huang, R. Cao, Y. Shen, K. Xu, W. Hao, Y. Wang, F. Dong, T. Zhang, Single-unit-cell layer established Bi_2WO_6 3D hierarchical architectures: Efficient adsorption, photocatalysis and dye-sensitized photoelectrochemical performance. Appl. Catal. B-Environ. **219**, 526–537 (2017). https://doi.org/10.1016/j.apcatb.2017.07.084
87. B.Y. Alfaifi, H. Bayahia, A.A. Tahir, Highly efficient nanostructured Bi_2WO_6 thin film electrodes for photoelectrochemical and environment remediation. Nanomaterials **9**(5), 755 (2019). https://doi.org/10.3390/nano9050755
88. M. Karbasi, F. Karimzadeh, K. Raeissi, S. Rtimi, J. Kiwi, S. Giannakis, C. Pulgarín, Insights into the photocatalytic bacterial inactivation by flower-like Bi_2WO_6 under solar or visible light, through in situ monitoring and determination of reactive oxygen species (ROS). Water **12**(4), 1099 (2020). https://doi.org/10.3390/w12041099
89. T. Wang, S. Liu, W. Mao, Y. Bai, K.S. Chiang, K. Shah, J. Paz-Ferreiro, Novel Bi_2WO_6 loaded N-biochar composites with enhanced photocatalytic degradation of rhodamine B and Cr(VI). J. Hazard. Mater. **389**, 121827 (2020). https://doi.org/10.1016/j.jhazmat.2019.121827
90. Y. Su, G. Tan, T. Liu, L. Lv, Y. Wang, X. Zhang, Z. Yue, H. Ren, A. Xia, Photocatalytic properties of Bi_2WO_6/$BiPO_4$ Z-scheme photocatalysts induced by double internal electric fields. Appl. Surf. Sci. **457**, 104–114 (2018). https://doi.org/10.1016/j.apsusc.2018.06.153
91. S. Yin, Y. Chen, Q. Hu, M. Li, Y. Ding, J. Di, J. Xia, Construction of NH_2-MIL-125(Ti) nanoplates modified Bi_2WO_6 microspheres with boosted visible-light photocatalytic activity. Res. Chem. Intermed. **46**(7), 3311–3326 (2020). https://doi.org/10.1007/s11164-020-04132-9
92. L. Cai, J. Gong, J. Liu, H. Zhang, W. Song, L. Ji, Facile preparation of nano-Bi_2MoO_6/diatomite composite for enhancing photocatalytic performance under visible light irradiation. Materials **11**(2), 267 (2018). https://doi.org/10.3390/ma11020267

93. S. Balasurya, A. Das, A.A. Alyousef, A. Alqasim, N. Almutairi, S.S. Khan, Facile synthesis of Bi_2MoO_6–Ag_2MoO_4 nanocomposite for the enhanced visible light photocatalytic removal of methylene blue and its antimicrobial application. J. Mol. Liq. **337**, 116350 (2021). https://doi.org/10.1016/j.molliq.2021.116350
94. A.M. Elgorban, A.A.A. Kheraif, A. Syed, Construction of Ag_2WO_4 decorated $CoWO_4$ nano-heterojunction with recombination delay for enhanced visible light photocatalytic performance and its antibacterial applications. Colloids Surf. A: Physicochem. Eng. Aspects **629**, 127416. https://doi.org/10.1016/j.colsurfa.2021.127416
95. K. Kusuma, M. Manju, C. Ravikumar, V. Dileepkumar, A. Kumar, A.N. Santosh, H.C.A. Murthy, K. Gurushantha, Probe Sonicated synthesis of bismuth oxide (Bi_2O_3): photocatalytic application and electrochemical sensing of ascorbic acid and lead. J. Nanomater. **2022**, 1–13 (2022). https://doi.org/10.1155/2022/3256611
96. Y. Li, J. Zhong, J. Li, S. Huang, S. Zhang, H. Yang, L. Ma, Enhanced visible light-driven photocatalytic destruction of decontaminants over $Bi_2O_3/BiVO_4$ heterojunctions with rich oxygen vacancies. Chem. Phys. Lett. **801**, 139722 (2022). https://doi.org/10.1016/j.cplett.2022.139722
97. X. Yang, S. Sun, Z. Shi, D. Yun, Y. Guo, C. Liu, B. Yang, M. Yang, Q. Yang, J. Cui, One-pot construction of highly efficient TaON/Bi_2O_3/S–BiOCl ternary photocatalysts: simultaneously integrating type-I with Z-scheme junctions for improved visible light-driven removal of organic pollutants. Chemosphere **307**, 135979 (2022). https://doi.org/10.1016/j.chemosphere.2022.135979
98. H. Peng, R. Guo, H. Lin, X. Liu, Synthesis of Bi_2O_3/g-C_3N_4 for enhanced photocatalytic CO_2 reduction with a Z-scheme mechanism. RSC Adv. **9**(64), 37162–37170 (2019). https://doi.org/10.1039/c9ra07485f
99. M. Gui, W. Zhang, S. Qingxi, C. Chen, Preparation and visible light photocatalytic activity of Bi_2O_3/Bi_2WO_6 heterojunction photocatalysts. J. Solid State Chem. **184**(8), 1977–1982 (2011). https://doi.org/10.1016/j.jssc.2011.05.057
100. Q. Que, Y. Xing, Z. He, Y. Yang, X. Yin, W. Que, Bi_2O_3/Carbon quantum dots heterostructured photocatalysts with enhanced photocatalytic activity. Mater. Lett. **209**, 220–223 (2017). https://doi.org/10.1016/j.matlet.2017.07.115
101. J. Ke, J. Liu, H. Sun, H. Zhang, X. Duan, P. Liang, X. Li, S. Liu, S. Wang, Facile assembly of $Bi_2O_3/Bi_2S_3/MoS_2$ n-p heterojunction with layered n-Bi_2O_3 and p-MoS_2 for enhanced photocatalytic water oxidation and pollutant degradation. Appl. Catal. B-Environ. **200**, 47–55 (2017). https://doi.org/10.1016/j.apcatb.2016.06.071
102. R. He, H. Liu, H. Liu, D. Xu, L. Zhang, S-scheme photocatalyst Bi_2O_3/TiO_2 nanofiber with improved photocatalytic performance. J. Mater. Sci. Technol. **52**, 145–151 (2020). https://doi.org/10.1016/j.jmst.2020.03.027
103. H. Lü, Q. Hao, T. Chen, L. Zhang, D. Chen, C. Ma, W. Yao, Y. Zhu, A high-performance Bi2O3/Bi2SiO5 p–n heterojunction photocatalyst induced by phase transition of Bi2O3. Appl. Catal. B-Environ. **237**, 59–67 (2018). https://doi.org/10.1016/j.apcatb.2018.05
104. X. Li, J. Chen, H. Li, J. Li, Y. Xu, Y. Liu, J. Zhou, Photoreduction of CO2 to methanol over Bi2S3/CdS photocatalyst under visible light irradiation. J. Nat. Gas Chem. **20**(4), 413–417 (2011). https://doi.org/10.1016/s1003-9953(10)60212-5
105. X. Wu, C.Y. Toe, C. Su, Y.H. Ng, R. Amal, J. Scott, Preparation of Bi-based photocatalysts in the form of powdered particles and thin films: a review. J. Mater. Chemistry. A. Mater. Energy Sustain. **8**(31), 15302–15318 (2020). https://doi.org/10.1039/d0ta01180k
106. A. Helal, S.M. El-Sheikh, J. Yu, A.I. Eid, S.A. El-Haka, S. Samra, Novel synthesis of $BiVO_4$ using homogeneous precipitation and its enhanced photocatalytic activity. J. Nanopart. Res. **22**(6) (2020). https://doi.org/10.1007/s11051-020-04861-3
107. L. Yosefi, M. Haghighi, Fabrication of nanostructured flowerlike p-BiOI/p-NiO heterostructure and its efficient photocatalytic performance in water treatment under visible-light irradiation. Appl. Catal. B-Environ. **220**, 367–378 (2018). https://doi.org/10.1016/j.apcatb.2017.08.028

108. P. Ju, P. Wang, B. Li, H. Fan, S. Ai, D. Zhang, Y. Wang, A novel calcined Bi_2WO_6 /$BiVO_4$ heterojunction photocatalyst with highly enhanced photocatalytic activity. Chem. Eng. J. **236**, 430–437 (2014). https://doi.org/10.1016/j.cej.2013.10.00
109. X. Quan, W. He, S. Liu, C. Li, J. Zhang, P.K. Wong, Bifunctional S-scheme g-C_3N_4/Bi/$BiVO_4$ hybrid photocatalysts toward artificial carbon cycling. Chin. J. Catal. **41**(1), 140–153 (2020). https://doi.org/10.1016/s1872-2067(19)63481-9
110. Y. Li, H. Jiang, X. Wang, X. Hong, B. Liang, Recent advances in bismuth oxyhalide photocatalysts for degradation of organic pollutants in wastewater. RSC Adv. **11**(43), 26855–26875 (2021). https://doi.org/10.1039/d1ra05796k
111. J. Lv, Q. Hu, C. Cao, Y. Zhao, Modulation of valence band maximum edge and photocatalytic activity of BiOX by incorporation of halides. Chemosphere **191**, 427–437 (2018). https://doi.org/10.1016/j.chemosphere.2017.09.149
112. K. Sharma, V. Dutta, S. Sharma, P. Raizada, A. Hosseini-Bandegharaei, P. Thakur, P. Singh, Recent advances in enhanced photocatalytic activity of bismuth oxyhalides for efficient photocatalysis of organic pollutants in water: a review. J. Ind. Eng. Chem. **78**, 1–20 (2019). https://doi.org/10.1016/j.jiec.2019.06.022
113. Y. Long, Q. Han, Z. Yang, Y. Ai, S. Sun, Y. Wang, Q. Liang, A novel solvent-free strategy for the synthesis of bismuth oxyhalides. J. Mater. Chem. A, Mater. Energy Sustain. **6**(27), 13005–13011 (2018). https://doi.org/10.1039/c8ta04529a
114. S. Wang, L. Wang, W. Huang, Bismuth-based photocatalysts for solar energy conversion. J. Mater. Chem. A, Mater. Energy Sustain. **8**(46), 24307–24352 (2020). https://doi.org/10.1039/d0ta09729b
115. M. Arumugam, M.Y. Choi, Recent progress on bismuth oxyiodide (BiOI) photocatalyst for environmental remediation. J. Ind. Eng. Chem. **81**, 237–268 (2020). https://doi.org/10.1016/j.jiec.2019.09.013
116. J.C. Durán-Álvarez, C. Martínez-Avelar, E. González-Cervantes, R.A. Gutiérrez-Márquez, M. Rodríguez-Varela, A.S. Varela, F.F. Castillón, R. Zanella, Degradation and mineralization of oxytetracycline in pure and tap water under visible light irradiation using bismuth oxyiodides and the effect of depositing Au nanoparticles. J. Photochem. Photobiol. A **388**, 112163 (2020). https://doi.org/10.1016/j.jphotochem.2019.112163
117. L. Kong, Z. Jiang, H.H. Lai, R.J. Nicholls, T. Xiao, M.O. Jones, P.P. Edwards, Unusual reactivity of visible-light-responsive AgBr–BiOBr heterojunction photocatalysts. J. Catal. **293**, 116–125 (2012). https://doi.org/10.1016/j.jcat.2012.06.011
118. X. Meng, Bismuth-based photocatalytic semiconductors: introduction, challenges and possible approaches. J. Mol. Catal. A-Chem. **423**, 533–549 (2016). https://doi.org/10.1016/j.molcata.2016.07.030
119. J. Zhang, Z. Liu, Z. Ma, Facile formation of $Bi_2O_2CO_3$/Bi_2MoO_6 nanosheets for visible light-driven photocatalysis. ACS Omega **4**(2), 3871–3880 (2019). https://doi.org/10.1021/acsomega.8b03699

Recent Advancements in Bismuth Oxyhalide Photocatalysis for Energy and Environmental Applications

Chandran Ashina, Loganathan Muruganandam, Ramalinga V. Mangalaraja, Chandrasekaran Rajasekaran, and Panneerselvam Sathishkumar

Abstract Energy crisis and environmental pollution are universal challenges affecting people and the environment. One effective strategy to deal with this global issue is the photocatalytic approach. Among them bismuth oxyhalides (BiOX) are a family of catalysts which show excellent solar energy conversion. The materials have a unique layered structure and have an inbuilt electric field. Moreover this material has high stability and electrical conductivity. These characteristics give the material more controllable qualities to boost efficiency. Many strategies like doping, defect engineering, facet exposure, solid solution preparation, etc., are employed for the enhancement of photocatalysis. Similarly, structural modifications like perovskite structure and, introduction of support can also enhance the photocatalytic property of the material. The bismuth oxyhalides are widely applied in many energy and environmental applications like CO_2 conversion, nitrogen fixation, water splitting, degradation of many organic and volatile compounds in air as well as water. This chapter discusses with recent modifications employed over BiOX and its application towards energy and environmental remediation.

C. Ashina · P. Sathishkumar (✉)
Department of Chemistry, School of Advanced Sciences, Vellore Institute of Technology (VIT), Vellore, India
e-mail: sathishkumar.p@vit.ac.in

L. Muruganandam
Department of Chemistry, Saranathan College of Engineering, Tiruchirapalli 620012, India

R. V. Mangalaraja
Faculty of Engineering and Sciences, Universidad Adolfo Ibáñez, Diagonal Las Torres, 2640 Peñalolén, Santiago, Chile

Instituto Universitario de Investigación y Desarrollo Tecnológico (IDT), Universidad Tecnológica Metropolitana (UTEM), Ignacio Valdivieso, 2409 San Joaquín, Santiago, Chile

C. Rajasekaran
Department of Biotechnology, School of Bioscience and Technology, Vellore Institute of Technology (VIT), Vellore, India

P. Sathishkumar (ed.), *Photocatalysis for Energy and Environmental Applications*, Green Energy and Technology, https://doi.org/10.1007/978-981-97-1939-6_8

Keywords Bismuth oxyhalides · Photocatalysis · Energy conversion · Environmental remediation

1 Introduction

Photocatalysis is a promising method for energy and environmental applications; TiO_2 is a traditional photocatalyst and is extensively investigated for UV light-irradiated photocatalysis. On keeping TiO_2 as a benchmark of several photocatalysts have been synthesized, among them bismuth oxyhalides are layered ternary photocatalysts with visible light absorption from V, VI and VII groups [1]. In bismuth oxyhalides the [X–Bi–O–Bi–X] layers are stacked with van der Waals forces [2]. Because of their unique structure bismuth oxyhalides develop an internal electric field, a peculiar property of bismuth oxyhalides which encourages the charge separation and mobility of ions from bulk to surface [3]. Small modifications in exposed facets and introduction of impurity can significantly alter the electric field and make modification in the photocatalyst and its efficiency [4]. Therefore, the BiOX nanophotocatalysts are employed in several applications including CO_2 reduction, heavy metal removal etc. The photocatalytic degradation of organic pollutant is one of the established application of bismuth oxyhalide photocatalyst and, several investigations have been reported [5]. Herein the chapter will discuss the recent advancements transpired in the bismuth oxyhalides as catalysts for the CO_2 conversion [6], nitrogen fixation [7], heavy metal ion reduction [8], air purification [9], organic pollutant degradation [5] and hydrogen generation [10].

1.1 CO_2 Conversion

The "Anthropogenic Greenhouse Effect" is a serious environmental problem and refers to the increased atmospheric CO_2 concentration caused by the burning of fossil fuels in stationary and mobile sources. Solar fuels are a diverse class of chemicals that can be applied to industrial, transportation, and power generation processes. The two primary types of solar fuels are hydrogen and carbon-based fuels like methanol (CH_3OH), carbon monoxide (CO), and methane (CH_4). So photocatalytic CO_2 conversion is a better green method for production of CH_3OH, CO and CH_4 [6]. The mechanisms involved are shown in Eqs. 1 and 2. The neutral cycle of carbon is shown in Scheme 1.

$$h\nu + 2H_2O \rightarrow 2H_2 + O_2 \tag{1}$$

$$h\nu + H_2 + CO_2 \rightarrow Fuel + O_2 \tag{2}$$

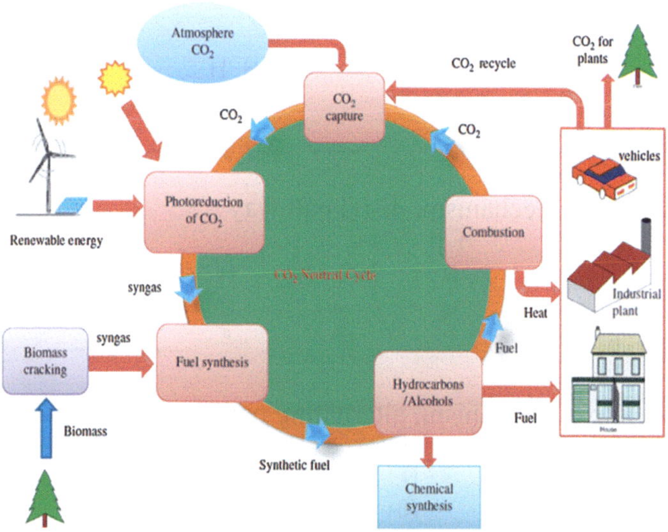

Scheme 1 Neutral cycle of CO_2 [6]. Reprinted with permission from Ref. [6]. Copyrights (2017) Elsevier Publications

1.2 Nitrogen Fixation

Production of ammonia through Haber- Bosch process consumes large amount of energy and are not eco-friendly, but biological nitrogen fixation are eco-friendly used by plants and not for industrial use [11]. During the artificial reduction of N_2 the solubility of N_2 creates hinderance. With the ionic liquids like [C_4mpyr][eFAP] the solubility can be increased but is highly expensive and is toxic to the environment [12]. Photocatalytic method is an eco-friendly method for nitrogen fixation and the mechanism pathway is schematically illustrated in Scheme 2. Nitrogen fixation is usually followed by the adsorption of N_2 over the surface of the catalyst but when compared with other bismuth based photocatalyst like Bi_2WO_6 adsorption of N_2 over the surface of bismuth oxyhalides are difficult. So bare bismuth oxyhalides shows only less efficiency or no conversion but when oxygen vacancies are introduced it shows a better conversion efficiency as it acts as an electron trapping agent.

1.3 Heavy Metal Ion Reduction

Metals like chromium, uranium, selenium etc. are found in large excess in natural water, and the presence of these metals implies carcinogenicity and toxicity to the environment. Conventionally the metals are removed by precipitation, ion exchange, co-precipitation methods etc. [13]. Then the hexavalent nature of chromium and the solubility of uranium (soluble U(VI) converts into insoluble U(IV)) hinders the

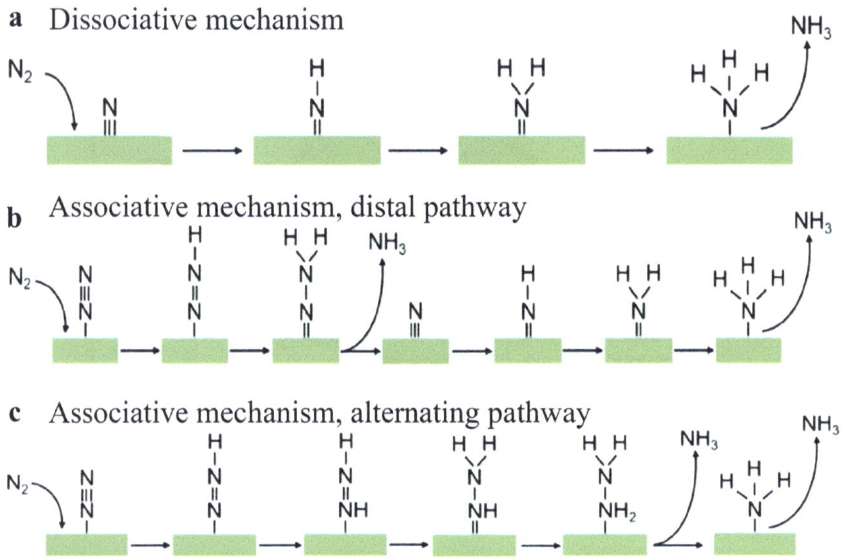

Scheme 2 The pathways of nitrogen fixation. Reprinted with permission from Ref. [12]. Copyrights (2014) Royal Society of Chemistry

efficient removal of the metals through conventional methods. Photocatalysis has shown a breakthrough in this process. Along with the photoreduction of metals in water, photocatalysis methods also oxidizes the organic contaminants present in water.

1.4 Air Purification

Volatile organic compounds (VOCs) are toxic in nature and create threat to the environment, especially indoor air, where the toxic indoor air may cause sick building syndrome (SBS) so it is important to remove VOCs from air. Catalytic oxidation, direct oxidation, combustion etc. were employed from the past decades for removal. An efficient and eco-friendly method has not been reported. Thus, the photocatalytic method reported a better degradation efficiency and is also an eco-friendly strategy.

1.5 Organic Pollutant Degradation

Most of the waterbodies are highly contaminated with various kinds of organic pollutants like drugs, phenol components, dyes etc. which are non-biodegradable and these are considered as emerging contaminants. Removal of these contaminants with the conventional water treatment method is not efficient and the visible light photocatalytic method showed a promising approach [14]. So several studies have been reported and the recent advancement achieved in organic contaminant degradation with BiOX catalyst are discussed in the following sections.

1.6 Hydrogen Generation

The major source of energy consumed by humans are from non-renewable fossil fuels which are on the verge of extinction and more over the expelling of CO_2 to the environment increases global warming [10]. So there is an urgent need to shift toward renewable source of energy like H_2 fuels and water splitting is a better method for H_2 production. Here also the photocatalytic method makes the process more green and efficient by absorbing the visible light. The basic mechanism for water splitting is shown in Eqs. (3–5).

$$H_2O(l) \rightarrow H_2(g) + \frac{1}{2}O_2(g) \tag{3}$$

$$H_2O + 2H^+ \rightarrow \frac{1}{2}O_2 + 4H^+ \tag{4}$$

$$2H^+ + 2e \rightarrow H_2 \tag{5}$$

The recent updates in photocatalytic water splitting with BiOX catalyst is summarized in the upcoming sections [15].

2 Bismuth Oxyhalides

Bismuth oxyhalides are generally represented as BiOX where X = Cl, Br, and I. These are P-type semiconductors with a band gap ranging from 3.2 to 1.7 eV which absorbs UV and visible light. As it has a layered structure there develops an internal electric field which helps in reducing the recombination effect. This material has a selective oxidation capacity as well. Various synthesis methods like Hydrothermal/Solvothermal, Calcination, Ionic liquid microemulsion can be employed for the successful synthesis of BiOX. Several modifications have been introduced to improve the catalytic property and stability of the material. The upcoming sections describe the various modifications applied to BiOCl, BiOBr and BiOI and their applications on various energy and environmental applications.

2.1 BiOCl and Modified BiOCl

BiOCl has an indirect band gap which ranges from 3.2 to 3.5 eV with a tetragonal malockite structure. The band gap response to UV absorption is in order to enhance its absorption to the visible region and also to improve the catalytic activities many modifications are carried out.

Lattice structure modification in BiOCl (BOC) was employed by Guan et al. who investigated the reduction of CO_2. For this, different concentrations of thiourea were added as BiOCl precursors. Addition of 0 mg of thiourea is denoted as BOC, 25 mg of thiourea as BOC-S1, 50 mg as BOC-S2, 100 mg as BOC-S3 and the formation of Bi–O bond with shrinkage was reported. The morphological change observed in bare BiOCl and modified BiOCl are the closed structures and open structures respectively, and also the surface area was increased from 0 to 100 mg on thiourea addition, where BOC-S3 is 45.327 cm^2/g [16]. The band gap energy remained unchanged despite the addition of thiourea. The absorbance peak and DRS plot is shown in Fig. 1a and b. So the improved efficiency when compared with BOC may be attributed to the shorter Bi-O bond due to heterojunction formation which reduces the recombination and enhances the charge transportation. In this investigation only one application is investigated but some multipollutant degradation can also be investigated with the structure modified catalyst. For instance Maimaitzi et al. [17] fabricated bare BiOCl, chlorophyll doped BiOCl (Chl/BiOCl) and chlorophyll Co-doped Mg BiOCl (Chl–Mg/BiOCl) and investigated the efficiency of the catalyst by reducing CO_2 and degrading ciprofloxacin [17]. There is a slight difference in the morphological structure of BiOCl and Chl–Mg/BiOCl when evaluated through SEM. The BiOCl have self-assembled hierarchical spheres whereas Chl–Mg/BiOCl has agglomerated particles which destroy the hierarchial structure. When comparing the optical properties BiOCl absorbs the UV light whereas modified BiOCl absorbs visible light. As mentioned above the author applied this set of photocatalyst for CO_2 reduction and antibiotic (ciprofloxacin) degradation. In these investigations the author confirmed that the efficiency and stability is in the following order (Chl–Mg/BiOCl) > (Chl/BiOCl) > BiOCl. The reason for increased efficiency in (Chl/BiOCl) than BiOCl may be due to chlorophyll acting as a photosensitizer whereas when comparing Chl-Mg/BiOCl and Chl/BiOCl the increased efficiency may be attributed to the electron trapping ability of Mg which reduces the recombination and uplifts the efficiency [17].

Regarding N_2 fixation, oxygen vacancy induced Bi rich $Bi_{12}O_{17}Cl_2$ (BOC-OV) were synthesized and NH_3 fixation was investigated. Even though the modifications hadn't reported an appreciable change in morphology, an increasing surface area was reported for the BOC-OV. Similarly the optical band gap energy is reduced from 2.47 to 2.09 eV. It enhances the visible light absorption, furthermore the charge recombination rate was observed by PL and reported a decrease in intensity in BOC-OV supporting the reduction of the recombination rate. The reasons for decreased rate may be due to the electron trapping capacity of oxygen vacancies, thereby enhancing the photocatalytic activity of BOC-OV than BOC. The reported yield of NH_3 conversion rate is 23.43 m mol g^{-1} h^{-1} in 3 h of solar light irradiation

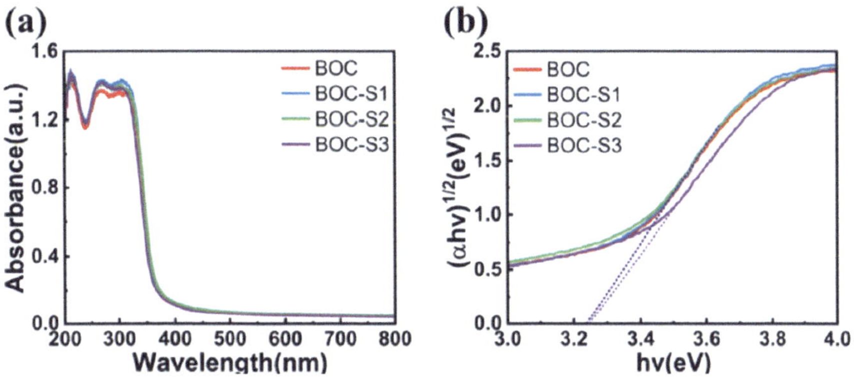

Fig. 1 **a** Absorbance peak. **b** DRS peak of BOC and BOC S. Reprinted with permission from Ref. [16]. Copyrights (2023) Elsevier Publications

[18]. In recent years modified BiOCl was investigated for air pollution, for instance, in order to remove gaseous toluene BiOCl/Bi_2WO_6 heterojuction was constructed. The optical properties of constructed heterojunction reported a band gap energy of 2.49 eV which is less than bare BiOCl and Bi_2WO_6. For further studies on valence band and conduction band XPS technique was analysed and the E_{VBXPS} was reported as 2.49 which is in between the E_{VB} of BiOCl and Bi_2WO_6 proving the successful synthesis of the heterojunction. The CB and VB potential of all syntheiszed catalysts are shown in Fig. 2. Similarly the time resolved spectroscopy reported a decrease in the recombination rate. The valence band potential and reduced recombination rate synergestically increases the photocatalytic activity leading to a removal efficiency of 99.73% in 180 min of simulated sunlight which is more than Bi_2WO_6 (21.4%) and BiOCl(16.38%) [19]. When exploring the other application of modified BiOCl the organic contaminant degradation is the one which is extensively reported from the last 3 years as a single pollutant as well as a multi pollutant.

Doping of W with BiOCl has changed the optical property of BiOCl by inducing an extra energy level of W in BiOCl structure and reducing the band gap energy from 3.29 to 3.00 eV. The efficiency of the catalyst was tested by the degradation of Rh-B. The bare BiOCl shows degradation in UV but is negligible in the visible region. But when it is doped with W the degradation efficiency in the visible region was reported as >90% in 180 min of visible light irradiation [20]. Apart from doping a minor change in the synthesis method can also influence the photocatalytic activity. For example Narenuch et al. [21] synthesized BiOCl by adding surfactants (SDS or CTAB) in the solvothermal method. The addition of the surfactant affects the activity of the photocatalyst by influencing the morphology and active sites. A complete degradation of Rhodamine- B was reported for SDS capped BiOCl in 40 min irradiation of sunlight, whereas 98% of degradation was reported for CTAB capped BiOCl. The increased efficiency in SDS capped BiOCl is due to the polar nature of the SDS head. Similarly the author investigated the degradation of antibiotics like ofloxacin, norfloxacin, and ciprofloxacin [21]. In another study, the band gap engineering was done by adjusting

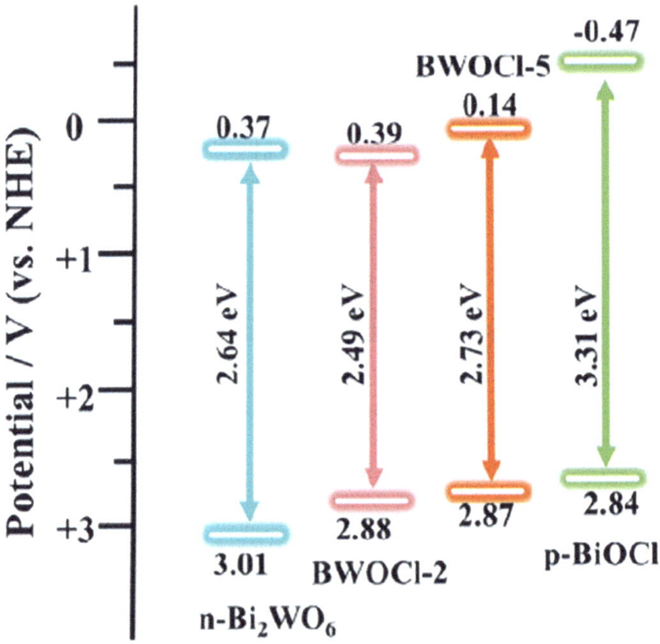

Fig. 2 The conduction and valence band potential of BiOCl/Bi_2WO_6. Reprinted with permission from Ref. [19]. Copyrights (2023) Elsevier Publications

the ratio of the solvents. In a solvothermal synthesis, the solvents ethylene glycol (EG) and ethyl alcohol (EA) are varied and fabricated with BiOCl. The highest photocatalytic activity reported for Rhodamine-B degradation is BOC-EG: 2Ea (99% in 10 min irradiation) whereas the other ratios reported are 36%, 84%, 94%, and 23% for BOC-EG, BOC-2EG:EA, BOC-EG:EA and BOC-EA respectively. This increased efficiency for BOC-EG: 2Ea has a large surface area (001) exposed facets, reduced band gap and creation of oxygen vacancies [22]. The (001) exposed facet usually reports a higher photocatalytic activity; so a single crystalline square shaped BiOCl was synthesized and the degradation of Rh-B was investigated. The (001) exposed facets are enhancing the photocatalytic activity by reducing the recombination [23].

2.2 *BiOBr and Modified BiOBr for Energy and Environmental Applications*

BiOBr semiconductor has a band gap ranging from 2.61 to 2.90 eV with a visible light response. So this nanophotocatalyst is widely investigated in many energy and environmental applications. In nitrogen fixation apart from oxygen vacancy induction, the addition of noble metals such as a co-catalyst is another effective method to improve the photocatalytic activity of the bismuth oxyhalide. Au metal supported as the co-catalyst with BiOBr and NH_3 fixation with N_2 is assessed. The author synthesized BiOBr and loaded Au nanoclusters and nanoparticles. The Au nanoparticles will produce plasmon effect and will increase the conversion whereas Au nanoclusters increase the reduction activity by increasing the active sites. The author optimized the catalyst by changing the weight % and confirmed BiOBr-1%Au-L15 (BiOBr/Au-Cp) as the optimized catalyst i.e. BiOBr/Au-Cp illuminated under xenon lamp for 15 min with 80.52 m mol g^{-1} h^{-1} nitrogen conversion [24]. Lan et al. also syntheized oxygen vacancy mediated bismuth rich Bi@ BiOBr nanophotocatalyst and reported a nitrogen reduction rate of 1350 m mol g^{-1} h^{-1} [24].

For volatile organic contaminant degradation (VOC's) $LaVO_4$/BiOBr heterojunctions were constructed through the hydrothermal method and the conversion of gaseous toluene and acetone was investigated. The photocatalytic reactor used for investigation is shown in Scheme 3. The author reported a conversion efficiency of 95.45% in 3 h of visible light irradiation for toluene and 87.1% in 4 h of irradiation for acetone and the intermediates formed were determined by FTIR [25]. A schematic representation of the conversion is shown in Scheme 4.

Construction of MoS_2 step scheme heterojunction with $Bi_4O_5Br_2$ and the degradation efficiency of Rhodamine-B (Rh-B) and disinfection of E.coli was investigated. The step scheme enhanced the charge separation and optimal catalyst (3 weight % of MoS_2) reported a 100% degradation of Rh-B in 24 min and 6.5 log E-coli bacteria in 180 min. The inactivation of E. coli was determined through the fluorescent stain test; the living bacteria displayed green fluorescence and dead bacteria displayed red and within 180 min of light irradiation only red fluorescence remained indicating the inactivation of bacteria and is depicted in Fig. 3.

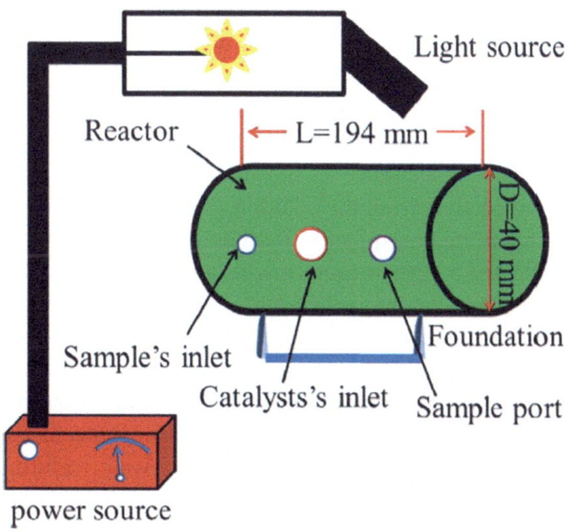

Scheme 3 Reactor used for conversion. Reprinted with permission from Ref. [25]. Copyrights (2020) Elsevier Publications

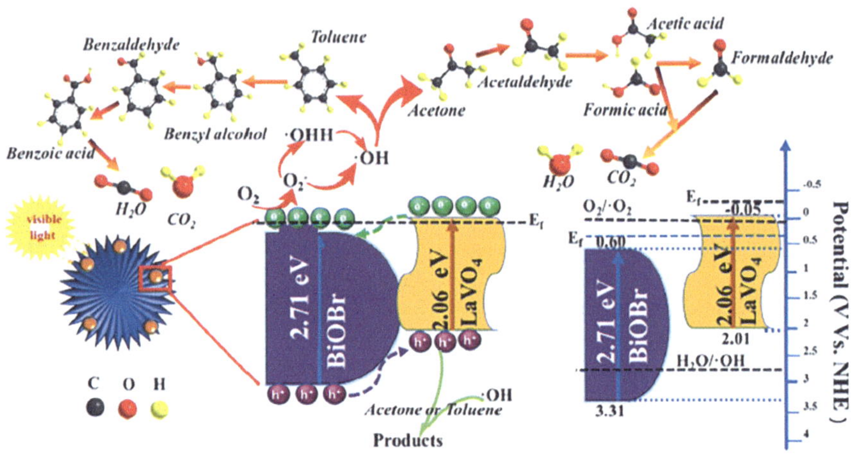

Scheme 4 Photocatalytic conversion of toluene and acetone. Reprinted with permission from Ref. [25]. Copyrights (2020) Elsevier Publications

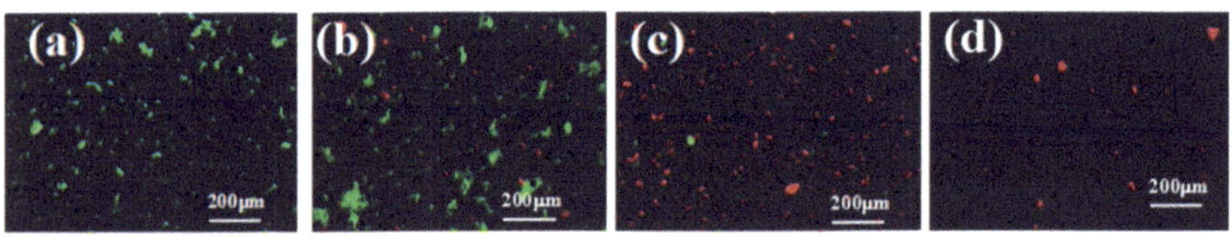

Fig. 3 Fluorescent stain test of bacteria. Reprinted with permission from Ref. [26]. Copyrights (2022) Elsevier Publications

Constructing a p–n junction with 2D-2D interfaces was investigated in BiOBr/(001-TiO_2) (Tb-*x*) for Rhodamine-B degradation. The better photocatalytic activity was reported due to the exposed facets and huge visible light absorption capacity [27].

Spinel ferrites have seeked much attention in photocatalysis due to its easy removal of the catalyst from the solution as it has magnetic properties. Several studies have been reported on spinel ferrite photocatalysts for water splitting recently. Bera et al. fabricated a heterostructure with p type $CuFe_2O_4$ (CFO) and n type BiOX. Type I, Type II and Z scheme heterojunctions are formed based on the electronic bands of BiOX and Z scheme reported the better efficiency in hydrogen generation, i.e. BiOCl/CFO reported better hydrogen generation which is 5 times higher than the bare. The contact between the BiOX and CFO promotes the charge transfer and visible light absorption along with the charge transport mechanism of the Z scheme [28]. The proposed mechanism in the Z scheme water splitting of BiOCl/CFO is shown in Scheme 5.

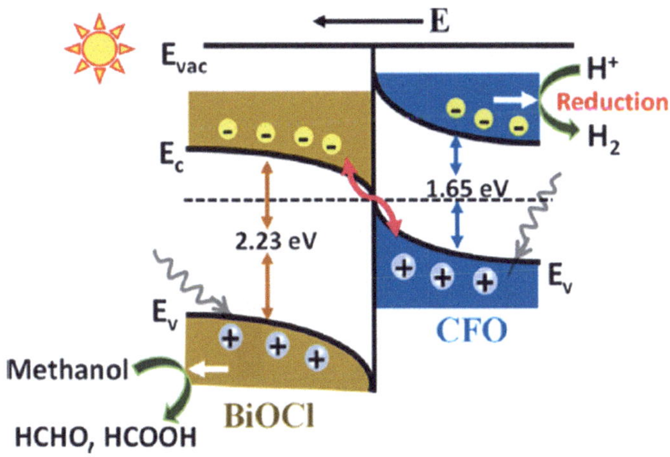

Scheme 5 The Z-scheme water splitting mechanism of BiOCl/CFO. Reprinted from Ref. [28]. Copyrights (2022) American Chemical Society

2.3 BiOI Modified BiOI for Energy and Environmental Applications

The BiOI nanophotocatalyst has a visible light response with a narrow band gap of 2.1–1.63 eV. Due to the narrow band gap, the recombination rate of the generated photocharges are also reported high. So some modifications are employed for improving the property of the nanophotocatalyst. Because of its high visible light response the BiOI as it is and the modified BiOI are investigated for wide applications. In CO_2 conversion Bi et al. synthesized mesoporous BiOI under vacuum through solid state chemistry strategy. The synthesized nanophotocatalyst has a mesoporous crystalline structure. The conversion of CO_2–CH_4 was investigated under dark, UV-Vis and solar radiations. The obtained results are shown in Fig. 4. The superior reduction of CO_2 (82.3%) was reported at 5 h of solar light irradiation under 2 bar pressure. And also the synthesized catalyst shows stability until 6 cycles. It is illustrated in Fig. 5.

BiOI and partially disordered BiOI (BiOI-D) were synthesized by Xiong et al. and photocatalytic CO_2 reduction and degradation efficiency of Rhodamine-B and ciprofloxacin [30] was investigated. The creation of disorder in BiOI reduced the band gap from 1.78 to 1.73 eV through Kubelk-monk equation which favours the visible light harvesting. Similarly the disordered BiOI reported an increase in surface area (14.77 $m^2\ g^{-1}$) when compared to pristine BiOI (2.83 $m^2\ g^{-1}$). The increased surface area increases the surface adsorption and enhances the catalytic activity. Moreover the disorder enhances the charge separation and improves the efficiency of photocatalytic reduction by stabilizing COOH* in intermediates and reduces the CO_2 activation energy barrier. It is illustrated in Fig. 6. The efficiency reported for BiOI-D is 15.1 m mol g^{-1} in 5 h which is 12.5 times more than BiOI nanosheets[30].

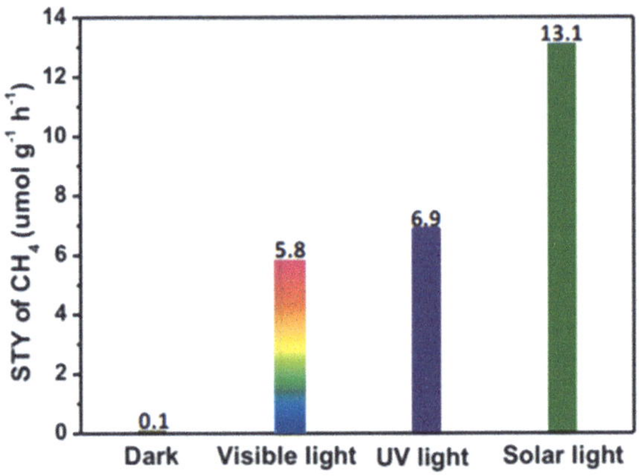

Fig. 4 Photocatalytic reduction of CO_2 with BiOI under dark, UV, visible and solar. Reprinted from Ref. [29]

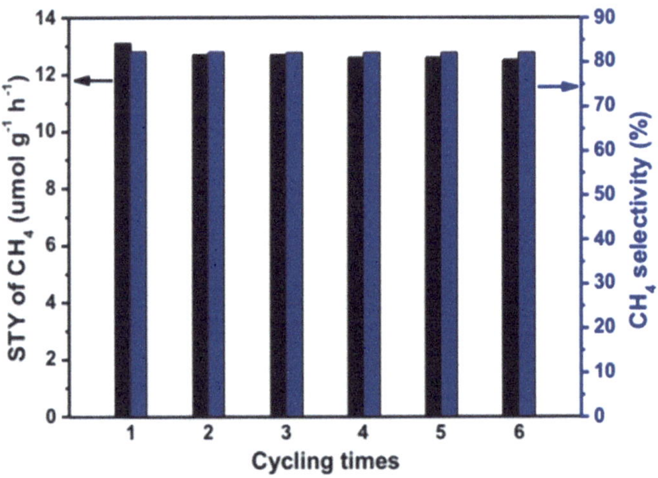

Fig. 5 Stability of catalyst till 6 runs. Reprinted from Ref. [29]

When investigating the degradation efficiency of Rhodamine-B and ciprofloxacin, 60 min of irradiation with xenon lamp 97.9% of efficiency is reported for Rhodamine-B and 47.3% for ciprofloxacin for the BiOI-D nanophotocatalyst whereas 52.4% and 4.9% respectively for the BiOI photocatalyst [30].

Zhou et al. synthesized BiOX through the solvothermal method where ethylene glycol acts as the solvent and the removal of gaseous mercury expelled from Zn smelting flue gas was investigated. A sphere like BiOCl and BiOBr and velvet shaped BiOI were reported. This change in morphology is the result of the chelating effect of ethylene glycol and it can be considered as a microreactor. As the change in

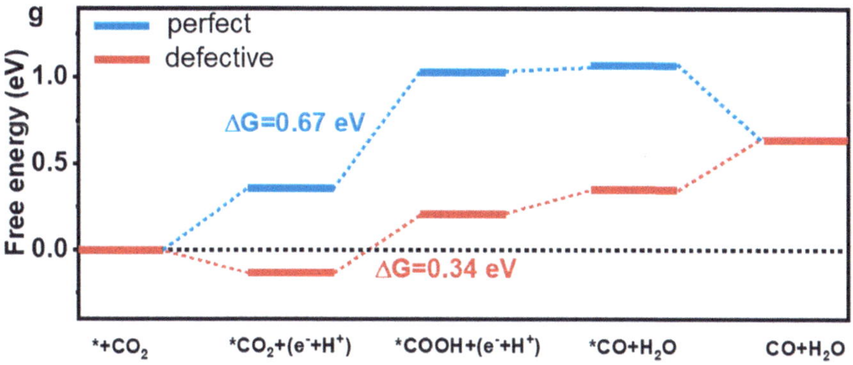

Fig. 6 Free energy diagram of CO_2–CO conversion with BiOI and BiOI-D. Reprinted with permission from Ref. [26]. Copyrights (2023) Elsevier publications

morphology there is a change in surface area also. The surface area is in the order of BiOI > BiOCl > BiOBr. The increase in the surface area and pore volume enhances the photocatalytic effect. And BiOI reported the highest removal efficiency of 96.83% under UV-Vis irradiation, moreover the absorption intensity and absorption edge was more for BiOI in the UV region. From the investigation the author came to the conclusion that morphology, surface hydroxyl group formation and optical property together are responsible for efficient photocatalytic efficiency [31].

2.4 *Bismuth-Rich Bismuth Oxyhalides*

By the addition of Bi the modifications can be made in the structural, optical and electronic properties of the nanocomposites. These kind of nanocomposites are called as bismuth-rich- bismuth oxyhalides and are represented as $Bi_aO_bX_c$. The X in $Bi_aO_bX_c$ influences the band structure along with Bi content. The increased Bi content with Cl increases the band gap whereas with Br reduces the band gap [32]. Moreover it influences the layered structure further influencing the internal electric field (IEF) [33]. Recent studies reported on bismuth-rich bismuth oxyhalides on various applications are discussed here. Cai et al. synthesized $BiOBr_{0.5}I_{0.5}$ through a hydrothermal method and investigated the degradation of Rhodamine-B and reported a degradation efficiency of 99.9% in 12 min of visible light irradiation. The rate of Rhodamine- B degradation reported with $BiOBr_{0.5}I_{0.5}$ is 5.28×10^{-2} min^{-1} whereas for BiOBr and BiOI is 0.57×10^{-2} min^{-1} and 0.11×10^{-2} min^{-1} respectively. The reported rate for $BiOBr_{0.5}I_{0.5}$ is 9.2 times and 48 times higher for BiOBr and BiOI respectively. Moreover $BiOBr_{0.5}I_{0.5}$ nanophotocatalyst is stable till 4 cycles [34].

Chachvalvutikul et al. [35] synthesized the $Bi_7O_9I_3$–$Bi_4O_5Br_2$ nanophotocatalyst by varying the ratio of I and Br, and investigated the degradation of phenolic compounds. The solid solution with a ratio of 0.3:0.7 for I:Br reported the maximum efficiency for 4-Nitrophenol (98.7%) in 360 min of visible light irradiation. Similarly the author investigated the degradation efficiency with 3-Nitrophenol and Bisphenol-A under 180 min of visible light irradiation. The following compounds reported 27.5% and 97.5% degradation efficiency respectively. $Bi_7O_9I_3$–$Bi_4O_5Br_2$ were sensitized with Rh-B and the degradation efficiency of 4-NP was observed but the efficiency was decreased from 98.7% to 25.7%. This is because Rh-B consumed all the generated photocharges and underwent degradation as its degradation efficiency of Rh-B is much faster than 4-NP but on BPA degradation the sensitization increased as the photodegradation efficiency is similar [35].

In a recent study the researchers tried to find the effect of built in electric field in BiOX. So, Dong et al. [36] synthesized BiOX and $BiOXO_3$ where X = Br and I and investigated the effect of the electric field in photocatalytic water splitting. The $BiOBrO_3$ reported the excellent efficiency in terms of stability and hydrogen production. The author states that the increased photocatalytic activity to $BiOBrO_3$ will be due to the polar built in electric field along the plane (001) which inhibits the recombination photogenerated charge carriers efficiently as the direction of the electric field has much influence on reducing the recombination [36]. Similarly the built in electric field in the bulk of BiOX has no impact on recombination as it has a mirror symmetery whereas BiOI shows better reduction of recombination due to the destruction of mirror symmetry and surface electric field. So the reported efficiency rate is $BiOBrO_3$ > BiOI > $BiOIO_3$ > BiOBr [36].

2.5 *Pervoskites*

Sillen aurivillius perovskite structure are $[Bi_2O_2][A_{n-1}B_nO_{3n+1}][Bi_2O_2][X]$ where X = halide and n is the number of the perovskite layer. For water splitting sillen aurivillus oxyhalides acts as an oxygen donating photocatalyst in a Z scheme system. For instance Ogawa et al. [37] synthesized $BaBi_5Ti_nOX$ where X = Cl, Br, I and n from 3 to 5 with 2 step-S, 2step-A, and 2step SA mechanism. $BaBi_5Ti_3O_{14}I$ reported better photocatalytic activity with a band gap of 2.48 eV. The polarizability of iodine destabilizes the oxygen in the perovskite layer and resulted in a negative valence band maxima. For the photocatalytic hydrogen evolution the catalyst was suspended in an aqueous methanol solution (electron donor) and for oxygen evolution the catalyst was suspended in $AgNO_3$ solution (electron acceptor). The gases evolved were analyzed through online gas chromatography. The higher oxygen evolution rate was shown in Table 1 (entry I-3) and also a gradual decrease in oxygen evolution rate with increasing irradiation time was observed due to Ag deposition [37].

Further the same author synthesized $BaMBi_3Nb_2O_{11}I$ with a double perovskite layer (n = 2) and M = Ba, Sr, Ca. The author explained that when the perovskite layer is 2 the introduction of I will not only effect the valence band but also influences the conduction band and also increases the interlayer distance between Bi–Bi bonds. This enhances the O_2 evolution capacity. The water oxidation activities of the prepared samples in the presence of Ag^+ is tabulated in Table 1 [38].

Table 1 O_2 evolution efficiencies of synthesized catalyst adopted from Ref. [38]

Slno	Compound	Initial rate of O_2 evolution (μmol h^{-1})	
		Cl	I
1	Bi_4NbO_8X ($n = 1$)	4.8	41.5
2	$Ba_2Bi_3Nb_2O_{11}X$ ($n = 2$)	4.7	37.6
3	$BaSrBi_3Nb_2O_{11}X$	7.7	62.0
4	$BaCaBi_3Nb_2O_{11}X$	6.2	32.3
5	$Bi_5BaTi_3O_{14}X$ ($n = 3$)	4.3	24.2
6	$Bi_6NbWO_{14}X$	13.8	89.6

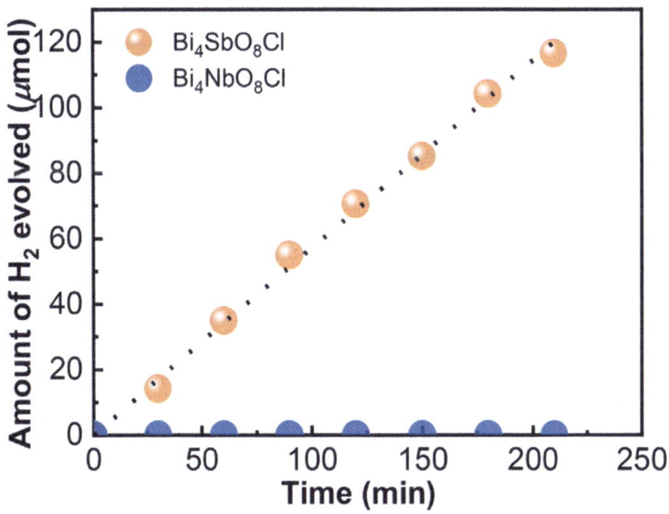

Fig. 7 The hydrogen evolution rate Bi_4SbO_8Cl compared with Bi_4NbO_8Cl. Reprinted with permissions from Ref. [39]. Copyrights (2022) Elsevier publications

Bi_4SbO_8Cl is another nanophotocatalyst belonging to the family of Sillen-Aurivillius oxyhalide. Here the transition material is substituted by the main group element. So the sp hybridization has been strengthened. Hence the optical band gap has been increased to 2.53 eV. The hydrogen evolution rate has been compared with Bi_4NbO_8Cl and shown in Fig. 7. The increased efficiency in Bi_4SbO_8Cl is attributed to the charge mobility and reducibility offered by Sb–O–Sb bonds [39].

2.6 *BiOX Modified with 2D Materials*

Boron nitride is an interesting 2D material and Xu et al. modified the BiOCl nanostructure with graphene like boron nitride (g-BN/BiOCl) and investigated the removal

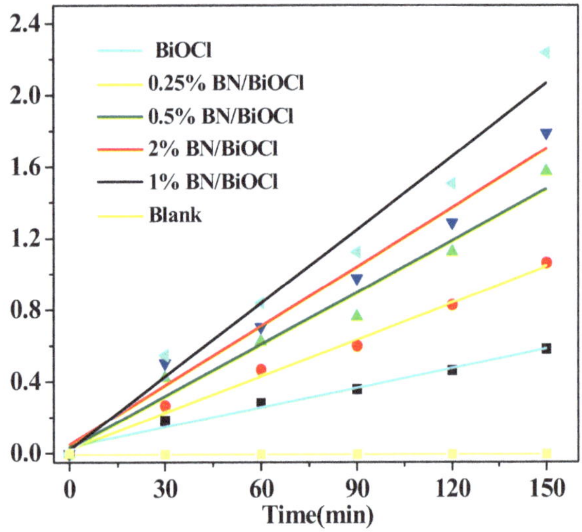

Fig. 8 Kinetic profile of Cr removal with respect to BiOCl and BN@BiOCl. Reprinted with permission from Ref. [40]. Copyrights (2015) Elsevier publications

efficiency of Cr(VI) under visible light irradiation. The doping of g-BN increases the specific surface area to 6.89 m^2/g from 3.918 m^2/g and so it enhances the adsorption capacity of the catalyst; Similarly the hydrophobic nature of the g-BN enhances the adsorption. The author varied the weight percentage of BN doping and investigated the removal efficiency and reported that the 1% g-BN/BiOCl has the highest removal efficiency than others, and 2% g-BN/BiOCl shows a slight decrease due to covering of the active site in BiOCl with overloading of g-BN. The kinetic profile of removal is indicated in Fig. 8. On other hand on doping the band gap energies were also reduced. 1% g-BN/BiOCl reported a band gap of 3.11 eV whereas bare BiOCl has 3.21 eV.

The absorption spectra and tauc's plot of all the prepared samples are shown in Fig. 9 [40].

The bismuth oxyhalides can be modified with materials like r-GO g-C_3N_4 etc., to enhance the activity and efficiency, for instance Cai et al. modified BiOX with r-GO nanosheets and removal efficiency of Hg^0 with BiOCl, BiOBr, BiOI were investigated and values of 0.90.97%, 92.99% and 100% respectively are reported, Under UV radiation and under visible light radiation the author reported 40.97%, 73.27% and 91.98% respectively in 120 min [41]. The reported efficiency is more than the bare BiOX and this increment is attributed to r-GO as it increases the surface area and pore volume and also enhances the charge separation [41]. A comparative study with other catalyst reported in literature is shown in Table 2.

Ti_3C_2 belongs to the family of mxene, an emerging 2D material, and is modified with BiOBr. Huang et al. fabricated BiOBr/Ti_3C_2 nanocomposites by electrostatically driven self-assembly method and investigated the removal efficiency of Cr(VI) and reported a 47.5% of removal in 40 min of UV-vis irradiation, whereas bare BiOBr reported only 8.7%. The improved removal efficiency is due to the high conductivity

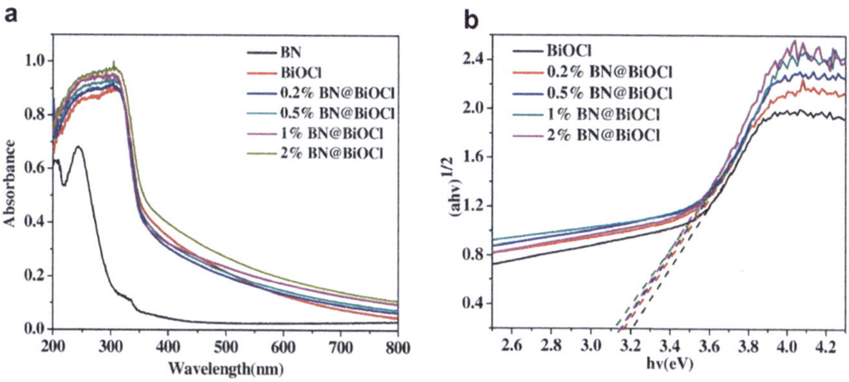

Fig. 9 **a** UV-vis absorption spectra and **b** Tauc's plot of synthesized BiOCl and g-BN/BiOCl. Reprinted with permission from Ref. [40]. Copyrights (2015) Elsevier publications

Table 2 Comparative study with the other catalyst reported in literature is shown in table adopted from Ref. [41] and the references are cited therein

Photocatalysts	Sample amount (mg)	Initial mercury (μg/m^3)	Maximum removal efficiency (%)
$BiOI/Bi_5O_7I^{(VL)}$	50	60	60
3D-BiOX (X = Cl, Br, I)$^{(UV)}$	200	1000	97
$BiOI/TiO_2{}^{(VL)}$	500	60	83
$Bi_4O_5I_2/Bi_5O_7I^{(VL)}$	50	60	79
$BiOIO_3$/graphene$^{(VL)}$	30	60	89
$CeO_2/BiOIO_3{}^{(VL)}$	30	65	87
$BiOI/BiOIO_3{}^{(UV)}$	50	60	99
rGO/BiOI$^{(UV)}$	50	400	100
rGO/BiOI$^{(VL)}$	50	400	91.98

VL: Visible light; UV: Ultraviolet light
Reprinted with permission from Ref. [41]. Copyrights (2022) Elsevier publications

of Ti_3C_2, capture of electrons from the Schottky barrier and creates separation of charges which hinders the charge recombination. Similarly the morphological structure of Ti_3C_2 provides large active sites and also it shows better redox reactivity. The same modified catalyst is used for the degradation of Rhodamine-B(Rh-B) and 2,4-dinitrophenol (2,4 DNP). The author reported around 99.4% degradation for Rh-B around 24 min of UV-Visible irradiation and for 2,4 DNP it is depicted in Fig. 10 [42].

Construction of heterojunction with 2D covalent organic frame work (COF) is another method of modification. In an investigation, Zhong et al. modified BiOX with the 2D covalent organic frame work for the removal of uranium and for

Fig. 10 Kinetic profile of 2,4-DNP degradation. Reprinted with permission from Ref. [42]. Copyrights (2019) Elsevier publications

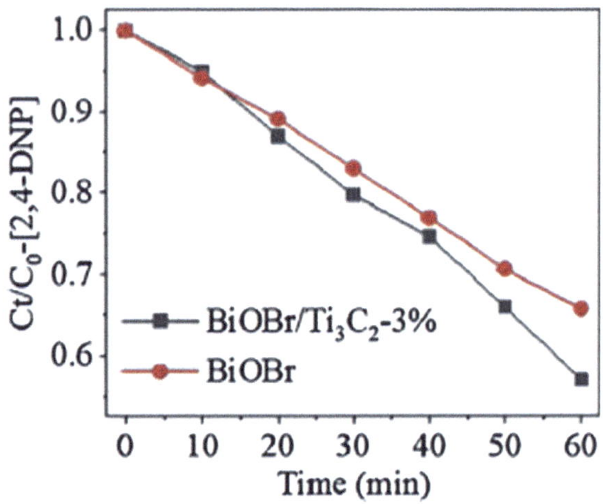

the degradation of Bisphenol A. The COF has an extended p bond conjugation system and it reduces the band gap as well as the rate of recombination by efficient separation of charge carriers acting as the migration path for charge carrier; similarly it reduces the band gap also. The change in specific surface area was also analyzed with BET when it was doped with COF (TpPa-1-x%). The better removal efficiency of uranium as well as BPA was reported by the photocatalyst BiOBr @ TpPa-1-x. The mechanism of the uranium removal and degradation of BPA is schematically illustrated in Scheme 6 [43].

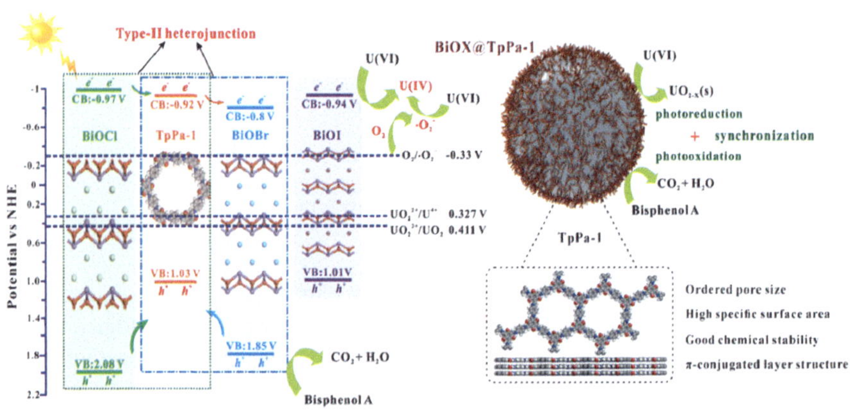

Scheme 6 Schematic mechanism of removal of uranium and BPA. Reprinted with permission from Ref. [43]. Copyrights (2022) Elsevier publications

2.7 Modifications with Quantum Dots

A novel class of discrete, quasi-spherical nanoparticles known as quantum dots (QDs) exists in the realm of nano-sized materials with diameters smaller than 10 nm [44]. In this family carbon quantum dots are prominent with exceptional properties like non toxicity, bio-compatiblity, and up-conversion ability. CQDs are a conjugated structural material made of sp2/sp3 hybridized carbon atoms, and they have outstanding electron transport and reservoir properties. It is understood that materials with delocalized conjugated structures can hasten the separation and transfer of photogenerated charges, thus enhancing photocatalytic performance [45]. When quantum dots are introduced in a photocatalyst it acts as photosensitizer [46], electron mediator [47–49], spectral converter [50] and photocatalyst as such [51, 52]. In this section, we discusses on various quantum dots embedded along with BiOX photocatalyst for different applications for the last 3 years. CQDS and Co modified BiOCl in which Co enhanced the charge separation and CQDS improved the visible light enhancement and reported a removal efficiency of NO_x around 49.5% within 30 min of visible light irradiation [53]. In another investigation, carbon dots modified BiOCl, BiOBr, BiOI (CQDS/BiOX) were synthesized and investigated its photocatalytic degradation efficiency under visible and near IR. Rhodamine-B and levofloxacin was selected as the pollutant. The degradation test was done under light and the catalyst is stable till 3 cycles. The CQDS/BiOBr is showing the maximum degradation efficiency (98.91%) in 60 min of irradiation. The photodegradation rate of Rhodamine B is shown in Fig. 11 and levofloxacin is shown in Fig. 12 and the degradation pathway of Levofloxacin is shown in Scheme 7. The addition of quantum dots have increased the surface area and the high absorption peak in the UV-Visible region was due to p–p* transition and n–p* transition. In CQDS/BiOCl the absorption edge was shifted from UV to visible whereas no significant change was observed for CQDS/BiOBr [54].

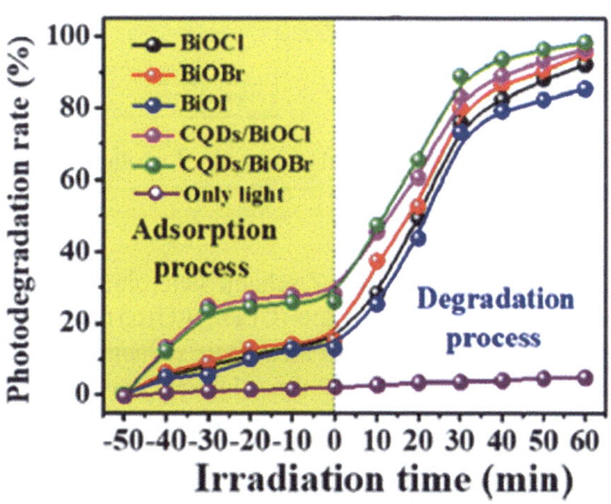

Fig. 11 Photocatalytic degradation of Rh-B under visible light. Reprinted with permission from Ref. [54]. Copyrights (2022) Elsevier publications

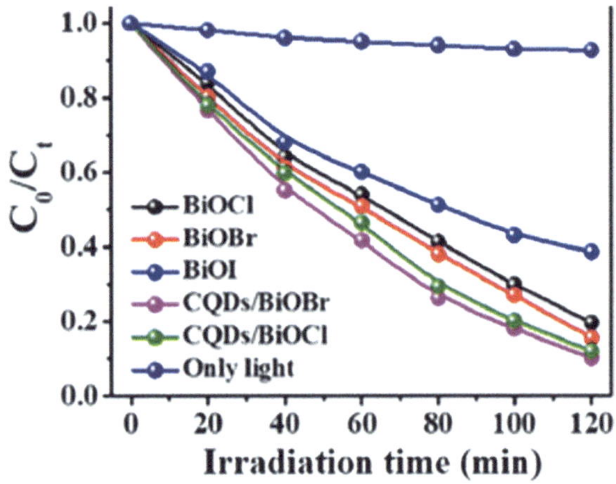

Fig. 12 Photocatalytic degradation of levofloxacin. Reprinted with permission from Ref. [54]. Copyrights (2022) Elsevier publications

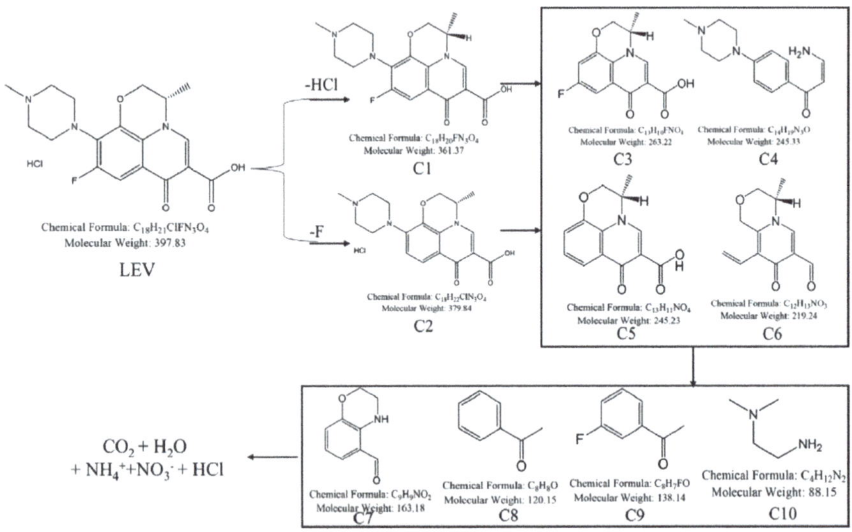

Scheme 7 Possible degradation pathway of levofloxacin. Reprinted with permission from Ref. [54]. Copyrights (2022) Elsevier publications

For CO_2 conversion a Z scheme heterojunction was constructed between BiOBr and carbonized polymer dots (CPDs/BiOBr) through solid state chemical method and was employed for the photocatalytic reduction of CO_2. The Z scheme heterojunction catalyst reported 2.02 times higher conversion efficiency than bare BiOBr. The comparative study with other reported BiOBr is shown in Table 3 [55]. The increased conversion efficiency may be attributed to the change in morphology and optical properties and the upconversion effect of CPDs. The bare BiOBr has a lamellar structure where as CPDs/BiOBr has a rosette structure; similarly the absorption maxima shows a red shift. The upconversion effect was confirmed through PL as it shows an emission peak at 460 nm when excitation wavelength is 750 nm and is shown in Fig. 13 [55].

In a study carbon quantum dots and Bi metal was decorated over BiOBr to form CQDS/Bi/BiOBr. The interface of the composite were held together by ohmic junction to form C-scheme. This mechanism reduces the electron migration path and also enables the smooth passage of electrons. The SPR property of the Bi metal/and O_2^- activation ability of CQDS together constitutes for the effective degradation of methyl orange under visible light [56]. The CO_2 conversion was investigated by fabricating CQDS/$Bi_{12}O_{17}Cl_2$/NiAl-LDH ternary nanocomposites. The 16.4 mol^{-1} g^{-1} h^{-1} of yield was reported for the optimal catalyst. The better conversion efficiency reported was due to the formation of Z scheme and upconversion effect offered by the QDS. Because of this unique feature of QDS the photoresponsive ability of the composite was improved and energy from near IR is absorbed [57].

Bismuth oxybromide quantum dots were synthesized by Shi et al. and was decorated over $Na_2Ti_3O_7$ and investigated the conversion of benzyl alcohol to benzaldehyde. Due to increased oxygen vacancy the nanophotocatalyst reported an efficiency

Table 3 Comparative study on CO_2 reduction on various BiOBr catalyst, adopted from Ref. [55] and the references are cited therein

Photocatalysts	Light sources	Reaction conditions	Main products	Evolution rate (μmol g^{-1} h^{-1})
CPDs/BiOBr	300 W Xe-lamp	Gas solid reaction	CO	144.6
AgBr/BiOBr	300 W Xe-lamp	Liquid solid reaction	CO	12.43
BiOBr/CdS	300 W Xe-lamp	Liquid solid reaction	CO	4.53
BiOBr/$Zn(OH)_2$	300 W Xe-lamp	Liquid solid reaction	CO	5.4
BiOBr/Bi_2 WO6	300 W Xe-lamp	Gas solid reaction	CO	55.17
Ov-BiOBr	300 W Xe-lamp	Liquid solid reaction	CO	2.03
Bi-CTS/BiOBr	300 W Xe-lamp	Liquid solid reaction	CO	50.13
BiOBr/HNb_3O_8	300 W Xe-lamp	Gas solid reaction	CO	32.92
BiOBr/ACSs	300 W Xe-lamp	Liquid solid reaction	CO	23.74
BiOBr/Bi_2S_3	300 W Xe-lamp	Liquid solid reaction	CO/CH_4	100.8/8.5
BiOBr/CoAl-LDH	300 W Xe-lamp	Liquid solid reaction	CO/CH_4	4.096/4.174

Reprinted with permission from Ref. [55]. Copyrights (2014) Elsevier publications

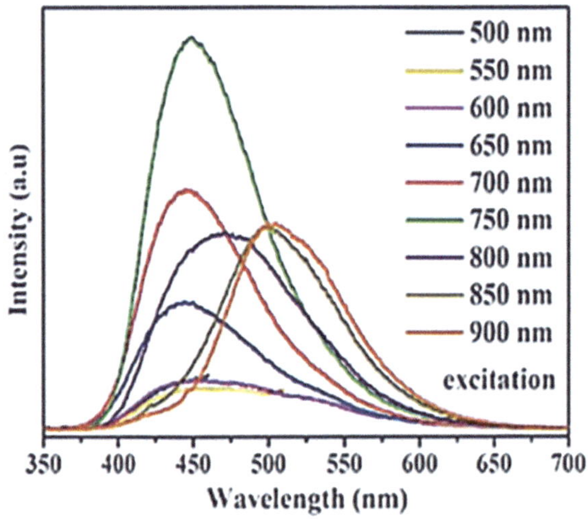

Fig. 13 The up conversion PL spectra. Reprinted with permission from Ref. [55]. Copyrights (2014) Royal society of chemistry

of 3.32 m mol g^{-1} h^{-1}. Through TEM analysis the author confirmed the size of quantum dots are between 2 and 4 nm and DFT calculation also reported superior photocatalytic activity due to electron density. The nanocomposite showed high selectivity towards benzaldehydes [58].

The tetracycline degradation was investigated by CN/CQDS/$BiOCl_{0.75}Br_{0.25}$ nanocomposite and reported a degradation efficiency of 84.3% which was better than the parent photocatalyst. The increased efficiency is attributed to the upconversion effect of CQDS [59].

2.8 *Luminescence Material*

A substance emits light after absorbing radiative energy, either directly or through processes that include the transfer of electrons or energy in an energy or a process called luminescence [60]. The upconversion (UC) luminous efficiency of layered bismuth oxyhalide doped with rare earth ions is adversely affected by internal lattice defects [61].

Zhang et al. synthesized the upconversion luminescence material (UCL) NaY F_4: Yb^{3+}, Tm^{3+}@ BiOCl heterostructured nanocomposite and investigated the degradation of Rhodamine-B under near IR light and stimulated solar light and reported a degradation efficiency of 98% in 2 h and 99.9% in 18 min respectively. The sensitizer Yb^{3+} and activator Tm^{3+} interaction enhances the upconversion. Bi_2GdO_4Cl:Yb^{3+}/Er^{3+} were doped with Li^+ and the degradation of tetracycline was investigated and a 62% degradation was reported under UV-Vis NIR radiation for 120 min. The defects and impurities usually quench the luminescence effect. So the introduction of a host material will reduce the oxygen vacancy and improves the luminescence effect [61].

In other studies BiOI was modified with Er^{3+} and Yb^{3+} and reported as a shift in band gap from 1.86 to 2.39 eV was observed and when doping is done along with optimum concentration $Bi_4O_5I_2 \cdot 6\%Yb^{3+}/Er^{3+}$ reported the maximum red emission and degradation study of BPA and Rhodamine-B under NIR was investigated [62].

2.9 Clay Materials as Support

Clay minerals are lamella red structured heterogeneous materials. Tetrahedral silicates and octahedral hydroxide sheets make up the fundamental components of clay minerals, which are members of the sheet silicate family known as phyllosilicates [63]. Clays have various mineralogical properties, which may influence how well they can load photocatalyst [64]. BiOCl was doped with Bi and montmorillonite clay was used as a supporter and the degradation of Rhodamine-B and Tetracycline was investigated. The addition of a supporter increased the adsorption capacity of the catalyst and the doping of bismuth enhanced the SPR property and synergistically increased the photocatalytic activity by enhancing the light absorption and separation of charge carriers [65]. Bentonite is another clay material and a TiO_2/BiOBr/bentonite composite was synthesized for the degradation of bentonite. The addition of clay minerals increased the surface area and reported a better degradation under visible light than bare TiO_2 and BiOBr [66].

In an investigation, illite a clay mterial has been used as a support over $Bi_4O_5I_2/Bi_5O_7I$ as the illite material has a large specific surface area and adsorbs the molecules more effectively, inhibiting self-aggregation. The degradation rate reported for BPA and Tetracycline hydrochloride under optimum calcination temperature (500) were 94% in 120 min of visible light irradiation and 73% in 60 min. Whereas the $Bi_4O_5I_2/Bi_5O_7I$, illite and the $Bi_4O_5I_2/Bi_5O_7I$ at various calcination temperatures reported lesser efficiency [67]. Sepiolite clay was used as a support in BiOCl/TiO_2 for the removal of formaldehyde. The TEM and SEM images revealed well dispersion of BiOCl and TiO_2 over sepiolite without any self-agglomeration. The BiOCl/TiO_2/Sepiolite shows a tight packing (Fig. 14) with exposed edges which facilitates transport and separation of photogenerated charge carriers. Similarly, the sepiolite shifted the absorption edge from UV to visible light. The removal efficiency was determined by the increasing concentration of CO_2 and decreasing concentration of HCHO monitored through a multigas analyzer. The decreased concentration of HCHO is 17.8 mg/m^3 from 66 mg/m^3 and increasing concentration of CO_2 is 9.42 mg/m^3 from 6.38 mg/m^3 in the presence of visible light irradiation for a time of 180 min proving the decomposition of HCHO [68].

Bi_5O_7I/Palygorskite composite was synthesized and the degradation of ciprofloxacin and tetracycline hydrochloride was investigated and a better degradation due to the efficient charge separation was reported that reduced the path distance of charge carriers and increased surface area [69]. In another investigation carbon quantum dots and clay mineral (saponite) were coupled with BiOBr together to form BiOBr/CQDS/Saponite and the degradation of ciprofloxacin was investigated. The

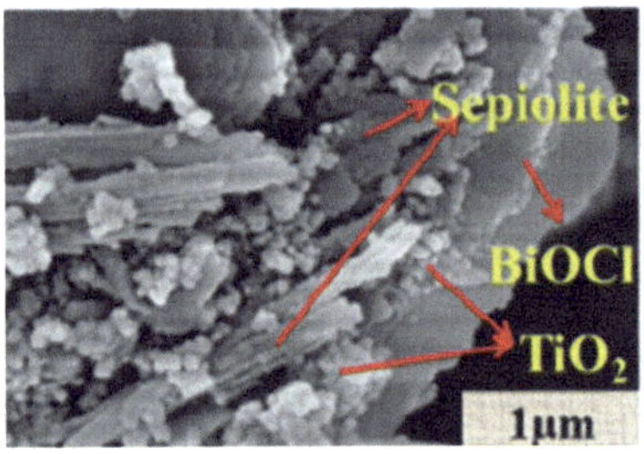

Fig. 14 TEM image of BiOCL/TiO_2/Sepiolite. Reprinted with permission from Ref. [68]. Copyrights (2020) Elsevier publications

uniform distribution of CQDS over saponite reduced the agglomeration providing the active surface [70].

Different clay materials have different layered structure and different minerals, so addition of various clay minerals enhances the photocatalytic properties through different mechanism finally leading to a better photocatalytic activity [71].

3 Summary and Conclusions

Bismuth oxyhalides are a wide range of photocatalysts with unique morphological structure and internal electric field and widely used in energy and environmental applications. Various strategies have been employed in this family due to its excellent tunability. In this chapter, we discussed by taking BiOX as the base material and modifications that has arised in each field of application. In CO_2 conversions and volatile organic contaminant degradation doping and heterojunctions are widely applied in the recent three years and most of these catalysts have efficiency to degrade multiple pollutants. The investigations reported on nitrogen fixation and heavy metal ions are less during these past three years and the nitrogen fixation mostly focusses on the creation of oxygen vacancies. Among the topic discussed organic contaminant degradations are the one in which several studies are reported. In this topic bismuth-rich bismuth oxyhalides photocatalysts are widely investigated. The other application which is discussed in this chapter is water splitting, where the catalyst which are widely applied are the sillenite aurvillous type. They are a kind of perovskites and these kind of materials are the ones which are widely advancing. The other modifications discussed are the addition of 2D and carbon materials, and introduction of clays as support. All these modifications improve the optical property, reduce the charge recombination, surface area and pore volume and also active sites and thereby enhance the photocatalytic activity.

When clay minerals are introduced it reduces the agglomeration as well as increases the surface area, apart from this the impurities in the clay may alter the optical property as well as the binding property of the nanocomposite. So introducing clay as a support is a ecofriendly modification for better photocatalytic activity.

4 Future Perspectives

The selectivity towards the product are still in bottle neck. The application of photocatalysts towards various area can be enhanced on increasing the selectivity towards the product which reduces the formation of toxic intermediates. Studies on the photocatalytic property of luminescent material has to be uphilled which paves the path to various sensing and biological applications. The application of photocatalyst in bioimaging, photodynamic therapy are some of the blooming field and more studies has to be encouraged. A non availability of proper synthetic method is another drawback of some higher ordered nanophotocatalyst. For instance, a proper synthetic method is not designed for perovskite oxyiodide structure where the perovskite layer is greater than 3. So more theoretical and experimental studies have to be carried out to fix the above problems.

References

1. G. Zhu, M. Hojamberdiev, S. Zhang, S.T.U. Din, W. Yang, Enhancing visible-light-induced photocatalytic activity of BiOI microspheres for NO removal by synchronous coupling with Bi metal and graphene. Appl. Surf. Sci. **467–468**, 968–978 (2019). https://doi.org/10.1016/j.apsusc.2018.10.246
2. D.S. Bhachu et al., Bismuth oxyhalides: synthesis, structure and photoelectrochemical activity. Chem. Sci. **7**(8), 4832–4841 (2016). https://doi.org/10.1039/c6sc00389c
3. K. Sridharan, S. Shenoy, S.G. Kumar, C. Terashima, A. Fujishima, S. Pitchaimuthu, Advanced two-dimensional heterojunction photocatalysts of stoichiometric and non-stoichiometric bismuth oxyhalides with graphitic carbon nitride for sustainable energy and environmental applications. Catalysts **11**(4) (2021). https://doi.org/10.3390/catal11040426
4. J. Di, J. Xia, H. Li, S. Guo, S. Dai, Bismuth oxyhalide layered materials for energy and environmental applications. Nano Energy **41**, 172–192 (2017). https://doi.org/10.1016/j.nanoen.2017.09.008
5. K. Sharma et al., Recent advances in enhanced photocatalytic activity of bismuth oxyhalides for efficient photocatalysis of organic pollutants in water: a review. J. Ind. Eng. Chem. **78**, 1–20 (2019). https://doi.org/10.1016/j.jiec.2019.06.022
6. A. Nikokavoura, C. Trapalis, Alternative photocatalysts to TiO2 for the photocatalytic reduction of CO2. Appl. Surf. Sci. **391**, 149–174 (2017). https://doi.org/10.1016/j.apsusc.2016.06.172
7. Y. Huang, N. Zhang, Z. Wu, X. Xie, Artificial nitrogen fixation over bismuth-based photocatalysts: fundamentals and future perspectives. J. Mater. Chem. A **8**(10), 4978–4995 (2020). https://doi.org/10.1039/c9ta13589h
8. T. A. Saleh, M. Mustaqeem, M. Khaled, Water treatment technologies in removing heavy metal ions from wastewater: a review. Environ. Nanotechnol. Monit. Manag. **17** (2022). https://doi.org/10.1016/j.enmm.2021.100617
9. S. Weon, F. He, W. Choi, Status and challenges in photocatalytic nanotechnology for cleaning air polluted with volatile organic compounds: visible light utilization and catalyst deactivation. Environ. Sci. Nano **6**(11), 3185–3214 (2019). https://doi.org/10.1039/c9en00891h
10. K. Maeda, Photocatalytic water splitting using semiconductor particles: History and recent developments. J. Photochem. Photobiol. C Photochem. Rev. **12**(237–268) (2011). http://ac.els-cdn.com.ezlibproxy1.ntu.edu.sg/S1389556711000505/1-s2.0-S1389556711000505-main.pdf?_tid=7a35fee6-2bb0-11e7-aac8-00000aab0f27&acdnat=1493342442_2527b72f57699efa726c6efef1b1ae6c

11. Y. Sun, Y. Ahmadi, K.H. Kim, J. Lee, The use of bismuth-based photocatalysts for the production of ammonia through photocatalytic nitrogen fixation. Renew. Sustain. Energy Rev. **170** (2022). https://doi.org/10.1016/j.rser.2022.112967
12. Q. Hao et al., Catalytic reduction of nitrogen to produce ammonia by bismuth-based catalysts: state of the art and future prospects. Mater. Horizons **7**(4), 1014–1029 (2020). https://doi.org/10.1039/c9mh01668f
13. M. Qamar, Z.H. Yamani, Bismuth oxychloride-mediated and laser-induced efficient reduction of Cr(VI) in aqueous suspensions. Appl. Catal. A Gen. **439–440**, 187–191 (2012). https://doi.org/10.1016/j.apcata.2012.07.004
14. X. Lv, F.L.Y. Lam, X. Hu, A review on bismuth oxyhalide (BiOX, X = Cl, Br, I) based photocatalysts for wastewater remediation. Front. Catal. **2** (2022). https://doi.org/10.3389/fctls.2022.839072
15. N. Fajrina, M. Tahir, A critical review in strategies to improve photocatalytic water splitting towards hydrogen production. Int. J. Hydrog. Energy **44**(2), 540–577 (2019). https://doi.org/10.1016/j.ijhydene.2018.10.200
16. C. Guan, T. Hou, W. Nie, Q. Zhang, L. Duan, X. Zhao, Boosting visible light driven gas–solid phase photocatalytic reduction of CO2 on BiOCl microspheres by enhanced carrier transportation through lattice structure modification. Sep. Purif. Technol. **306** (2023). https://doi.org/10.1016/j.seppur.2022.122654
17. H. Maimaitizi, A. Abulizi, D. Talifu, Y. Tursun, Nanoarchitectonics of chlorophyll and Mg co-modified hierarchical BiOCl microsphere as an efficient photocatalyst for CO2 reduction and ciprofloxacin degradation. Adv. Powder Technol. **33**(5) (2022). https://doi.org/10.1016/j.apt.2022.103562
18. S.H.W. Kok et al., Bismuth-rich Bi12O17Cl2 nanorods engineered with oxygen vacancy defects for enhanced photocatalytic nitrogen fixation. J. Alloys Compd. **952** (2023). https://doi.org/10.1016/j.jallcom.2023.170015
19. J. Yang et al., A compact Z-scheme heterojunction of BiOCl/Bi2WO6 for efficiently photocatalytic degradation of gaseous toluene. J. Colloid Interface Sci. **631**, 44–54 (2023). https://doi.org/10.1016/j.jcis.2022.11.023
20. F. Mokhtari, N. Tahmasebi, Hydrothermal synthesis of W-doped BiOCl nanoplates for photocatalytic degradation of rhodamine B under visible light. J. Phys. Chem. Solids **149** (2021). https://doi.org/10.1016/j.jpcs.2020.109804
21. T. Narenuch, T. Senasu, T. Chankhanittha, S. Nanan, Solvothermal synthesis of CTAB capped and SDS capped BiOCl photocatalysts for degradation of rhodamine B (RhB) dye and fluoroquinolone antibiotics. J. Solid State Chem. **294** (2021). https://doi.org/10.1016/j.jssc.2020.121824
22. H. Wang, X. Jiang, Y. Qi, J. Du, T. Guo, Engineering bandgap structure of BiOCl nanoplates with oxygen vacancies for accelerated photocatalytic degradation of Rhodamine B. J. Alloys Compd. **910** (2022). https://doi.org/10.1016/j.jallcom.2022.164860
23. Z. Hao, X. Lv, W. Hou, X. Liu, Facile synthesis of BiOClsingle-crystal photocatalyst with high exposed (0 0 1) facets and its application in photocatalytic degradation. Inorg. Chem. Commun. **134** (2021). https://doi.org/10.1016/j.inoche.2021.109038
24. C. Wang, J. Wang, X. Yu, Enhancement of photocatalytic ammonia production over BiOBr nanosheets with photo-assembled Au cocatalysts. Colloids Surf. Physicochem. Eng. Asp. **662**, 131055 (2023). https://doi.org/10.1016/j.colsurfa.2023.131055
25. X. Zou, C. Yuan, Y. Dong, H. Ge, J. Ke, Y. Cui, Lanthanum orthovanadate/bismuth oxybromide heterojunction for enhanced photocatalytic air purification and mechanism exploration. Chem. Eng. J. **379** (2020). https://doi.org/10.1016/j.cej.2019.122380
26. H. Shi et al., In-situ construction of step-scheme MoS2/Bi4O5Br 2 heterojunction with improved photocatalytic activity of Rhodamine B degradation and disinfection. J. Colloid Interface Sci. **623**, 500–512 (2022). https://doi.org/10.1016/j.jcis.2022.04.148
27. L. Yang et al., 2D/2D BiOBr/(001)-TiO2 heterojunction toward enhanced photocatalytic degradation activity of Rhodamine B. J. Alloys Compd. **884** (2021). https://doi.org/10.1016/j.jallcom.2021.161064

28. S. Bera, S. Ghosh, T. Maiyalagan, R.N. Basu, Band edge engineering of BiOX/CuFe2O4 heterostructures for efficient water splitting. ACS Appl. Energy Mater. **5**(3), 3821–3833 (2022). https://doi.org/10.1021/acsaem.2c00296
29. Q. Bi, Y. Fang, R. Luo, F. Huang, One-step solid-state-chemistry synthesized layered bismuth oxyiodide crystal for efficient solar-driven CO2 photoreduction. Catal. Commun. **175** (2023). https://doi.org/10.1016/j.catcom.2023.106600
30. J. Xiong, X. Zhu, J. Xia, J. Di, Partial disorder structured BiOI atomic layers boosting excitons dissociation for photocatalytic CO2 reduction and pollutant removal. Appl. Surf. Sci. **627** (2023). https://doi.org/10.1016/j.apsusc.2023.157338
31. Y. Nan Zhou, R. Li, L. Tao, R. Li, X. Wang, P. Ning, Solvents mediated-synthesis of 3D-BiOX (X = Cl, Br, I) microspheres for photocatalytic removal of gaseous Hg0 from the zinc smelting flue gas. Fuel **268** (2020). https://doi.org/10.1016/j.fuel.2020.117211
32. J. Xiong, P. Song, J. Di, H. Li, Bismuth-rich bismuth oxyhalides: a new opportunity to trigger high-efficiency photocatalysis. J. Mater. Chem. A **8**(41), 21434–21454 (2020). https://doi.org/10.1039/d0ta06044e
33. J.J. Foo, S.F. Ng, W.J. Ong, Dimensional heterojunction design: the rising star of 2D bismuth-based nanostructured photocatalysts for solar-to-chemical conversion. Nano Res. **16**(4), 4310–4364 (2023). https://doi.org/10.1007/s12274-021-4045-0
34. W. Cai et al., BiOBrxI(1–x) solid solution with oxygen vacancies boosting the photocatalytic degradation of multiple pollutants. Mater. Res. Bull. **169**, 112506 (2024). https://doi.org/10.1016/J.MATERRESBULL.2023.112506
35. A. Chachvalvutikul, T. Luangwanta, B. Inceesungvorn, S. Kaowphong, Bismuth-rich oxyhalide (Bi7O9I3–Bi4O5Br2) solid-solution photocatalysts for the degradation of phenolic compounds under visible light. J. Colloid Interface Sci. **641**, 595–609 (2023). https://doi.org/10.1016/j.jcis.2023.03.063
36. X.D. Dong, Y.M. Zhang, Z.Y. Zhao, Role of the polar electric field in bismuth oxyhalides for photocatalytic water splitting. Inorg. Chem. **60**(12), 8461–8474 (2021). https://doi.org/10.1021/acs.inorgchem.0c03220
37. M. Ogawa, H. Suzuki, K. Ogawa, O. Tomita, R. Abe, Synthesis of multi-layered perovskite oxyiodides: impact of number of perovskite layers and type of halide layer for band levels and photocatalytic properties. Solid State Sci. **141** (2023). https://doi.org/10.1016/j.solidstatesciences.2023.107221
38. K. Ogawa, H. Suzuki, A. Walsh, R. Abe, Orbital engineering in Sillén-Aurivillius phase bismuth oxyiodide photocatalysts through interlayer interactions. Chem. Mater. (2023). https://doi.org/10.1021/acs.chemmater.3c00932
39. W. Fang, Y. Mi, Y. Yang, Y. Jiang, Y. Liu, W. Shangguan, Conduction band tuning by strengthening s-p hybridization of novel layered oxyhalide Bi4SbO8Cl for efficient visible-light photocatalytic water splitting. Mater. Today Chem. **26** (2022). https://doi.org/10.1016/j.mtchem.2022.101175
40. H. Xu, Z. Wu, M. Ding, X. Gao, Microwave-assisted synthesis of flower-like BN/BiOCl composites for photocatalytic Cr(VI) reduction upon visible-light irradiation. Mater. Des. **114**, 129–138 (2017). https://doi.org/10.1016/j.matdes.2016.10.057
41. J. Cai et al., rGO-modified BiOX (X = Cl, I, Br) for enhanced photocatalytic eradication of gaseous mercury. Appl. Surf. Sci. **594** (2022). https://doi.org/10.1016/j.apsusc.2022.153502
42. Q. Huang, Y. Liu, T. Cai, X. Xia, Simultaneous removal of heavy metal ions and organic pollutant by BiOBr/Ti3C2 nanocomposite. J. Photochem. Photobiol. A Chem. **375**, 201–208 (2019). https://doi.org/10.1016/J.JPHOTOCHEM.2019.02.026
43. X. Zhong, Q. Ling, S. Wang, B. Hu, Visible-light-driven 2D/2D Bismuth oxyhalides/covalent organic framework heterojunctions for synchronous photocatalytic U(VI) reduction and bisphenol a degradation. J. Environ. Chem. Eng. **10**(4) (2022). https://doi.org/10.1016/j.jece.2022.108097
44. H. Yu et al., ZnO/carbon quantum dots nanocomposites: one-step fabrication and superior photocatalytic ability for toxic gas degradation under visible light at room temperature. New J. Chem. **36**(4), 1031–1035 (2012). https://doi.org/10.1039/c2nj20959d

45. J. Di et al., Novel visible-light-driven CQDs/Bi2WO6 hybrid materials with enhanced photocatalytic activity toward organic pollutants degradation and mechanism insight. Appl. Catal. B Environ. **168–169**, 51–61 (2015). https://doi.org/10.1016/j.apcatb.2014.11.057
46. H. Yu et al., Carbon quantum dots/TiO2 composites for efficient photocatalytic hydrogen evolution. J. Mater. Chem. A **2**(10), 3344–3351 (2014). https://doi.org/10.1039/c3ta14108j
47. J. Di et al., Carbon quantum dots modified BiOCl ultrathin nanosheets with enhanced molecular oxygen activation ability for broad spectrum photocatalytic properties and mechanism insight. ACS Appl. Mater. Interfaces **7**(36), 20111–20123 (2015). https://doi.org/10.1021/acsami.5b05268
48. J. Di et al., The synergistic role of carbon quantum dots for the improved photocatalytic performance of Bi2MoO6. Nanoscale **7**(26), 11433–11443 (2015). https://doi.org/10.1039/c5nr01350j
49. J. Pan et al., Fabrication of carbon quantum dots modified granular SnO2 nanotubes for visible light photocatalysis. Mater. Lett. **170**, 187–191 (2016). https://doi.org/10.1016/j.matlet.2016.02.033
50. Y. Guo, P. Yao, D. Zhu, C. Gu, A novel method for the development of a carbon quantum dot/carbon nitride hybrid photocatalyst that responds to infrared light irradiation. J. Mater. Chem. A **3**(25), 13189–13192 (2015). https://doi.org/10.1039/c5ta02262b
51. P. Yang, J. Zhao, J. Wang, H. Cui, L. Li, Z. Zhu, Pure carbon nanodots for excellent photocatalytic hydrogen generation. RSC Adv. **5**(27), 21332–21335 (2015). https://doi.org/10.1039/c5ra01924a
52. M. Pirsaheb, A. Asadi, M. Sillanpää, N. Farhadian, Application of carbon quantum dots to increase the activity of conventional photocatalysts: a systematic review. J. Mol. Liq. **271**, 857–871 (2018). https://doi.org/10.1016/j.molliq.2018.09.064
53. W. Li et al., CQDs modified Co-BiOCl nanosheets with improved effective light absorption and charge separation for photocatalytic CIP degradation and NOX removal. Surf. Interfaces **27** (2021). https://doi.org/10.1016/j.surfin.2021.101541
54. Y. Zhao et al., Effective photodegradation of rhodamine B and levofloxacin over CQDs modified BiOCl and BiOBr composite: mechanism and toxicity assessment. J. Colloid Interface Sci. **627**, 180–193 (2022). https://doi.org/10.1016/j.jcis.2022.07.046
55. J. Xie et al., Up-conversion effect boosted the photocatalytic CO2 reduction activity of Z-scheme CPDs/BiOBr heterojunction. Inorg. Chem. Front. (2023). https://doi.org/10.1039/d3qi00701d
56. Y. Guan et al., C-scheme electron transfer mechanism: an efficient ternary heterojunction photocatalyst carbon quantum dots/Bi/BiOBr with full ohmic contact. J. Colloid Interface Sci. **624**, 168–180 (2022). https://doi.org/10.1016/j.jcis.2022.05.091
57. R. Tang Guo et al., Carbon quantum dots-modified Z-scheme Bi12O17Cl2/NiAl-LDH for significantly boosting photocatalytic CO2 reduction. J. Colloid Interface Sci. **627**, 343–354 (2022). https://doi.org/10.1016/j.jcis.2022.07.078
58. Q. Shi, A. Raza, L. Xu, G. Li, Bismuth oxyhalide quantum dots modified sodium titanate necklaces with exceptional population of oxygen vacancies and photocatalytic activity. J. Colloid Interface Sci. **625**, 750–760 (2022). https://doi.org/10.1016/j.jcis.2022.06.066
59. M. Hussein Abdurahman et al., Tunable band structure of synthesized carbon dots modified graphitic carbon nitride/bismuth oxychlorobromide heterojunction for photocatalytic degradation of tetracycline in water. J. Colloid Interface Sci. **629**, 189–205 (2023). https://doi.org/10.1016/j.jcis.2022.08.172
60. É. Whelan, F.W. Steuber, T. Gunnlaugsson, W. Schmitt, Tuning photoactive metal–organic frameworks for luminescence and photocatalytic applications. Coord. Chem. Rev. **437** (2021). https://doi.org/10.1016/j.ccr.2020.213757
61. Y. Peng et al., Enhanced upconversion luminescence in layered Bi2GdO4Cl:Yb3+/Er3+ by reducing OVs and its application in photocatalysis. J. Lumin. **252** (2022). https://doi.org/10.1016/j.jlumin.2022.119269
62. Y. Li, Z. Cheng, Y. Peng, S. Yang, Y. Zhang, Modulation of single-band red upconversion luminescence in Er3+/Yb3+ codoped bismuth oxyiodide nanoplates by bandgap engineering

and their application in NIR photocatalysis. Ceram. Int. **46**(5), 6351–6359 (2020). https://doi.org/10.1016/j.ceramint.2019.11.111
63. A. Heidari, A. Shahbazi, T.M. Aminabhavi, D. Barceló, S. Rtimi, A systematic review of clay-based photocatalysts for emergent micropollutants removal and microbial inactivation from aqueous media: status and limitations. J. Environ. Chem. Eng. **10**(6) (2022). https://doi.org/10.1016/j.jece.2022.108813
64. Z. Cao, Q. Wang, H. Cheng, Recent advances in kaolinite-based material for photocatalysts. Chinese Chem. Lett. **32**(9), 2617–2628 (2021). https://doi.org/10.1016/j.cclet.2021.01.009
65. H.H. Naing, Y. Li, J.B. Ghasemi, J. Wang, G. Zhang, Enhanced visible-light-driven photocatalysis of in-situ reduced of bismuth on BiOCl nanosheets and montmorillonite loading: synergistic effect and mechanism insight. Chemosphere **304** (2022). https://doi.org/10.1016/j.chemosphere.2022.135354
66. M.C. Dlamini et al., Photocatalytic abatement of phenol on amorphous TiO2-BiOBr-bentonite heterostructures under visible light irradiation. J. Ind. Eng. Chem. **111**, 419–436 (2022). https://doi.org/10.1016/j.jiec.2022.04.023
67. L. Gao, B. Zhou, W. Yu, N. Tian, H. Huang, Y. Zhang, Efficient mineralization of organic pollutants by in-situ synthesized illite/bismuth-based heterojunction photocatalytic composites. Appl. Surf. Sci. **612** (2023). https://doi.org/10.1016/j.apsusc.2022.155882
68. X. Hu, C. Li, Z. Sun, J. Song, S. Zheng, Enhanced photocatalytic removal of indoor formaldehyde by ternary heterogeneous BiOCl/TiO2/sepiolite composite under solar and visible light. Build. Environ. **168** (2020). https://doi.org/10.1016/j.buildenv.2019.106481
69. Y. Luo, K. Wang, T. Hu, X. Liu, Controlled synthesis of palygorskite/Bi5O7I hybrid microspheres with high efficient photodegradation of antibiotics. Colloids Surf. Physicochem. Eng. ASP. **616** (2021). https://doi.org/10.1016/j.colsurfa.2021.126225
70. C. Chuaichamet al., Efficient photocatalytic degradation of emerging ciprofloxacin under visible light irradiation using BiOBr/carbon quantum dot/saponite composite. Environ. Res. **212** (2022). https://doi.org/10.1016/j.envres.2022.113635
71. C. Chuaicham et al., Recent clay-based photocatalysts for wastewater treatment. Separations **10**(2) (2023). https://doi.org/10.3390/separations10020077

Heterojunction Perovskites for Hydrogen Production via Photocatalytic Water Splitting

Mano Ganapathy and Viswanathan Alagan

Abstract Photocatalysis is a very promising and efficient method with significant potential for energy and environmental applications. Photoreaction and chemical reaction occur upon the illumination of light on a photocatalyst, leading to numerous benefits in terms of degradation and hydrogen generation. Various types of metal oxide-based photocatalysts, including ZnO, TiO_2, $SrTiO_3$, WO_3, and others, have been utilized in the field of photocatalytic applications. Zinc oxide (ZnO), titanium dioxide (TiO_2), and strontium titanate ($SrTiO_3$) are metal oxides that have garnered significant attention in several fields due to their promising qualities. Recent year's perovskite materials have become great attention for the photocatalytic application due to their potential properties. However, the majority of the perovskites exhibited low light absorption as a result of their bandgap and the recombination of charge carriers. To address the aforementioned limitations in the perovskites, strategies including doping and narrow bandgap semiconductor coupling have been employed. The utilization of narrow bandgap semiconductor coupling is a viable approach to address these limitations, while preserving the intrinsic characteristics of the perovskite materials. In recent years, researchers have explored the various combinations of narrow bandgap metal sulfides and metal oxides, including CdS, CdO, ZnO, TiO_2, and $SrTiO_3$ with perovskite material. The process of semiconductor coupling results in the creation of a heterojunction, which serves to improve the sensitivity of visible light detection and minimize the recombination of photogenerated charge carriers. This chapter will elucidate the concept of perovskite based heterostructures in relation to their photocatalytic activity when exposed to visible light irradiation.

Keywords Perovskite · Semiconductor · Heterojunction · Hydrogen production

M. Ganapathy
Department of Physics, Bharathiyar Institute of Engineering for Women, Deviyakuruchi, Salem, Tamil Nadu, India

V. Alagan (✉)
Department of Physics, Anna University (UCE) - BIT Campus, Tiruchirappalli, Tamil Nadu, India
e-mail: alaganviswa@gmail.com

P. Sathishkumar (ed.), *Photocatalysis for Energy and Environmental Applications*, Green Energy and Technology, https://doi.org/10.1007/978-981-97-1939-6_9

1 Introduction

Energy sources play a crucial role in sustaining our contemporary way of life. Fuel has been shown to have significant implications in various domains, including transportation systems such as vehicles, airplanes, and ships, as well as in the generation of electricity, among other applications [1]. The phenomenon of population growth is closely associated with the process of fast industrialization, which in turn necessitates substantial use of fuel or electricity. This phenomenon will result in a significant increase in energy demand in the future. The classification of main energy resources encompasses two distinct categories: non-renewable and renewable energy sources [2]. The non-renewable energy sources encompass coal, oil, natural gas, and nuclear power plants [3–5]. The renewable energy resources encompass hydropower, geothermal energy, wind power, solar energy, biomass, and hydrogen [6, 7]. Electricity generation is derived from several sources such as, fossil fuels, nuclear power, hydropower, geothermal power, biofuel, wind, and solar energy contributing approximately 60%, 10%, 16%, 0.33%, 1.6%, 4%, and 2% respectively [8]. In the majority of instances, energy sources have exhibited certain restrictions, including the utilization of fossil fuels such as oil and natural gas, which result in the production of detrimental byproducts like carbon dioxide (CO_2) [9]. Additionally, the transportation of these fuels is subject to certain constraints. In recent years, many governments have opted to decrease the number of nuclear power plants in response to concerns regarding the disposal of radioactive waste, which poses significant threats to safety and public health [10]. Hydropower exhibits geographical constraints. Hence, it is imperative to identify energy resources that are cost-effective. The increasing focus on hydrogen production has been driven by its lack of hazardous waste production and its cleaner nature compared to other conventional energy options [8].

Hydrogen fuel is a versatile and clean energy carrier that has gained significant attention as a potential solution for a wide range of applications, including transportation, power generation, and industrial processes [11, 12]. Hydrogen is used in fuel cells, it produces no harmful emissions. This can help to reduce air pollution and greenhouse gas emissions, especially when hydrogen is produced using renewable energy sources. Hydrogen has a high energy-to-weight ratio, making it a viable option for long-range transportation and energy storage applications. Hydrogen can be used in various sectors, including transportation, industry, and power generation, making it a versatile energy carrier [13]. Electrolysis is commonly adopted method for the hydrogen production, which involves using electricity to split water (H_2O) into oxygen (O_2) and hydrogen (H_2) [14]. This process can use renewable electricity sources like solar or wind power, making it environmentally friendly. Steam Methane Reforming (SMR) is a process that extracts hydrogen from natural gas by reacting it with steam [14]. However, this method produces carbon dioxide (CO_2) as a byproduct. Thermochemical Processes is another method, such as high-temperature water splitting and biomass gasification, can also be used to produce hydrogen [15].

Producing and distributing hydrogen can be energy-intensive and costly, especially when not produced from renewable sources [1]. Hydrogen has a low energy density per unit volume, which can make storage and transportation challenging. Developing the necessary infrastructure for hydrogen production, distribution, and refueling is a significant challenge [1]. Hydrogen is flammable so, the safety measures are crucial when storing, transporting, and using hydrogen. The use of hydrogen as a clean energy carrier is an active area of research and development, with ongoing efforts to improve its production methods, storage solutions, and infrastructure. It is seen as a potential component of a sustainable energy future, particularly when produced using renewable energy sources and when integrated into a broader energy system.

Photocatalytic water splitting is a chemical process that uses light energy to split water molecules (H_2O) into hydrogen gas (H_2) and oxygen gas (O_2) [6, 16, 17]. This process is of significant interest because hydrogen is a clean and efficient fuel source, and the production of hydrogen through water splitting using renewable energy sources can be environmentally friendly.

1.1 Photocatalytic Water Splitting

Photocatalytic water splitting has several advantages, including the use of abundant and renewable sunlight as an energy source and the production of clean hydrogen fuel [18]. However, it also faces challenges and limitations. Achieving high hydrogen conversion efficiency is still challenging. Researchers are continually working on improving the efficiency of photocatalysts and optimizing reaction conditions. Photocatalysts may degrade over time due to exposure to harsh reaction conditions and accumulated damage from repeated charge carrier generation. Finding photocatalysts that are efficient, cost-effective, and stable remains a focus of research [7, 19]. Transitioning from laboratory-scale experiments to large-scale production for practical applications is a significant challenge. Despite these challenges, photocatalytic water splitting holds great promise as a clean and sustainable method for hydrogen production, which can be used in fuel cells and other applications for clean energy storage and utilization. The ongoing research aims to address these challenges and makes this technology more practical and widespread (Fig. 1).

A photocatalyst is a substance that can initiate or accelerate a chemical reaction when exposed to light, typically in the form of photons. Photocatalysts are widely used in various applications, including environmental remediation, energy production, and chemical synthesis [20]. They work by absorbing light energy and using it to facilitate a chemical reaction, but they are not consumed in the process, making them reusable. Photocatalysts have the ability to absorb photons from light sources, including sunlight or artificial light. When photons are absorbed, photocatalysts can generate electron-hole pairs within their structure [21]. This separation of charges is a crucial step in many photocatalytic reactions. Photocatalysts facilitate redox (reduction-oxidation) reactions by transferring electrons from one reactant to another. This allows them to drive various chemical reactions, including the degradation of pollutants or the production of energy carriers like hydrogen gas. Photocatalysts can

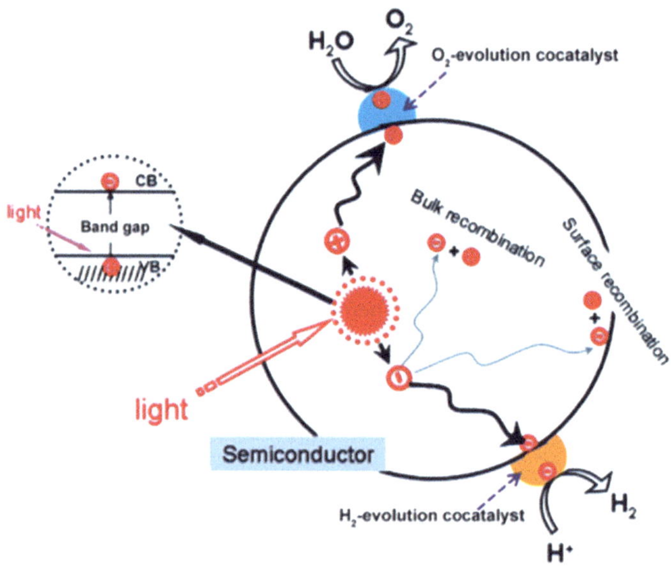

Fig. 1 Schematic diagram of photocatalytic water splitting (Applied Surface Science, Volume 351, 2015, 779–793, https://doi.org/https://doi.org/10.1016/j.apsusc.2015.05.171)

be designed or selected for specific reactions, offering a level of control and selectivity in chemical transformations. Common photocatalytic materials include semiconductors like titanium dioxide (TiO_2) and various metal oxides [18] are widely used in photocatalytic applications. Researchers are continually exploring new materials and improving existing ones to enhance the efficiency and versatility of photocatalysts for various applications, with a focus on sustainability and environmental benefits. Photocatalytic water splitting using perovskite materials is a promising area of research in the field of renewable energy and hydrogen production.

Perovskite is a type of mineral and a crystal structure that has gained significant attention and popularity in various scientific and technological applications, particularly in the fields of materials science, physics, and renewable energy. The name "perovskite" originally referred to a specific mineral called calcium titanate ($CaTiO_3$), which was named after Russian mineralogist Lev Perovski [22]. Perovskite materials have a unique crystalline structure characterized by the general chemical formula ABX3, where "A" represents a larger cation (positively charged ion), "B" represents a smaller cation, and "X" represents an anion (negatively charged ion) (Fig. 2) [23]. The distinctive perovskite structure consists of a three-dimensional network of corner-sharing BX6 octahedra, forming a cubic or tetragonal lattice. Perovskite is a class of materials that have gained significant attention in various fields, including photovoltaics (solar cells), optoelectronics, and catalysis [24, 25]. The name "perovskite" is derived from the mineral perovskite, which has a similar crystal structure.

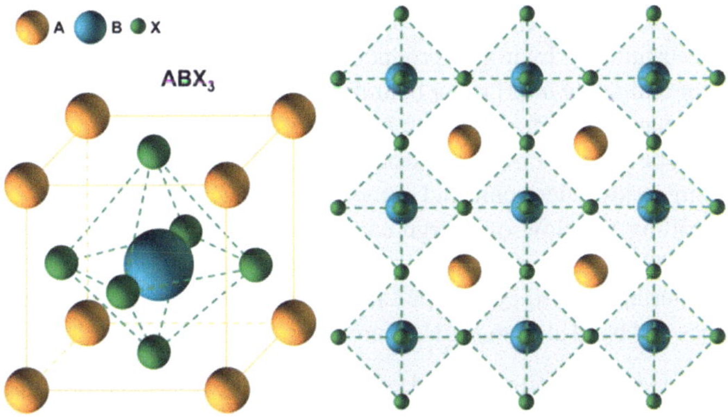

Fig. 2 Schematic diagram of perovskite structure (ACS Energy Lett. 2022, 7, 8, 2490–2514)

1.2 Perovskite Based Photocatalyst

Perovskite materials have a specific crystal structure known as the perovskite structure. It consists of a cubic arrangement of oxygen atoms, with a metal atom at the center of the cube and other metal or organic cations located at the corners of the cube. Many perovskite compounds are semiconductors, meaning they can conduct electricity under certain conditions [26]. This property makes them attractive for use in solar cells and other electronic devices. Perovskite materials have excellent optoelectronic properties, such as a high absorption coefficient, which makes them efficient at capturing and converting sunlight into electricity. Perovskite solar cells have shown great promise in achieving high power conversion efficiencies [27]. The bandgap of perovskite materials can be engineered by changing the composition of the cations and anions. This tunability allows researchers to design perovskite materials with specific optical and electronic properties for various applications. Perovskite materials can be fabricated using solution-based processes, which are relatively low-cost and suitable for large-scale production. This makes them attractive for commercial applications in the solar industry.

Perovskite materials for photocatalytic water splitting show efficient light absorption and charge separation [28]. High conversion efficiency and tunability of perovskite compositions for specific photocatalytic applications [29]. Utilization of renewable energy is (solar) for hydrogen production. Perovskite materials can be sensitive to moisture and environmental factors, which can affect their long-term stability. Some perovskite materials may contain toxic elements, such as lead. Efforts are ongoing to develop lead-free perovskites for environmental safety. Scaling up perovskite-based photocatalytic systems for practical, large-scale hydrogen production is a challenge [28]. Developing perovskite compositions and structures that optimize performance and durability is an active area of research. Research in the field of perovskite-based photocatalytic water splitting is ongoing, and scientists are working to address these challenges while harnessing the potential of perovskite materials for sustainable hydrogen production and other clean energy applications.

A heterojunction is an interface or boundary between two different semiconductor materials that have distinct electronic properties, such as their bandgap, conductivity type (n-type or p-type), and electron affinity [30]. Heterojunctions are essential components in various electronic and optoelectronic devices, and they play a crucial role in controlling the flow of charge carriers (electrons and holes) across semiconductor interfaces. Heterojunctions are interfaces between two different semiconductor materials, and they can be classified into several types based on their characteristics and band alignment. Here are some common types of heterojunctions:

1. Straddling Heterojunction (Type-I): In a straddling heterojunction, the conduction band and valence band edges of one semiconductor material lie above those of the other material. In this particular alignment, the conduction and valence bands of narrow or small band gap materials are positioned within the band locations of a wider band gap material (as seen in Fig. 3). Within this particular alignment, the charge carriers that are created by the incident light encounter a potential barrier at the interface between the two semiconductors [31].
2. Staggered Heterojunction (Type-II): In the context of type-II alignment, the conduction band of a material with a smaller band gap is positioned between the conduction and valence bands of another material with a wider band gap.

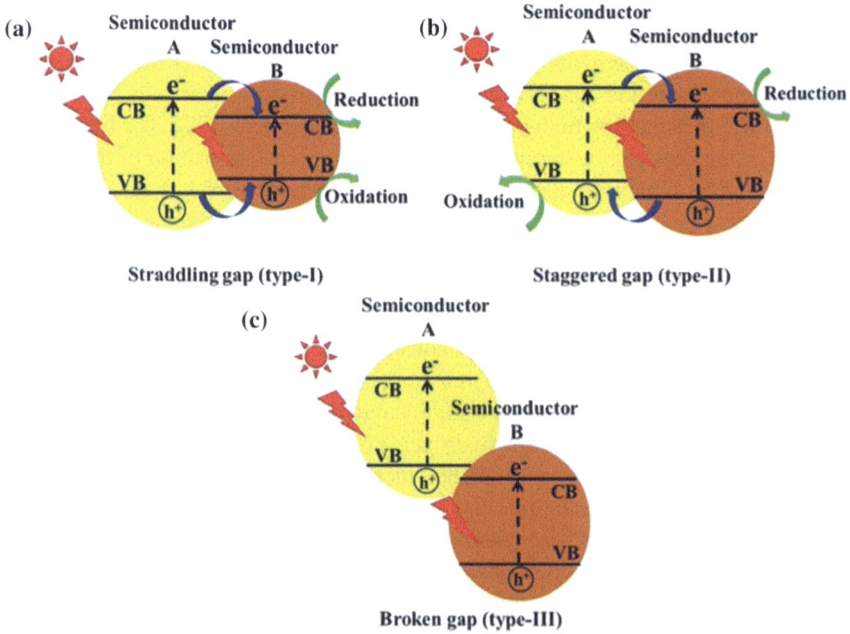

Fig. 3 Schematic diagram of types of heterojunction (Environ Chem Lett 19, 4563–4601 (2021). https://doi.org/https://doi.org/10.1007/s10311-021-01295-8)

This particular band alignment is mostly utilized in photovoltaic applications in order to mitigate the recombination of photo-generated charge carriers. Furthermore, this particular form of alignment significantly improves the breadth of spectral response [32].

3. Broken-Gap or Offset Heterojunction (Type-III): In a broken-gap or offset heterojunction, the conduction band and valence band edges of one material are not aligned with those of the other material. This creates a discontinuity in the band structure. The resulting band alignment can affect the flow of charge carriers across the junction and is often used to control electron and hole confinement [33].

Heterojunctions can confine charge carriers to specific regions, reducing electron-hole recombination and enhancing device efficiency. Different materials in a heterojunction can have varying bandgap energies, allowing engineers to design devices with specific electronic and optical properties [34]. Heterojunctions often result in devices with superior performance characteristics compared to homojunction devices, making them valuable in modern electronics. Heterojunctions are fundamental to the operation of many semiconductor devices and are a key element in advancing technology in areas such as electronics, photonics, and renewable energy. Researchers continue to explore new materials and design strategies to optimize heterojunctions for various applications [35].

In recent years, various types of heterojunctions have been synthesized with the aim of augmenting photocatalytic hydrogen production efficiency. This chapter discusses about the metal oxide perovskite based heterojunction and halide based heterojunction in detail.

2 Metal Oxide Perovskite Based Heterojunction

Juan et al., prepared $Bi_4Ti_3O_{12}/ZnIn_2S_4$ s-type heterojunction with different weight ratios of $Bi_4Ti_3O_{12}$ (10%, 20%, 30% and 50%) which were labeled as 10% BTO/ZIS, 20% BTO/ZIS, 30% BTO/ZIS and 50% BTO/ZIS, respectively and it measures the photocatalytic hydrogen production efficiency (Fig. 4). The low-temperature water bath reaction method was adopted for the preparation of $Bi_4Ti_3O_{12}/ZnIn_2S_4$ heterojunction. The enhanced absorption of visible light, efficient charge transfer channel, and improved generation, separation, and migration of photogenerated charges are achieved through the construction of a sheet-sheet-bonded microstructure and S-scheme heterojunction. Consequently, these factors contribute to the overall enhancement of the photocatalytic reduction reaction activity under visible light response. The findings of the performance test indicate that the BTO/ZIS composite with a composition of 20% exhibits the most favorable catalytic performance, reaching a value as high as 19.8 mmol h^{-1} g^{-1} and its performance is 2.1 times more than that of pure ZIS [36].

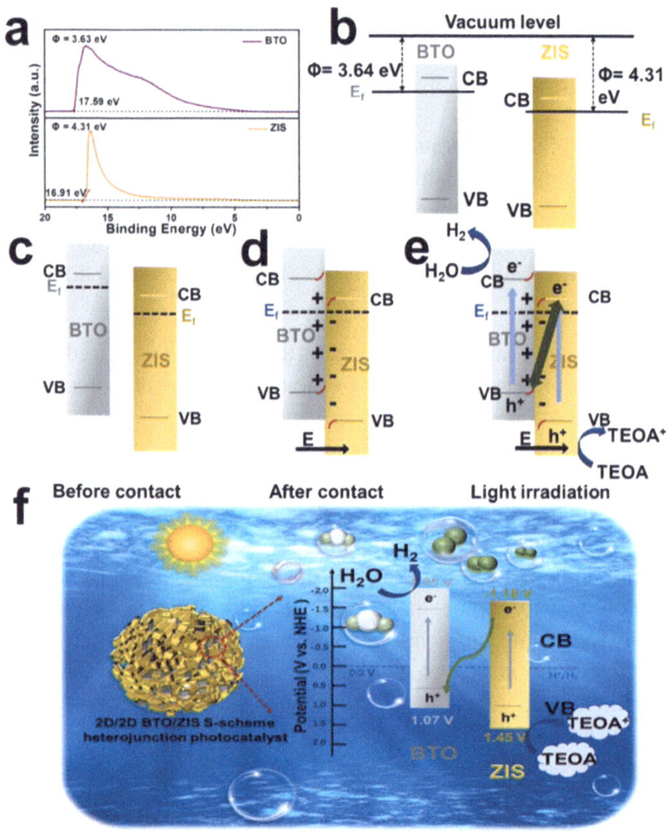

Fig. 4 UPS spectra of BTO and ZIS. Deduced E_f of **a** BTO and **b** ZIS based on work function. heterojunction mechanism, **c** before contact, **d** after contact and **e** under light irradiation. **f** photocatalytic H_2 production reaction mechanism of BTO/ZIS heterojunction

$CdSe/SrTiO_3$ nanocomposites were synthesized by varying concentrations of Cd (0.001 M, 0.01 M, and 0.1 M) denoted as $CdSe/SrTiO_3$-1, $CdSe/SrTiO_3$-2, and $CdSe/SrTiO_3$-3, respectively. The nanocomposites were investigated for their potential in photcatalytic hydrogen production. The solution was prepared by dispersing 0.025 g of the photocatalyst in 100 mL of an aqueous solution containing 0.35 M Na_2S and 0.25 M Na_2SO_3 as sacrificial reagents. The catalyst $CdSe/SrTiO_3$-3 produce H_2 evolution of 3350.53 μmol over a period of 8 h. The electrochemical impedance spectrum (EIS) of $CdSe/SrTiO_3$-3 demonstrates the presence of a tiny arc radius. Additionally, the photocurrent density of $CdSe/SrTiO_3$-3 is measured to be 0.26 $\mu A\ cm^{-2}$. $CdSe/SrTiO_3$ nanocomposites have significantly enhanced photocatalytic hydrogen production performance due to their increased absorption of visible light, efficient separation and transmission of photogenerated electrons, and reduced recombination of electron-hole pairs [37].

Naveen et al., were effectively constructed novel $BaTiO_3/CaFe_2O_4$ heterojunction composites utilizing the sonication-calcination method. The composites were made with varying weight ratios of $CaFe_2O_4$–$BaTiO_3$. The $BaTiO_3$ and $CaFe_2O_4$ powders used in the fabrication process were synthesized separately using hydrothermal and sol-gel processes, respectively. The composites obtained in this study were designated as BT/CF10, BT/CF20, BT/CF30, BT/CF40, BT/CF50, and BT/CF60. These labels correspond to the weight ratios of $BaTiO_3/CaFe_2O_4$ at 10%, 20%, 30%, 40%, 50%, and 60%, respectively. The significant cooperative effect observed between BaTiO3 and $CaFe_2O_4$ is a crucial factor contributing to the notable improvement in the photocatalytic efficiency of the heterojunction. This enhancement can be attributed to the increased absorption of visible light and the efficient separation of electron-hole pairs at the interfacial junction of $BaTiO_3$ and $CaFe_2O_4$, facilitated by the electric field. The band edges of the $BaTiO_3$ and $CaFe_2O_4$ constituents inside the composite material are arranged in a manner that results in the formation of a type-II heterojunction. This configuration generates a driving force that facilitates the separation of the charge carriers [38].

Jiaqi et al., were prepared fusiform $g\text{-}C_3N_4/CaTiO_3$ nano-heterojunctions by the hydrothermal co-deposition approach. The samples were modified by g-C3N4 with the different molar ratios ($g\text{-}C_3N_4/CaTiO_3$) of 1/13.5, 2/13.5, and 3/13.5 and were labeled as CaeC-1, CaeC-2 and CaeC-3, respectively. The fusiform nano-heterojunction exhibits noteworthy improvement in photocatalytic performance for hydrogen production, exhibiting a catalytic activity that is approximately 18 times more than that of unmodified $CaTiO_3$. The primary factor contributing to the enhancement of photocatalytic hydrogen production is the presence of a distinctive nano-heterojunction. This nano-heterojunction facilitates the transfer of photon-generated carriers by utilizing the internal electric field of the heterojunction. As a result, the separation of photon-generated carriers is promoted, leading to an increased lifetime of photon-generated electrons. This extended lifetime reduces the recombination of Hþ ions, ultimately resulting in an improved efficiency of photocatalytic hydrogen production [22].

Xueyon et al., were synthesized type II heterojunction including $PbTiO_3/CdS$ composites that was successfully achieved and afterwards utilized in the field of piezo-photocatalytic hydrogen production, marking the first instance of such application. When subjected to light irradiation alone, the $PbTiO_3/CdS$ system experiences complete shielding of the spontaneous polarization field of $PbTiO_3$. Consequently, this shielding impedes the effective separation of carriers, resulting in suboptimal photocatalytic activity. Following the implementation of ultrasonication, the $PbTiO_3$/CdS-10% composite demonstrated a significantly enhanced rate of hydrogen (H_2) generation by piezo-photocatalysis, reaching 849.0 $\mu mol\ h^{-1}\ g^{-1}$. This value surpasses the H_2 production rates achieved through individual piezocatalysis (400.6 $\mu mol\ h^{-1}\ g^{-1}$) and photocatalysis (98.9 $\mu mol\ h^{-1}\ g^{-1}$). The enhanced hydrogen production of $PbTiO_3/CdS$ through piezo-photocatalysis can be ascribed to the presence of alternating piezoelectric potential. This potential facilitates the separation of charge carriers and hinders the complete screening of the polarization electric field by the carriers. The utilization of a binary piezoelectric semiconductor material combination has the potential to effectively impede the recombination of charge carriers, hence leading to a notable improvement in its catalytic performance. In addition, it is noteworthy that the piezo-photo coupling activity exhibited by the $PbTiO_3$/CdS-10% composite material is approximately 7.2, 2.5, and 2.7 times greater than that observed in the $PbTiO_3$-c/CdS-10% composite, pure $PbTiO_3$, and CdS individually, respectively [39].

3 Halide Based Heterojunction

Metal Halide perovskites (MHPs) have garnered considerable scholarly attention in the field of photocatalysis in recent times. This is mostly owing to their remarkable optoelectronic characteristics, including favorable band structures, extended lifetimes of charge carriers, high quantum yields of photoluminescence, and various other notable attributes [25, 40, 41]. It has been found that the properties of MHPs exhibit a wide range of diversity and distinctiveness, suggesting that their potential applications should be further expanded. The A+ cation, represented by methylammonium (MA+), formamidinium (FA+), and cesium (Cs+), along with the B2+ cations (such as lead and tin), and X anions (chlorine, bromine, and iodine), participate in the formation of metal halide perovskites [42]. These perovskites exhibit a corner-sharing arrangement of [BX6]− [4] units, with the A+ cation exhibiting a coordination number of 12. The bonding between the B-site and hybridized X orbitals, characterized by the valence and conduction bands of high-performing metal halide perovskites (MHPs), distinguishes them from conventional tetrahedral linked semiconductors like gallium arsenide (GaAs) [35]. The heterojunction can be established through the spatial separation of semiconductors, which enables the completion of charge transfer. This can be achieved by employing an approximate redox shuttle or by utilizing a heterojunction to connect two semiconductors in a strong and cohesive manner. Through precise control of the interfacial contact, a proficient exchange of electrons in the heterojunction is achieved, resulting in a notable enhancement of the conversion of H_2O–H_2 under the influence of visible light [43].

Yixuan et al., were prepared the $CsCu_2I_3$ (CCI) and graphite carbon nitride (CN) heterojunction by the in-situ hot-injection technique. The CCI/CN composite exhibits a notably improved performance in terms of photocatalytic hydrogen evaluation production through the hydroiodic acid splitting process. The photocatalytic hydrogen evaluation rate achieved by the CCI/CN composite is measured at 2212 umol/g/h, which is four times greater than that of CCI and nearly ten times greater than that of CN. The interface between CN and CCI facilitates the formation of a distinct N–I bond, which enables the rapid transfer of photogenerated electrons. This N–I bond plays a crucial role in establishing a type II mechanism, hence enhancing the performance of the photogenerated hot electron (PHE) process [44].

Yunqi et al., were adopted solvent engineering strategy to achieve a successful in situ formation of heterojunctions at the interface of (methylammonium bismuth iodide) $MA_3Bi_2I_9$ and tri(dimethylammonium)hexa-iodobismuthate (DMA_3BiI_6). The valence band (VB) and conduction band (CB) locations of $MA_3Bi_2I_9$ and DMA_3BiI_6 exhibited a closely aligned type-II heterostructure, which facilitated the development of an interfacial charge transfer pathway at the interface of $MA_3Bi_2I_9$ and DMA_3BiI_6. This resulted in an elongated lifespan for exciton diffusion, with a time constant of around 38 ns. This effect facilitated enhanced electron transport, leading to a reduction in H+ ions and so increasing the process of photocatalytic hydrogen evolution. The electrochemical reactivity of the as-precipitated samples was assessed using electrochemical impedance spectroscopy (EIS). The results showed that the $MA_3Bi_2I_9/DMA_3BiI_6$ sample exhibited the lowest charge transfer resistance (Rct) value of 1.15 kΩ cm^2. This indicates a relatively high electrochemical performance, which ultimately contributes to the significantly improved efficiency and stability of the solar-induced hydrogen evolution in the $MA_3Bi_2I_9$/DMA_3BiI_6 perovskite heterojunctions. Notably, these enhancements were achieved without the need for any noble metal co-catalyst, even when exposed to visible light irradiation. Under visible-light illumination ($\lambda > 420$ nm) at an intensity of 100 mW cm^{-2}, the powdered $MA_3Bi_2I_9/DMA_3BiI_6$ heterostructured composite (referred to as BBP-5) exhibits a hydrogen (H_2) evolution rate of 198.2 μmol h^{-1} g^{-1}, even in the absence of platinum (Pt) co-catalysts [29].

Tian et al., were prepared the compound $Cs_2AgBiBr_6$/xRGO (referred to as CABB/xRGO) which was synthesized using a photo-reduction method. In particular, the sample consists of 200 mg of CABB and varying amounts of GO, specifically 2, 5, and 10 mg, corresponding to mass ratios of 1%, 2.5%, and 5% respectively. The hydrogen (H_2) production achieved with CABB/2.5%RGO was found to be 80 times more compared to the bare CABB material. This observation was made in a solution containing saturated HBr and H_3PO_2, with a volume ratio of 5:1. In addition, it was observed that the CABB/2.5%RGO composite exhibited remarkable stability in a saturated solution and retained its high catalytic activity for hydrogen evolution over a continuous period of 120 h. It has been demonstrated that the utilization of reduced graphene oxide (RGO) can enhance the process of separating photogenerated excitons and facilitating the transfer of electrons from carbon-based catalyst (CABB) to RGO. As a result, this leads to a reduction in charge recombination and an improvement in the production of hydrogen gas (H_2) [45].

Zhuo et al., were prepared nano-flower structure $Cs_2AgBiBr_6$/Ag_3PO_4 (CA) heterojunction for photocatalytic hydrogen production and organic pollutants degradation. CA was found to have a high crystalline quality and a significant interface effect. This remarkable photocatalyst possesses a high redox capacity, efficient catalysis behavior, and electron-hole separation, in addition to being highly adsorptive, optically active, and stable. Within six hours, approximately 4915.4 mol/g of photocatalytic H_2 evolution (PHE) were produced [46].

Bianca et al., were prepared binary $Cs_3Bi_2I_9$/g-C_3N_4 heterojunction prepared by the wet chemical method for the photocatalytic process. The weight proportion of $Cs_3Bi_2I_9$ in relation to g-C_3N_4 was either 1 or 10 wt%. These respective samples are identified as CNCSBI001 and CNCSBI01. The photocatalytic activity of the pure g-C_3N_4 sample under solar-light conditions, where the hydrogen evolution process

(HER) occurred with a rate of 496.85 μmol g^{-1} h^{-1}. This activity can be attributed to the modest bandgap and unique electronic structure of g-C_3N_4. The introduction of metal halide perovskite resulted in a notable enhancement of around 46% in the photocatalytic activity of the CNCSBI001 sample. This improvement was observed by a hydrogen evolution rate (HER) of approximately 920.76 μmol g^{-1} h^{-1}. The increase in the efficiency of separating and transporting photoinduced electron-hole pairs in the photocatalyst was confirmed using photoelectrochemical experiments. The observed phenomena of suppressed charge recombination and efficient charge transfer can be attributed to the creation of heterojunctions. The improved photocatalytic performance can be ascribed to the effective direct scheme heterojunction with a significantly large redox capacity. The findings derived from this study validate the potential for enhancing the photocatalytic efficacy of individual semiconductors by means of composite creation [25].

4 Summary

This chapter presents a comprehensive summary of the perovskite structure and its heterojunctions with different materials and analyzed its photocatalytic hydrogen evaluation performances. In recent years, the formation of heterojunctions provides great attention due to its potential properties. However, certain obstacles continue to persist in heterojunction materials, including the recombination of photogenerated charge carriers, light absorption qualities, corrosion, separation and recycling. In order to address the aforementioned issues, it is imperative to explore the vast research opportunities that exist in the field of heterojunction materials.

References

1. I. Staffell et al., The role of hydrogen and fuel cells in the global energy system. Energy Environ. Sci. **12**, 463–491 (2019)
2. P.J. Megia, A.J. Vizcaino, J.A. Calles, A. Carrero, Hydrogen production technologies: from fossil fuels toward renewable sources. A mini review. Energy Fuels **35**, 16403–16415 (2021)
3. A. de Klerk, *Transport Fuel: biomass-, Coal-, Gas- and Waste-to-Liquids Processes. Future Energy: improved, Sustainable and Clean Options for Our Planet* (Elsevier, 2020). https://doi.org/10.1016/B978-0-08-102886-5.00010-4
4. Y. Xu et al., Facile fabrication of molybdenum compounds (Mo2C, MoP and MoS2) nanoclusters supported on N-doped reduced graphene oxide for highly efficient hydrogen evolution reaction over broad pH range. Chem. Eng. J. **417**, 129233 (2021)
5. Z. Abdin et al., Hydrogen as an energy vector. Renew. Sustain. Energy Rev. **120**, 109620 (2020)
6. W. Wang, M.O. Tadé, Z. Shao, Research progress of perovskite materials in photocatalysis- and photovoltaics-related energy conversion and environmental treatment. Chem. Soc. Rev. **44**, 5371–5408 (2015)
7. J. Ran, J. Zhang, J. Yu, M. Jaroniec, S.Z. Qiao, Earth-abundant cocatalysts for semiconductor-based photocatalytic water splitting. Chem. Soc. Rev. **43**, 7787–7812 (2014)

8. Electricity in the U.S.—U.S. Energy Information Administration (EIA). https://www.eia.gov/energyexplained/electricity/electricity-in-the-us.php
9. T. Su, Z. Qin, H. Ji, Y. Jiang, G. Huang, Recent advances in the photocatalytic reduction of carbon dioxide. Environ. Chem. Lett. **14**(1), 99–112 (2015)
10. W. Mu et al., Porous three-dimensional reduced graphene oxide merged with WO3for efficient removal of radioactive strontium. Appl. Surf. Sci. **423**, 1203–1211 (2017)
11. K. Maeda, Photocatalytic water splitting using semiconductor particles: history and recent developments. J. Photochem. Photobiol. C **12**, 237–268 (2011)
12. K.H. Ng, S.Y. Lai, C.K. Cheng, Y.W. Cheng, C.C. Chong, Photocatalytic water splitting for solving energy crisis: myth, fact or busted? Chem. Eng. J. **417**, 128847 (2021)
13. Q. Yao, Y. DIng, Z.H. Lu, Noble-metal-free nanocatalysts for hydrogen generation from boron- and nitrogen-based hydrides. Inorg. Chem. Front. **7**, 3837–3874 (2020)
14. A.I. Osman et al., Hydrogen production, storage, utilisation and environmental impacts: a review. Environ. Chem. Lett. **20**(1), 153–188 (2021)
15. J.E. Funk, Thermochemical hydrogen production: past and present. Int. J. Hydrog. Energy **26**, 185–190 (2001)
16. Y. Wei et al., Efficient sequential harvesting of solar light by heterogeneous hollow shells with hierarchical pores. Natl. Sci. Rev. **7**, 1638–1646 (2020)
17. G.I.N. Waterhouse et al., Hydrogen production by tuning the photonic band gap with the electronic band gap of TiO2. Sci. Rep. **3**, 1–5 (2013)
18. H. Ahmad, S.K. Kamarudin, L.J. Minggu, M. Kassim, Hydrogen from photo-catalytic water splitting process: a review. Renew. Sustain. Energy Rev. **43**, 599–610 (2015)
19. H. Hou, X. Zhang, Rational design of 1D/2D heterostructured photocatalyst for energy and environmental applications. Chem. Eng. J. **395**, 125030 (2020)
20. A.A. Ismail, D.W. Bahnemann, Photochemical splitting of water for hydrogen production by photocatalysis: a review. Sol. Energy Mater. Sol. Cells **128**, 85–101 (2014)
21. F.T. Wagner, G.A. Somorjai, Photocatalytic hydrogen production from water on Pt-free SrTiO3 in alkali hydroxide solutions. Nature **285**, 559–560 (1980)
22. J. Pan et al., The enhanced photocatalytic hydrogen production of the fusiform g-C3N4 modification CaTiO3 nano-heterojunction. Int. J. Hydrogen Energy **43**, 19019–19028 (2018)
23. P. Jayabal, V. Sasirekha, J. Mayandi, K. Jeganathan, V. Ramakrishnan, A facile hydrothermal synthesis of SrTiO3 for dye sensitized solar cell application. J. Alloy. Compd. **586**, 456–461 (2014)
24. Y. Ma et al., Synthesis of nanocrystalline strontium titanate by a sol-gel assisted solid phase method and its formation mechanism and photocatalytic activity. CrystEngComm **21**, 3982–3992 (2019)
25. B.M. Bresolin, P. Sgarbossa, D.W. Bahnemann, M. Sillanpää, Cs3Bi2I9/g-C3N4 as a new binary photocatalyst for efficient visible-light photocatalytic processes. Sep. Purif. Technol. **251**, 117320 (2020)
26. J. Miao, F. Zhang, Recent progress on highly sensitive perovskite photodetectors. J. Mater. Chem. C **7**, 1741–1791 (2019)
27. S.K. Karuturi et al., Perovskite photovoltaic integrated CdS/TiO2 photoanode for unbiased photoelectrochemical hydrogen generation. ACS Appl. Mater. Interfaces **10**, 23766–23773 (2018)
28. A. Bloesser, R. Marschall, Layered perovskite nanofiber heterojunctions with tailored diameter to enhance photocatalytic water splitting performance. ACS Appl. Energy Mater. **1**, 2520–2525 (2018)
29. Y. Tang et al., In situ formation of bismuth-based perovskite heterostructures for high-performance cocatalyst-free photocatalytic hydrogen evolution. Adv. Funct. Mater. **30**, 2006919 (2020)
30. J.Y. Tang et al., Two-dimensional interface engineering of g-C3N4/g-C3N4 nanohybrid: synergy between isotype and p-n heterojunctions for highly efficient photocatalytic CO2 reduction. Chem. Eng. J. **466**, 143287 (2023)

31. Y. Lin et al., The evolution from a typical type-I CdS/ZnS to type-II and Z-scheme hybrid structure for efficient and stable hydrogen production under visible light. ACS Sustain. Chem. Eng. **8**, 4537–4546 (2020)
32. A. Das, M. Patra, P.M. Kumar, M. Bhagavathiachari, R.G. Nair, Role of type II heterojunction in ZnO–In2O3 nanodiscs for enhanced visible-light photocatalysis through the synergy of effective charge carrier separation and charge transport. Mater. Chem. Phys. **263** (2021)
33. C. Lei et al., Broken-gap type-III band alignment in WTe2/HfS2 van der Waals heterostructure. J. Phys. Chem. C **123**, 23089–23095 (2019)
34. K. Afroz, M. Moniruddin, N. Bakranov, S. Kudaibergenov, N. Nuraje, A heterojunction strategy to improve the visible light sensitive water splitting performance of photocatalytic materials. J. Mater. Chem. A **6**, 21696–21718 (2018)
35. R. Mishra et al., A review on Z/S—Scheme heterojunction for photocatalytic applications based on metal halide perovskite materials. Appl. Surf. Sci. Adv. **9**, 100241 (2022)
36. J. Wang, Y. Shi, H. Sun, W. Shi, F. Guo, Fabrication of Bi4Ti3O12/ZnIn2S4 S-scheme heterojunction for achieving efficient photocatalytic hydrogen production. J. Alloy. Compd. **930**, 167450 (2023)
37. J. Han et al., Synthesis of CdSe/SrTiO3 nanocomposites with enhanced photocatalytic hydrogen production activity. Appl. Surf. Sci. **467–468**, 1033–1039 (2019)
38. N.K. Veldurthi, N. KrishnaRao Eswar, S.A. Singh, G. Madras, Cooperative effect between BaTiO3 and CaFe2O4 in a cocatalyst-free heterojunction composite for improved photochemical H2 generation. Int. J. Hydrog. Energy **43**, 22929–22941 (2018)
39. X. Huang et al., Insight into the piezo-photo coupling effect of PbTiO3/CdS composites for piezo-photocatalytic hydrogen production. Appl. Catal. B **282**, 119586 (2021)
40. S. Li, J. Luo, J. Liu, J. Tang, Self-trapped excitons in all-inorganic halide perovskites: fundamentals, status, and potential applications. J. Phys. Chem. Lett. **10**, 1999–2007 (2019)
41. K.M. McCall, C.C. Stoumpos, S.S. Kostina, M.G. Kanatzidis, B.W. Wessels, Strong electron-phonon coupling and self-trapped excitons in the defect halide perovskites A3M2I9 (A = Cs, Rb; M = Bi, Sb). Chem. Mater. **29**, 4129–4145 (2017)
42. D.J. Kubicki et al., Local structure and dynamics in methylammonium, Formamidinium, and Cesium Tin(II) mixed-halide perovskites from 119Sn solid-state NMR. J. Am. Chem. Soc. **142**, 7813–7826 (2020)
43. J. Sun et al., Emerging two-dimensional organic semiconductor-incorporated perovskites–A fascinating family of hybrid electronic materials. J. Am. Chem. Soc. (2023). https://doi.org/10.1021/JACS.3C02143
44. Y. Lv et al., In-situ construction of lead-free halide perovskite CsCu2I3/g-C3N4 heterojunction for photocatalytic H2 generation. Sep. Purif. Technol. **316**, 123813 (2023)
45. T. Wang, D. Yue, X. Li, Y. Zhao, Lead-free double perovskite Cs2AgBiBr 6/RGO composite for efficient visible light photocatalytic H2 evolution. Appl. Catal. B **268**, 118399 (2020)
46. Z. Chen et al., S-scheme Cs2AgBiBr6/Ag3PO4 heterojunction with efficient photocatalysis performance for H2 production and organic pollutant degradation under visible light. Sep. Purif. Technol. **295**, 121250 (2022)

Photocatalytic and Photoelectrochemical Technologies for Hydrogen Production: Commercialization Aspect

Rajendran Lakshmi Priya, Boopathi Shagunthala Hariprasad, Chettipalayam Arunasalam Dhayanithi, Ponnusamy Paunkumar, Chellapandi Bhuvaneswari, and Sundaram Ganesh Babu

Abstract Sustainable solar energy and photoactive materials are harnessed to efficiently utilize naturally occurring water resources for the production of environmentally friendly green fuel. Globally, conventional sources primarily rely on fossil fuels, which not only contribute to energy depletion but also pose significant environmental threats such as air pollution and ozone layer depletion. To address these challenges and transition away from conventional fossil fuels, eco-friendly hydrogen fuels emerge as a compelling alternative. Hydrogen offers several advantages, including being carbon-free, possessing a high energy density, and enabling rapid replenishment. The global effort to produce over 50 million tons of hydrogen per year aims to replace fossil fuels in various applications. Ongoing research focuses on enhancing the performance of low-cost photocatalytic materials and ensuring their long-term stability to meet the demands of large-scale hydrogen production. This chapter focuses on the profitable aspects of ultraviolet and visible light-based single & multi-component photocatalysts, cocatalyst-loaded composites, as well as carbon-based support composites for photocatalytic and photoelectrochemical water splitting. The effective engineering of these photocatalytic materials, characterized by high photocatalytic rates, holds the potential for widespread adoption in industrial hydrogen gas production. This approach aligns with the growing need for sustainable energy solutions and contributes to a greener and more environmentally responsible energy landscape.

Keywords Photocatalytic · Photoelectrochemical · Water splitting · Large-scale · Electron-hole pairs · Commercialization

---

R. L. Priya · B. S. Hariprasad · C. A. Dhayanithi · P. Paunkumar · C. Bhuvaneswari · S. G. Babu (✉)
Nano-Catalysis Research Lab, Department of Chemistry, School of Advanced Sciences, Vellore Institute of Technology, Vellore 632014, Tamilnadu, India
e-mail: ganeshbabu.s@vit.ac.in; babuchemist007@gmail.com

P. Sathishkumar (ed.), *Photocatalysis for Energy and Environmental Applications*, Green Energy and Technology, https://doi.org/10.1007/978-981-97-1939-6_10

1 Introduction

Over the past several decades, fossil fuels have been a tremendous source of energy for mankind. Besides the fossil fuels like petroleum, coal, natural gas, and other mineral oils which were formed millions of years ago with the help of fossil matter, the extraction of fossil fuels is a bit tedious [1]. Evolution of fossil fuels for limited periods on the earth such as petroleum for 40 years, natural gas for 60 years, and coal for 156 years, respectively [2–4]. Furthermore, the usage of non-renewable fossil fuels leads to global warming and environmental pollution. The global climatic changes are due to the contamination of the earth by consuming high energies. The increase in population leads to high energy demand. About 1.3% of energy demand increases every year. To overcome these disadvantages of fossil fuels, researchers expand their view on the production of green renewable fuels. The existing energy demand can be replaced by eco-friendly fuels such as hydrogen gas, synthetic natural gas, and green methanol which are termed as green fuels [5]. Among these green fuels, hydrogen gas is considered as one of the best clean energy sources and supports zero carbon emission strategies with high calorific value. The hydrogen fuel can be produced from several sources like coal, biomass, and water [6, 7]. Environmental benign hydrogen occurs in pure form without any mixture of other toxic gases [8, 9]. Currently, the researchers mainly target sustainable solar hydrogen production using the water-splitting method. The source of water in the earth is more abundant and the presence of sunlight over the earth in every season leads to the production of hydrogen fuels. Generally, hydrogen gas is used in industries such as steel, petrochemicals, and fertilizer. It is also used as fuel for rockets and power generation for electric vehicles to reduce carbon monoxide emissions [10]. Solar hydrogen production involves many methods such as electrolytic, thermochemical, mechanocatalytic, plasmolytic, photocatalytic, photoelectrochemical, photobiological, and solar thermal [4, 6, 11]. The conventional method of fossil fuel extraction involves high thermal energy is required. The naturally abundance resources of water and sunlight are feasible for the photocatalytic and photoelectrochemical production of a large quantity of hydrogen gas.

The photocatalytic water splitting helps to generate renewable energy [12, 13]. The main mechanistic processes involved the photocatalytic water splitting are light absorption, charge separation, and migration, and surface redox reaction. Oxidation and reduction reactions take place in the interface of photocatalytic material [14, 15]. There are two different approaches in the photocatalytic system single-material and multicomponent material [16, 17]. Materials possess narrow bandgaps, suitable thermodynamic potential, and stable in photocatalytic reactions. The significant characteristics such low cost and easy separation of hydrogen during the reaction [18]. Both ultraviolet and visible light can be used for the photocatalytic system according to the bandgap nature of the material.

The photoelectrochemical hydrogen production using various photoactive catalysts was coated on the surface of the electrode. Generally, transition metal oxides that behave as semiconductor materials, metal-organic frameworks, 2D materials, perovskite, and carbon materials were preferably used as the photo electrocatalyst. The engineering of the well-defined catalysts with the help of support or cocatalyst reduces the recombination rate. Photoelectrochemical water splitting involves the utilization of light energy to form the chemically desired output [19]. The task of the catalyst is to absorb the light at a particular wavelength and excite the photon from the valance band to the conduction band [20]. Significantly, semiconductor materials play a vital role in the high production of hydrogen using water. It possesses low-cost materials, low overpotential, suitable band edge, and high stability in photochemical, photogenerated electron-hole pairs [4, 21, 22]. Generation of the electron-hole pair in the catalyst helps to split the water into hydrogen and oxygen with the help of an oxidation-reduction reaction. Pilot-scale production of hydrogen from the various single and multi-component systems is widely described.

In this perspective, we mainly focus on the commercial aspects of ultraviolet light and visible light-based single & multi-composite, cocatalyst-loaded composite, carbon-based support composites of photocatalytic and photoelectrochemical water splitting.

2 Photocatalytic Techniques

The synthesis of clean and renewable energy like hydrogen with the help of solar sources is termed photocatalytic hydrogen production. Photocatalyst plays a predominant role in the functioning of the photocatalytic water splitting by converting the solar energy to chemical energy [23]. Significantly, 173,000 terawatts of solar energy reaches the earth continuously [24]. A semiconductor with a band gap value of 1.23 eV is suitable for photocatalytic techniques. The energy of the photocatalytic materials possesses high or equal energy to its band gap [25, 26]. The energy helps to pave the way for producing the electron-hole pair uniformly [27]. A single photocatalyst produces the one-step excitation whereas another way two different photocatalysts show the two-step excitation mechanism looks like the Z-scheme pathway of photosynthesis and it is also termed as artificial photosynthesis [28–32].

2.1 Ultraviolet Light-Assisted System

CuO-modified titania was prepared by the conventional impregnation and followed by the calcination. The cluster of CuO was deposited on the surface of the titania by the impregnation in Degussa P25 TiO_2 powder and copper nitrate. The bare TiO_2 shows less efficiency, but with the help of a co-catalyst CuO cluster the efficiency of Hydrogen production increases and reduces the recombination rate. The optimal production of the H_2 rate was 2061 μ mol h^{-1} g^{-1} and QE of 13.4% at 365 nm. On comparing the rate of pure TiO_2, the CuO–TiO_2 shows 129 times higher efficiency. The quantum size effect of the CuO cluster possesses a lower energy level between the conduction and valance bands paving the way for higher efficiency as shown in Fig. 1. Low-cost material of $CuO–TiO_2$ can be used commercially for the production of hydrogen [33].

Three solid photocatalysts of Bi_2MNbO_7 (M = Al^{3+}, Ga^{3+}, and In^{3+}) were synthesized by the solid-state reactions and active in the UV light irradiation. Hydrogen production can be done using a mixture of methanol/water and pure water. The contribution of the M^{3+} ionic radii decreases with increases in hydrogen production. The order of ionic radii $Al^{3+} < Ga^{3+} < In^{3+}$. The rate of hydrogen production 0.71, 0.3, and 0.18 mmol g^{-1} h^{-1} were obtained for Bi_2MNbO_7 (M = Al^{3+}, Ga^{3+}, and In^{3+}) respectively. Furthermore, the diffuse reflection spectra were used to calculate the band gap photocatalysts 2.9, 2.75, and 2.7 eV of Bi_2MNbO_7 (M = Al^{3+}, Ga^{3+}, and

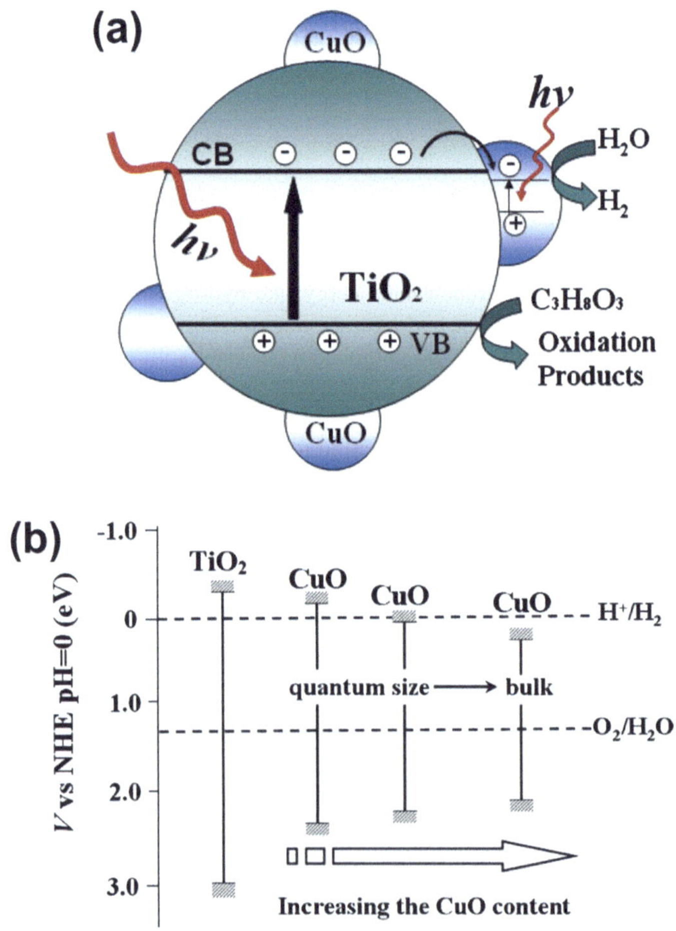

Fig. 1 **a** Systematic representation of CuO-modified titania and **b** The energy level diagram of $CuO–TiO_2$. Reprinted with permission from Ref. [33]. Copyright (2011), Elsevier

In^{3+}). The ionic radii play an important role in trimetallic photocatalytic hydrogen production. As in the commercial scale, the high surface material of TiO_2 added one percent of the conventional photocatalyst may lead to the high production of hydrogen [34].

The photocatalyst of $NiO/Ni/La_2Ti_2O_7$ was used for the hydrogen evolution reaction with the help of UV irradiation from the natural seawater. $La_2Ti_2O_7$ was synthesized by the conventional solid-state reaction. $NiO/Ni/La_2Ti_2O_7$ was prepared by incipient-wetness impregnation method. $NiO/Ni/ La_2Ti_2O_7$ was used to produce the hydrogen using pure water, the natural seawater, and the $MgCl_2$ rate was 137.2, 69.6, and 59.4 μ mol h^{-1}. As the hard water contains the Mg^{2+} ion, it restricts the production of hydrogen in the natural seawater. Removal of the hardness present in the natural seawater may lead to a high rate of hydrogen production. A surplus amount of seawater was present in the environment compared to the pure water. So, commercially valuable production of green fuel can be produced in the natural seawater source [35]. The efficient photocatalytic system of silver-loaded mesoporous-assembled TiO_2–ZrO_2 mixed oxide nanocrystal. TiO_2–ZrO_2 was prepared by the sol-gel method, and Ag was loaded with mixed oxide via the photochemical deposition method (PCD). The mesoporous nature of the high surface area material of TiO_2–ZrO_2 shows high efficiency towards hydrogen production. Compared with the commercially procured TiO_2, the synthesized titania shows a more activity as a catalyst. The co-catalyst of the Ag was loaded to induce the Schottky barrier of the mixed oxide, and it captured the electron on the surface of the Ag to promote the hydrogen production. The (wt%) of the Ag loaded was optimized to the concentration of the Ag increases with increases in high efficiency. The finally loaded mixed oxide catalyst (Ag loaded TiO_2–ZrO_2) rate of hydrogen production was 4.16 cm^3 h g_{cat}. As compared with the bare metal oxide, Ag (0.5 wt%) loaded shows enhanced activity up to 358.2% of hydrogen production. This shows that the highly active catalyst of a UV-irradiated photocatalyst can be used for the commercial production of hydrogen [36]. Barium titanate nanoparticles was synthesized by the sol-gel method. The UV-active barium titanate was utilized for hydrogen production from water. 26 μ mol g^{-1} after 5 h. The size of the barium is very small to form the nanoparticles on the surface. As it forms the nanoparticles Schottky barriers the interface of the nanoparticles decreasing the rate of hydrogen production [37].

The 2D hexagonal structure of mesoporous Tantalum oxide (Ta_2O_5) catalyst was used to produce the hydrogen with the help of UV radiation. The different co-catalysts were used for hydrogen production such as NiO_x, RuO_2, or $Rh_{2-y}Cr_yO_3$. The amorphous nature of the Ta_2O_5 became crystalline once after the photocatalytic reaction as shown in Fig. 2. NiO_x loaded on the Ta_2O_5 catalyst shows a high rate of hydrogen production up to 3360 μ mol h^{-1}. Alkaline treatment on the catalyst helps to change the surface properties and may be the reason for the electron-hole migration in hydrogen production [38]. Various modified TiO_2-based catalysts under UV and visible light irradiation was used for hydrogen production reaction. The suspension of TiO_2 (20.9 mmol/g_{cat} h) was three times higher efficient than the bare TiO_2 (7.7 mmol/g_{cat} h) under UV light irradiation. The macroporous nature of the TiO_2

in the visible light irradiation shows 0.64 mmol/g_{cat} h (Fig. 3). The presence of the co-catalyst enhances the visible light active catalyst up to 0.82 mmol/g_{cat} h. Up on discussion of the UV irradiated photocatalytic hydrogen production is more efficient than the visible light active photocatalyst [39].

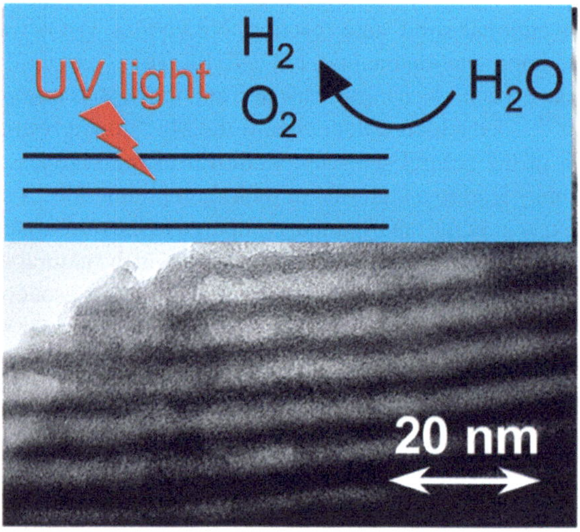

Fig. 2 TEM image of the 2D-hexagonal structure of mesoporous Ta_2O_5 after reinforcement with SiO_2, UV active photocatalyst. Reprinted with permission from Ref. [38]. Copyright (2008), ACS Publications

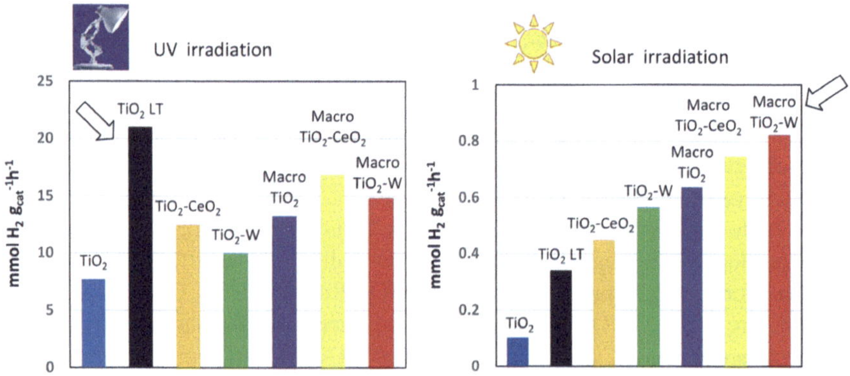

Fig. 3 Systematic representation of three different approaches to modify the TiO_2-based photocatalyst in UV and solar light irradiation. Reprinted with permission from Ref. [39]. Copyright (2019), Elsevier

2.2 Visible Light-Assisted System

The nature of the active sites mainly depends on the physiochemical properties and reaction conditions. Co-catalysts reduce the recombination rate of the photocatalytic material and enhance the activity [40]. Few researchers successfully reported that, without the use sacrificial agent in the water splitting [41, 42]. Recently, the evolution of the cocatalyst and reaction mechanism of the visible light active photocatalytic water splitting was elaborated [30, 43, 44]. In the last decade number of visible-light active catalysts have investigated a huge role in water splitting.

2.2.1 Single Materials

Single-semiconductor materials grasp the photo energy from sunlight to split the water into a major source of H_2. Hydrogen gas replaces fossil non-renewable resources. In the absence of the sacrificial agent single semiconductor material was used for the effective photocatalyst for the water splitting. The 3D graphitic carbon nitride nanosheets (3D g-C_3N_4 NS) were a metal-free and organic photocatalyst synthesized by a facile bottom-up supramolecular self-assembly route. Figure 4a represents the synthesis of 3D g-C_3N_4 NS from melamine and cyanuric acid as the major precursor. The formed aerogel material exhibits a low volume fraction of about 18.3 mg cm^{-3}. Prepared 3D g-C_3N_4 NS lattice fringes show 0.329 nm, and this indicates the π–π inter-layer stacking motif as shown in Fig. 4b, c. Figure 4d represents the AFM image of 3D g-C_3N_4 NS layer thickness 2.3 nm. This is attributed to 5 g-C_3N_3 layer units. Hydrogen evolution rate up to 101.4 μ mol h^{-1} g^{-1} under visible light. The hydrogen production rate of the g-C_3N_4 nanosheet was 11.8 times higher than the bulk g-C_3N_4. Because NS possesses a highly crystalline ultrathin nanosheet unit creates the charge carrier transport and the 3D structure g-C_3N_4 prevents the agglomeration of the nanosheets. These fine properties enhance the stability of the catalyst for more than 100 h in a hydrogen evolution reaction [45].

2.2.2 Co-catalyst Loaded Materials

To replace fossil fuels, which have significantly harmed the environment, with renewable energy that can be found, produced, and used, numerous efforts have been launched. A possible method for producing hydrogen (H_2) while utilizing free solar energy is photocatalytic water splitting. Simple processes and minimal environmental impact are two major advantages of photocatalytic systems for the production of H_2. To fabricate the essential materials for photocatalytic water splitting, in particular, there are numerous difficulties for commercial use [46, 47]. One of many potential semiconductor materials is used for a minor photocatalytic function. It is anticipated that by 2025, advanced renewable photocatalytic water splitting technologies will have been developed with low estimated hydrogen production costs at the plant gate,

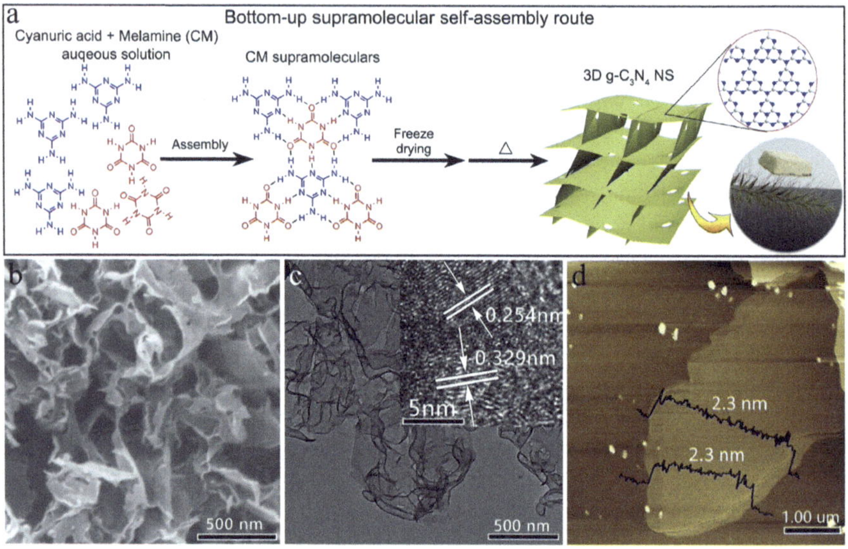

Fig. 4 **a** Bottom-up supramolecular self-assembly route for synthesizing 3D g-C_3N_4 NS. **b** SEM image, **c** TEM image (HRTEM image inset), and **d** AFM image of 3D g-C_3N_4 NS-2. Reprinted with permission from Ref. [45]. Copyright (2019), Elsevier

which are less expensive than the hydrogen production costs from photoelectrochemical processes. Utilizing semiconductors as co-catalyst photocatalytic technology powered by sunlight, splitting water into hydrogen (H_2) is the best solution to the energy deficit issue. However, the main reason for lower activity over photocatalysts is significant carrier recombination [48, 49]. In this research investigation, Sahar and the group presented a novel method for producing Pd/Rb_2O@TiO_2 catalysts for photocatalytic hydrogen production from water-splitting processes (Fig. 5). Palladium and rubidium contents were in-situ decreased and placed on the R-TiO_2 surfaces for this reason. By adding dual co-catalysts (Pd and Rb_2O) that reduce the barrier of activation energy for the two half-reactions, the current study aimed to improve the performance of rutile TiO_2. As the ideal and most suitable sacrificial reagent, just 5% C_2H_5OH was utilized. $Pd_{1.8}$/$RbO_{0.2}$@TiO_2 was discovered to be the most effective catalyst, capable of delivering hydrogen at a rate of 22.74 mmol h^{-1} g^{-1}. Pd and Rb_2O cocatalysts were thought to be responsible for the higher H_2 production activities. To achieve equilibrium during photoreactions, Rb_2O elevated fermi energy levels to drive electrons from TiO_2 to Pd active sites. Here, the mechanistic technique for greater charge separation and higher electron transit has been illustrated and addressed. Based on the findings, it can be said that the work presented here is trustworthy for applications involving sustainable energy and will be more beneficial for the highly commercialized component [50].

Carrier separation can be efficiently aided by the creation of an internal electric field (IEF) by linking a semiconductor with a metal co-catalyst. Here, Co@C is

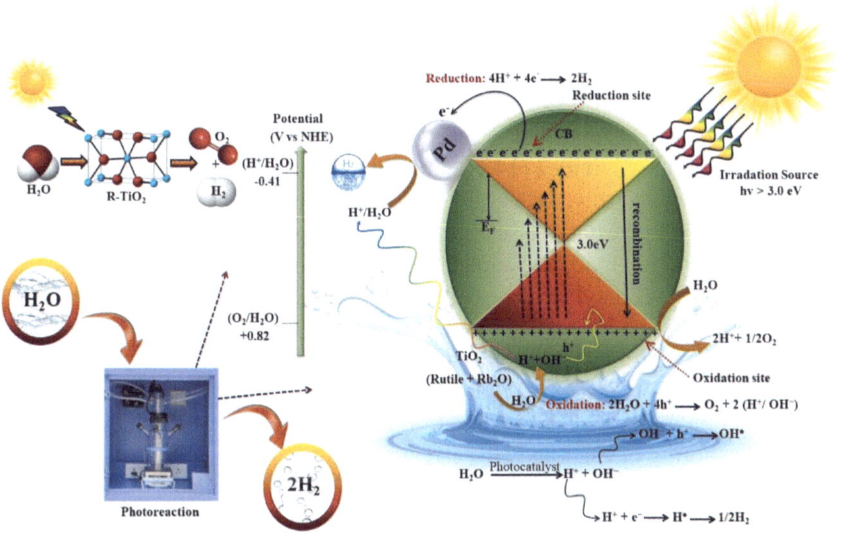

Fig. 5 The photocatalytic water splitting mechanism for hydrogen production using Pd/ $Rb_2O@TiO_2$. Reprinted with permission from Ref. [50]. Copyright (2023), Elsevier

used as a co-catalyst anchored on the surface of $ZnIn_2S_4$ nanosheets under simulated solar irradiation (AM 1.5G) with a production rate of 18.1 mmol h^{-1} g^{-1}, which is 109.7 times higher than that of bare $ZnIn_2S_4$ without assisted by Pt. The development of a huge IEF (4.6 times greater than $ZnIn_2S_4$) and an environment conducive to hydrogen adsorption and desorption (GH* 0), which offers the following benefits, are primarily responsible for the enhancement of the photocatalytic H_2 evolution activity of $Co@C/ZnIn_2S_4$. Increasing the kinetics of hydrogen release and (i) hastening the migration and separation of photogenerated charges. Shi et al. broaden the use of transition metal@carbon as a co-catalyst in energy photocatalysis and offer a design concept for an easy way to make a highly efficient composite photocatalyst [51]. Tian and their team members reported the optimal rate of hydrogen evolution is as high as 5.69 mmol h^{-1} g^{-1}. In this study, when CoS_2 nanoparticles are used as cocatalysts to grow on $Zn_3In_2S_6$ nanosheets with abundant sulfur vacancies. This rate is 2.87 and 2.29 times higher than that of $CoS_2/Zn_3In_2S_6$ (with few sulfur vacancies) and $Zn_3In_2S_6$ (with rich sulfur vacancies). Additionally, the $CoS_2/Zn_3In_2S_6$ composites hydrogen generation rate is 24.17 mmol h^{-1} g^{-1}, which is 4.25 and 1.90 times that of $CoS_2/Zn_3In_2S_6$ (with rich sulfur vacancies) and 1%-Pt/ $Zn_3In_2S_6$ (with rich sulfur vacancies), respectively. $CoS_2/Zn_3In_2S_6$ composite (with abundant sulfur vacancies and 1 weight percent Pt) exhibits an apparent quantum efficiency (AQE) of 66.20% when exposed to light at a wavelength of 370 nm. In particular, they show how sulfur vacancies and dual cocatalysts (CoS_2 and Pt) can help $Zn_3In_2S_6$ nanosheets perform hydrogen evolution efficiently. For the advancement of photocatalytic hydrogen production technology, this work offered fresh perspectives on commercial reactors and ideas [52]. According to Saleem et al. the hydrothermal method was used to synthesize a variety of catalysts, including CdS, Cu@CdS, Ni@CdS, and Cu/Ni@CdS. By using a chemical reduction technique, Cu/Ni cocatalysts were in-situ coated over CdS surfaces to increase activity and stability. Results show that a cocatalyst ratio that is tuned (2% Cu and 1% Ni) over CdS exhibits greater H_2 evolution activities (14.16 mmol g^{-1} h^{-1} with 72% quantum efficiencies). The synergism between Cu and Ni cocatalysts was blamed for the higher activity. Considering the findings, it is possible to conclude that the study brought new possibilities for highly marketed applications [53]. Similarly, Yao and group improvised CdS@ZnNi–C photocatalyst has an apparent quantum efficiency (AQE) of 31.4% and a hydrogen evolution rate of 15.7 mmol g^{-1} h^{-1} at 420 nm. The rate is, respectively, 2.4 and 3.7 times higher than that of Zn^{2+} and Ni^{2+} modified carbon cocatalysts on CdS (CdS@Zn–C) and (CdS@Ni–C). According to studies using density functional theory, CdS@ZnNi–C's improved catalytic efficiency is caused by the co-anchoring of Zn^{2+} and Ni^{2+} in carbon layers, which results in a favorable Gibbs free energy (GH*) for hydrogen evolution [54]. Only tungsten oxides are used by Thangamuthu et al. to build a sturdy Z-scheme system for fully visible-driven water splitting in both neutral and alkaline solutions, where sodium tungsten oxide bronze ($Na_{0.56}WO_{3-x}$) is used as an H_2 evolution photocatalyst. With a remarkably high apparent quantum yield of 6.06% at 420 nm under neutral conditions, this system effectively generates H_2 (14 mol h^{-1}) at an optimum molar ratio of 2:1 in an aqueous solution driven by light (Fig. 6). Both observations and a calculation using density functional theory show that this exceptionally selective H_2 is caused by the preferential adsorption of iodate (IO^{3-}) on WO_3 and iodide (I^-) on $Na_{0.56}WO_{3-x}$ [55].

According to Saleh et al. the greatest value of the hydrogen produced was seen at a mass loading of 0.3 weight percent for Pt and Cu co-catalysts. Cu/TiO_2 synthesis generated an initial rate of 24 mmol h^{-1} g^{-1}, which is 3.5 and 1.4 times greater than

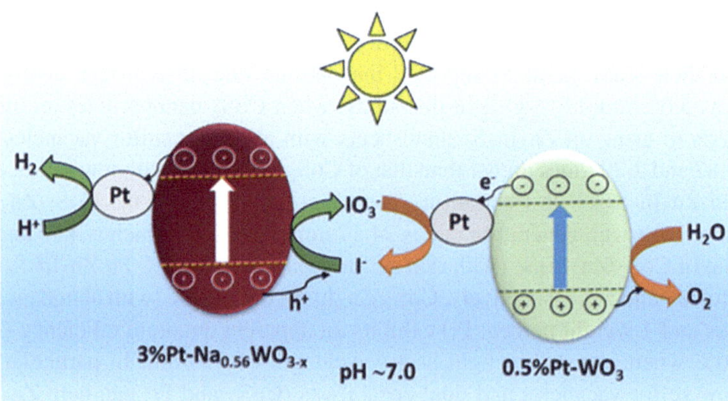

Fig. 6 The photocatalytic water splitting mechanism for hydrogen production using 3% Pt-$Na_{0.56}WO_{3-x}$. Reprinted with permission from Ref. [55]. Copyright (2023), Elsevier

the composite even without a valuable metal like Pt. The maximum rate for Pt co-catalyst, on the other hand, was observed and was 58 mmol h^{-1} g^{-1}, which was 1.6 and 1.1 times greater than the composite produced, respectively. This work provides fresh insights into the major impact of the co-catalyst deposition method on photocatalysis's effectiveness in creating hydrogen through water splitting [56]. Jayachitra et al. produced nanosized $NiSe_2$ particles that were utilized as a co-catalyst with TiO_2 to increase the rate of H_2 creation. TiO_2 produced H_2 at a significantly higher rate in the presence of $NiSe_2$, reaching a maximum activity of 219.2 mmol h^{-1} g^{-1} with an STH of 9%, which is close to the activity of Pt-loaded TiO_2 (9.7% for 235.5 mmol h^{-1} g^{-1}). $NiSe_2$ strong conductivity enables the quick transport of photogenerated electrons from TiO_2 to $NiSe_2$ [57]. To prepare highly conductive $NiSe_2$ as an electron sink for effective H_2 generation, this work validates a quick response. This finding inspires and creates a new field of study for researchers to investigate other co-catalysts as a distinctly marketable element of photocatalytic water splitting.

2.2.3 Supported Materials

Through highly competitive, commercially feasible renewable solutions for green hydrogen, sustainable energy innovation is leading the path to decarbonization. Due to its promotion of the direct conversion of solar energy to chemical energy and hydrogen generation, photocatalytic water splitting has attracted attention on a global scale with the help of supporting materials [58]. By using reduced graphene oxide (rGO) nanosheets as charge separators for visible-light-driven photocatalytic water splitting, an effective supporting material like graphene-based nanocomposite has been created to get beyond its restriction of quick charge recombination. In this study, Samajdar et al. presented a carefully crafted Lanthanum orthoferrite/Reduced Graphene oxide ($LaFeO_3$/rGO) nanocomposites and investigated its photocatalytic efficacy. Comparing the nanocomposites to pure $LaFeO_3$, the photocurrent density of the latter showed to be about 21 times higher. When exposed to visible light, the nanocomposites have been used successfully as photocatalysts for H_2 creation by water reduction. As one of the highest values obtained utilizing noble-metal-free graphene-based photocatalytic nanocomposites, H_2 production has been seen to increase significantly for nanocomposites (82 mmol h^{-1} g^{-1}) compared to that of bare $LaFeO_3$ (9 mmol h^{-1} g^{-1}) as shown in (Fig. 7). The method for creating simple yet very effective supporting materials for photocatalysis applications is provided in this paper [59].

Covalent organic frameworks (COFs) have become a new breed of photocatalytic materials. The high recombination rate of photogenerated electron-hole pairs, however, limits their photocatalytic capabilities. Yan et al. successfully construct a novel metal-free 2D/2D vander Waals heterojunction here, which consists of a 2D COF with ketoenamine linkage (TpPa-1-COF using 1,3,5-triformylphloroglucinol (Tp) and paraphenylenediamine ligands) and 2D defective hexagonal boron nitride (h-BN). Additionally, the structural changes of TpPa-1-COF combined with defective h-BN have the potential to reduce electron backflow by widening the space

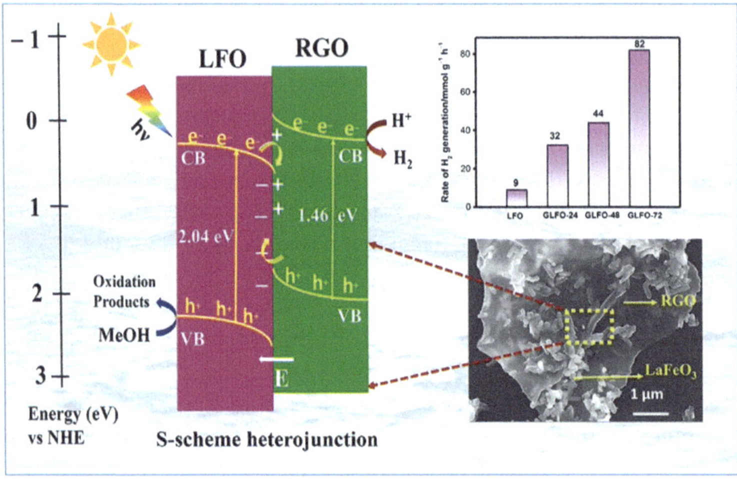

Fig. 7 Charge-transfer mechanism of the $LaFeO_3$/rGO nanocomposites. Reprinted with permission from Ref. [59]. Copyright (2023), Elsevier

between their conduction band positions. This is supported by both experimental data and the results of calculations using density functional theory. Accordingly, the resulting porous h-BN/TpPa-1-COF metal-free VDW heterojunction displays outstanding solar energy catalytic activity for water splitting without co-catalysts, and the H_2 evolution rate can reach up to 3.15 mmol h^{-1} g^{-1}, which is about 67 times greater than that of pristine TpPa-1-COF, also surpassing that of state-of-the-art metal-free-based photocatalysts reported to date. It is the first study to use h-BN to build COFs-based heterojunctions, which may open up new possibilities for creating extremely effective metal-free photocatalysts for H_2 evolution [60]. Ali et al. worked on an innovative paper on the production of MoS_2–BN/TiO_2 envisions advancements in the applicability of sustainable H_2 energy by using the same TiO_2 with different cocatalysts. MoS_2-BN loaded concentration was adjusted, and 2.5 wt% with an H_2 production rate of 2.6 mmol h^{-1} g^{-1} and 6.94%, MoS_2–BN/TiO_2 displayed the best photocatalytic performance. The regulated loading of sp^2 hybridized h-BN effectively accelerates the electron mobility of MoS_2 and reduces its band edges, which aid in the reduction of the Schottky barriers, corroborating the potential of MoS_2-BN nanoflowers in boosting the H_2 production rate of TiO_2 [61]. A hybrid photo-catalyst that resembles a spherical bead and is composed of reduced graphene oxide, a highly effective electron transporter, Sodium alginate, a water adsorbent, and cadmium sulphide was tested for hydrogen synthesis from water in the presence of a 45 W LED. CdS nanoflowers and sodium alginate nanofibrils were found scattered on the surface of graphene sheets during the morphological analysis of the produced composite photocatalyst. The photocatalytic activity was improved by the addition of graphene, going from 3169 mmol h^{-1} g^{-1} for pure CdS to 5001 mmol h^{-1} g^{-1} for rGO-CdS-Sodium Alginate (Fig. 8). The idea behind the bead-like Alginate photocatalyst is successful because the hydrogen bonds that form between the water molecules and the functional groups of the Alginate entrap the water molecules, binding them and enhancing the retention time, increasing the contact time of the photocatalyst with water, and producing a higher yield of hydrogen (5 mmol h^{-1} g^{-1}) investigated by Sarkar et al. [62].

The catalyst (Pd–Rb_2O@g-C_3N_4/TiO_2) created in this study by Rafiq et al. is highly efficient. Performances for charge separation and photocatalysis are improved by the presence of Pd metal over g-C_3N_4/TiO_2. Increased electron transport to the metal active sites (Pd co-catalysts) was the result of Rb_2O consumption of holes, which also reduced the semiconductors Fermi energy level during the photoreaction. Once the holes had been consumed, Rb species were once more deposited on the semiconductors surface. The findings show that the type II heterojunction between g-C_3N_4 and TiO_2 and the development of a Schottky barrier between the semiconductor and Pd are what caused the greater H_2 production activities. To produce 19.7 mmol g^{-1} h^{-1} of hydrogen, Pd–Rb_2O@g-C_3N_4/TiO_2 was discovered to be the most effective photocatalyst. The authors anticipated that their work play a pioneering role in the creation of this type of new structure that produces commercially viable H_2 efficiently and facilitates its conversion using sustainable methods [63]. A $NiCo_2O_4$/

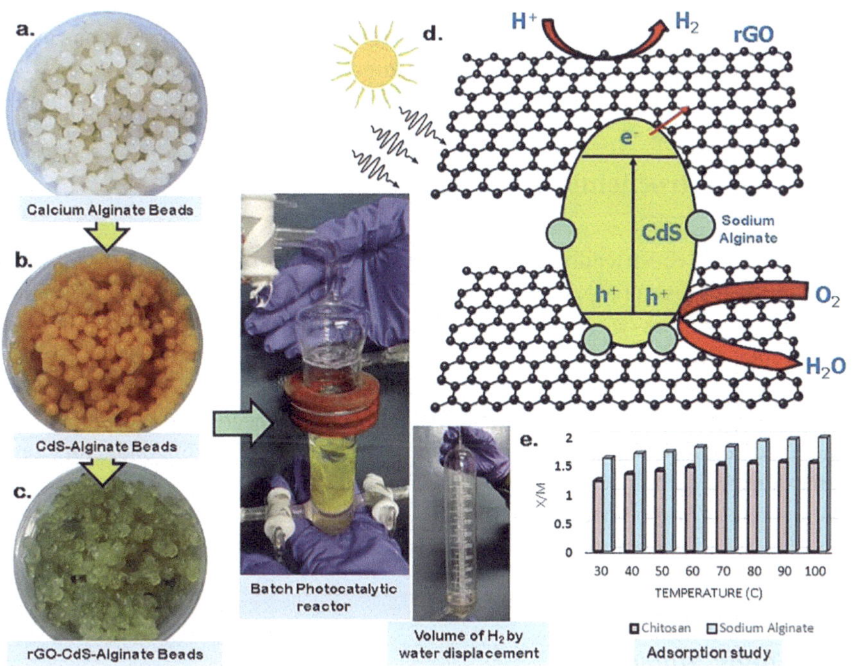

Fig. 8 Schematic of synthesis procedures of **a** Calcium alginate, **b** CdS-Sodium alginate, **c** rGO-CdS-Sodium alginate (bead form), **d** catalytic application of bead-based structured catalyst, and **e** comparison of water adsorption capabilities of Chitosan and alginate. Reprinted with permission from Ref. [62]. Copyright (2023), Elsevier

g-C_3N_4 S-scheme heterojunction is built to increase H_2 evolution under visible light irradiation, according to research by Ding et al. In comparison to pure $NiCo_2O_4$ and pure g-C_3N_4, the photocatalytic activity of the $NiCo_2O_4$/g-C_3N_4 hybrid is greater. In comparison to pure g-C_3N_4 (5.1 mol/h), the activity of the 3% $NiCo_2O_4$/g-C_3N_4 catalyst is the highest, reaching 21.2 mol/h of H_2 yield over 3% $NiCo_2O_4$/g-C_3N_4. The resulting heterojunction effectively isolates the charges caused by photoinduction, broadens the range of the light response, and limits the recombination of photogenerated carriers. Furthermore, the $NiCo_2O_4$/g-C_3N_4 catalyst is quite stable. In contrast, Rafig et al. use the $NiCo_2O_4$/g-C_3N_4 catalyst to maximize H_2 generation [64]. Nevertheless, Yuvarani et al. employed the identical supporting photocatalyst, g-C_3N_4 with 5 weight percentage of $NiTe_2$. With an apparent quantum yield of 9.8%, $NiTe_2$ displayed an H_2 evolution rate of 3155 mol h^{-1} g^{-1} at 420 nm monochromatic light, which is much greater than that of a bare $NiTe_2$ photocatalyst. The significant photocatalytic activity of g-C_3N_4 nanosheets in hydrogen (H_2) production is attributed to the extensive surface area they offer [65]. The photocatalytic systems advantages for commercial marketing and research in the area of photocatalytic hydrogen generation include being environmentally friendly, reusable, economical, and well-designed. In light of this, photocatalytic water splitting may be a desirable and practical method for solar hydrogen synthesis. High efficiency, long-term and sustainable stability, and low capital costs are desirable developments that should be realized in order to boost the viability of practical production of hydrogen gas using photocatalytic water splitting systems for commercial uses.

3 Photoelectrochemical Techniques

An attractive and eco-friendly technique for producing hydrogen energy without CO_2 emissions is through photoelectrochemical (PEC) water splitting [66]. This innovative technology relies on specialized materials, often semiconductors, to capture and convert solar energy into chemical energy. Extensive research efforts are underway to produce semiconductor photoelectrodes using materials such as TiO_2, $BiVO_4$, WO_3, Fe_2O_3, ZnO, and SnO_2. Notably, Honda and Fujishima pioneered the use of TiO_2 as a photoanode for PEC water splitting in 1972 [67]. While n-type semiconductors have been extensively employed in photoelectrochemical (PEC) hydrogen production, the development of p-type oxide materials for photocathodes has been relatively rare, primarily due to the limited availability of such materials [68].

3.1 Single Materials

In 2014, Jin and his team reported a drop casting 1T-MoS_2 nanosheet solutions directly onto Si is the most straightforward method for determining that heterostructures of 1T-MoS_2 and Si are viable for PEC hydrogen production [69]. By chemically exfoliating MoS_2 nanoflakes produced using the chemical vapor deposition (CVD) method. Using a three-electrode system, the photocurrent density was recorded of the Si photocathode with different volumes of drop casted 1T-MoS_2 (0–16 μL) in 0.5 M H_2SO_4 under sunlight irradiation and current density and limiting current density of 1T MoS_2 at 0 V (Fig. 9a, b). From PEC investigations, photocurrent density was increased to 17.6 mA cm^{-2} at 0 V versus RHE. The authors claim the highest photocurrent density was achieved compared to other reported chalcogenides (1T polymorphs) for PEC solar energy conversion.

In 2015, Li et al. synthesized a scalable route of a single crystalline Cu_2ZnSnS_4 (CZTS) nanosheet for PEC water splitting. The photocurrent density was found to be 1.32 mA/cm^2 at 0 V versus RHE under 1.5 AM. The highest photocurrent density was reported without a co-catalyst, and this work could be used as a reality practical application of using abundant materials for efficient hydrogen production [70].

In 2015, Qurashi and co-workers synthesized TiO_2 nanospheres at room temperature for PEC water splitting under light (AM 1.5G) [71]. The photocurrent density was observed around 0.95 mA cm^{-2} at 1.23 V, and photocurrent conversion efficiency was found to be 0.89%. The use of organometallic precursor to synthesize nanosphere is a further development of an energy conversion system. In the same year [72], Tan and his team fabricated Ru-doped hematite nanorods as photoanode for PEC water splitting. Figure 10 displays, that the photocurrent transients of the 2Ru–Fe_2O_3 samples are notably decreased, and those of the 5Ru–Fe_2O_3 samples completely vanish, in comparison to the undoped and 10Ru–Fe_2O_3 photoanodes which show substantial positive and negative current transients. The highest photocurrent density was observed around 5.7 mA cm^{-2} at 1.23 V versus RHE.

In 2020, Cho et al. reported a dual heterostructure (core-shell) of CdS@CdSe NP and 1D ITO@ TiO_2 NWs was prepared by electrochemical deposition method for PEC hydrogen production (materials preparation and hydrogen production are presented as schematic illustration in Fig. 11 A and B. As the synthesized photoanode was investigated the photocurrent density and the highest photocurrent were achieved

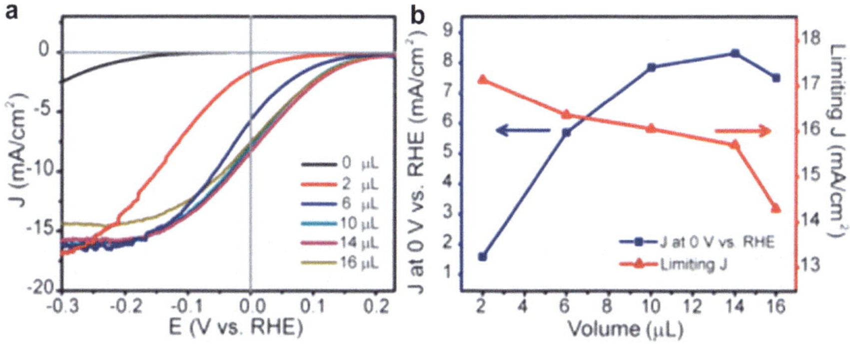

Fig. 9 **a** J E curve of p-Si photocathodes measured with the increasing volume of drop casted 1T-MoS_2 solutions, **b** limiting current density and current density at 0 V as a function of 1T-MoS_2 solution volume. Reprinted with permission from Ref. [69]. Copyright (2014), Elsevier

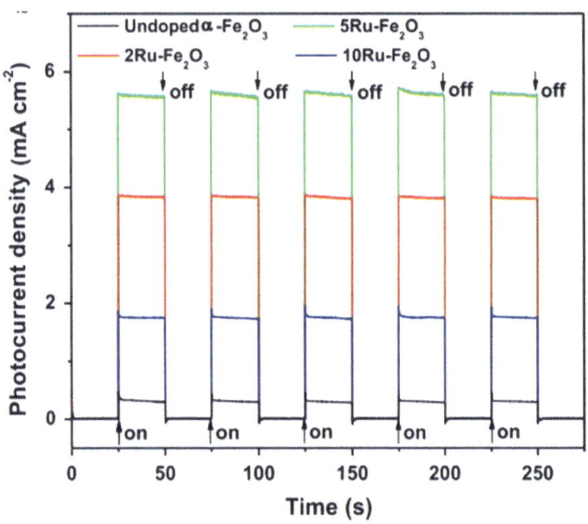

Fig. 10 Transients photocurrent density versus time of undoped and doped hematite at 1.2 V (applied potential). Reprinted with permission from Ref. [72]. Copyright (2015), Elsevier

at 20.11 mA cm^{-2} (at 1.23 V vs. RHE). The prepared material proved an efficient catalyst for energy conversion material [73].

In 2023, Amin et al. reported a spin coating and spray deposition were used to fabricate $LaFeO_3$ photocathodes using a low-cost commercialized disinfectant spray gun [68]. Improved visible light absorption was achieved through morphological modifications, and the band gap energy was lowered from 2.45 to 2.42 eV as a result. The PEC was performed in 0.1 M NaOH (pH 13) as an electrolyte using a 1.5 AM solar simulator. From linear sweep voltammetry (LSV), steady-state current

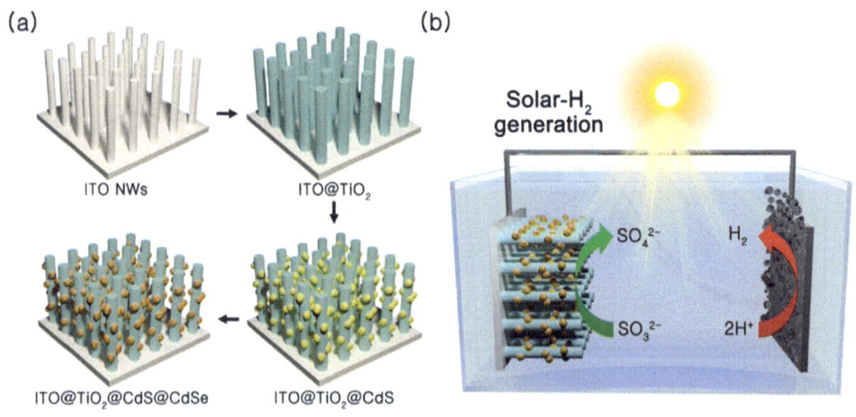

Fig. 11 **a** Fabrication process and **b** solar H_2 production process of ITO@TiO_2@CdS@CdS. Reprinted with permission from Ref. [73]. Copyright (2020), Elsevier

density-potential (J-V) of as prepared electrodes was measured at 6 mV s^{-1} (scanning potential direct is positive to negative). Inclusive, the fabricated $LaFeO_3$ via both spin and the combined spin and spray technique exhibited good photocathode responses under light on and off mode. The photocurrent density of spray deposition applied on top of the $LaFeO_3$ spin-coated layer (27 μA cm^{-2} at 0.6 V) was twice as high as that of the spin-coated $LaFeO_3$ layer (15.2 μA cm^{-2} at 0.36 V). As a result, this study demonstrated that a simple technique for modifying the surface structure using cheap tools may be used to build efficient photosensitive devices, proving their practicality for PEC water-splitting applications.

In 2023, Zheng et al. created CuO nanorod arrays using a gas phase cation and a sacrificial template (ZnO) for adequate photoelectrochemical water splitting [74]. To obtain a pure form of CuO from ZnO, the cation exchange process was investigated further at various temperatures; ZnO was completely transformed into CuO from 350 to 600 °C. To study the photoelectrochemical behavior of the synthesized CuO, it developed into a photoelectrode that was deposited on ITO glass and exposed to 100 mW cm^{-2} AM 1.5 solar irradiation in 1 M Na_2SO_4 solution as an electrolyte. The optimized reaction temperature was discovered to be 450 °C, due to the photocurrent at 0 V versus RHE reaching 2.1 mA cm^{-2}. By using linear sweep voltammetry with chopped light, the photoelectrochemical properties of the optimized CuO electrode were further examined. One of the highest values in the pure CuO system reported for the photocurrent is 2.42 mA cm^{-2} at 0 V versus RHE. The authors claim that this technique offers a fresh strategy for the development of CuO-based nanodevices. In the same year [75], Kuritka and co-workers demonstrated the nanoforest morphology of ZnO nanorods deposited onto ITO glass using a hydrothermal method. The highest photocurrent density was found to be 0.06 mA cm^{-2} at 0.5 V. It could be used in energy-related technologies.

3.2 *Co-catalyst-Loaded Materials*

The introduction of an adequate co-catalyst is an effective method to improve the performance of semiconductor nanomaterials that are in photoelectrochemical (PEC) water splitting, where low photoconversion efficiency, high charge transfer resistance, and fast recombination rate are their complications [76]. Cost-effective co-catalysts need to be developed to increase the efficiency and stability of the semiconducting photoanodes in a PEC cell [77].

In 2015, Sudhagar and his team demonstrated a 1-T MoS_2 nanosheet on TiO_2–CdS as a co-catalyst with TiO_2–CdS–ZnS as a photoanode for hydrogen production [78]. It was feasible to control the charge recombination through exposed CdS sites on TiO_2 by treating the interface with 1-T MoS_2 sheets. This tandem electrode (TCZM) displayed higher hydrogen production of 1.47 ml h^{-1} cm^{-2} at 1 V, which is 150 percent better than utilizing Pt as CE.

Pareek et al. [77] synthesized $Co(OH)_2$, Co_3O_4, $Ni(OH)_2$, and NiO was deposited onto the surface of CdS photo-anodes using a chemical impregnation process [77]. Interestingly, the co-catalysts modification increases the stability of CdS electrodes (>8 h) and enhances driven H_2 production in PEC cells and the PEC explored highest H_2 production was about 2.5 mmol h^{-1}. From this investigation, the development of a cost-effective co-catalyst for PEC hydrogen production.

In 2018, Wang et al. described the role of co-catalyst and cost-effectiveness in photoelectrochemical hydrogen production, due to the replacement of expensive and rare abundant metals [79]. The prepared 1D CoP nanowires (NW) as co-catalysts have excellent properties than bulk CoP. Followed by, $Zn_{0.5}Cd_{0.5}S$ (ZCS) was prepared and the assembled hybrid ZCS/10% CoP NWs exhibited photocatalytic activity of 12,175.8 μmol h^{-1} g^{-1} much higher than ZCS/10% CoP NPs than that of pristine ZCS and hydrogen rate displayed by different wt% CoP/ZCS is given in Fig. 12. The development of an efficient co-catalyst is cost-effective for hydrogen production, due to commercialization in the future perspective.

In 2019, Peng and co-workers designed a CoS_x nanosheet loaded onto a CdS nanorod by a simple method [80]. This CdS–CoS_x composite was used as a photoanode in PEC hydrogen production. Compared to CdS, the composite was given the H_2 production of 168.6 mmol cm^{-2} (7.77 L m^{-2} h^{-1}). In addition to that Na_2S–Na_2SO_4 system is feasible for PEC hydrogen production.

In 2022, Yuan et al. introduced iron phosphates (FePi) onto Fe_2O_3 by surface phosphorization to enhance catalytic activity in PEC hydrogen production [81]. The highest photocurrent density showed 11.9 mA cm^{-2} for FePi/Fe_2O_3/p-Si. This strategy could be employed in PEC hydrogen production by modifying the co-catalyst as an efficient.

In 2023, Alharthi and co-workers revealed the effect of the co-catalyst on Bi_2O_3 such as ZrO_2 and Bi_2S_3. The highest photocurrent of Bi_2O_3:Zr/Bi_2S_3 was found to be around 3.55 mA cm^{-2} at 1.23 V RHE [82]. From Fig. 13a, b, after ion conversion

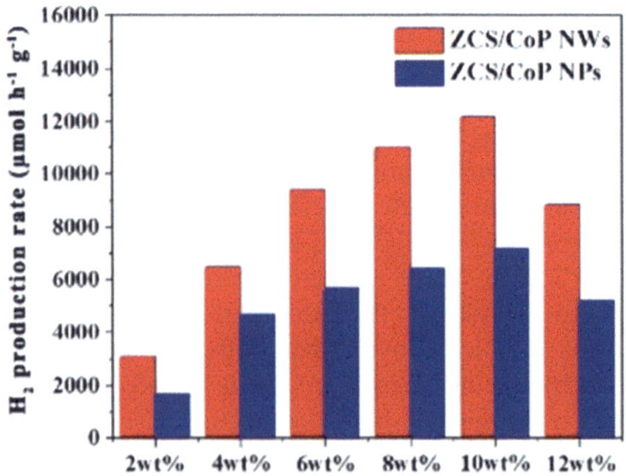

Fig. 12 Different wt % of CoP/ZCS in PEC hydrogen production. Reprinted with permission from Ref. [79]. Copyright (2018), Elsevier

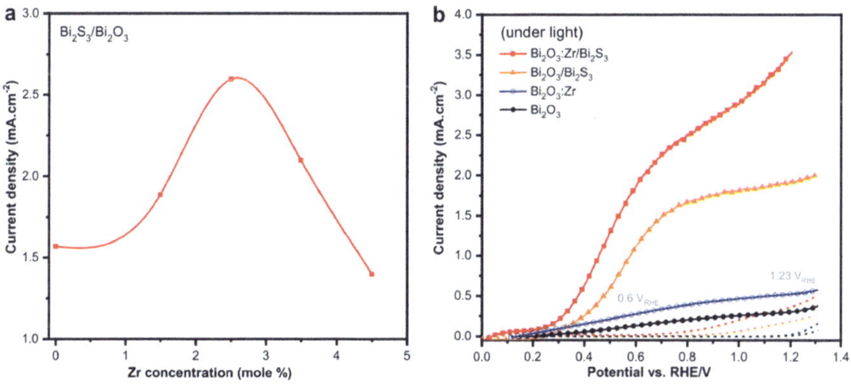

Fig. 13 **a** Photocurrents were observed for Bi_2S_3/Bi_2O_3 for different Zr concentrations, **b** polarization curves for Bi_2O_3 and Zr:Bi_2O_3 in 0.1 M Na_2SO_4, Bi_2O_3/Bi_2S_3 and Bi_2O_3:Zr/Bi_2S_3 in 0.1 M Na_2SO_4/0.1 M Na_2S. Reprinted with permission from Ref. [82]. Copyright (2023), Elsevier

in Bi_2O_3/Bi_2S_3, the photocurrent density increased from 0.3 to 1.97 mA cm^{-2} due to better optical properties, which successfully increased the photogenerated production of carriers. Additionally, with Bi_2S_3 trans-formation and Zr content, the photocurrent density significantly increased and reached 3.55 mA cm^{-2} at 1.23 V RHE measured in 0.1 M Na_2SO_3/Na_2S (pH 12.8). Bi_2O_3/Bi_2S_3 photo-electrodes are a prospective replacement for traditional precious metals water-splitting electrodes due to their outstanding PEC properties and stability as well as their quick, simple, and organized for commercial development.

In 2023, Yan et al. prepared nickel phosphate ($Ni_3(PO_4)_2$) decorated onto Si nanowires using chemical vapor deposition method (CVD) [83]. The ($Ni_3(PO_4)_2$) nanoparticles were uniformly distributed on Si NWs (($Ni_3(PO_4)_2$)-Si NWs). The highest photocurrent density of that composite was exhibited at about 16 mA cm^{-2} at 0 V versus RHE under SAM 1.5G solar irradiation.

In 2023, Chen et al. described three MOFs or derivatives (MIL(Fe), ZIF67, and oxidized ZIF67) that were coupled with $BiVO_4$ (BVO) as a catalyst for hydrogen production [84]. For heterojunction behavior, (MIL(Fe)/W doped $BiVO_4$ material was prepared. Similarly, the other two MOFs were also prepared. The highest photocurrent density was recorded at about 4.14 mA cm^{-2} at 1.23 V versus RHE. The stability of MIL/W: BVO was measured from chronoamperometry (Fig. 14). More than 90% of the photocurrent was retained throughout the full 5000 s period of time. The remarkable stability of the MIL/W: BVO electrode is indicated by the significant retention of the photocurrent.

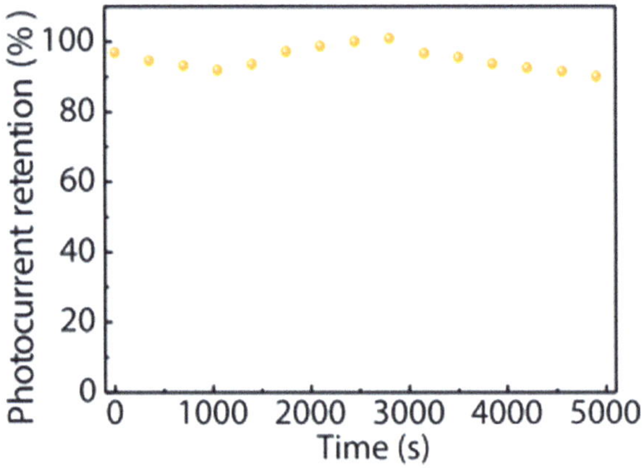

Fig. 14 Stability test of the MIL/W: BVO. Reprinted with permission from Ref. [84]. Copyright (2023), Elsevier

3.3 Supported Materials

As the field of PEC hydrogen production continues to advance, the use of graphene, graphene oxide (GO), reduced graphene oxide (rGO), and graphitic carbon nitride (g-C_3N_4) as support materials holds promise for improving the overall efficiency and commercial viability of this sustainable hydrogen production technology. Researchers are actively exploring these materials to bring photoelectrochemical (PEC) hydrogen production closer to widespread commercialization [85].

Photoelectrodes, primarily consisting of semiconductors, have a crucial role in absorbing light, generating electron-hole pairs, and facilitating charge transfer within photoelectrochemical cells. Researchers have delved into the thorough examination of heterostructures, considering the synergistic effects between various semiconductors and support materials [86]. For instance, Weidong Shi and co-workers [87] designed a novel heterostructure from TiO_2, a widely researched photocatalyst material with a wide bandgap of 3.2 eV. Furthermore, a limitation in the context of TiO_2 is the recombination of electron-hole pairs within its matrix [88]. To address this issue and mitigate the draws associated with pure TiO_2, an effective approach involves sensitizing TiO_2 by incorporating Cu_2O photoelectrode material which has a narrow bandgap (1.9–2.2 eV) p-type semiconductor to create heterostructures. Nonetheless, the limited interfacial area between distinct semiconductors directly impacts the effectiveness of electron-hole separation. Therefore, optimizing the interfaces within heterostructures by introducing reduced graphene oxide which has outstanding ability to establish connections between diverse semiconductor nanoparticles further enhances the efficiency of photoelectrochemical conversion. The hydrogen production apparatus was maintained at a potential of 0.6 V relative to a calomel electrode, as illustrated in Fig. 15. The TiO_2/RGO/Cu_2O electrodes within this system played a crucial role in accumulation holes. These holes were primarily utilized to oxidize methanol to formaldehyde, thereby substantially enhancing the separation of photogenerated electrons and holes. Simultaneously, the electrons were

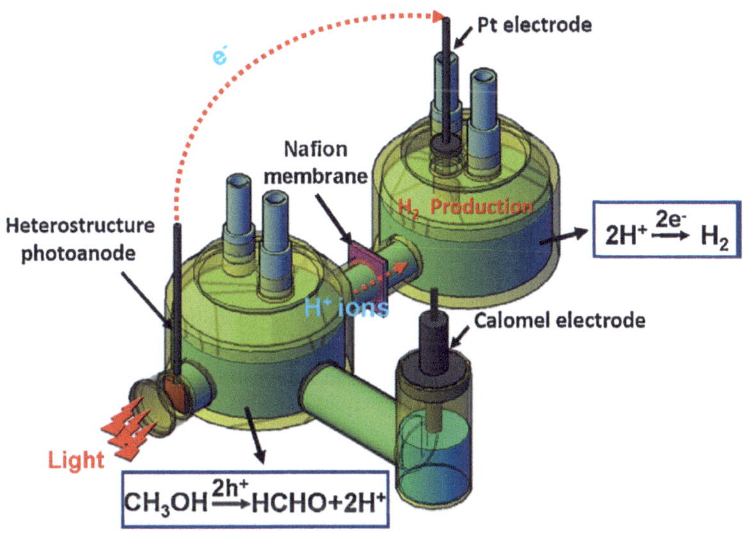

Fig. 15 Schematic diagram of equipment for photoelectrochemical hydrogen production. Reprinted with permission from Ref. [87]. Copyright (2016), Elsevier

directed towards the Pt electrode to support the production of hydrogen. The TiO_2/RGO/Cu_2O heterostructure displayed an impressive hydrogen production rate of 631.6 μmol/h m^2 under a radiation intensity of 47 mW/cm^2.

Further studies were developed by designing a ternary monolithic photo membrane using ZnS/CdS heterojunction nanorods inside the interlayers of the multilayer-stacked rGO sheets [89]. The presence of dual heterojunction interfaces (ZnS/CdS and CdS/rGO) along with optimized pathways effectively enhances the separation and transfer of photogenerated electron-hole pairs. This, in turn, leads to a substantial improvement in the visible-light-driven photochemical and photoelectrochemical activity of the ZnS/CdS/rGO photo membrane. The ZnS/CdS/rGO membrane was directly used as a photoanode in hydrogen production. It obtained an IPCE of 25% at 450 nm and a good photo-stability of ~5.7 mA cm^{-2} after 120 min.

The production of hydrogen gas via water-splitting processes serves as a crucial energy resource, especially for application in aviation and industrial sectors where hydrogen fuel is essential. Abdullah Almohammedi et al. prepared a TiN/TiO_2/Al_2O_3 template as a photoelectrode for hydrogen production from sewage water [90]. The Al_2O_3 template has been fabricated with precision and a well-structured arrangement, featuring hexagonal pores, through the Ni imprinting technique. The deposition of TiN/TiO_2 inside these pores is accomplished using Atomic Layer Deposition (ALD) and DC sputter-coating methods. The use of TiN/TiO_2/Al_2O_3 materials as a photocathode was conducted with and without the coating of Au nanoparticles. Integrating Au nanoparticles (NPs) decreases electron-hole recombination and enhances light absorption, thereby leading to an enhancement in photocatalytic performance.

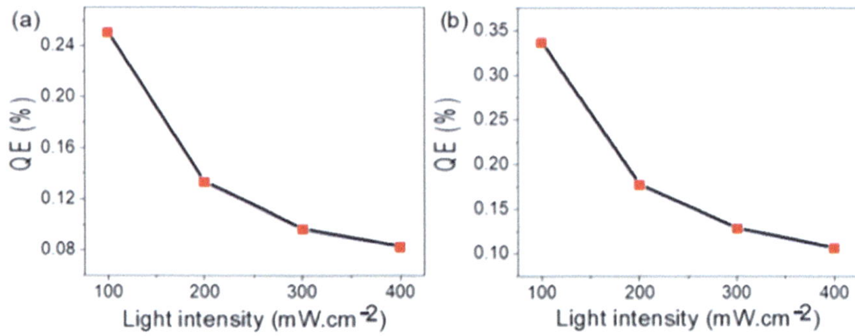

Fig. 16 The relation between quantum efficiency and light intensity for **a** $TiN/TiO_2/Al_2O_3$ and **b** $Au/TiN/TiO_2/Al_2O_3$. Reprinted with permission from Ref. [90]. Copyright (2021), MDPI

Quantum efficiency (QE) measures the effectiveness of converting absorbed photons into useful photochemical or electrochemical reactions. It can be measured by the Eq. (1).

$$QE = \frac{J_{ph} \cdot h \cdot C \cdot c}{P \cdot \lambda} \quad (1)$$

Here, J_{ph} is the current density, h is Plank's constant, c is the speed of light, P is the light intensity, and λ is the wavelength of light. Figure 16 shows the difference between the QE with and without Au coating of 0.25% and 0.34% at 400 mW cm^{-2}, respectively.

The activation energy for electrodes $TiN/TiO_2/Al_2O_3$ and $Au/TiN/TiO_2/Al_2O_3$ is found to be 27.22 and 18.84 kJ mol^{-1}, respectively using the Arrhenius equation. Due to its excellent stability and cost-effectiveness, the fabricated photoelectrode holds promise for potential utilization in industrial settings.

Support materials like graphene, GO, rGO, and $g\text{-}C_3N_4$ are integral in photoelectrochemical (PEC) hydrogen production. They enhance electron transfer kinetics, stability, and charge separation, improving PEC efficiency. These materials offer tailored properties, durability, and scalability, making them promising for commercialization. They can be customized to specific PEC requirements and integrated into various configurations. As PEC technology advances, these support materials play a key role in enhancing hydrogen production.

4 Conclusion and Future Perspective

In conclusion, the book chapter provides a comprehensive exploration of the commercialization aspects surrounding photocatalytic and photoelectrochemical technologies for hydrogen production. These innovative approaches offer great promise in the transition towards sustainable and clean energy solutions. As we continue to bridge the gap between research and practical implementation, the prospects for widespread adoption in the industrial, residential, and transportation sectors are becoming increasingly tangible. Throughout this exploration, we have highlighted the significance of material innovation, efficiency optimization, and scalability in bringing these technologies closer to practical industrial applications. The future of photocatalytic and photoelectrochemical hydrogen production holds immense promise in the global pursuit of sustainable energy solutions. Photocatalytic processes, utilizing semiconductor materials to harness solar energy for hydrogen production, continue to evolve with the development of advanced materials and innovative reactor designs. Likewise, photoelectrochemical methods, which rely on specialized electrodes to directly convert sunlight into hydrogen, are advancing rapidly. As the world seeks to reduce carbon emissions and reliance on fossil fuels, the future of hydrogen production through photocatalysis and photo-electrochemistry appears brighter than ever.

References

1. Z. Haider, H.W. Yim, H.W. Lee, H.-I. Kim, Surface and bulk modification for advanced electrode design in photoelectrochemical water splitting. Int. J. Hydrog. Energy **45**(10), 5793–5815 (2020)
2. I. Iordache, A.V. Gheorghe, M. Iordache, Towards a hydrogen economy in Romania: statistics, technical and scientific general aspects. Int. J. Hydrog. Energy **38**(28), 12231–12240 (2013)
3. X. Luo, J. Wang, M. Dooner, J. Clarke, Overview of current development in electrical energy storage technologies and the application potential in power system operation. Appl. Energy **137**, 511–536 (2015)
4. P.M. Rajaitha, S. Hajra, K. Mistewicz, S. Panda, M. Sahu, D. Dubal et al., Multifunctional materials for photo-electrochemical water splitting. J. Mater. Chem. A **10**(30), 15906–15931 (2022)
5. B. Sørensen, A history of renewable energy technology. Energy Policy **19**(1), 8–12 (1991)
6. G. Liu, Y. Sheng, J.W. Ager, M. Kraft, R. Xu, Research advances towards large-scale solar hydrogen production from water. EnergyChem. **1**(2), 100014 (2019)
7. P. Muthukumar, A. Kumar, M. Afzal, S. Bhogilla, P. Sharma, A. Parida et al., Review on large-scale hydrogen storage systems for better sustainability. Int. J. Hydrog. Energy (2023)
8. B. Sørensen, Renewable energy: a technical overview. Energy Policy **19**(4), 386–391 (1991)
9. I. Dincer, C. Acar, Review and evaluation of hydrogen production methods for better sustainability. Int. J. Hydrog. Energy **40**(34), 11094–11111 (2015)
10. K.S. Gabriel, R.S. El-Emam, C. Zamfirescu, Technoeconomics of large-scale clean hydrogen production—A review. Int. J. Hydrog. Energy **47**(72), 30788–30798 (2022)
11. S.G. Babu, R. Vinoth, D. Praveen Kumar, M.V. Shankar, H.-L. Chou, K. Vinodgopal et al., Influence of electron storing, transferring and shuttling assets of reduced graphene oxide at the interfacial copper doped TiO_2 p–n heterojunction for increased hydrogen production. Nanoscale **7**(17), 7849–7857 (2015)
12. K. Fujishima, A. Honda, A. Fujishima, K. Honda, Nature publishing group. Nature **1972**(238), 37–38 (1972)
13. H. Lv, Y.V. Geletii, C. Zhao, J.W. Vickers, G. Zhu, Z. Luo et al., Polyoxometalate water oxidation catalysts and the production of green fuel. Chem. Soc. Rev. **41**(22), 7572–7589 (2012)
14. Z. Han, R. Eisenberg, Fuel from water: the photochemical generation of hydrogen from water. Acc. Chem. Res. **47**(8), 2537–2544 (2014)

15. Z. Wang, C. Li, K. Domen, Recent developments in heterogeneous photocatalysts for solar-driven overall water splitting. Chem. Soc. Rev. **48**(7), 2109–2125 (2019)
16. L. Jiao, Y. Dong, X. Xin, L. Qin, H. Lv, Facile integration of Ni-substituted polyoxometalate catalysts into mesoporous light-responsive metal-organic framework for effective photogeneration of hydrogen. Appl. Catal. B **291**, 120091 (2021)
17. M. Zhang, H. Li, J. Zhang, H. Lv, G.-Y. Yang, Research advances of light-driven hydrogen evolution using polyoxometalate-based catalysts. Chin. J. Catal. **42**(6), 855–871 (2021)
18. F.E. Osterloh, B.A. Parkinson, Recent developments in solar water-splitting photocatalysis. MRS Bull. **36**(1), 17–22 (2011)
19. J. Zhang, M. Zhang, Y. Dong, C. Bai, Y. Feng, L. Jiao et al., CdTe/CdSe-sensitized photocathode coupling with Ni-substituted polyoxometalate catalyst for photoelectrochemical generation of hydrogen. Nano Res. **15**(2), 1347–1354 (2022)
20. K. Prakash, V. Selvam, S.G. Babu, S. Meena, S. Karuthapandian, Rational design of novel 3D flower-like praseodymium molybdate anchored graphitic carbon Nitride: an efficient and sustainable photocatalyst for mitigation of carcinogenic pollutants. Appl. Surf. Sci. **569**, 151104 (2021)
21. M. Ni, M.K. Leung, D.Y. Leung, K. Sumathy, A review and recent developments in photocatalytic water-splitting using TiO_2 for hydrogen production. Renew. Sustain. Energy Rev. **11**(3), 401–425 (2007)
22. P. Kasirajan, S. Karunamoorthy, M. Velluchamy, K. Subramaniam, C.M. Park, G.B. Sundaram, Fabrication of copper molybdate nanoflower combined polymeric graphitic carbon nitride heterojunction for water depollution: synergistic photocatalytic performance and mechanism insight. Environ. Res. **233**, 116428 (2023)
23. K. Maeda, K. Domen, Photocatalytic water splitting: recent progress and future challenges. J. Phys. Chem. Lett. **1**(18), 2655–2661 (2010)
24. A. Gupta, B. Likozar, R. Jana, W.C. Chanu, M.K. Singh, A review of hydrogen production processes by photocatalytic water splitting—From atomistic catalysis design to optimal reactor engineering. Int. J. Hydrog. Energy **47**(78), 33282–33307 (2022)
25. A.J. Bard, Photoelectrochemistry and heterogeneous photo-catalysis at semiconductors. J. Photochem. **10**(1), 59–75 (1979)
26. P. Ganguly, M. Harb, Z. Cao, L. Cavallo, A. Breen, S. Dervin et al., 2D nanomaterials for photocatalytic hydrogen production. ACS Energy Lett. **4**(7), 1687–1709 (2019)
27. S.G. Babu, V. Ramalingam, B. Neppolian, D.D. Dionysiou, M. Ashokkumar, Diffused sunlight driven highly synergistic pathway for complete mineralization of organic contaminants using reduced graphene oxide supported photocatalyst. J. Hazard. Mater. **291**, 83–92 (2015)
28. P. Senthil Kumar, M. Selvakumar, S.G. Babu, S.K. Jaganathan, S. Karuthapandian, S. Chattopadhyay, Novel CuO/chitosan nanocomposite thin film: facile hand-picking recoverable, efficient and reusable heterogeneous photocatalyst. RSC Adv. **5**(71), 57493–57501 (2015)
29. K. Maeda, T. Takata, M. Hara, N. Saito, Y. Inoue, H. Kobayashi et al., GaN: ZnO solid solution as a photocatalyst for visible-light-driven overall water splitting. J. Am. Chem. Soc. **127**(23), 8286–8287 (2005)
30. Y. Lee, H. Terashima, Y. Shimodaira, K. Teramura, M. Hara, H. Kobayashi et al., Zinc germanium oxynitride as a photocatalyst for overall water splitting under visible light. J. Phys. Chem. C **111**(2), 1042–1048 (2007)
31. Y. Sasaki, H. Nemoto, K. Saito, A. Kudo, Solar water splitting using powdered photocatalysts driven by Z-schematic interparticle electron transfer without an electron mediator. J Phys Chem C **113**(40), 17536–17542 (2009)
32. S.G. Babu, P. Karthik, M.C. John, S.K. Lakhera, M. Ashokkumar, J. Khim et al., Synergistic effect of sono-photocatalytic process for the degradation of organic pollutants using $CuO–TiO_2$/rGO. Ultrason. Sonochem. **50**, 218–223 (2019)
33. J. Yu, Y. Hai, M. Jaroniec, Photocatalytic hydrogen production over CuO-modified titania. J. Colloid Interface Sci. **357**(1), 223–228 (2011)
34. Z. Zou, J. Ye, H. Arakawa, Photocatalytic water splitting into H_2 and/or O_2 under UV and visible light irradiation with a semiconductor photocatalyst. Int. J. Hydrog. Energy **28**(6), 663–669 (2003)

35. S.M. Ji, H. Jun, J.S. Jang, H.C. Son, P.H. Borse, J.S. Lee, Photocatalytic hydrogen production from natural seawater. J. Photochem. Photobiol. A **189**(1), 141–144 (2007)
36. S. Onsuratoom, S. Chavadej, T. Sreethawong, Hydrogen production from water splitting under UV light irradiation over Ag-loaded mesoporous-assembled TiO_2–ZrO_2 mixed oxide nanocrystal photocatalysts. Int. J. Hydrog. Energy **36**(9), 5246–5261 (2011)
37. K. Karthik, C.V. Reddy, K.R. Reddy, R. Ravishankar, G. Sanjeev, R.V. Kulkarni et al., Barium titanate nanostructures for photocatalytic hydrogen generation and photodegradation of chemical pollutants. J. Mater. Sci.: Mater. Electron. **30**, 20646–20653 (2019)
38. Y. Noda, B. Lee, K. Domen, J.N. Kondo, Synthesis of crystallized mesoporous tantalum oxide and its photocatalytic activity for overall water splitting under ultraviolet light irradiation. Chem. Mater. **20**(16), 5361–5367 (2008)
39. R. Fiorenza, S. Sciré, L. D'Urso, G. Compagnini, M. Bellardita, L. Palmisano, Efficient H_2 production by photocatalytic water splitting under UV or solar light over variously modified TiO_2-based catalysts. Int. J. Hydrogen Energy **44**(29), 14796–14807 (2019)
40. M.I. Fadlalla, S.G. Babu, Chapter 5—Role of graphene in photocatalytic water splitting for hydrogen production, in *Graphene-Based Nanotechnologies for Energy and Environmental Applications*, eds. by M. Jawaid, A. Ahmad, D. Lokhat (Elsevier, 2019). pp. 81–108
41. X. Wang, K. Maeda, A. Thomas, K. Takanabe, G. Xin, J.M. Carlsson et al., A metal-free polymeric photocatalyst for hydrogen production from water under visible light. Nat. Mater. **8**(1), 76–80 (2009)
42. K. Maeda, M. Higashi, D. Lu, R. Abe, K. Domen, Efficient nonsacrificial water splitting through two-step photoexcitation by visible light using a modified oxynitride as a hydrogen evolution photocatalyst. J. Am. Chem. Soc. **132**(16), 5858–5868 (2010)
43. K. Sayama, K. Mukasa, R. Abe, Y. Abe, H. Arakawa, Stoichiometric water splitting into H_2 and O_2 using a mixture of two different photocatalysts and an IO^{3-}/I^- shuttle redox mediator under visible light irradiation. Chem. Commun. **23**, 2416–2417 (2001)
44. M. Higashi, R. Abe, K. Teramura, T. Takata, B. Ohtani, K. Domen, Two step water splitting into H_2 and O_2 under visible light by $ATaO_2N$ (A = Ca, Sr, Ba) and WO_3 with IO^{3-}/I^- shuttle redox mediator. Chem. Phys. Lett. **452**(1–3), 120–123 (2008)
45. X. Chen, R. Shi, Q. Chen, Z. Zhang, W. Jiang, Y. Zhu et al., Three-dimensional porous g-C_3N_4 for highly efficient photocatalytic overall water splitting. Nano Energy **59**, 644–650 (2019)
46. T. Puangpetch, T. Sreethawong, S. Yoshikawa, S. Chavadej, Hydrogen production from photocatalytic water splitting over mesoporous-assembled $SrTiO_3$ nanocrystal-based photocatalysts. J. Mol. Catal. A: Chem. **312**(1–2), 97–106 (2009)
47. C. Acar, I. Dincer, G.F. Naterer, Review of photocatalytic water-splitting methods for sustainable hydrogen production. Int. J. Energy Res. **40**(11), 1449–1473 (2016)
48. D. Jing, L. Guo, L. Zhao, X. Zhang, H. Liu, M. Li et al., Efficient solar hydrogen production by photocatalytic water splitting: from fundamental study to pilot demonstration. Int. J. Hydrog. Energy **35**(13), 7087–7097 (2010)
49. C.S. Gopinath, N. Nalajala, A scalable and thin film approach for solar hydrogen generation: a review on enhanced photocatalytic water splitting. J. Mater. Chem. A **9**(3), 1353–1371 (2021)
50. K.U. Sahar, K. Rafiq, M.Z. Abid, A. Rauf, U. ur Rehman, M.A. Nadeem et al., Sun-light driven hydrogen generation by Pd/Rb_2O cocatalysts: Escalate the utility of rutile TiO_2 for photocatalytic water splitting. Colloids Surf. A: Physicochem. Eng. Aspects **674**, 131942 (2023)
51. Y. Shi, L. Li, Z. Xu, X. Qin, Y. Cai, W. Zhang et al., Coupled internal electric field with hydrogen release kinetics for promoted photocatalytic hydrogen production through employing carbon coated transition metal as co-catalyst. J. Colloid Interface Sci. **630**, 274–285 (2023)
52. T. Liu, Y. Xiong, X. Wang, Y. Xue, W. Liu, J. Tian et al., Dual cocatalysts and vacancy strategies for enhancing photocatalytic hydrogen production activity of $Zn_3In_2S_6$ nanosheets with an apparent quantum efficiency of 66.20%. J. Colloid Interface Sci. **640**, 31–40 (2023)
53. F. Saleem, M.Z. Abid, K. Rafiq, A. Rauf, K. Ahmad, S. Iqbal et al., Synergistic effect of Cu/Ni cocatalysts on CdS for sun-light driven hydrogen generation from water splitting. Int. J. Hydrog. Energy (2023)

54. L. Xue, Y. Shi, C. Huang, Q. Wu, B. Chen, W. Yao et al., Zn and Ni dual hydrogen evolution sites integrated onto CdS for effective photocatalytic hydrogen production. J. Colloid Interface Sci. **635**, 72–82 (2023)
55. M. Thangamuthu, K. Vankayala, L. Xiong, S. Conroy, X. Zhang, J.J.A.C. Tang, Tungsten oxide-based Z-scheme for visible light-driven hydrogen production from water splitting. ACS Catal. **13**, 9113–9124 (2023)
56. M. Saleh, H.N. Abdelhamid, D.M. Fouad, H.M.J.F. El-Bery, Enhancing photocatalytic water splitting: comparative study of TiO_2 decorated nanocrystals (Pt and Cu) using different synthesis methods. Fuel **354**, 129248 (2023)
57. S. Jayachitra, D. Mahendiran, P. Ravi, P. Murugan, M. Sathish, Highly conductive $NiSe_2$ nanoparticle as a co-catalyst over TiO2 for enhanced photocatalytic hydrogen production. Appl. Catal. B: Environ. **307**, 121159
58. L. Yang, X. Li, G. Zhang, P. Cui, X. Wang, X. Jiang et al., Combining photocatalytic hydrogen generation and capsule storage in graphene based sandwich structures. Nat. Commun. **8**(1), 16049 (2017)
59. S. Samajdar, S. Bera, P.S. Das, H. Finch, V.R. Dhanak, S. Chakraborty et al., Exploration of 1D–2D $LaFeO_3$/RGO S-scheme heterojunction for photocatalytic water splitting. Int. J. Hydrog. Energy **48**(47), 17838–17851 (2023)
60. G. Yan, X. Sun, Y. Zhang, H. Li, H. Huang, B. Jia et al., Metal-free 2D/2D van der Waals Heterojunction based on covalent organic frameworks for highly efficient solar energy catalysis. Nano-Micro Lett. **15**(1), 132 (2023)
61. S.A. Ali, T.J. Ahmad, Enhanced hydrogen generation via overall water splitting using novel MoS_2-BN nanoflowers assembled TiO_2 ternary heterostructures. Int. J. Hydrog. Energy (2023)
62. A. Sarkar, S. Chaule, S. Mandal, S. Saha, S. Ganguly, D. Banerjee et al., Enhanced photocatalytic Hydrogen generation by splitting water using Sodium alginate decorated rGO-CdS hybrid photo-catalyst. Mater. Today: Proc. (2023)
63. K.U. Sahar, K. Rafiq, M.Z. Abid, U. ur Rehman, A. Rauf, E. Hussain et al., Surface sensitization of g-C_3N_4/TiO_2 via Pd/Rb_2O co-catalysts: accelerating water splitting reaction for green fuel production in the absence of organic sacrificial agents. React. Chem. Eng. (2023)
64. Q. Ding, X. Zou, J. Ke, Y. Dong, Y. Cui, G. Lu et al., S-scheme 3D/2D $NiCo_2O_4$@g-C_3N_4 hybridized system for boosting hydrogen production from water splitting. Renew. Energy **203**, 677–685 (2023)
65. S. Yuvarani, S. Kannan, M. Raja Saravanan, P. Thamaraiselvan, M. Shkir, H. Algarni et al., Facile strategy to fabricate $NiTe_2$/g-C_3N_4 heterojunction photocatalyst with enhanced visible-light photocatalytic hydrogen production. J. Mater. Sci.: Mater. Electron. **34**(17), 1342 (2023)
66. B.A. Pinaud, J.D. Benck, L.C. Seitz, A.J. Forman, Z. Chen, T.G. Deutsch et al., Technical and economic feasibility of centralized facilities for solar hydrogen production via photocatalysis and photoelectrochemistry. Energy Environ. Sci. **6**(7), 1983–2002 (2013)
67. A. Fujishima, K. Honda, Electrochemical photolysis of water at a semiconductor electrode. Nature **238**(5358), 37–38 (1972)
68. A. Aadenan, N.A. Arzaee, M.F.M. Noh, I.N. Nawas Mumthas, N.A. Mohamed, M.N.M. Daud et al., Photoelectrochemical water splitting performance of $LaFeO_3$ thin film photocathode prepared by spray gun deposition. Thin Solid Films **777**, 139892 (2023)
69. Q. Ding, F. Meng, C.R. English, M. Cabán-Acevedo, M.J. Shearer, D. Liang et al., Efficient photoelectrochemical hydrogen generation using heterostructures of Si and chemically exfoliated metallic MoS_2. J. Am. Chem. Soc. **136**(24), 8504–8507 (2014)
70. B.-J. Li, P.-F. Yin, Y.-Z. Zhou, Z.-M. Gao, T. Ling, X.-W. Du, Single crystalline Cu_2ZnSnS_4 nanosheet arrays for efficient photochemical hydrogen generation. RSC Adv. **5**(4), 2543–2549 (2015)
71. V. Avasare, Z. Zhang, D. Avasare, I. Khan, A. Qurashi, Room-temperature synthesis of TiO_2 nanospheres and their solar driven photoelectrochemical hydrogen production. Int. J. Energy Res. **39**(12), 1714–1719 (2015)
72. X. Guo, L. Wang, Y. Tan, Hematite nanorods Co-doped with Ru cations with different valence states as high performance photoanodes for water splitting. Nano Energy **16**, 320–328 (2015)

73. Y. Cho, M. Park, J.K. Kim, S. Kim, H. Suk Jung, J.H. Park, Electrochemically controlled CdS@CdSe nanoparticles on ITO@TiO_2 dual core–shell nanowires for enhanced photoelectrochemical hydrogen production. Appl. Surf. Sci. **505**, 144569 (2020)
74. Z. Zheng, M. Morgan, P. Maji, X. Xia, X. Zu, W. Zhou, CuO nanorod arrays by gas-phase cation exchange for efficient photoelectrochemical water splitting. RSC Adv. **13**(6), 3487–3493 (2023)
75. A.C. Guler, J. Antos, M. Masar, M. Urbanek, M. Machovsky, I. Kuritka, Comprehensive evaluation of photoelectrochemical performance dependence on geometric features of ZnO nanorod electrodes. Nanosc. Adv. **5**(11), 3091–3103 (2023)
76. Y. Pang, W. Zang, Z. Kou, L. Zhang, G. Xu, J. Lv et al., Assembling of Bi atoms on TiO_2 nanorods boosts photoelectrochemical water splitting of semiconductors. Nanoscale **12**(7), 4302–4308 (2020)
77. A. Pareek, P. Paik, P.H. Borse, Stable hydrogen generation from Ni- and Co-based co-catalysts in supported CdS PEC cell. Dalton Trans. **45**(27), 11120–11128 (2016)
78. R. Raja, P. Sudhagar, A. Devadoss, C. Terashima, L.K. Shrestha, K. Nakata et al., Pt-free solar driven photoelectrochemical hydrogen fuel generation using 1T MoS2 co-catalyst assembled CdS QDs/TiO_2 photoelectrode. Chem. Commun. **51**(3), 522–525 (2015)
79. P. Wang, S. Zhan, H. Wang, Y. Xia, Q. Hou, Q. Zhou et al., Cobalt phosphide nanowires as efficient co-catalyst for photocatalytic hydrogen evolution over $Zn_{0.5}Cd_{0.5}S$. Appl. Catal. B: Environ. **230**, 210–219 (2018)
80. L. Chu, Y. Lin, Y. Liu, H. Wang, Q. Zhang, Y. Li et al., Preparation of CdS-CoSx photocatalysts and their photocatalytic and photoelectrochemical characteristics for hydrogen production. Int. J. Hydrog. Energy **44**(51), 27795–27805 (2019)
81. Y. Yuan, B. Zhong, F. Li, H. Wu, J. Liu, H. Yang et al., Surface phosphorization for the enhanced photoelectrochemical performance of an Fe_2O_3/Si photocathode. Nanoscale **14**(31), 11261–11269 (2022)
82. A.I. Alharthi, M.N. Shaddad, T.F. Qahtan, M.A. Alotaibi, A.A. Alanazi, P. Arunachalam, Cooperative catalytic behavior of Bi_2S_3 and ZrO_2 nanoparticles on Bi_2O_3 photoanodes for promoted stability and solar driven photoelectrochemical hydrogen production. J. Alloy. Compd. **967**, 171733 (2023)
83. C. Yan, Y. Wang, H. Wang, Y. Zhang, Nickel phosphate improved silicon nanowires arrays hybrid nanostructures for solar-driven photoelectrochemical hydrogen production. J. Photochem. Photobiol. A **445**, 115094 (2023)
84. C.Y. Chen, H. Pai, H.M. Chen, S. Yougbaré, S. Husain, Y.C. Hsiao et al., Distinguished heterojunction and co-catalyst behaviors on tungsten doped $BiVO_4$ for photoelectrochemical catalytic water splitting reaction. Mater. Today Sustain. **24**, 100488 (2023)
85. I. Hamdani, A. Bhaskarwar, Recent progress in material selection and device designs for photoelectrochemical water-splitting. Renew. Sustain. Energy Rev. **138**, 110503 (2021)
86. W. Fan, H. Bai, W. Shi, Semiconductors with NIR driven upconversion performance for photocatalysis and photoelectrochemical water splitting. CrystEngComm **16**(15), 3059–3067 (2014)
87. W. Fan, X. Yu, H.-C. Lu, H. Bai, C. Zhang, W. Shi, Fabrication of TiO_2/RGO/Cu_2O heterostructure for photoelectrochemical hydrogen production. Appl. Catal. B **181**, 7–15 (2016)
88. W. Ho, J.C. Yu, J. Lin, J. Yu, P. Li, Preparation and photocatalytic behavior of MoS2 and WS2 nanocluster sensitized TiO2. Langmuir **20**(14), 5865–5869 (2004)
89. W. Zhang, S. Huo, S. Yang, Y. Zhong, S. Zhang, H. Wang et al., Ternary monolithic ZnS/CdS/rGOphotomembrane with desirable charge separation/transfer routes for effective photocatalytic and photoelectrochemical hydrogen generation. Chem. Asian J. **14**(19), 3431–3441 (2019)
90. A. Almohammedi, M. Shaban, H. Mostafa, M. Rabia, Nanoporous TiN/TiO_2/alumina membrane for photoelectrochemical hydrogen production from sewage water. Nanomaterials **11**(10), 2617 (2021)

Mesocrystals for Photocatalysis: A Review of Functional Modification

Kalaiselvi Chandiran and Krishna Chandar Nagamuthu Raja

Abstract Mesocrystal is originally a term for describing the superstructure of nanocrystals with common crystallographic orientation. Mesocrystal photocatalysts have many special properties, such as high specific surface area, high crystallinity, availability of pores, strong electrical conductivity, and thermal stability. It is predicted that the superstructure of mesocrystal photocatalysts plays an important role in improving the performance of materials in many applications, such as photocatalysis, Li-ion batteries, sensing, solar-energy conversion, and supercapacitors. In the past, the majority of attention was concentrated on the mechanism of mesocrystal formation, properties, and applications. Currently, remarkable progress is being described with a particular focus on mesocrystal photocatalysts. However, so far there is no review report on effective functional modifications of mesocrystal photocatalysts. The mini-review focuses on different functional modifications (including metal, non-metal doping, and carbon-based mesocrystal) photocatalytic application, and formation mechanisms of mesocrystal photocatalysts.

Keywords Mesocrystal photocatalyst · Metal doped mesocrystal · Non-metal doped mesocrystal · Carbon-based mesocrystal · Visible light

1 Introduction

Water pollution caused by dye discharges from various sectors (for example, textile dyeing, cosmetics, paints, paper-making food industry, etc.) has recently received a lot of attention owing to the dangers it poses to public health and ecosystems. The present global pigment market is 38 billion USD, and it is predicted to grow up to 48 billion USD by 2023 [1–3]. The annual quantity supplied of synthetic dyes exceeds 700,000 tonnes. Each year, more than 15% of these synthetic dyes are released into water. The dye pollutants in water are toxic, cytotoxic, and genotoxic [4].

K. Chandiran · K. C. N. Raja (✉)
Department of Physics, School of Advanced Sciences, Vellore Institute of Technology, Vellore 632014, India
e-mail: nkchandar@vit.ac.in

P. Sathishkumar (ed.), *Photocatalysis for Energy and Environmental Applications*, Green Energy and Technology, https://doi.org/10.1007/978-981-97-1939-6_11

Dyes are categorized as reactive dyes, solvent dyes, basic dyes, direct dyes, and vat dyes based on their chromophore structure, as illustrated in Fig. 1. Only around 47% of synthetic dyes are biodegradable [5, 6]. Traditional wastewater treatment technologies (e.g., adsorption, coagulation/flocculation, and precipitation) often need a prolonged processing period and generate secondary waste, which is expensive to dispose. Advanced oxidation processes are a rapidly developing efficient technology for removing organic pollutants from water bodies. Due to its low cost, eco-friendly nature, and excellent physical and chemical stability, it can perform as an effective catalyst [7–9]. Currently, the rapid recombination of photogenerated charge carriers and low consumption of visible light are the two blocks that limit the practical application of photocatalysts. Generally, photocatalytic material with morphology-control or facet-control properties can also achieve photodegradation under visible light regions [10, 11].

Mesocrystal superstructure photocatalysts can increase the surface area, pore size, and pore volume and enhance the diffusion mass transfer of the substrate molecule. Currently, there is a need to improve research in the field of mesocrystal photocatalysts under UV-visible light illumination. The effective separation of photogenerated electron-hole pair in a photocatalyst. The ultraviolet region occupies only a small fraction (4–5%) of the entire solar spectrum, while (45%) of the solar spectrum is in the visible light range [12, 13]. Achieving a photocatalytic activity under the visible light region has greater significance in the minimization of the environmental issue.

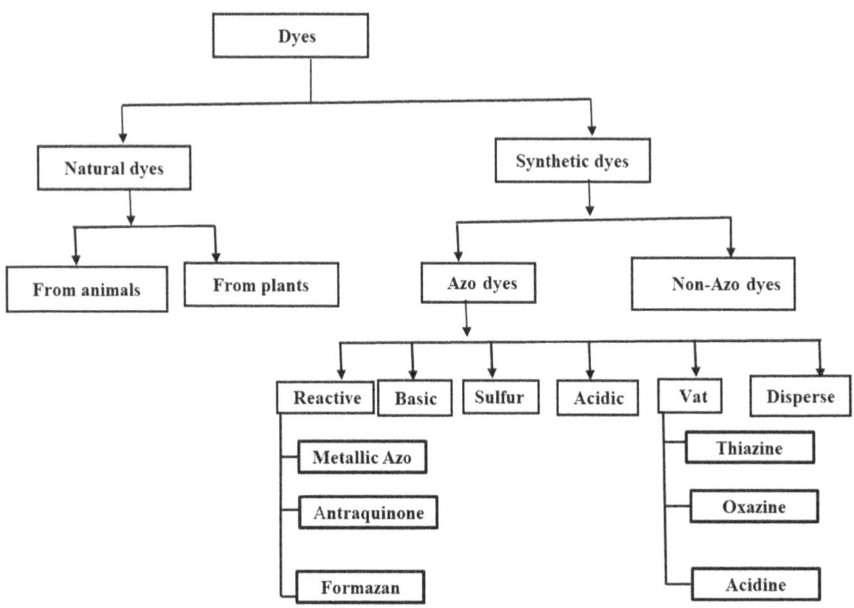

Fig. 1 Flow chart indicating dye classifications based on the chemical constitution

Further, the surface morphology has an impact on the photocatalytic dye degradation through numerous factors namely on the absorption capacity, electron-hole pair separation efficiency, and migration of charge. As a reason, it is necessary to develop and synthesize novel hierarchical architectures sensibly.

1.1 Mechanism of Photo-Catalytic Degradation of Dyes

The photocatalytic dye degradation mechanism is illustrated in Fig. 2. Synthetic dyes are transported from the aqueous phase surrounding the photocatalyst surface. The dye molecules are adsorbed onto the photocatalyst surface by superoxide radicals and hydroxyl radicals. This radical is formed by incident photons [14]. The products released from the surface are returned to the liquid phase. The basic equation involved in the dye degradation mechanism is

$$\text{Photocatalyst} + h\nu(\text{UV or visible irradiations}) \rightarrow \text{photocatalyst} + h^+(\text{VB}) + e^-(\text{CB}) \quad (1)$$

$$H_2O \rightarrow H^+ + OH^- \quad (2)$$

$$OH^- + h^+(\text{VB}) \rightarrow OH^\circ(\text{Hydroxyl radicals}) \quad (3)$$

$$O_2 + e^-(\text{CB}) \rightarrow O_2{}^\circ - (\text{Anionic superoxide radicals}) \quad (4)$$

$$OH^\circ + \text{dye} \rightarrow CO_{2\_} + H_2O \quad (5)$$

$$O_2^{\circ -} + \text{dye} \rightarrow CO_2 + H_2O \quad (6)$$

The above mechanism is a general equation and can be varied under different experimental conditions [15, 16].

2 Mesocrystal Photocatalysts

Mesocrystal is a new class of hierarchical nanostructure that consists of crystallographically oriented nanocrystals with three-dimensional superstructures [17]. In 1969, Peters et al. discovered a mesocrystal of $BaSO_4$ porous with an intermediate structure. In 2003 Colfen and Mann described that growing single crystals are supported by mesoscale assembly of nanocrystals [18, 19]. In 2005 Colfen et al. observed the mesocrystal superstructures and formation mechanisms in biominerals and their mimetics. They further claimed that it is a special type of colloidal solid

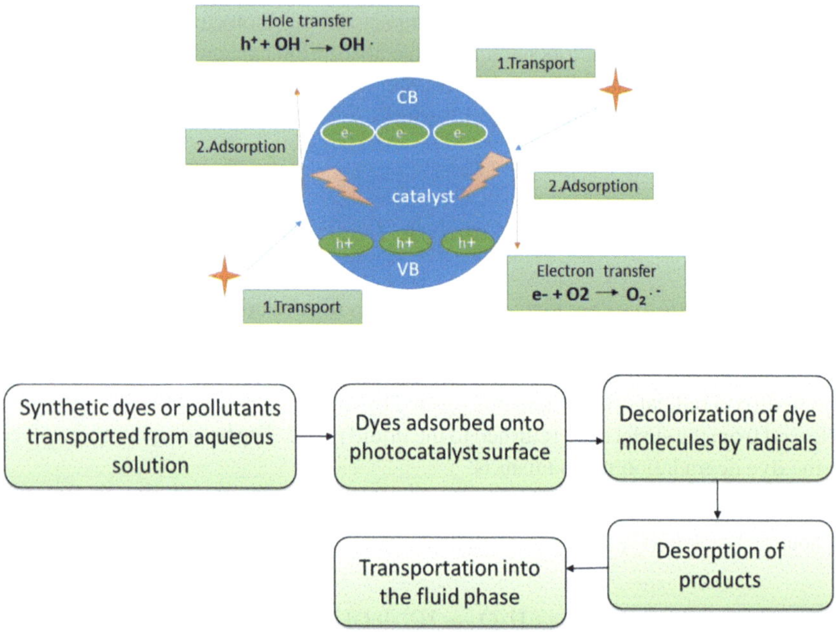

Fig. 2 Basic photocatalytic dye degradation mechanism

and collected aligned nanocrystals with specific mutual crystallographic orientation. It is interesting to note that both man-made and natural mesocrystal superstructures are available such as red coral and sea urchin spine and bones. Mesocrystals superstructures are designed in Fig. 3a: the atoms, ions, or molecules interfere and grow as single crystals by the process of oriented attachment. Such a group of single crystals from this process is typically called classical crystallization [20]. In non-classical crystallization process of mesocrystal superstructures is illustrated in Fig. 1b, c.

The three-dimensional superstructure is designed by nanoparticle aggregation in a common crystallographic direction, which is crystallographically combined to form a single crystal by the process of oriented attachment (OA), with spatially separated nanoparticles in a common crystallographic register (Fig. 3c). In general, mesocrystals are made up of two types of crystalline nanoparticles namely monodisperse and polydisperse crystalline nanoparticles [20]. The building units of type I mesocrystal are monodisperse nanoparticles arranged in an array with long-range order and small-angle diffraction pattern. Type II mesocrystals show mutually oriented polydisperse colloidal nanoparticles with a probable degree of alignment. This is demonstrated by Grazing-Incidence Small-Angle X-ray scattering (GISAXS) as well as in situ TEM and small/wide-angle X-ray scattering (SAXS/WAXS). The following methods explain the formation of mesocrystals superstructure [19]:

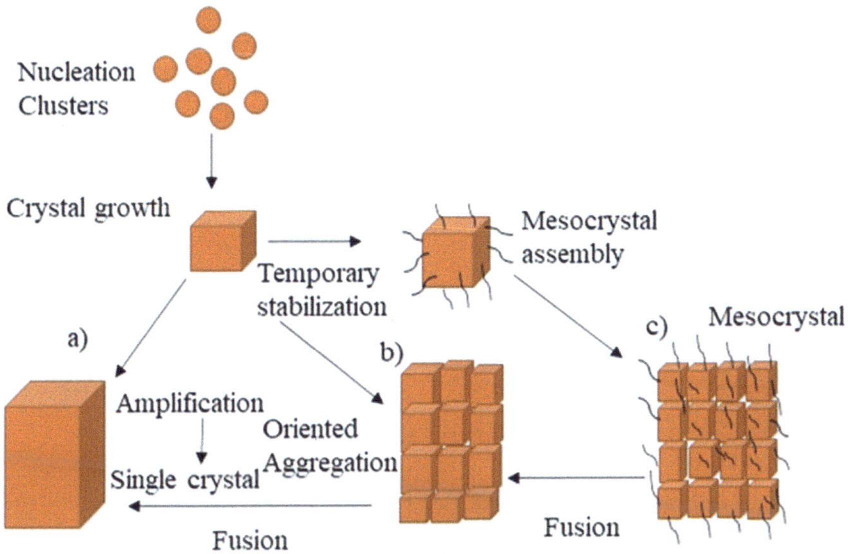

Fig. 3 Classical and nonclassical crystallization of mesocrystal. **a** Classical crystallization via atoms/ions/molecules addition versus single crystal formation. **b** Forming an iso-oriented crystal by oriented aggregation of primary nanoparticles, **c** mesocrystal formation by the self-assembly of primary nanoparticles

(a) due to the alignment of an organic matrix,
(b) physical fields of nanoparticles (such as electronic, magnetic, and optical fields or mutual alignment of identical crystal faces,
(c) the connection of nanoparticles employing miner bridge to form a topotactic (epitaxial) growth of the nanoparticle.

The three-dimensional (3D) superstructure of mesocrystalline materials has received significant research attention, which essentially reduces the recombination rate of electron-hole pairs. The structural and morphological aspects have received a lot of attention over the last few decades. The synthesis and formation mechanism of nanocrystals has been investigated in a few review articles. In 2019, remarkable progress has been reported, with particular focus on the formation mechanism of nonclassical crystallization pathways of mesocrystal [20]. In the literature, an increasing number of papers have been published. Most of the researchers have concentrated on the properties of mesocrystal superstructures as well as their formation mechanism. Besides in 2016, TiO_2 mesocrystal photocatalyst has been predominantly studied [21]. To the best of our knowledge, there are no review articles published on the various types of mesocrystal photocatalysts with their efficient functional modification and their application in dye degradation under visible light regions [20].

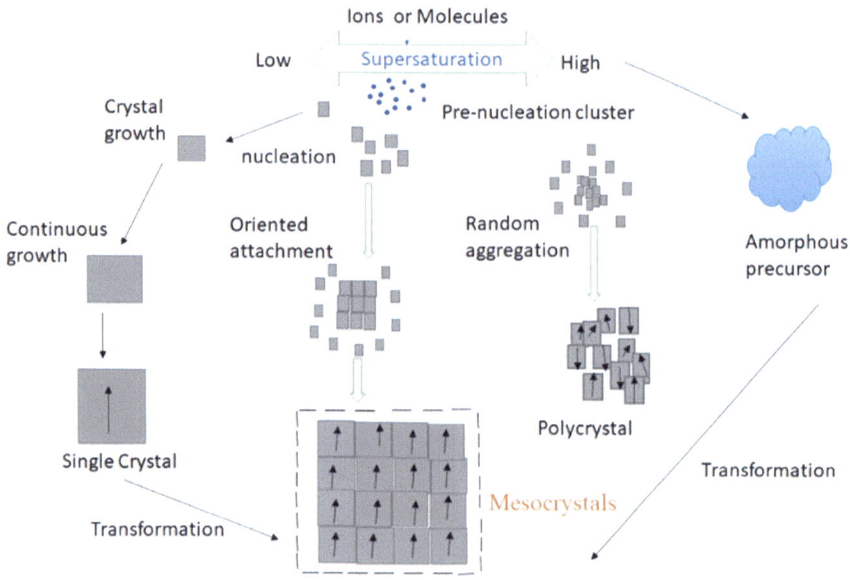

Fig. 4 Single crystal, polycrystal, and mesocrystal formation paths of in solution systems. The arrows indicate the crystallographic direction

In this review, we mostly recapitulate the significant advances in the expansion of monocrystal-oriented photocatalysis in various facets. Mesocrystal-based photocatalyst and their dye degradation efficiency under visible light irradiation by some functional modification are described [14]. To date, several changes have been made to expand the optical response of mesocrystal photocatalysts from ultraviolet to visible light by functional modification of nanocrystals such as metal-doped mesocrystal, non-metal doped mesocrystal and carbon-based mesocrystals (Fig. 4).

3 Application Potential of Mesocrystals

Although several semiconductors with favorable structure performance relationships have been researched for enhancing photocatalysis's poor efficiency, the presence of charge recombination at the surface or within the volume cannot be avoided successfully. The highly ordered mesocrystal superstructures could reduce the electron-hole pairs recombination, due to the anisotropic electron flow along the interparticle [22]. As a result, the surface of nanocrystals can achieve rapid and effective collection of photogenerated electron-hole pairs on catalytic sites. Over the last few decades, serious environmental pollution has begun to have a major impact on modern society, and the rapid development of industrialization and population growth. In addition to the lack of clean water sources due to the long-term droughts have become global problems. Hierarchical superstructure photocatalysis is used for removing organic pollutants from wastewater [22].

3.1 Functionally Modification of Mesocrystal Photocatalysts

The correlation between the altered interfacial/electronic structures and improved photocatalytic performances is crucial for developing novel monocrystal-based photocatalysts. Mesocrystals photocatalysts of functional modifications can be classified into two types namely doping and composites.

Anil Kumar et al. have prepared anatase square bipyramidal TiO_2 mesocrystal with various morphologies by simple one-pot hydrothermal method. The different morphologies were simply controlled by changing urea concentration in the aqueous solution. The different morphologies such as rods, rhombohedral, square pyramidal, and truncated square pyramidal structures [23]. The decomposition of urea generates amines and carboxyl groups, which are deposited on the surface of TiO_2 anatase mesocrystal during crystal formation.

Figure 5 shows the photocatalytic performance of the different morphologies of TiO_2, which are mentioned by ST-0, ST-0.025M, ST-0.05M, and ST-0.1. The rod-like morphology (ST-0) shows the decolorization efficiency is 40% within 20 min. The photocatalytic performance of ST-0.025M and ST-0.05M increased to 60% and 70%, due to urea content increased. For the same time interval, ~95% of the methylene blue dye was degraded for the high concentration of urea (0.15 M) sample, and it is due to crystal facts (001) orientation [23]. In particular, the photocatalytic degradation ability is that the high surface energy of the (001) plane is greater than that of the lower surface energy (101) facets [23]. The surface-active oxygen atoms and Penta coordinated Ti_5C as well as the superior mobility of the photogenerated electron-hole pairs towards the (001) facets are responsible for the enhanced photocatalytic performance of the truncated square bipyramidal TiO_2 mesocrystal (Fig. 6).

3.1.1 Doping Mesocrystal

It is well known that doping is a well-demonstrated strategy for effectively enhancing the physicochemical properties by changing the electronic structure of semiconductors; therefore, studies on doped nanocrystals have attracted increased attention.

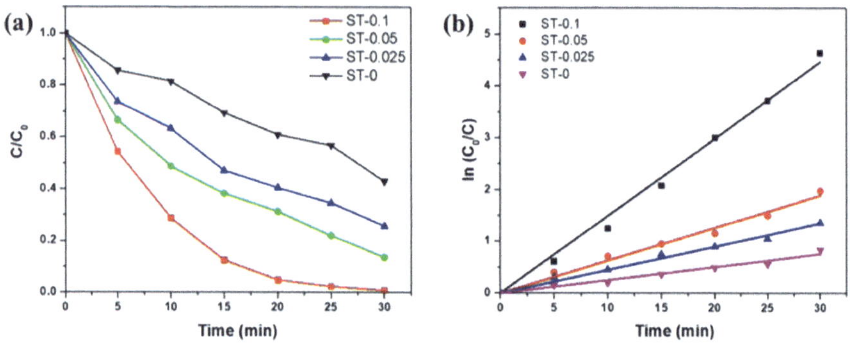

Fig. 5 **a** Photocatalytic dye degradation of MB in aqueous solution. **b** Pseudo-first-order kinetics **b** of the 0-ST, 0.025-ST-, 0.05-ST-, and 0.1-ST samples degradation

In general, the addition of metallic components intensifies new binding functions, resulting in unique photocatalytic capabilities in the doped system by reducing the bandgap and boosting visible light absorption.

To import metal ions into the framework of the mesocrystal, the relevant soluble salt is always uniformly mixed with the mesocrystal precursors; hence, metallic impurities are doped into the mesocrystal at the same time.

3.1.2 Metal Doping

José Miguel Calatayud et al. have reported a successful synthesis of TiO_2 mesocrystal without aggregation of nanoparticles in the range of 20 nm. Also, the doped TiO_2 mesocrystal contains Cr and Sb-codoped mesocrystal to form the ellipsoidal cross-section mesocrystal with a size less than 100 nm × 50 nm of dimension. By preventing the recombination of the electron-hole pair process, the photocatalytic activity is strengthened. Electron hole pair recombination interference is accomplished by doping with metals (Ag, Cd, Cu, Co, Fe, or Mn) or a nonmetal (N, S, or F) [24].

The self-assembly of the primary nanoparticles was formed by the interactions of the organic ligands and the interface-energy effect. The ellipsoidal mesocrystal shows an excellent photocatalytic degradation of organic compounds. Figure 7 shows photocatalytic degradation of MO by Ax photocatalyst under (a) visible and (b) UV light irradiation. The photocatalyst with a low dopant load, Cr0.03Sb0.03Ti0.94, i.e. A0.03, achieved a degradation degree of roughly 80%. Under UV light irradiation, the doped titania samples also show improved MO degrading efficiency [24]. For Cr0.05Sb0.05Ti0.9 and Cr0.1Sb0.1Ti0.9 samples, i.e. A0.05 and A0.1, respectively, degradation was nearly complete. Mesocrystals are superstructures constructed of crystalline nanoparticles that are reciprocally oriented between nanocrystal-building units and have the potential to significantly improve charge separation when exposed to visible light, Ax with nominal compositions $Cr_xSb_xTi1{-}2xO_2$ $(0 \leq x \leq 0.1)$.

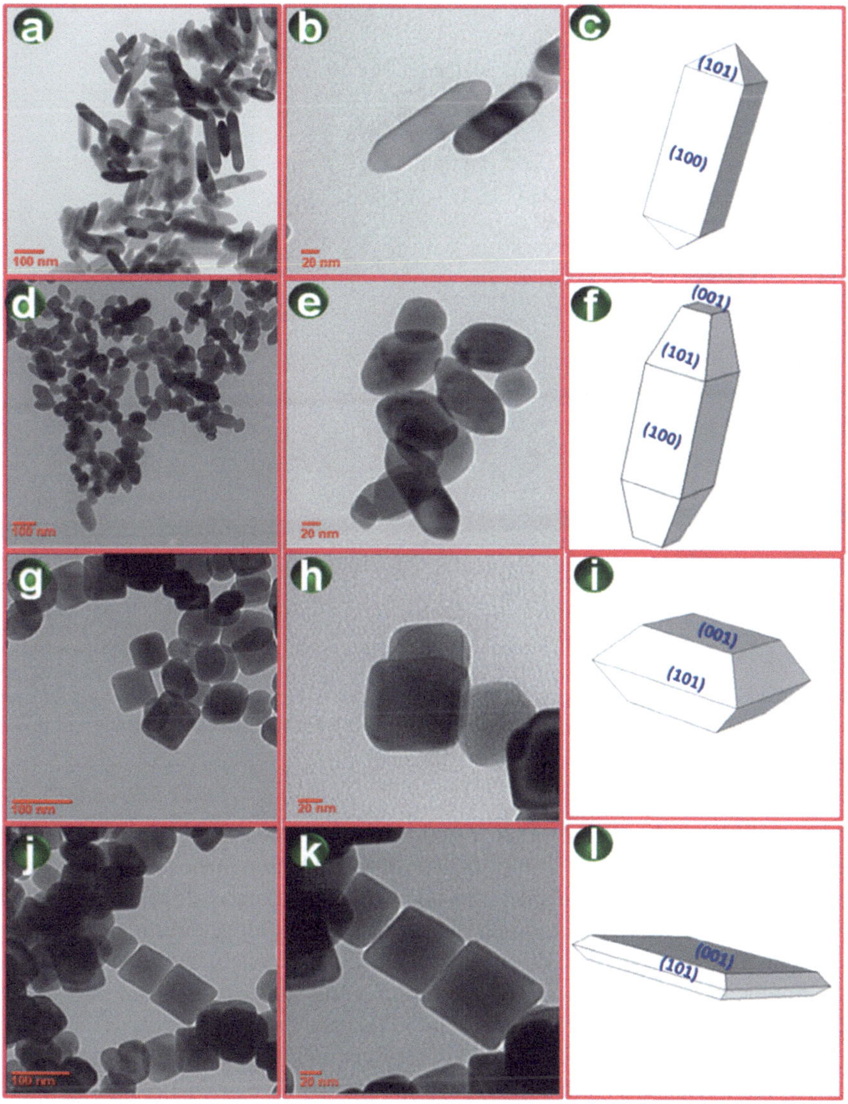

Fig. 6 TEM morphology of the 0-ST (**a**, **b**), 0.025-ST (**d**, **e**), 0.05-ST (**g**, **h**), and 0.1 ST (**j**, **k**) and their respective crystallites (**c**, **f**, **i**, and **l**)

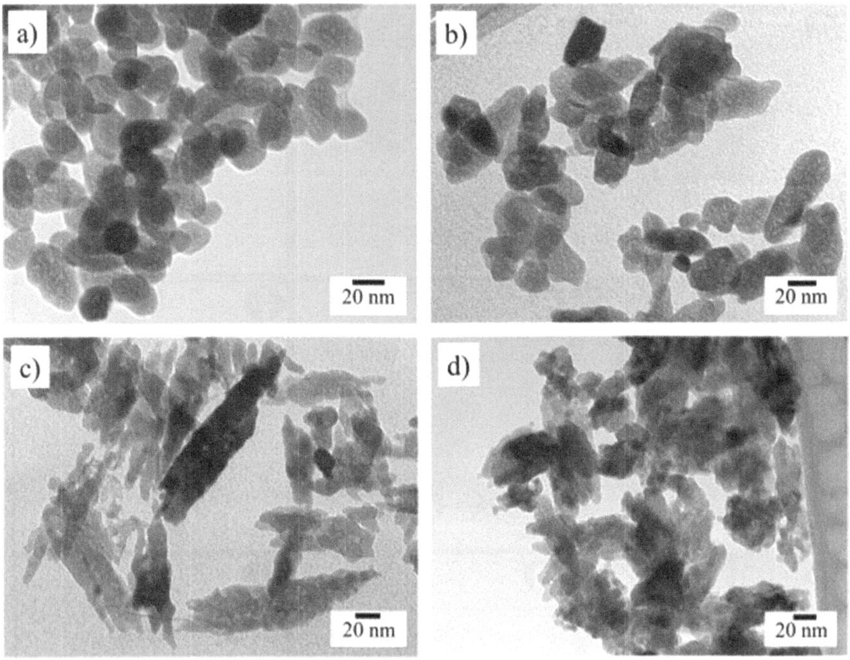

Fig. 7 SEM morphology of samples Ax: **a** $x = 0$, **b** $x = 0.03$, **c** $x = 0.05$, and **d** $x = 0.1$. Bars on insets are 50 nm and decolorization of methyl orange by photocatalyst under, **a** visible and **b** UV light irradiation

Figure 7 shows the photocatalytic reaction under UV and visible light irradiation. The respectable photocatalytic ability can be described through the pore size, surface area, and pore volume. It is also established by the occurrence of the antimony and chromium-doped mesocrystal, which are composed of building units of nanocrystals with mutual crystallographic direction [24]. The nanocrystals with a mutual crystallographic direction beneficial to expand the segregation of charge carriers under illumination and also to produce are found to be long-lived charge carriers after the photocatalytic performance under the visible light region.

TiO_2 has been doped with a foreign element to make a new energy level near the conduction band, which allows photons with some lower energy levels to excite the higher energy level. However, the wide bandgap of TiO_2 has limited its application in the visible light region. To rectify this limitation, the band gap of TiO_2 is modified by the process of metal doping (Table 1).

Table 1 Mesocrystal photocatalyst properties and their applications

Photocatalyst	Dyes	Light source	Reason for enhanced property	References
Anatase TiO	Decolorization of methylene blue	UV irradiation	Crystal plane led to a relatively large crystallite size	[32]
TiO_2 photocatalyst Anatase structure	Decolorization of methylene blue	UV irradiation	Largely exposed (001) plane improved photodegradation	[33]
TiO_2 photocatalyst anatase structure	Photodegradation of rhodamine B	Visible light	In this photocatalytic reaction, a significant number of surface-active sites enhanced the degradation	[34]
Anatase TiO_2	Photodegradation of methyl orange	UV irradiation	The high crystalline and surface area of the oxidative photocatalytic sample results in more active sites and superior adsorptive activity	[35]
ZnO nanosheet assembly	Photodegradation of methyl orange	UV-vis visible irrational	Exposed nonpolar facets of ZnO have good photocatalytic performance	[36]
ZnO nanosheet assemblies	Photodegradation of methylene blue	UV lamp	To improve photodegradation, the morphology of mesocrystal enhanced specific surface area and pore volume	[37]
CuO spindle	Photocatalytic dye decolorization of Rhodamine B	Visible light irradiation	3D mesostructure spindles exhibited more pores to adsorb molecules and could	[38]
$SrTiO_3$	Photodegradation of rhodamine B	Visible light irradiation	The well-defended superstructure can deliver photo-charges more efficiently	[39]
$Nb_3O_7(OH)$	Decolorization of methylene blue, rhodamine B, and indigo carmine	UV lamp	To enhance photocatalytic performance by vacancies Large surface area and high crystallinity	[40]

(continued)

Table 1 (continued)

Photocatalyst	Dyes	Light source	Reason for enhanced property	References
AgIn (WO_4)	Decolorization of eosin Y, rhodamine B, and methyl orang	UV-vis	Oriented attachment, PH, and phase formation	[41]
Nb_5 and Sb_{5+}TiO2 mesocrystals	Decolorization of methylene blue and rhodamine B	UV irradiation	Doping and co-doping Sb-doping make photocatalytic activity independent of the crystal's facets and thus of the particle's morphology	[42]

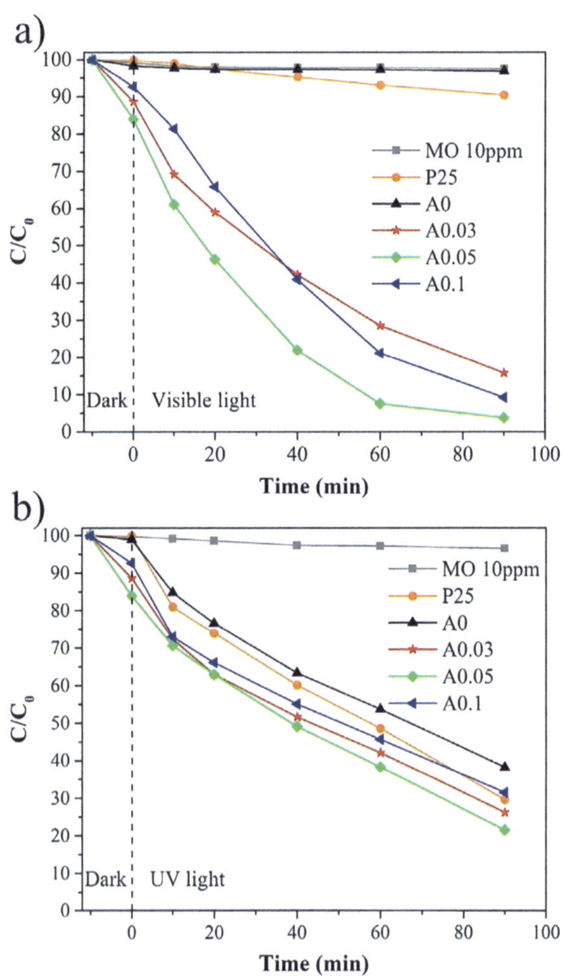

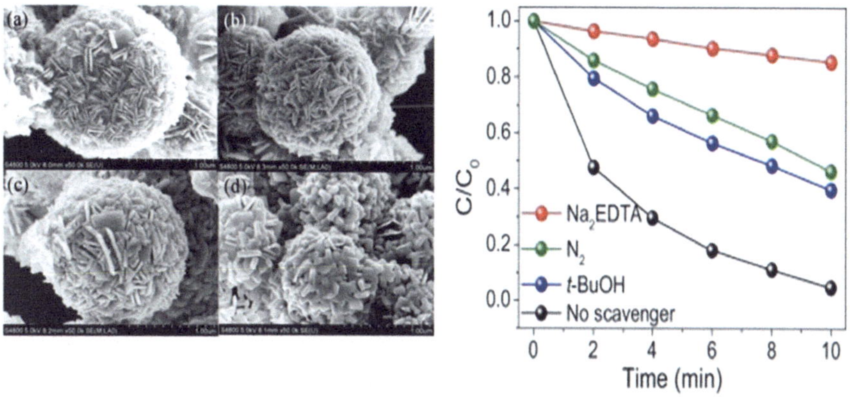

Fig. 8 SEM images of samples annealed at **a** 200 °C, **b** 300 °C, **c** 700 °C, **d** 800 °C

Longfei Liu et al. have prepared hollow microspheres mesocrystal by Au modified TiO_2 mesocrystal (with diameter 3–4 nm) [25]. The mesocrystal was synthesised by using the methodology of ultrasound-assisted aerosol spray, by controlling the spray drying temperatures. Figure 8 shows the SEM images of the samples retain the hierarchical microsphere morphology with increasing calcination temperatures from 200 to 900 °C. Figure 8 shows that the microsphere mesocrystal exhibits good photocatalytic capabilities for degradation of 4-CP oxidation reaction under visible light irradiation [25]. TiO_2 hollow microsphere mesocrystals show 76% photodegradation efficiency, which is greater than the nanosheet of TiO_2 mesocrystals.

The authors have suggested that crystallinity, surface area, and hollow superstructure are accountable for the photocatalytic application. Owing to the oriented superstructures and porous properties (more active species) significantly enhanced the charge separation. The photocatalytic activity is improved by the creation of heterojunctions between the mixed phases. TiO_2 mesocrystal modified with Au nanoparticles enhances the photocatalytic decolorization performance under visible light irradiation [25]. It is due to the Au nanoparticles, which have a surface plasmon resonance (SPR) and absorb light in the visible region.

Microspheres photodegradation of methyl orange under the UV light illumination using the scavengers.

3.1.3 Carbon-Based Mesocrystal

Graphene oxide, reduced graphene oxide, carbon dots, and carbon nanotubes are some of the excellent carbon nanomaterials. The carbon support-based mesocrystal would act as good electron acceptors for charge separation and migration in mesocrystal [26]. Moreover, graphene has received much research attention from theoretical and experimental science communities. Because of its unique properties, such as high electron mobility, mechanical strength, large surface area, and high electrical conductivity. Moreover, the graphene can decrease the photogenerated rate of electron-hole pairs recombination, and chemical molecules are adsorbed on their surface through the п–п* interactions [27].

Xiaofei Yanga et al. have synthesized spindle-shaped mesocrystals by modifying TiO_2 with graphene using the method of facile solvothermal process. The mesocrystal shows an excellent photocatalytic performance under UV/Visible irradiation for the degradation of Rhodamine B dye molecules [28]. Composites made with varying concentrations of GO all had stronger photocatalytic activity than pure TiO_2 mesocrystals. Graphene-TiO_2 composites show much greater photocatalytic performance than pure TiO_2 mesocrystal [27]. Figure 9 demonstrates the modification with 20 mg of graphene oxide. TiO_2-Graphene mesocrystal can lead to a 100% increased degradation efficiency [28]. The authors attributed the TiO_2 mesocrystal hybridized with graphene sheets that can improve the photocatalytic performance due to the number of chemical molecules adsorbed on the surface and the recombination rate of charge carriers reduced [27]. The amount of graphene oxide (GO) added dramatically influences the morphologies of TiO_2 mesocrystals as well as the photocatalytic activity of as-prepared nanocomposites. Figure 9 shows the function of graphene, in addition to assisting the morphological control of TiO_2 particles, which significantly improves the photocatalytic activity.

Chae Hee Park et al. developed a TiO_2 mesocrystal onto multiwalled carbon nanotubes by using a facile one-pot chemical method. The mesocrystal was formed by using different calcination temperatures. The authors described that the TiO_2 mesocrystal has transformed into porous single crystals at above 400 °C [29]. The

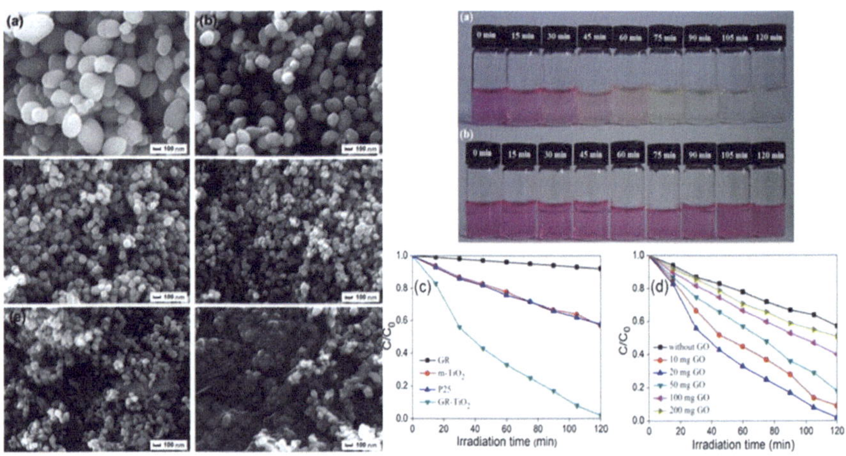

Fig. 9 SEM morphology of synthesized samples in the presence of different weight percentages of graphene oxide and digital photos of decolorization of Rhodamine B after visible light irradiation. **a** Graphene-TiO_2 composites and **b** TiO_2 as catalysts; **c**, **d** decolorization curves of RhB solution with as-prepared TiO_2, GR, P25, and GR-TiO_2 samples

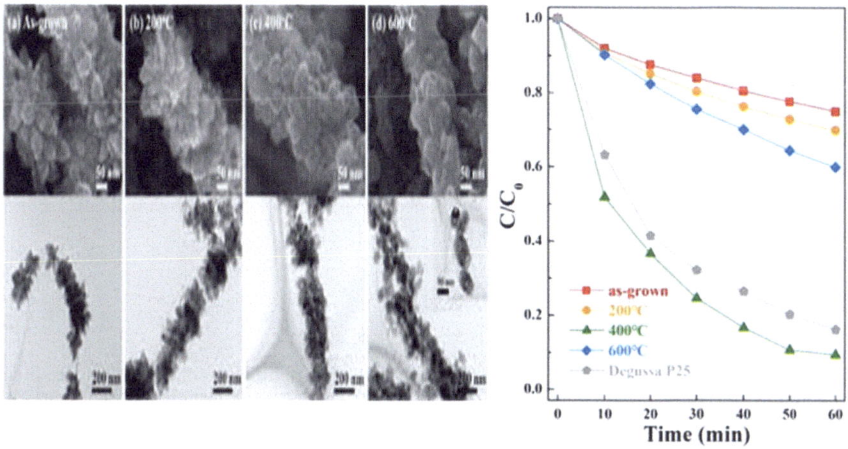

Fig. 10 FE-SEM morphology and HR-TEM morphology **a** as-synthesized and calcined TiO_2/CNT composites at **b** 200 °C, **c** 400 °C, **d** 600 °C and the graph represents photocatalytic degradation of the MB using TiO_2/CNT composites and commercial Degussa P_{25}

authors attributed this to improving surface diffusion at high temperatures. The photocatalytic experiment shows the decolorization process of MB under ultraviolet irradiation. From Fig. 10, it is clear that more than 90% of methylene blue was degraded by the TiO_2/CNT (calcined at 400 °C) sample. The authors ascribed strong covalent bonds between TiO_2 and carbon atoms. The author mainly described the (CNTs) have been one-dimensional structure with an increase in the delocalization of electrons and a reduction in the probability of charge carrier recombination [29]. However, the main drawback of titanium photocatalysts is a large band gap semiconductor with values over the range of 3.0–3.2 eV due to their photocatalytic reaction under UV light illumination.

Penghu Dong and Yao Yin et al. synthesized tetrahedral and spherical Ag_3PO_4 mesocrystal. The homogeneous Ag_3PO_4 tetrahedron mesocrystal with an average apex length of 1.1 μm. The photocatalytic performance is performed under visible light illumination. The photodegradation experiment indicates that the photodegradation of methylene blue and methyl orange. The photocatalytic dye degradation efficiency in tetrahedral mesocrystal Ag_3PO_4 is higher (25%) than that of spherical Ag_3PO_4 mesocrystal for the Methylene blue dye molecules [30]. Figure 11. The photocatalytic degradation rate of tetrahedral Ag_3PO_4 mesocrystal shows a much higher (32%) efficiency than spherical Ag_3PO_4 for methyl orange dye molecules. The authors stated that Ag_3PO_4 tetrahedral mesocrystals enhance the adsorption of dye molecules with a higher specific surface area. The superstructure of mesocrystal has both characteristics such as porous and high crystallinity [30]. The Ag_3PO_4 mesocrystal was a well-crystallized tetrahedral phase, which dramatically decreased the probability of electron-hole pair recombination and improved photocatalytic performance [30].

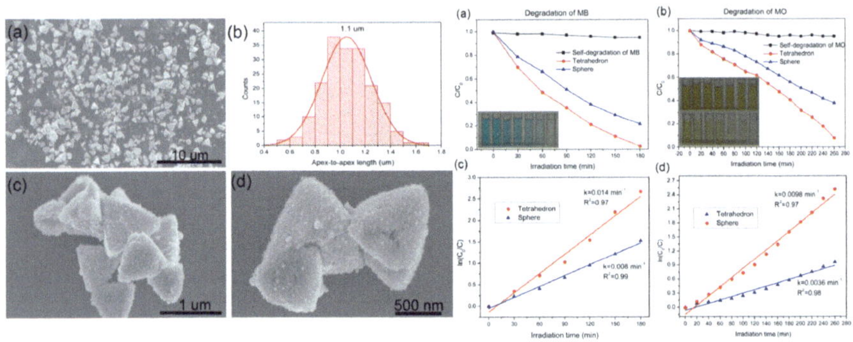

Fig. 11 **a** SEM image of Ag_3PO_4 tetrahedron synthesized at optimized condition: 1.4 mL oleic acid, H3PO4, DMF/H2O = 9:1, 25 °C, 1 h; **b** the apex-to-apex length distribution of these tetrahedrons; and **c**, **d** Ag_3PO_4 tetrahedron at large magnifications Fig. 12. Variation of MB (**a**) and MO (**b**) concentration against irradiation time in the presence of spherical Ag_3PO_4, tetrahedral Ag_3PO_4 mesocrystals and without photocatalyst in the dark and under visible light irradiation

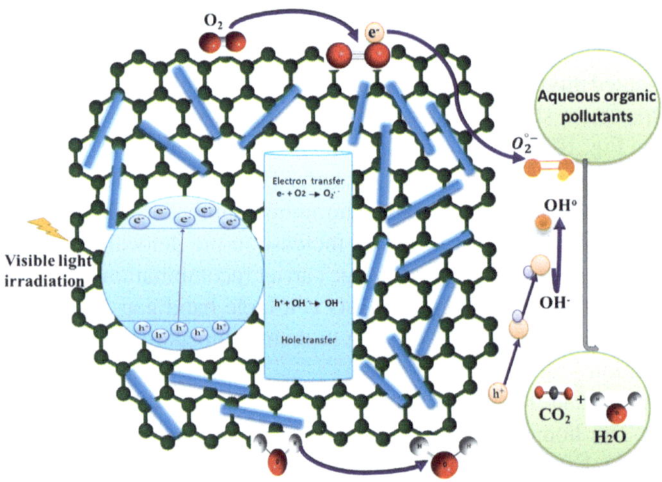

Fig. 12 Formation mechanism of carbon-based mesocrystal photocatalyst

3.1.4 Non-metal Doping

Besides selective photocatalyst microsphere mesocrystal synthesized by Ying Xiong and Bing Wang et al. using the method of facile co-precipitation route, Yttrium doped $PBWO_4$ mesocrystal with an average diameter of 1.5 and thickness of 0.7 μm and the author showed different types of mesocrystal [31]. The photocatalytic performance of the Y-doped $PbWO_4$ mesocrystal was examined in terms of the photocatalytic dye degradation of the orange II under UV irradiation. In Fig. 13 microspheres show poor photocatalytic ability, which is partly ascribed to the poor crystallinity of mesocrystals. It is well known that the absorption ability of photocatalysts also a significant factor in photocatalytic activity is closely related to the separation of photo-generated electron-hole pairs. As a result, the photocatalytic activity is expected to be reduced when compared to $PbWO_4$ microspheres [31].

Besides Lu-Lu Lai and Jin-Ming Wu et al. synthesized hollow rutile mesocrystal nanorods of TiO_2 microspheres. Using the method of Low-temperature one-pot synthesis, the hollow microspheres consisted of an average diameter of 200 nm and a length of 500 nm [32]. The dye adsorption for commercial P25 was calculated to be 7%, whereas the synthetic photocatalyst adsorbed Rhodamine B dye molecules were destroyed at a rate of 17%. The authors compared the photocatalytic dye degradation of commercial Degussa P_{25} TiO_2 nanocrystals and synthesized rutile mesocrystal hollow microspheres [32]. The unique morphology of the

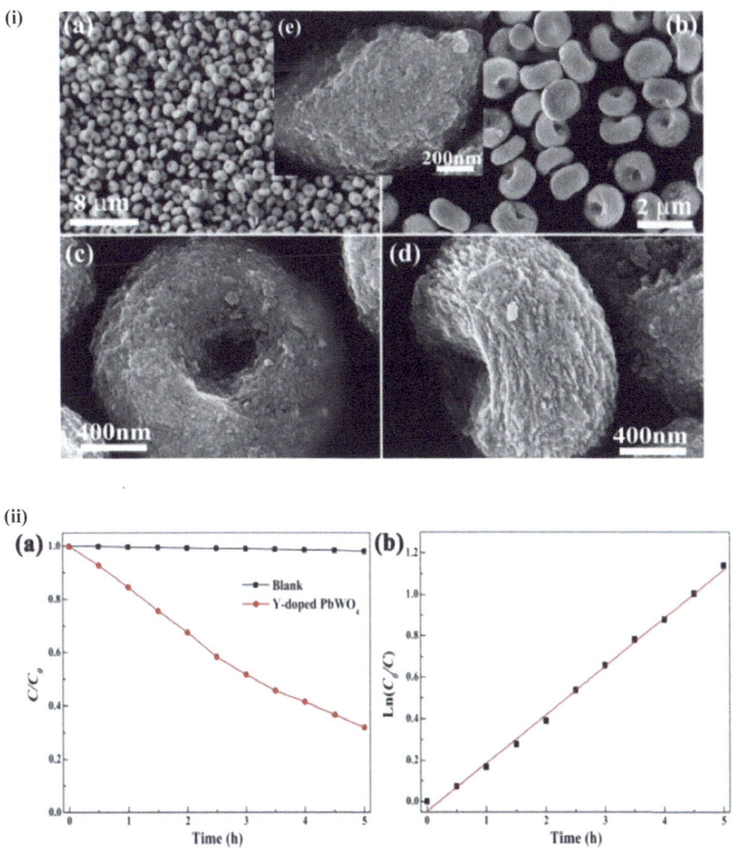

Fig. 13 (i) FESEM morphology of Y-doped $PbWO_4$ samples with different reaction times. **a** 2 h, **b** 16 h, **c** 32 h, **d** 38 h. (ii) Dynamic curve of photodegradation (**a**), plot of *ln*(C_0/C) versus time (**b**) for acid orange II over the erythrocyte-like Y-doped PbWO4 mesocrystals

hollow microspheres mesocrystal exhibits excellent photocatalytic performance in decolorization of rhodamine B under UV light irradiation. The author found that microspheres adsorb more dye molecules due to their relatively small particle size and high specific surface area (Fig. 14 and Table 2).

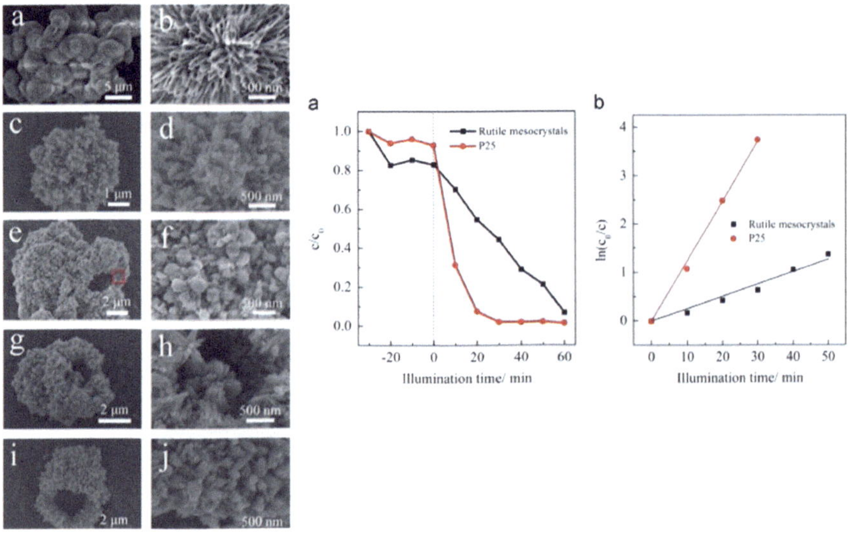

Fig. 14 FESEM morphologies of the powders derived using different reaction times of **a**, **b** 12 h, **c**, **d** 24 h, **e**, **f** 36 h, **g**, **h** 48 h, and **i**, **j** 60 h

Table 2 An estimation regarding the number of publications on the mesocrystal photocatalyst for dye degradation. The estimation is based on science directly taking the words mesocrystal photocatalyst specific application of dye degradation

Mesocrystal photocatalysts for dye degradation publication	Years
08	2021
18	2020
14	2019
12	2018
15	2017
11	2016
08	2015
10	2014
11	2013
7	2012
6	2011
2	2010

4 Conclusion

This paper clearly shows that increased efforts have been performed in the wastewater. Some of the outcomes are quite fascinating. Mesocrystal photocatalysts are widely described in various biominerals, compared to single crystal and polycrystal which is a new class of intermediate crystalline materials. Mesocrystal superstructures are also synthesized with specific crystal fact-oriented that are produced by additives-controlled crystallization through particle-based nonclassical routes or ion-based classical routes. Mesocrystal superstructures have many physical and physicochemical properties for example crystallinity in subunits, subunit alignment, high porosity, and the elegant three-dimensional network superstructure. These are distinct from regular nanoparticulate and single-crystal materials and as a result are projected to exhibit high photocatalytic efficiency. Discussed the above functional modification of mesocrystal photocatalysts for wastewater treatment with better photocatalytic ability. At the same time, developments in analytical characterization namely in-situ techniques like Grazing-Incidence Small-Angle X-ray Scattering (GISAXS)/Grazing-Incidence Wide-Angle X-ray Scattering (GIWAXS), TEM, and High Resolution (HR)-TEM. As well as their in-situ microscopic and diffraction techniques, namely SAXS/WAXS, will allow for the observation of the mechanisms and structuration principles of mesocrystal formation.

In the bottom-up view, a mesocrystal is made of crystallographically oriented nanoparticles. Individual nanoparticles have numerous properties, for example, superparamagnetic, photoluminescence photocatalytic activity behavior, etc. The crystallized portions of the mesocrystals are closely related to the synergetic effect. Mesocrystal photocatalyst has many fusion sites between the nanoparticles. In a top-down view, a mesocrystal can be thought of as a bulk single crystal, with some components being replaced by an inorganic/organic substance to form a network structure. However, the bandgap of TiO_2 corresponds to the ultraviolet wavelength, which only accounts for a small fraction (3–5%) of the solar spectrum. Several recent studies have shown that graphene-based nanocomposites and doped mesocrystal photocatalysts can produce better performances than the impartial mesocrystal complements in photocatalytic dye degradation application. The addition of graphene and its derivatives as well as of metallic and non-metallic dopants has been analyzed to enhance photocatalytic performance under visible light region. Another approach to improving photocatalytic efficiency is to reduce the recombination of electron-hole pairs. Mesocrystal superstructures have gathered immense interest because of their inherent anisotropic nature and tuneable spatial distribution, Although pore size, pore volume, high crystallinity, and specific surface area. Which gives rise to extraordinary optical, magnetic, and electronic properties. However, the surface morphology of the materials is also an important factor in photocatalytic application.

References

1. H. Anwer, A. Mahmood, J. Lee, K.H. Kim, J.W. Park, A.C.K. Yip, Photocatalysts for degradation of dyes in industrial effluents: opportunities and challenges. Nano Res. **12**, 955–972 (2019). https://doi.org/10.1007/s12274-019-2287-0
2. C. Xu, G.P. Rangaiah, X.S. Zhao, Photocatalytic degradation of methylene blue by titanium dioxide: experimental and modeling study. Ind. Eng. Chem. Res. **53**, 14641–14649 (2014). https://doi.org/10.1021/ie502367x
3. D. Sudha, P. Sivakumar, Review on the photocatalytic activity of various composite catalysts. Chem. Eng. Process. **97**, 112–133 (2015). https://doi.org/10.1016/j.cep.2015.08.006
4. V.K. Gupta, I. Ali, T.A. Saleh, A. Nayak, S. Agarwal, Chemical treatment technologies for waste-water recycling—An overview. RSC Adv. **2**, 6380–6388 (2012). https://doi.org/10.1039/c2ra20340e
5. A. Ajmal, I. Majeed, R.N. Malik, H. Idriss, M.A. Nadeem, Principles and mechanisms of photocatalytic dye degradation on TiO2 based photocatalysts: a comparative overview. RSC Adv. **4**, 37003–37026 (2014). https://doi.org/10.1039/c4ra06658h
6. V.S. Kirankumar, N. Mayank, S. Sumathi, Photocatalytic performance of cerium doped copper aluminate nanoparticles under visible light irradiation. J. Taiwan Inst. Chem. Eng. **95**, 602–615 (2019). https://doi.org/10.1016/j.jtice.2018.09.020
7. U.G. Akpan, B.H. Hameed, Parameters affecting the photocatalytic degradation of dyes using TiO2-based photocatalysts: a review. J. Hazard. Mater. **170**, 520–529 (2009). https://doi.org/10.1016/j.jhazmat.2009.05.039
8. M.N. Chong, B. Jin, C.W.K. Chow, C. Saint, Recent developments in photocatalytic water treatment technology: a review. Water Res. **44**, 2997–3027 (2010). https://doi.org/10.1016/j.watres.2010.02.039
9. E.S. Karafas, M.N. Romanias, V. Stefanopoulos, V. Binas, Effect of metal doped and co-doped TiO2 photocatalysts oriented to degrade indoor/outdoor pollutants for air quality improvement. A kinetic and product study using acetaldehyde as a probe molecule. J. Photochem. Photobiol. A: Chem. **371**, 255–263 (2019). https://doi.org/10.1016/j.jphotochem.2018.11.023
10. Y.P. Yuan, L.W. Ruan, J. Barber, S.C. Joachim Loo, C. Xue, Hetero-nanostructured suspended photocatalysts for solar-to-fuel conversion. Energy Environ. Sci. **7**, 3934–3951 (2014). https://doi.org/10.1039/c4ee02914c
11. S. Zhu, D. Wang, Photocatalysis: basic principles, diverse forms of implementations and emerging scientific opportunities. Adv. Energy Mater. **7**, 1–24 (2017). https://doi.org/10.1002/aenm.201700841
12. L. Bergström, E.V. Sturm née Rosseeva, G. Salazar-Alvarez, H. Cölfen, Mesocrystals in biominerals and colloidal arrays. Accounts Chem. Res. **48**, 1391–1402 (2015). https://doi.org/10.1021/ar500440b
13. N.K.C.R. Jayavel, C 14 TAB-assisted CeO2 mesocrystals: self-assembly mechanism and its characterization, 263–269 (2013). https://doi.org/10.1007/s13204-012-0131-7
14. H. Imai, Mesostructured crystals: growth processes and features. Prog. Cryst. Growth Charact. Mater. **62**, 212–226 (2016). https://doi.org/10.1016/j.pcrysgrow.2016.04.011
15. Y. Deng, R. Zhao, Advanced oxidation processes (AOPs) in wastewater treatment, 167–176 (2015). https://doi.org/10.1007/s40726-015-0015-z
16. J. Peng, X. Chen, W.J. Ong, X. Zhao, N. Li, Surface and heterointerface engineering of 2D MXenes and their nanocomposites: insights into electro- and photocatalysis. Chem **5**, 18–50 (2019). https://doi.org/10.1016/j.chempr.2018.08.037
17. N. Singh, J. Prakash, R.K. Gupta, Design and engineering of high-performance photocatalytic systems based on metal oxide-graphene-noble metal nanocomposites. Mol. Syst. Des. Eng. **2**, 422–439 (2017). https://doi.org/10.1039/c7me00038c
18. M.G. Ma, H. Cölfen, Mesocrystals—Applications and potential. Curr. Opin. Colloid Interface Sci. **19**, 56–65 (2014). https://doi.org/10.1016/j.cocis.2014.03.001

19. H. Cölfen, M. Antonietti, Mesocrystals: inorganic superstructures made by highly parallel crystallization and controlled alignment. Angew. Chem. Int. Ed. **44**, 5576–5591 (2005). https://doi.org/10.1002/anie.200500496
20. E.V. Sturm, H. Cölfen, Mesocrystals: past, presence, future, crystals (vol. 7, 2017). https://doi.org/10.3390/cryst7070207
21. P. Zhang, M. Fujitsuka, T. Majima, Development of tailored TiO2 mesocrystals for solar driven photocatalysis. J. Energy Chem. **0**, 1–10 (2016). https://doi.org/10.1016/j.jechem.2016.11.012
22. S. Sun, X. Yu, Q. Yang, Z. Yang, S. Liang, Mesocrystals for photocatalysis: a comprehensive review on synthesis engineering and functional modifications. Nanosc. Adv. **1**, 34–63 (2019). https://doi.org/10.1039/c8na00196k
23. A. Kumar, R. Police, With exposed (001) facets. Ceram. Int. **45**, 2178–2184 (2019). https://doi.org/10.1016/j.ceramint.2018.10.128
24. J. Miguel, J. Balbuena, M. Cruz-yusta, F. Martín, P. Pardo, L. Sánchez, J. Alarcón, Mesocrystalline Cr and Sb-codoped anatase visible-light-driven photocatalyst. Ceram. Int. 0–1 (2018). https://doi.org/10.1016/j.ceramint.2018.02.003
25. C. Tang, L. Liu, Y. Li, Z. Bian, Aerosol spray assisted assembly of TiO2 mesocrystals into hierarchical hollow microspheres with enhanced photocatalytic performance. Appl. Catal. B **201**, 41–47 (2017). https://doi.org/10.1016/j.apcatb.2016.08.006
26. S. Singhal, R. Sharma, C. Singh, S. Bansal, Enhanced photocatalytic degradation of methylene blue using ZnFe2O4 /MWCNT composite synthesized by hydrothermal method. Indian J. Mater. Sci. **2013**, 1–6 (2013). https://doi.org/10.1155/2013/356025
27. S. Sagadevan, J.A. Lett, G.K. Weldegebrieal, S. Garg, W. Oh, Enhanced photocatalytic activity of rGO-CuO nanocomposites (2021)
28. X. Yang, J. Qin, Y. Li, R. Zhang, H. Tang, Graphene-spindle shaped TiO2 mesocrystal composites: facile synthesis and enhanced visible light photocatalytic performance. J. Hazard. Mater. **261**, 342–350 (2013). https://doi.org/10.1016/j.jhazmat.2013.07.044
29. C. Hee, C. Min, J. Won, G. Chul, J. Joo, Enhanced photocatalytic activity of porous single crystal TiO2/CNT composites by annealing process. Ceram. Int. 0–1 (2017). https://doi.org/10.1016/j.ceramint.2017.10.086
30. P. Dong, Y. Yin, N. Xu, R. Guan, G. Hou, Y. Wang, Facile synthesis of tetrahedral Ag3PO4mesocrystals and its enhanced photocatalytic activity. Mater. Res. Bull. **60**, 682–689 (2014). https://doi.org/10.1016/j.materresbull.2014.09.047
31. Y. Xiong, B. Wang, W. Hu, L. Dai, Facile synthesis and characterization of erythrocyte-like Y-doped PbWO4 mesocrystals and their photocatalytic activity. Mater. Charact. **105**, 24–29 (2015). https://doi.org/10.1016/j.matchar.2015.04.016
32. L.L. Lai, J.M. Wu, Hollow TiO2 microspheres assembled with rutile mesocrystals: low-temperature one-pot synthesis and the photocatalytic performance. Ceram. Int. **41**, 12317–12322 (2015). https://doi.org/10.1016/j.ceramint.2015.06.060
33. L. Zhou, L. Zhou, J. Chen, C. Ji, P. O'Brien, A facile solid phase reaction to prepare TiO2 mesocrystals with exposed 001 facets and high photocatalytic activity. CrystEngComm **15**, 5012–5015 (2013). https://doi.org/10.1039/c3ce27095e
34. R.O. Da Silva, R.H. Gonalves, D.G. Stroppa, A.J. Ramirez, E.R. Leite, Synthesis of recrystallized anatase TiO2 mesocrystals with Wulff shape assisted by oriented attachment. Nanoscale **3**, 1910–1916 (2011). https://doi.org/10.1039/c0nr01016b
35. L. Li, C.Y. Liu, Organic small molecule-assisted synthesis of high active TiO2 rod-like mesocrystals. CrystEngComm **12**, 2073–2078 (2010). https://doi.org/10.1039/b924170a
36. J.Y. Dong, W.H. Lin, Y.J. Hsu, D.S.H. Wong, S.Y. Lu, Ultrafast formation of ZnO mesocrystals with excellent photocatalytic activities by a facile Tris-assisted antisolvent process. CrystEngComm **13**, 6218–6222 (2011). https://doi.org/10.1039/c1ce05503h
37. M. Wang, Y. Zhang, Y. Zhou, F. Yang, E.J. Kim, S.H. Hahn, S.G. Seong, Rapid room-temperature synthesis of nanosheet-assembled ZnO mesocrystals with excellent photocatalytic activity. CrystEngComm **15**, 754–763 (2013). https://doi.org/10.1039/c2ce26660a

38. S. Sun, X. Zhang, J. Zhang, L. Wang, X. Song, Z. Yang, Surfactant-free CuO mesocrystals with controllable dimensions: green ordered-aggregation-driven synthesis, formation mechanism and their photochemical performances. CrystEngComm **15**, 867–877 (2013). https://doi.org/10.1039/c2ce26216a
39. S.B. Betzler, A. Wisnet, B. Breitbach, C. Mitterbauer, J. Weickert, L. Schmidt-Mende, C. Scheu, Template-free synthesis of novel, highly-ordered 3D hierarchical Nb 3O7(OH) superstructures with semiconductive and photoactive properties. J. Mater. Chem. A **2**, 12005–12013 (2014). https://doi.org/10.1039/c4ta02202e
40. B. Hu, L.H. Wu, S.J. Liu, H. Bin Yao, H.Y. Shi, G.P. Li, S.H. Yu, Microwave-assisted synthesis of silver indium tungsten oxide mesocrystals and their selective photocatalytic properties. Chem. Commun. **46**, 2277–2279 (2010). https://doi.org/10.1039/b921455k
41. D. Primc, G. Zeng, R. Leute, M. Walter, L. Mayrhofer, M. Niederberger, Chemical substitution—Alignment of the surface potentials for efficient charge transport in nanocrystalline TiO2 photocatalysts. Chem. Mater. **28**, 4223–4230 (2016). https://doi.org/10.1021/acs.chemmater.6b00846
42. J. Chen, G. Li, H. Zhang, P. Liu, H. Zhao, T. An, Anatase TiO2 mesocrystals with exposed (0 0 1) surface for enhanced photocatalytic decomposition capability toward gaseous styrene. Catal. Today **224**, 216–224 (2014). https://doi.org/10.1016/j.cattod.2013.10.073
43. Z. Hong, M. Wei, T. Lan, G. Cao, Self-assembled nanoporous rutile TiO2 mesocrystals with tunable morphologies for high rate lithium-ion batteries. Nano Energy **1**, 466–471 (2012). https://doi.org/10.1016/j.nanoen.2012.02.009

The manufacturer's authorised representative in the EU is Springer Nature Customer Service Centre GmbH, Europaplatz 3, 69115 Heidelberg, Germany. If you have any concerns regarding our products, please contact ProductSafety@springernature.com

Printed and bound by CPI Group (UK) Ltd, Croydon, CR0 4YY
01/10/2026
02225864-0014